An Introduction to Radiative Transfer

Methods and applications in astrophysics

Astrophysicists have developed several very different methodologies for solving the radiative transfer equation. *An Introduction to Radiative Transfer* presents these techniques as applied to stellar atmospheres, planetary nebulae, supernovae and other objects with similar geometrical and physical conditions. Accurate methods, fast methods, probabilistic methods and approximate methods are all explained, including the latest and most advanced techniques. The book includes the different methods used for computing line profiles, polarization due to resonance line scattering, polarization in magnetic media and similar phenomena. Exercises at the end of each chapter enable these methods to be put into practice, and enhance understanding of the subject. This textbook will be of great value to graduates, postgraduates and researchers in astrophysics.

ANNAMANENI PERAIAH obtained his doctorate in radiative transfer from Oxford University. He was formerly a Senior Professor at the Indian Institute of Astrophysics, Bangalore, India. He has held positions in India, Canada, Germany and the Netherlands. His research interests include developing solutions to the radiative transfer equation in stellar atmospheres and line formation in expanding atmospheres with different physical and geometrical conditions.

An Introduction to Radiative Transfer

Methods and applications in astrophysics

Annamaneni Peraiah

CAMBRIDGE
UNIVERSITY PRESS

CAMBRIDGE UNIVERSITY PRESS
Cambridge, New York, Melbourne, Madrid, Cape Town, Singapore,
São Paulo, Delhi, Dubai, Tokyo

Cambridge University Press
The Edinburgh Building, Cambridge CB2 8RU, UK

Published in the United States of America by Cambridge University Press, New York

www.cambridge.org
Information on this title: www.cambridge.org/9780521779890

First published 2002

A catalogue record for this publication is available from the British Library

Library of Congress Cataloguing in Publication data

Peraiah, Annamaneni, 1937–
An introduction to radiative transfer:
Methods and applications in astrophysics / Annamaneni Peraiah.
p. cm.
Includes bibliographical references and index.
ISBN 0 521 77001 7 – ISBN 0 521 77989 8 (pb.)
1. Radiative transfer. 2. Stars–Radiation. I. Title.

QB817.P47 2001 523.8′2–dc21 2001025557

ISBN 978-0-521-77001-9 Hardback
ISBN 978-0-521-77989-0 Paperback

Transferred to digital printing 2009

Contents

Preface

Astrophysicists analyse the light coming from stellar atmosphere-like objects with widely differing physical conditions using the solution of the equation of radiative transfer as a tool. A method of obtaining the solution of the transfer equation developed to suit a given physical condition need not necessarily be useful in a situation with different physical conditions. Furthermore, each individual has his/her preferences to a particular type of methodology. These factors necessitated the development of several widely differing methods of solving the transfer equation.

In the second half of the twentieth century several books were written on the subject of radiative transfer: one each by Chandrasekhar, Kourganoff and Sobolev, two books by Mihalas, two by Kalkofen and more recently two books by Sen and Wilson. These books, which describe the developments of the transfer theory, will remain milestones. They will be of great value to the researcher in this field. A beginner needs to understand the basic concepts and the initial development of the subject to proceed to use the latest advances. It is felt that it is necessary to have a book on radiative transfer which presents a comprehensive view of the subject as applied in astrophysics or more particularly in stellar atmospheres and objects with similar geometrical and physical conditions. This book serves such a purpose. Several methods are presented in the book so that the students of radiative transfer can familiarise themselves with the techniques old and new.

It became a daunting task to include all the existing techniques in the book as there is a restriction on its size. This resulted in leaving out a few methods that are of equal interest as those that appear in the book. I apologize to the authors of these methods in advance. The subject matter of the book assumes of the student a knowledge of basic mathematics and physics at the undergraduate level. This book is intended to be included in the advanced course work of undergraduate students, and the course work of graduate students. Several exercises have been included at the end of each chapter for practising the concepts described in the chapter. These problems are straightforward and can be solved by direct application of the theory. Some of them involve just supplying the intermediate steps in the derivations of the chapter.

The material in the book is largely drawn from the books mentioned earlier and from various other references cited at the end of each chapter. If there are any errors these are mine and I shall be grateful if these are brought to my attention. Any suggestions for improvements and corrections are welcome.

It is a pleasure to thank Dr W. Kalkofen for a brief discussion on the subject matter of the book. I am grateful to Professor K. K. Sen for not only giving a few tips on writing books but also for going through the first draft and pointing out several typographical errors and adding a few conceptual points. This book would not have been possible without the active help from Mr Baba Anthony Varghese who very patiently typed the text. His phenomenal computer expertise enabled the book to rapidly and easily take its present form. It is pleasure to thank him for all this. I thank Drs A. Vagiswari and Christina Louis for their magnanimous and kind help in securing me any reference that I needed. Further, I thank Mr M. Srinivasa Rao, Mr S. Muthukrishnan and Mrs Pramila Kaveriappa for helping me in various ways during the writing of the book.

There is one person whose memory always lingers on in my mind – that of Professor M. K. Vainu Bappu. From him I have learnt several aspects not only of science but also of life. I fondly cherish the memory of my association with him.

I am grateful to my wife Jayalakshmi and my children Rajani (Vaidhyanathan), Chandra (Edith) and Usha (Madhusudan) – spouses in brackets – for the love and affection shown to me.

Finally I thank the staff of Cambridge University Press who have been connected with the publication of the book, especially Dr Simon Mitton and Miss Jacqueline Garget for clearing my doubts from time to time and Ms Maureen Storey, who very patiently went through the manuscript and suggested several corrections.

Bangalore
October 2000

Annamaneni Peraiah

Chapter 1

Definitions of fundamental quantities of the radiation field

1.1 Specific intensity

This is the most fundamental quantity of the radiation field. We shall be dealing with this quantity throughout this book.

Let dE_ν be the amount of radiant energy in the frequency interval $(\nu, \nu + d\nu)$ transported across an element of area ds and in the element of solid angle $d\omega$ during the time interval dt. This energy is given by

$$dE_\nu = I_\nu \cos\theta \, d\nu \, d\sigma \, d\omega \, dt, \tag{1.1.1}$$

where θ is the angle that the beam of radiation makes with the outward normal to the area ds, and I_ν is the *specific intensity* or simply *intensity* (see figure 1.1).

The dimensions of the intensity are, in CGS units, erg cm^{-2} s^{-1} hz^{-1} ster^{-1}. The intensity changes in space, direction, time and frequency in a medium that absorbs

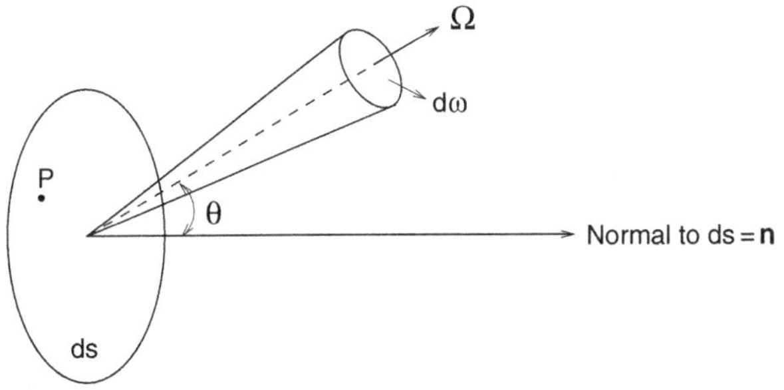

Figure 1.1 Schematic diagram which shows how the specific intensity is defined.

and emits radiation. I_ν can be written as

$$I_\nu = I_\nu(\mathbf{r}, \Omega, t), \tag{1.1.2}$$

where $\mathbf{r}$ is the position vector and Ω is the direction. In Cartesian coordinates it can be written as

$$I_\nu = I_\nu(x, y, z; \alpha, \beta, \gamma; t), \tag{1.1.3}$$

where x, y, z are the Cartesian coordinate axes and α, β, γ are the direction cosines. If the medium is stratified in plane parallel layers, then

$$I_\nu = I_\nu(z, \theta, \varphi; t), \tag{1.1.4}$$

where z is the height in the direction normal to the plane of stratification and θ and φ are the polar and azimuthal angles respectively. If I_ν is independent of φ, then we have a radiation field with axial symmetry about the z-axis. Instead of z, we may choose symmetry around the x-axis.

In spherical symmetry, I_ν is

$$I_\nu = I_\nu(r, \theta; t), \tag{1.1.5}$$

where r is the radius of the sphere and θ is the angle made by the direction of the ray with the radius vector.

The radiation field is said to be isotropic at a point, if the intensity is independent of direction at that point and then

$$I_\nu = I_\nu(\mathbf{r}, t). \tag{1.1.6}$$

If the intensity is independent of the spatial coordinates and direction, the radiation field is said to be homogeneous and isotropic. If the intensity I_ν is integrated over all the frequencies, it is called the integrated intensity I and is given by

$$I = \int_0^\infty I_\nu \, d\nu. \tag{1.1.7}$$

There are other parameters that characterize the state of polarization in a radiation field. These are studied in chapters 11 and 12.

1.2 Net flux

The flux F_ν is the amount of radiant energy transferred across a unit area in unit time in unit frequency interval. The amount of radiant energy in the area ds in the direction θ (see figure 1.1) to the normal, in the solid angle $d\omega$, in time dt and in the frequency interval $(\nu, \nu + d\nu)$ is equal to $I_\nu \cos\theta \, d\omega \, d\nu \, ds \, dt$. The net flow in all directions is

$$d\nu \, ds \, dt \int I_\nu \cos\theta \, d\omega,$$

or

$$F_\nu = \int I_\nu \cos\theta \, d\omega. \tag{1.2.1}$$

The integration is over all solid angles. This is the net flux and is the rate of flow of radiant energy per unit area per unit frequency.

In polar coordinates, where the outward normal is in the z-direction, we have

$$d\omega = \sin\theta \, d\theta \, d\varphi, \tag{1.2.2}$$

where φ is the azimuthal angle. The net flux F_ν then becomes

$$F_\nu = \int_0^{2\pi} \int_0^{\pi} I_\nu \cos\theta \sin\theta \, d\varphi \, d\theta. \tag{1.2.3}$$

The dimensions of flux are erg cm^{-2} s^{-1} hz^{-1}. Equation (1.2.3) can also be written as

$$\begin{aligned} F_\nu &= \int_0^{2\pi} d\varphi \int_0^{\pi/2} I_\nu \cos\theta \sin\theta \, d\theta + \int_0^{2\pi} d\varphi \int_{\pi/2}^{\pi} I_\nu \cos\theta \sin\theta \, d\theta \\ &= F_\nu(+) - F_\nu(-), \end{aligned} \tag{1.2.4}$$

where

$$F_\nu(+) = \int_0^{2\pi} \int_0^{\pi/2} I_\nu \cos\theta \sin\theta \, d\theta \, d\varphi \tag{1.2.5}$$

and

$$F_\nu(-) = \int_0^{2\pi} \int_\pi^{\pi/2} I_\nu \cos\theta \sin\theta \, d\theta \, d\varphi. \tag{1.2.6}$$

The physical meaning of equation (1.2.4) is as follows: $F_\nu(+)$ represents the radiation illuminating the area from one side and $F_\nu(-)$ represents the radiation illuminating the area from another side. Therefore F_ν, the flux of radiation transported through the area, is the difference between these illuminations of the area. The flux depends on the direction of the normal to the area. The dependence of the flux on direction shows that flux is of vector character. In the Cartesian coordinate system, let the angles made by the direction of radiation with the axes x, y and z be α_1, β_1 and γ_1 respectively, then the flux or radiation along the coordinate axes is given by

$$F_\nu(x) = \int I_\nu \cos\alpha_1 \, d\omega, \tag{1.2.7}$$

$$F_\nu(y) = \int I_\nu \cos\beta_1 \, d\omega, \tag{1.2.8}$$

$$F_\nu(z) = \int I_\nu \cos\gamma_1 \, d\omega. \tag{1.2.9}$$

Furthermore, if α_2, β_2 and γ_2 are the angles made by the coordinate axes and the normal to the area and θ is the angle between the normal and the direction of the radiation, then

$$\cos\theta = \cos\alpha_1 \cos\alpha_2 + \cos\beta_1 \cos\beta_2 + \cos\gamma_1 \cos\gamma_2. \tag{1.2.10}$$

Substituting equation (1.2.10) into equation (1.2.1), we get

$$F_\nu = \cos\alpha_2 \, F_\nu(x) + \cos\beta_2 \, F_\nu(y) + \cos\gamma_2 \, F_\nu(z). \tag{1.2.11}$$

The integrated flux over frequency is

$$F = \int_0^\infty F_\nu \, d\nu. \tag{1.2.12}$$

If the radiation field is symmetric with respect to the coordinate axes, then the net flux across the surface oriented perpendicular to that axis is zero as the oppositely directed rays cancel each other. In a homogeneous planar geometry, $F_\nu(x)$ and $F_\nu(y)$ are zeros and only $F_\nu(z)$ exists. In such a situation, we have

$$F_\nu(z, t) = 2\pi \int_{-1}^{+1} I(z, \mu, t)\mu \, d\mu, \tag{1.2.13}$$

where $\mu = \cos\theta$.

The astrophysical flux $F_{A\nu}(z, t)$ normally absorbs the π on the RHS of equation (1.2.13) and is written as

$$F_{A\nu}(z, t) = 2 \int_{-1}^{+1} I(z, \mu, t)\mu \, d\mu \tag{1.2.14}$$

and the Eddington flux $F_{E\nu}$ is defined as

$$F_{E\nu}(z, t) = \frac{1}{2} \int_{-1}^{+1} I(z, \mu, t)\mu \, d\mu. \tag{1.2.15}$$

1.2.1 Specific luminosity

The specific luminosity was suggested by Rybicki (1969) and Kandel (1973). We define it following Collins (1973).

From figure 1.2, we define the specific luminosity $\mathcal{L}(\psi, \xi)$ in terms of the orientation variables ψ and ξ as

$$\mathcal{L}(\psi, \xi) = 4\pi \int_A I(\theta, \phi)\hat{n}(\theta, \phi) \cdot \hat{o}(\theta, \phi) \, dA(\theta, \phi), \tag{1.2.16}$$

where $\hat{n}(\theta, \phi)$ and $\hat{o}(\theta, \phi)$ are position dependent unit vectors normal to the surface and in the direction of the observer respectively. The area A over which the specific intensity $I(\theta, \phi)$ is to be integrated is the 'observable' surface and is defined by the orientation angles ψ and ξ. It is obvious from equation (1.2.16) that $\mathcal{L}(\psi, \xi)$ is a function of the orientation of the object with respect to the observer and is measured per unit solid angle; the total luminosity L is given in terms of $\mathcal{L}(\psi, \xi)$ as

$$L = \frac{1}{4\pi} \int_{4\pi} \mathcal{L}(\psi, \xi)\, d\Omega(\psi, \xi). \tag{1.2.17}$$

1.3 Density of radiation and mean intensity

Let V and Σ be two regions (see figure 1.3) the latter being larger than the former in linear dimensions but sufficiently small for a pencil not to have its intensity changed appreciably in transit. The radiation travelling through V must have crossed the region Σ through some element; let $d\Sigma$ be such an element with normal $\mathbf{N}$. The

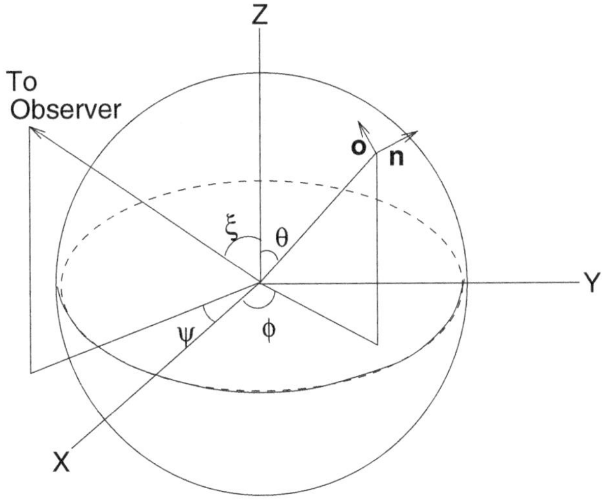

Figure 1.2 The angles θ and ϕ are the angular coordinates of a point on the stellar surface, and therefore represent a local structure. The angles ψ and ξ represent the orientation of the stellar body (from Collins (1973), with permission).

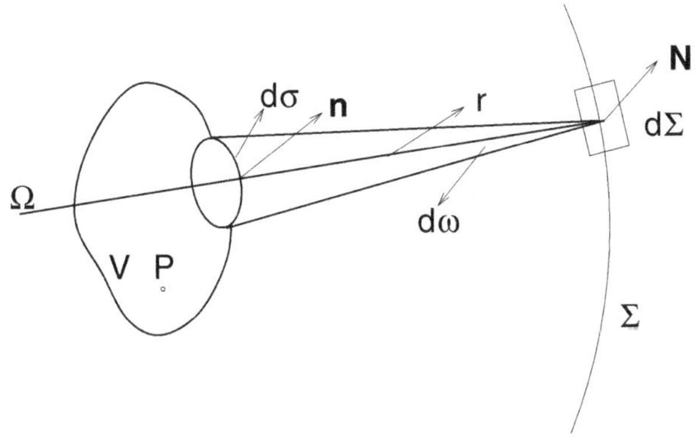

Figure 1.3 Schematic diagram to define density of radiation.

energy passing through $d\Sigma$ which also passes through $d\sigma$ with normal $\mathbf{n}$ on V per unit time is

$$I_\nu(\mathbf{\Omega}, \mathbf{N})\, d\Sigma\, d\omega'\, d\nu, \tag{1.3.1}$$

where

$$d\omega' = (\mathbf{\Omega} \cdot \mathbf{n})\, d\sigma / r^2. \tag{1.3.2}$$

If l is the length travelled by the pencil in V, then an amount of energy

$$\frac{I_\nu(\mathbf{\Omega} \cdot \mathbf{n})(\mathbf{\Omega} \cdot \mathbf{N})\, d\sigma\, d\Sigma\, d\nu}{r^2} \frac{l}{c} \tag{1.3.3}$$

will have travelled through the element in time l/c, where c is the velocity of light.

The solid angle $d\omega$ subtended by $d\Sigma$ at P is $(\Omega \cdot \mathbf{N})\, d\Sigma / r^2$ and the volume intercepted in V by the pencil is given by

$$dV = l(\mathbf{\Omega} \cdot \mathbf{n})\, d\sigma. \tag{1.3.4}$$

This amount of energy is given by

$$\frac{1}{c} I_\nu\, d\nu\, dV\, d\omega. \tag{1.3.5}$$

Therefore, the contribution to the energy per unit volume per unit frequency range (in the interval ν, $\nu + d\nu$) coming from the solid angle $d\omega$ about the direction Ω is $I_\nu\, d\omega / c$ and the energy density is defined as

$$U_\nu = \frac{1}{c} \int I_\nu\, d\omega. \tag{1.3.6}$$

The average intensity or mean intensity J_ν is

$$J_\nu = \frac{1}{4\pi} \int I_\nu\, d\omega, \tag{1.3.7}$$

so that

$$U_\nu = \frac{4\pi}{c} J_\nu. \tag{1.3.8}$$

For an axially symmetric radiation field, J_ν is given by

$$\begin{aligned} J_\nu &= \frac{1}{2} \int_0^\pi I_\nu \sin\theta\, d\theta \\ &= \frac{1}{2} \int_{-1}^{+1} I(\mu)\, d\mu. \end{aligned} \tag{1.3.9}$$

The integrated energy density U is

$$U = \int_0^\infty U_\nu\, d\nu = \frac{1}{c} \int I\, d\omega. \tag{1.3.10}$$

The dimensions of energy density are erg cm^{-3} hz^{-1} and those of the integrated energy density are erg cm^{-3}. The dimensions of the mean intensity are erg cm^{-2} s^{-1} hz^{-1}.

1.4 Radiation pressure

A quantum of energy $h\nu$ will have a momentum of $h\nu/c$, where c is the velocity of light in the direction of propagation. The pressure of radiation at the point P (see figure 1.1) is calculated from the net rate of transfer of momentum normal to an area ds, which contains the point P. The amount of radiant energy in the frequency range $(\nu, \nu + d\nu)$ incident on ds making an angle θ with the normal to ds traversing the solid angle $d\omega$ in time dt is

$$I_\nu \cos\theta \, d\omega \, d\nu \, ds \, dt. \tag{1.4.1}$$

The momentum associated with this energy in the direction I_ν is

$$\frac{1}{c} I_\nu \cos\theta \, d\omega \, d\nu \, ds \, dt. \tag{1.4.2}$$

Therefore the normal component of the momentum transferred across ds by the radiation is

$$\frac{1}{c} \, d\sigma \, dt \, I_\nu \cos^2\theta \, d\omega \, dt. \tag{1.4.3}$$

The net transfer of momentum across ds by the radiation in the frequency interval $(\nu, \nu + d\nu)$ is

$$\frac{d\sigma \, dt}{c} \int I_\nu \cos^2\theta \, d\omega \, d\nu, \tag{1.4.4}$$

where the integration is over the whole sphere. The pressure at the point P is the net rate of transfer of momentum normal to the element of the surface area containing P in the unit area; the pressure $p_r(\nu) \, d\nu$ can be written in the frequency interval as

$$p_r(\nu) = \frac{1}{c} \int_0^{2\pi} \int_0^{\pi} I_\nu \cos^2\theta \sin\theta \, d\theta \, d\varphi. \tag{1.4.5}$$

If the radiation field is isotropic, then

$$p_r(\nu) = \frac{2\pi}{c} I_\nu \int_0^{\pi} \mu^2 \, d\mu = \frac{4}{3} \frac{\pi}{c} I_\nu \qquad (\mu = \cos\theta) \tag{1.4.6}$$

or in terms of energy density U_ν

$$p_r(\nu) = \frac{1}{3} U_\nu. \tag{1.4.7}$$

The radiation pressure integrated over all frequencies is

$$p_r = \int_0^{\infty} p_r(\nu) \, d\nu \tag{1.4.8}$$

or

$$p_r = \frac{1}{c} \int I \cos^2\theta \, d\omega, \tag{1.4.9}$$

where I is the integrated intensity. Furthermore

$$p_r = \frac{1}{3}U. \tag{1.4.10}$$

It can be seen that the dimensions of radiation pressure are the same as those of energy density, that is, erg cm^{-3} hz $^{-1}$ and the integrated radiation pressure has the dimensions of erg cm^{-3}.

1.5 Moments of the radiation field

Moments are defined in such a way that the nth moment over the radiation field is given by

$$M_n(z, n) = \frac{1}{2}\int_{-1}^{+1} I_\nu(z, \mu)\mu^n \, d\mu. \tag{1.5.1}$$

Following Eddington, we can have the zeroth, first and second moments as:

1. Zeroth moment (mean intensity):

$$J_\nu(z) = \frac{1}{2}\int_{-1}^{+1} I(z, \mu) \, d\mu. \tag{1.5.2}$$

2. First moment (Eddington flux):

$$H_\nu(z) = \frac{1}{2}\int_{-1}^{+1} I(z, \mu)\mu \, d\mu. \tag{1.5.3}$$

3. Second moment (the so called K-integral):

$$K_\nu(z) = \frac{1}{2}\int_{-1}^{+1} I(z, \mu)\mu^2 \, d\mu. \tag{1.5.4}$$

1.6 Pressure tensor

The rate of transfer of the x-component of the momentum across the element of surface normal to the x-direction by radiation in the solid angle dw per unit area in the direction whose direction cosines are l, m, n is

$$\frac{1}{c} Il \, d\omega \, l, \tag{1.6.1}$$

where I is the integrated radiation. If monochromatic radiation is considered, then I should be replaced by $I_\nu \, d\nu$. The total rate of x-momentum transfer across the element per unit area is $p_r(xx)$:

$$p_r(xx) = \frac{1}{c}\int Il^2\,d\omega. \tag{1.6.2}$$

Similarly the y- and z-components are given by

$$p_r(xy) = \frac{1}{c}\int Ilm\,d\omega \quad \text{and} \quad p_r(xz) = \frac{1}{c}\int Iln\,d\omega. \tag{1.6.3}$$

The quantities $p_r(yx)$, $p_r(yy)$, $p_r(yz)$, $p_r(zx)$, $p_r(zy)$ and $p_r(zz)$ are similarly defined for elements of the surfaces normal to the y- and z-directions. These nine quantities constitute the 'stress tensor'.

One can see that $p_r(xy) = p_r(yx)$, $p_r(xz) = p_r(zx)$ and $p_r(yz) = p_r(zy)$ or that the tensor is symmetrical. The mean pressure $\bar{p}$ is defined by

$$\bar{p} = \frac{1}{3}\left[p_r(xx) + p_r(yy) + p_r(zz)\right], \tag{1.6.4}$$

and

$$\bar{p} = \frac{1}{3c}\int I\omega = \frac{1}{3}U, \tag{1.6.5}$$

as $l^2 + m^2 + n^2 = 1$.

In the case of an isotropic radiation field

$$\bar{p} = p_r(xx) = p_r(yy) = p_r(zz) = \frac{1}{3}U, \tag{1.6.6}$$

and

$$\left.\begin{aligned} p_r(xy) &= p_r(yx) = 0,\\ p_r(xz) &= p_r(zx) = 0,\\ p_r(yz) &= p_r(xy) = 0. \end{aligned}\right\} \tag{1.6.7}$$

1.7 Extinction coefficient: true absorption and scattering

A pencil of radiation of intensity I_ν is attenuated while passing through matter of thickness ds and its intensity becomes $I_\nu + dI_\nu$, where

$$dI_\nu = -I_\nu \kappa_\nu\,ds. \tag{1.7.1}$$

The quantity κ_ν is called the mass extinction coefficient or the mass absorption coefficient. κ_ν comprises two important processes: (1) true absorption and (2) scattering. Therefore we can write

$$\kappa_\nu = \kappa_\nu^a + \sigma_\nu, \tag{1.7.2}$$

where κ_ν^a and σ_ν are the absorption and scattering coefficients respectively. Absorption is the removal of radiation from the pencil of the beam by a process which involves changing the internal degrees of freedom of an atom or a molecule. Examples of these processes are: (1) photoionization or bound–free absorption by which the photon is absorbed and the excess energy, if any, goes into the kinetic energy of the electron thermalizing the medium; (2) the absorption of a photon by a freely moving electron that changes its kinetic energy which is known as free–free absorption; (3) the absorption of a photon by an atom leading to excitation from one bound state to another bound state, which is called bound–bound absorption or photoexcitation; (4) the collision of an atom in a photoexcited state which will contribute to the thermal pool; (5) the photoexcitation of an atom which ultimately leads to fluorescence; (6) negative hydrogen absorption, etc. The reversal of the above processes may contribute to the emission coefficient (see section 1.8).

The coefficient κ_ν^a depends on the thermodynamic state of the matter at (pressure p, temperature T, chemical abundances α_i) any given point in the medium. At the point r the coefficient is given by

$$\kappa_\nu^a(r, T) = \kappa_\nu^a\,[p(r, T), T(r), \alpha_i(r, T), \dots, \alpha_\kappa(r, T)], \tag{1.7.3}$$

when there is local thermodynamic equilibrium (LTE). This kind of situation does not exist in reality and one needs to determine the κ_ν^a in a non-LTE situation. In static media κ_ν^a is isotropic while in moving media it is angle and frequency dependent due to Doppler shifts.

Another process by which energy is lost from the beam is the scattering of radiation which is represented by the mass scattering coefficient κ_ν^s. Scattering changes not only the photon's direction but also its energy. If we define the *albedo for single scattering* as ω_ν, then

$$\omega_\nu = \frac{\sigma_\nu}{\kappa_\nu}, \tag{1.7.4}$$

is the ratio of scattering to the extinction coefficients.

The extinction coefficient is the product of the atomic absorption coefficients or scattering coefficients (cm^2) and the number density of the absorbing or scattering particles (cm^{-3}). The dimension of κ_ν is cm^{-1} and $1/\kappa_\nu$ gives the photon mean free path which is the distance over which a photon travels before it is removed from the pencil of the beam of radiation.

1.8 Emission coefficient

Let an element of mass with a volume element dV emit an amount of energy dE_ν into an element of solid angle $d\omega$ centred around $\mathbf{\Omega}$ in the frequency interval ν to $\nu + d\nu$ and time interval t to $t + dt$. Then

$$dE_\nu = j_\nu\, dV\, d\omega\, d\nu\, dt, \tag{1.8.1}$$

where j_ν is called the macroscopic emission coefficient or emissivity. The emissivity has dimensions erg cm^{-3} sr^{-1} hz^{-1} s^{-1}. Emission is the combination of the reverse of the physical processes that cause true absorption. These processes are: (a) radiative recombination: when a free electron occupies a bound state creating a photon whose energy is the sum of the kinetic energy of the electron and the binding energy; (b) bremsstrahlung: a free electron moving in one hyperbolic orbit emits a photon by moving into a different hyperbolic orbit of lower energy; (c) photo de-excitation or collisional de-excitation: a bound electron changes to another bound state by emitting a photon through collision; (d) collisional recombination: a photoexcited atom contributes photon energy by collisional ionization; the reverse of this is called (three-body) collisional recombination; and (e) fluorescence: if a photon is absorbed by an atom and it is excited from bound state p to another bound state r, decays to an intermediate bound state q and then to the original state p, this process is called fluorescence. The energy from the original absorbed photon is re-emitted in two photons each of different energy.

A true picture of the occupation numbers is obtained only when the statistical equilibrium equation, which describes all necessary processes that are to be taken into account, is written. When LTE exists, the emission coefficient is given by

$$j_\nu^a(LTE) = \kappa_\nu^a B_\nu(T), \tag{1.8.2}$$

where $B_\nu(T)$ is the Planck function:

$$B_\nu(T) = \frac{2h\nu^3}{c^2}\left[\exp\left(\frac{h\nu}{kT}\right) - 1\right]^{-1}. \tag{1.8.3}$$

Equation (1.8.2) is known as Kirchhoff–Planck relation. In a non-LTE situation one has to consider stimulated emission due to the presence of the radiation field and spontaneous emission and the Einstein transition coefficients involved.

Emission of radiation can also be from the scattered photons. One can write

$$j_\nu^s(\mathbf{r}, \mathbf{\Omega}) = \frac{1}{4\pi}\int\int \sigma_{\nu'}^s(\mathbf{r}, t)p(\nu, \mathbf{\Omega}; \nu', \mathbf{\Omega}'; \mathbf{r}, t)I_{\nu'}(\mathbf{r}, \mathbf{\Omega}', t)\, d\nu'\, d\omega'. \tag{1.8.4}$$

The phase function p can be normalized in such a way that

$$\int\int p(\nu', \mathbf{\Omega}'; \nu, \Omega, ; \mathbf{r}, t)\, d\nu'\, d\omega' = 4\pi. \tag{1.8.5}$$

This is the manifestation of the conservation of radiation flux, that is, the emitted radiation balances that removed from the beam.

Equation (1.8.2) should be corrected for the stimulated scattering by multiplying it by the correction factor

$$\left\{1 + \frac{c^2}{2h\nu^3} I_\nu(r, \mathbf{\Omega}, t)\right\}. \tag{1.8.6}$$

This makes the transfer equation non-linear in I_ν. Particles, such as ions, atoms, molecules, electrons, solid particles, etc., scatter radiation and contribute to the scattering coefficient.

1.9 The source function

The source function is defined as the ratio of the emission coefficient to the absorption coefficient:

$$S_\nu = j_\nu / \kappa_\nu. \tag{1.9.1}$$

From equations (1.7.4), (1.8.2) and (1.8.4), we can write the source function as

$$\begin{aligned} S_\nu(\mathbf{r}, \mathbf{\Omega}, t) &= [1 - \omega_\nu(r, t)] B_\nu(r, t) \\ &+ \frac{\omega_\nu(r, t)}{4\pi} \int \int p(\nu', \Omega'; \nu, \mathbf{\Omega}; r, t) I_{\nu'}(\mathbf{r}, \mathbf{\Omega}'; t)\, d\nu'\, d\omega'. \end{aligned} \tag{1.9.2}$$

1.10 Local thermodynamic equilibrium

The state of the gas (the distribution of atoms over bound and free states) in thermodynamic equilibrium is uniquely specified by the thermodynamic variables – the absolute temperature T and the total particle density N. The assumption of LTE gives us the freedom to use (in a stellar atmosphere) the local values of T and N in spite of the gradients that exist in the atmosphere. In LTE, the same temperature is used in the velocity distribution of atoms, ions, electrons, etc. Thus the implications of its assumption are drastic. The velocity distribution of the particles is Maxwellian and the degrees of ionization and excitation are determined by the Saha Boltzmann equation (see Mihalas (1978), Sen and Wilson (1998)).

The principle of detailed balance holds good for every transition. This means that the number of radiative transitions $i \to j$ is balanced by the photoexcitation $j \to i$ transitions, where i and j are the upper and lower levels respectively. Thus,

$$n_i \left[A_{ij} + B_{ij} B_{ij}(\nu, T) \right] = n_j B_{ji} B_{ji}(\nu, T) \quad j < i, \; i = 2, \ldots, \tag{1.10.1}$$

where A_{ij}, B_{ij} and B_{ji} are the Einstein coefficients and $B_{ij}(\nu, T)$ and $B_{ji}(\nu, T)$ are the Planck functions given by

$$B_{ij}(\nu, T) = \frac{2h\nu_{ij}^3}{c^2} \left[\exp\left(\frac{h\nu_{ij}}{kT} \right) - 1 \right]^{-1}. \tag{1.10.2}$$

The radiative ionization from level i is balanced by radiative recombination to i. This gives us

$$n_e\,[A_{ci} + B_{ci}B_{ic}(\nu_{ic}, T)] = n_i B_{ic} B_{ic}(\nu_{ic}, T), \quad i = 1, 2, \ldots, \tag{1.10.3}$$

for collisional transition, with the detailed balance transitions given by the relations

$$n_i C_{ij} = n_j C_{ji}, \quad i, j = 1, 2, \ldots \quad i \neq j, \tag{1.10.4}$$

where the Cs are collisional rates and the subscript c denotes the continuum.

In the LTE situation, the radiative transitions are negligible compared to collisional transitions. This is an important consideration in treating non-LTE conditions in stellar atmospheres.

1.11 Non-LTE conditions in stellar atmospheres

In LTE conditions the particle distribution is Maxwellian. Every transition is exactly balanced by its inverse transition, that is, the principle of detailed balance holds good in LTE. Generally, the excitation and de-excitation of the atomic levels is caused by radiative and collisional processes. In the interior of the stars collisions dominate over the radiative processes and LTE prevails. Near the surface of the atmosphere, the radiative rates are not in detailed balance and there is a strong departure from the LTE situation and then the non-LTE situation exists and one should adopt a joint detailed balancing of the excitation and de-excitation of atomic levels. The LTE condition can be determined by the comparative contribution of collisional rates and radiative rates – dominance of the former prevails in the LTE situation, while the opposite situation leads to a non-LTE situation. In stellar atmospheres, non-LTE predominates and this should be taken into account in any transfer calculations.

Statistical equilibrium equations describe the equilibrium among various processes leading to the establishment of an equilibrium state. The state of the gas is assumed to be described by its kinetic temperature, the degrees of excitation and the ionization of each atomic level. The equations of statistical equilibrium (or rate equations) are used to calculate the occupation numbers of bound and free states of atoms assuming complete redistribution (that is, the emission and absorption profiles are identical) in a steady atmosphere.

Consider the changes in time of the number of particles in a given state i of a chemical species α in a given volume element of a moving medium. The net rate at which particles are brought to state i by radiative and collisional processes is given by

$$\left(\frac{\partial n_{i\alpha}}{\partial t}\right) = \sum_{j \neq i} n_{j\alpha} P^{\alpha}_{ji} - n_{i\alpha} P^{\alpha}_{i} + \nabla \cdot (n_{i\alpha} \cdot \mathbf{V}), \tag{1.11.1}$$

where $\mathbf{V}$ is the velocity of the moving medium and P_{ji} represents the total rate of transfer from level j to level i (radiative and collisional). The second term on the RHS gives the total number of particles entering and leaving the volume element, through the divergence theorem. The total number of particles of type α, N_α, is given by the sum over all states of species α:

$$N_\alpha = \sum_i n_{i\alpha}. \tag{1.11.2}$$

Then we have the continuity equation

$$\left(\frac{\partial N_\alpha}{\partial t}\right) + \nabla \cdot S\,(N_\alpha \mathbf{V}) = 0. \tag{1.11.3}$$

If m_α is the mass of each particle of type α, then by multiplying equation (1.11.3) by m_α and summing over all species of particles in this volume element, we get

$$\rho = \sum_\alpha m_\alpha N_\alpha \tag{1.11.4}$$

and

$$\frac{\partial \rho}{\partial t} + \nabla \cdot (\rho \mathbf{V}) = 0. \tag{1.11.5}$$

If the flow is steady, then

$$\sum_{j \neq i} \left(n_{j\alpha} P_{ji}^{\alpha} - n_{i\alpha} P_{ij}^{\alpha}\right) = \nabla \cdot (n_{i\alpha} \mathbf{V}). \tag{1.11.6}$$

If the atmosphere is static, then equation (1.11.6) becomes

$$n_i \sum_{j \neq i} P_{ij} - \sum_{j \neq i} n_j P_{ji} = 0. \tag{1.11.7}$$

We will write a simple model of the statistical equilibrium equation (see Mihalas and Mihalas (1984), pages 386–398 for a detailed account or Mihalas (1978), chapter 5). The equation for the population n_i is

$$\sum_{k=n+1}^{c} n_k (A_{ki} + B_{ki} \bar{J}_{ik} + n_e C_{ki}) + \sum_{j=1}^{i-1} n_j \left(B_{ji} \bar{J}_{ji} + n_e C_{ji}\right)$$
$$= n_i \left[\sum_{j=1}^{i-1} \left(A_{ij} + B_{ij} \bar{J}_{ji} + n_e C_{ij}\right) + \sum_{k=i+1}^{c} \left(B_{ik} \bar{J}_{ik} + n_e C_{ik}\right)\right], \tag{1.11.8}$$

where $\bar{J}$ is the line profile weighted mean intensity. The terms on the LHS of equation (1.11.8) represent different physical quantities: $\sum n_k (A_{ki} + B_{ki} \bar{J}_{ik})$ represents the spontaneous and stimulated radiative transitions from higher discrete levels; $\sum n_k n_e C_{ki}$ represents the collision induced transitions from upper levels; $\sum n_j B_{ji} \bar{J}_{ji}$ represents the photoexcitation from lower levels; and $\sum n_e n_j C_{ji}$ represents the collisional excitation. Similarly the terms on the RHS of (1.11.8) have the following meanings: $n_i \sum (A_{ij} + B_{ij} \bar{J}_{ji})$ represents the spontaneous and stimulated transitions to lower levels; $n_e n_i \sum C_{ij}$ represents the downward transitions induced by collisions (second kind); $n_i \sum B_{ik} \bar{J}_{ik}$ represents the photoexcitation into higher levels; and $n_e n_i \sum C_{ik}$ represents the upward transitions due to collisions with electrons.

Equation (1.11.8) specifies the gas at a given point in the medium if the radiation field (through $\bar{J}$), temperature and electron density n_e are specified.

1.12 Line source function for a two-level atom

This is one of the most useful quantities in the study of line transfer and has been studied extensively.

Consider two levels 1 and 2 (lower and upper respectively) of an atom. The principle of detailed balance gives us (see Mihalas and Mihalas (1984))

$$g_2 B_{21} = g_1 B_{12} \tag{1.12.1}$$

and

$$A_{21} = \frac{2h\nu_{12}^3}{c^2} B_{21}, \tag{1.12.2}$$

where g_1 and g_2 are the statistical weights, $h\nu_{12}$ is the energy difference between levels 1 and 2 measured relative to the ground state and A and B are the Einstein coefficients. The line absorption coefficient in terms of a convenient width Δs is

$$\kappa_l(\nu) = \frac{h\nu_0}{4\pi \Delta s}(N_1 B_{12} - N_2 B_{21}), \tag{1.12.3}$$

where N_1 and N_2 are the population densities of levels 1 and 2 respectively and ν_0 is the central frequency of the line. The line source function S_L (see Grant and Peraiah (1972)) is now written as

$$S_L = \frac{A_{21} N_2}{(B_{12} N_1 - B_{21} N_2)}. \tag{1.12.4}$$

We will use the following statistical equilibrium equation for a two-level atom:

$$N_1 \left[B_{12} \int_{-\infty}^{+\infty} \phi(x) J(x)\, dx + C_{12} \right] = N_2 \left[A_{21} + C_{21} + B_{21} \int_{-\infty}^{+\infty} \phi(x) J(x)\, dx \right], \tag{1.12.5}$$

where

$$x = \frac{(\nu - \nu_0)}{\Delta s} \tag{1.12.6}$$

and $\phi(x)$ is the line profile function (see below) and then combining (1.12.4) and (1.12.5) we obtain

$$S_L = (1-\epsilon)\int_{-\infty}^{+\infty} \phi(x) J(x)\, dx + \epsilon B, \tag{1.12.7}$$

where

$$\epsilon = \frac{C_{21}}{C_{21} + A_{21}\left[1 - \exp(h\nu_0/kT)\right]^{-1}} \tag{1.12.8}$$

is the probability per scatter that a photon will be destroyed by collisional de-excitation. When $\epsilon = 1$, LTE prevails and if $\epsilon \ll 1$, a non-LTE situation occurs. In equations (1.12.7) and (1.12.8), B is the Planck function, k is the Boltzmann constant and T is the temperature. Sometimes the line source function is written as

$$S_L = \frac{\bar{J} + \epsilon' B}{1 + \epsilon'}, \tag{1.12.9}$$

where

$$\epsilon' = \epsilon/(1-\epsilon) \tag{1.12.10}$$

and

$$\bar{J} = \int_{-\infty}^{+\infty} \phi(x) J(x)\, dx. \tag{1.12.11}$$

The line profiles are given by (Mihalas 1978):

$$\text{Doppler:} \quad \phi(x) = \pi^{-\frac{1}{2}} \exp(-x^2), \tag{1.12.12}$$

$$\text{Lorentz:} \quad \phi(x) = \frac{1}{\pi}\frac{1}{1+x^2}, \tag{1.12.13}$$

$$\text{Voigt:} \quad \phi(x) = a\pi^{-\frac{3}{2}} \int_{-\infty}^{+\infty} \exp(-x^2)\left[(x-y)^2 + a^2\right] dy, \tag{1.12.14}$$

where a is the ratio of the damping width to the Doppler width ($\Gamma/4\pi\,\Delta\nu_D$). The profile $\phi(x)$ is normalized such that

$$\int_{-\infty}^{+\infty} \phi(x)\, dx = 1. \tag{1.12.15}$$

1.13 Redistribution functions

In the process of the formation of spectral lines, we assume that scattering is either coherent or completely redistributed over the profile of the line. These assumptions are ideal and not achieved in real stellar atmospheres. It is necessary to find out how after scattering the photons are redistributed in angle and frequency across the line profile. These calculations are described in the form of partial redistribution functions. First, we consider an atom in its own frame of reference and find the redistribution that happens within the substructure of the bound states. We need to take into account the Doppler redistribution in the frequency produced by the atom's motion. Generally, the directions of the incident and emergent photons are different, therefore the projection of the atom's velocity vector along the propagation vectors will be different for the two photons and a different Doppler shift occurs. This gives rise to the Doppler redistribution. One needs to average over all possible velocities to obtain the final redistribution function. This redistribution function will be used in the line transfer calculation to obtain the correlation (if any) between the incoming and outgoing photons. In what follows, we will give the redistribution functions that will be useful in line transfer (see Hummer (1962), Mihalas (1978)).

The probability of emission of a photon after absorption is

$$R(\nu, \mathbf{q}, \nu', \mathbf{q}')\, d\nu'\, d\Omega'\, d\nu\, d\Omega, \tag{1.13.1}$$

where ν and $\mathbf{q}$ are the frequency and direction of the absorbed photon and ν' and $\mathbf{q}'$ are the frequency and direction of the emitted photon. This probability is subject to the condition

$$\iiiint R(\nu, \mathbf{q}; \nu', \mathbf{q}')\, d\nu'\, d\Omega'\, d\nu\, d\Omega = 1. \tag{1.13.2}$$

Here $d\Omega$ and $d\Omega'$ are the real elements normal to directions $\mathbf{q}$ and $\mathbf{q}'$ respectively. If $\phi(\nu')\, d\nu'$ is the probability that a photon with a frequency in the interval $(\nu, \nu + d\nu)$ is emitted in the interval $(\nu', \nu' + d\nu')$, then

$$4\pi \iint R(\nu', \mathbf{q}'; \nu, \mathbf{q})\, d\nu\, d\Omega = \phi(\nu', \mathbf{q}'), \tag{1.13.3}$$

where $\phi(\nu', \mathbf{q}')$ is the profile function, which is again subjected to the normalization condition that

$$\iint \phi(\nu'\mathbf{q}')\, d\nu'\, d\Omega' = 4\pi. \tag{1.13.4}$$

The redistribution functions are given as follows (the roman subscripts are due to Hummer (1962)):

(a) If we have two perfectly sharp upper and lower states in a bound–bound transition, the photons follow a Doppler redistribution. This does not apply to any real line. This redistribution function is given by (see Hummer (1962) and Mihalas (1978))

$$R_{I-AD}(x, \mathbf{q}; x', \mathbf{q}) = \frac{g(\mathbf{q}, \mathbf{q}')}{4\pi^2 \sin\gamma} \exp\left[-x'^2 - \left(x - x'\cos\gamma\right)^2 \operatorname{cosec}^2\gamma\right], \tag{1.13.5}$$

where R_{I-AD} is the angle dependent redistribution function, the x's are the normalized frequencies (see equation (1.12.6)) and γ is the angle between the vectors $\mathbf{q}$ and $\mathbf{q}'$. For isotropic scattering, the phase function is

$$g_{iso}(\mathbf{q}, \mathbf{q}') = \frac{1}{4\pi}, \tag{1.13.6}$$

and for dipole scattering

$$g_{dip}(\mathbf{q}, \mathbf{q}') = \frac{3}{16\pi}\left(1 + \cos^2 \gamma\right). \tag{1.13.7}$$

The redistribution function for isotropic scattering was first obtained by Thomas (1947).

The angle-averaged redistribution function R_{I-A} is given by

$$R_{I-A}(x, x') = \frac{1}{2}\,\text{erfc}\,|\bar{x}|\,, \tag{1.13.8}$$

where

$$\text{erfc}(x) = 2\pi^{-\frac{1}{2}} \int_x^{\infty} \exp\left(-t^2\right) dt \tag{1.13.9}$$

and

$$|\bar{x}| = \max\left(x, x'\right). \tag{1.13.10}$$

(b) In this case, we have an atom with a perfectly sharp lower state and an upper state broadened by radiative decay or an upper state whose finite life time against radiative decay (back to the lower state) leads to a Lorentz profile. This applies to resonance lines in media of low densities in which collisional broadening of the upper state is negligible, for example, the Lyman alpha line of hydrogen in the interstellar medium. The angle dependent redistribution function is given by

$$R_{II-AD}(x, \mathbf{q}; x', \mathbf{q}') = \frac{g(\mathbf{q}, \mathbf{q}')}{4\pi^2 \sin \gamma} \exp\left[-\left(\frac{x - x'}{2}\right)^2 \text{cosec}^2\left(\frac{\gamma}{2}\right)\right]$$
$$\times H\left(\sigma \sec \frac{\gamma}{2}, \frac{x + x'}{2} \sec \frac{\gamma}{2}\right), \tag{1.13.11}$$

where $\sigma = \delta/\Delta$, $4\pi\delta$ being the sum of the transition probabilities from the concerned states and Δ the Doppler width given by

$$\Delta = \nu_0\left(\frac{v}{c}\right), \quad v = \left(\frac{2kT}{m}\right)^{\frac{1}{2}}, \tag{1.13.12}$$

and H is the Voigt function given by

$$H(a, u) = \frac{a}{\pi} \int_{-\infty}^{+\infty} \exp(-y^2)\left[(u - y)^2 + a^2\right]^{-1} dy. \tag{1.13.13}$$

The function R_{II} was first introduced by Henyey (1941).

The angle-averaged R_{II} function is given by

$$R_{II-A}(x,x') = \pi^{-\frac{3}{2}} \int_{\frac{1}{2}|\bar{x}-\underline{x}|}^{\infty} \exp(-u^2) \left[\tan^{-1} \frac{\underline{x}+u}{\sigma} - \tan^{-1} \frac{\bar{x}-u}{\sigma} \right] du, \tag{1.13.14}$$

where $\bar{x} = \max(|x|, |x|')$ and $\underline{x} = \min(|x|, |x|')$. R_{II-A} was first obtained by Unno (1952) and later by Sobolev (1955). Furthermore,

$$\phi(x) = \int_{-\infty}^{+\infty} R_{II-A(iso)}(x,x')\,dx' = H(a,x), \tag{1.13.15}$$

a being the damping constant.

(c) The atom has a perfectly sharp lower state and a collisionally broadened upper state. All the excited electrons are randomly distributed over the substates of the upper states before emission occurs. In this case, the absorption profile is Lorentzian. The damping comprises radiative and collisional rates and represents the full width of the upper state. The redistribution function R_{III} is given by

$$R_{III-AD}(\nu',\mathbf{q}';\nu,\mathbf{q}) = \frac{g(\mathbf{q}',\mathbf{q})}{\pi^2 \sin\gamma} a \times \int_{-\infty}^{+\infty} \frac{\exp(-u^2) H(a\,\mathrm{cosec}\,\gamma, (x-u\cos\theta)\,\mathrm{cosec}\,\theta)}{(x-u)^2+a^2} du, \tag{1.13.16}$$

where a is the damping constant of the upper level. Heinzel (1981) gives an R_{III} in laboratory frame which is different from that of Hummer (1962):

$$R_{III-AD}(\nu',\mathbf{q}';x,\mathbf{q}) = \frac{g(\mathbf{q}',\mathbf{q})}{4\pi^2 \sin\gamma} \left[H\left(a_j\,\mathrm{cosec}\,\frac{\gamma}{2}, \frac{x-x'}{2}\,\mathrm{cosec}\,\frac{\gamma}{2}\right) \times \exp\left(-\frac{x+x'}{2}\sec^2\frac{\theta}{2}\right) + E_{III}(x',x,\gamma)\right]; \tag{1.13.17}$$

see Heinzel (1981) for $E_{III}(x',x,\gamma)$.

The angle-averaged R_{III-A} is given by

$$R_{III-A}(x',x) = \pi^{-\frac{5}{2}} \int_0^{\infty} \exp(-u^2) \left[\tan^{-1}\left(\frac{x'+u}{a}\right) - \tan^{-1}\left(\frac{x'-u}{a}\right)\right] \times \left[\tan^{-1}\left(\frac{x+u}{a}\right) - \tan^{-1}\left(\frac{x-u}{a}\right)\right] du. \tag{1.13.18}$$

(d) This function applies when a line is formed by an absorption from a broadened state i to a broadened upper state j, followed by a radiative decay to state i. It applies to scattering in subordinate lines. This was derived by several authors with some controversy but we will quote from Hummer (1962):

$$R_{IV-AD}(x', \mathbf{q}'; x, \mathbf{q}) = \frac{g(\mathbf{q}', \mathbf{q})}{2\pi^2 \sin\gamma} \frac{a_i \sec\frac{\gamma}{2}}{\pi}$$
$$\times \int_{-\infty}^{+\infty} \frac{\exp(-y^2) H\left(a_j \operatorname{cosec}\frac{\gamma}{2}, y\cot\frac{\gamma}{2} - x\operatorname{cosec}\frac{\gamma}{2}\right)}{\left[(x - x')\sec\frac{\gamma}{2} - 2y\right]^2 + \left(a_i \sec\frac{\gamma}{2}\right)^2} dy, \quad (1.13.19)$$

and the angle-averaged R_{IV} is

$$R_{IV-A}(x', x) = \pi^{-\frac{5}{2}} a_j \int_0^{+\infty} \exp(-u^2)\, du$$
$$\times \int_{-1}^{+1} \left[\tan^{-1}\left(\frac{x' - x + u(1-\mu)}{a_i}\right) - \tan^{-1}\left(\frac{(x' - x - u(1-\mu)}{a_i}\right)\right]$$
$$\times \frac{d\mu}{(x - \mu u)^2 + a_j^2} du, \quad (1.13.20)$$

where

$$\mathbf{q} \cdot \mathbf{u} = \mu. \quad (1.13.21)$$

(e) Heinzel (1981) has given R_V, which becomes R_I, R_{II} and R_{III} in special cases. R_V is given in the laboratory reference frame by

$$R_V(x', \mathbf{q}'; x, \mathbf{q}) = \frac{g(\mathbf{q}', \mathbf{q})}{4\pi^2 \sin\gamma} \left[H\left(a_j \sec\frac{\gamma}{2}, \frac{x + x'}{2} \sec\frac{\gamma}{2}\right)\right]$$
$$\times H\left(a_i \operatorname{cosec}\frac{\gamma}{2}, \frac{x - x'}{2} \operatorname{cosec}\frac{\gamma}{2}\right) + E_V(x', x, \gamma), \quad (1.13.22)$$

where

$$E_V(x', x, \gamma) = \frac{4}{\pi} \int_{v=0}^{\infty} \int_{u=\epsilon v}^{\infty} \exp\left[-u^2 - v^2 - 2A_j u\right]$$
$$\times \left[\exp(-2A_j u) - \exp(-2A_i \epsilon v)\right] \cos Cu \cos Du\, du\, dv, \quad (1.13.23)$$

with

$$\left.\begin{aligned} &A_j = \alpha' a_j, \quad A_i = \alpha' a_i, \quad \alpha' = \frac{1}{\alpha} = \sec\left(\frac{\gamma}{2}\right), \\ &\beta' = \frac{1}{\beta} = \operatorname{cosec}\left(\frac{\gamma}{2}\right), \quad \epsilon = \frac{\alpha}{\beta}, \\ &C = \alpha'(x + x'), \quad D = \beta'(x - x'), \end{aligned}\right\} \quad (1.13.24)$$

a_j, a_i being the damping parameters. A detailed study is given in Heinzel (1981, 1982), Hubený (1982), Heinzel and Hubený (1983), Hubený *et al.* (1983).

The angle-averaged R_V is given by,

$$R_{V-A}(x', x) = 8\pi^2 \int_0^{\pi} R_V(x', x, \gamma) \sin\gamma \, d\gamma. \tag{1.13.25}$$

The corresponding absorption profile is

$$\phi(x) = \int_{-\infty}^{+\infty} R_V(x', x)\, dx = H(a_i + a_j, x). \tag{1.13.26}$$

The function R_{V-A} has been calculated by Mohan Rao *et al.* (1984).

(f) The redistribution due to electron scattering (see Chandrasekhar (1960), Mihalas (1978)) is given by

$$R_e(\nu', \mathbf{q}'; \nu, \mathbf{q}) = g(\mathbf{q}', \mathbf{q}) \left[\frac{mc^2}{4\pi kT(1-\cos\gamma)^2}\right]^{\frac{1}{2}} \exp\left[\frac{-mc^2(\nu-\nu')^2}{4kT\nu'^2(1-\cos\gamma)}\right]. \tag{1.13.27}$$

From the above equation, the width of the 'line' for an incident monochromatic light scattered in the direction γ is

$$\left[\frac{4kT}{mc^2}\lambda^2(1-\gamma)\right]^{\frac{1}{2}}. \tag{1.13.28}$$

Rangarajan *et al.* (1991) computed the line profiles using the electron redistribution function in the framework of discrete space theory (see chapter 6) (see figure 1.4).

(g) The redistribution function developed by Domke and Hubený (1988) and Streater *et al.* (1988) represents the radiative and collisional redistribution of an arbitrarily polarized radiation in resonance lines. This function is given by (see Nagendra (1994))

$$\begin{aligned} R_{DH}(x, \mu; x', \mu') = {} & W\alpha R^A_{II}(x, \mu; x', \mu')\hat{P}^A(\mu, \mu') \\ & + R^B_{II}(x, \mu; x', \mu')\hat{P}^B(\mu, \mu') + R^C_{II}(x, \mu; x', \mu')\hat{P}^C(\mu, \mu') \\ & + (1-W)\alpha\left[R^A_{II}(x, \mu; x', \mu')\hat{P}^I(\mu, \mu')\right] \\ & + W\beta^{(2)}\left[R^A_{III}(x, \mu; x', \mu')\hat{P}^A(\mu, \mu')\right. \\ & \left. + R^B_{III}(x, \mu; x'\mu')\hat{P}^B(\mu, \mu') + R^C_{III}(x, \mu; x', \mu')\hat{P}^C(\mu, \mu')\right] + \beta^{(0)} \\ & - W\beta^{(2)}\left[R^A_{III}(x, \mu; x', \mu')\hat{P}^I(\mu, \mu')\right], \end{aligned} \tag{1.13.29}$$

where

$$\hat{P}^A(\mu, \mu') = -\frac{3}{2}\begin{pmatrix} 0 & 1 \\ 1 & 0 \end{pmatrix}, \quad \hat{P}^B(\mu, \mu') = \frac{3}{2}\begin{pmatrix} 1 & 1 \\ 1 & 1 \end{pmatrix}. \tag{1.13.30}$$

The frequency redistribution functions in equation (1.13.29) are

$$R^{A,B,C}_{II,III}(x,\mu;x',\mu') = \frac{1}{2\pi}\int_0^{2\pi} R_{II,III}(x,\mu;x',\mu',\Delta)$$
$$\times\left[1, \frac{3}{4}(+\cos^2\gamma), \sin^2\Delta\right] d\Delta \qquad (1.13.31)$$

($\Delta = (\phi - \phi')$) and α is the probability that re-emission of radiation occurs before any type of collision, $\beta^{(0)}$ is the probability that re-emission occurs after an elastic collision but before an inelastic quenching collision, $\beta^{(2)}$ is the probability that re-emission occurs after an inelastic collision changing the phase of the oscillating atomic dipole without changing the alignment and W is the probability that intrinsic level depolarization does not occur during scattering. Nagendra (1994) used the redistribution function R_{DH} (equation (1.13.29)) to study the radiation field in spherical atmospheres.

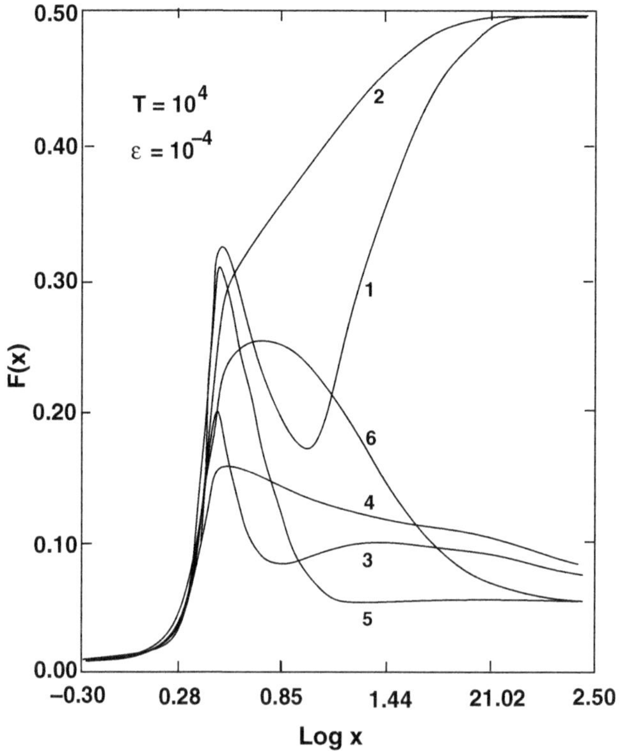

Figure 1.4 Emergent flux is plotted for a line with total line centre optical depth $T = 10^4$ and $\epsilon = 10^{-4}$. Odd numbers in the figure represent partial redistribution (PRD) results and even numbers represent those of CRD (complete redistribution). The curves labelled 1 and 2 are the results without electron scattering and those numbered 3 and 4 represent non-coherent scattering with $\beta_e = 10^{-5}$, where β_e is the ratio of electron scattering to the line absorption coefficient. Curves 5 and 6 represent the results for coherent electron scattering with the same β_e value (from Rangarajan *et al.* (1991), with permission).

(h) The Rayleigh redistribution matrix given by (see Chandrasekhar (1960))

$$\hat{R}_{Rayl}(x,\mu;x',\mu') = \delta(x-x')\frac{3}{16} \times \begin{pmatrix} 3-\mu^2-\mu'^2+3\mu^2\mu'^2 & (1-3\mu'^2)(1-\mu^2) \\ (1-\mu^2)(1-3\mu'^2) & 3(1-\mu^2)(1-\mu'^2) \end{pmatrix}, \quad (1.13.32)$$

is the coherent limit of the Compton scattering redistribution matrix for photon energies $x \ll 1$. The Compton redistribution matrix is given by (Nagirner and Poutanen 1994)

$$\hat{R}(x,\mu;x',\mu') = \frac{3}{16\pi}\frac{x}{x'}\int_0^{2\pi} d\varphi\,\delta\left[x'-x-xx'(1-\cos\Theta)\right] \times \hat{P}(x,\mu;x',\mu';\varphi), \quad (1.13.33)$$

where

$$\hat{P}(x,\mu;x',\mu';\varphi) = \begin{pmatrix} 1+\cos^2\Theta+\omega_c & \cos^2\Theta-1+2(1-\mu^2)\sin^2\varphi \\ \cos^2\Theta-1+2(1-\mu'^2)\sin^2\varphi & 1+\cos^2\Theta-2(\mu^2+\mu'^2)\sin^2\varphi \end{pmatrix}, \quad (1.13.34)$$

where $\omega_c(=xx'(1-\cos\Theta)^2)$ is the Compton depolarization factor and Θ is the scattering angle given by

$$\cos\Theta = \mu\mu' + (1-\mu^2)^{\frac{1}{2}}(1-\mu'^2)^{\frac{1}{2}}\cos\varphi. \quad (1.13.35)$$

The Dirac δ-function in equation (1.13.32) retains the momentum in Compton scattering. Integrating equation (1.13.33) over φ, we get

$$\hat{\mathbf{R}}_{Comp}(x,\mu;x',\mu') = \frac{3}{8\pi x'^2}\frac{1}{|\sin\varphi_0|}\hat{\mathbf{P}}(x,\mu;x',\mu',\varphi_0), \quad |\cos\varphi_0| \le 1, \quad (1.13.36)$$

where

$$\cos\phi_0 = \left(\cos\Theta - \mu\mu'\right)(1-\mu^2)^{-\frac{1}{2}}(1-\mu'^2)^{-\frac{1}{2}},$$
$$\cos\Theta = 1 - \frac{1}{x} + \frac{1}{x'}. \quad (1.13.37)$$

$\hat{R}_{Comp}(x,\mu;x'\mu') = 0$ for $|\cos\varphi_0| > 1$ – a condition of cut-offs in the redistribution matrix at scattering angles given by

$$\cos\Theta_{\pm} = \mu\mu' \pm (1-\mu^2)^{\frac{1}{2}}(1-\mu'^2)^{\frac{1}{2}}. \quad (1.13.38)$$

The elements of the $\hat{\mathbf{R}}_{Comp}$ matrix satisfy certain symmetry relations (see Poutanen *et al.* (1990)). The fluorescent line redistribution matrix (isotropic and unpolarized) is given by

$$\hat{\mathbf{R}}_{fluor}(x, \mu; x', \mu') = \frac{1}{2}\left[Y_\alpha \delta(x - x_\alpha) + Y_\beta \delta(x - x_\beta)\right] \frac{\sigma_{Fe}(x')}{\sigma_T} \times H(x' - x_c)\left(1 - \frac{1}{J_{Fe}}\right)\begin{pmatrix} 1 & 0 \\ 0 & 0 \end{pmatrix}, \quad (1.13.39)$$

where the fluorescent yields are $Y_\alpha = 0.035$ and $Y_\beta = 0.035$ (Kikoin 1976) for the 6.4 keV K_α and the 7.06 keV k_β lines of Fe I respectively, x_α, x_β are the corresponding centroid energies, $\sigma_{Fe}(x)$ is the photoelectric absorption cross section for Fe I and J_e is the absorption-edge jump (see Fernández *et al.* (1993)). $H(x' - x)$ is the Heaviside function which accounts for the absorption threshold at x_c corresponding to 7.1 keV for Fe I K lines.

The above redistribution functions have been used in Compton scattering problems by Poutanen *et al.* (1990).

Rangarajan *et al.* (1990) studied non-LTE line transfer with stimulated emission. They obtained the ratio of emission to absorption profiles $\psi(x)/\phi(x)$ using the R_{II}

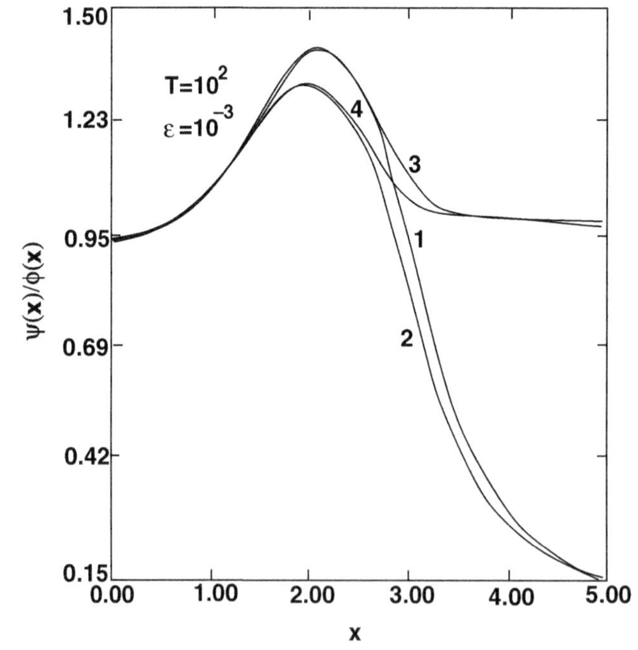

Figure 1.5 Ratio of the emission profile to the absorption profile for a self emitting plane parallel medium. The curves labelled 1 and 2 denote the results for R_{II} function with stimulated emission parameter $\rho = 0$ and 2 respectively where $\rho = [\exp(h\nu/kT) - 1]^{-1}$. Corresponding results for R_{III} function are shown by the curves labelled 3 and 4 (from Rangarajan *et al.* (1990), with permission).

and R_{III} functions. The function R_{III} gives the same profile for the emission as for absorption and is similar to that of the complete redistribution (CRD) in the core and wings except at a few intermediate frequency points whether stimulated emission exists or not. The emission and absorption profiles are different by several factors in the case of the R_{II} redistribution function (see figure 1.5).

1.14 Variable Eddington factor

The quantity $f_\nu(r,t) = K_\nu(r,t)/J_\nu(r,t)$ is called the Eddington factor. This depends on the isotropy of the radiation field. It changes normally from 1/3 to 1 in a stellar atmosphere and is therefore also called the variable Eddington factor.

Exercises

1.1(a) Derive Snell's law from the principle that a light ray travels in the path that requires least time. (Hint: use figure 1.6.)

(b) If n is the refractive index of the medium and I is the specific intensity, show that $n^{-2}I$ is constant along the path of the ray.

(c) Show that the specific intensity is invariant along the path of the ray in free space.

1.2(a) Show that the density of radiation on the surface of a star is $(2\pi/c)I_\nu$.

(b) If I is constant in the interval $0 < \theta \leq \pi/2$, show that the flux is equal to πI_ν.

(c) If I_ν is constant, show that the energy density of radiation at a distance r from the centre of the star is given by

$$\frac{2\pi I_\nu}{c}\left[1 - \sqrt{1-(r_*/r)^2}\right],$$

where c is the velocity of light and r_* is the radius of the star. The quantity $W = 1 - \sqrt{1-(r_*/r)^2}$ is called the dilution factor. Show that it is equal to 1/2 on the surface of the star and to $(r_*/2r)^2$ far away from the star.

(d) With constant I_ν, show that the flux is given by $\pi I_\nu (r_*/r)^2$.

1.3 Show that the direction cosines of the direction of propagation of radiation in spherical polar coordinates (with $d\omega = \sin\theta\, d\theta\, d\varphi$) are $(1-\mu^2)^{1/2}\cos\varphi$, $(1-\mu^2)^{1/2}\sin\varphi$ and μ, where $\mu = \cos\theta$.

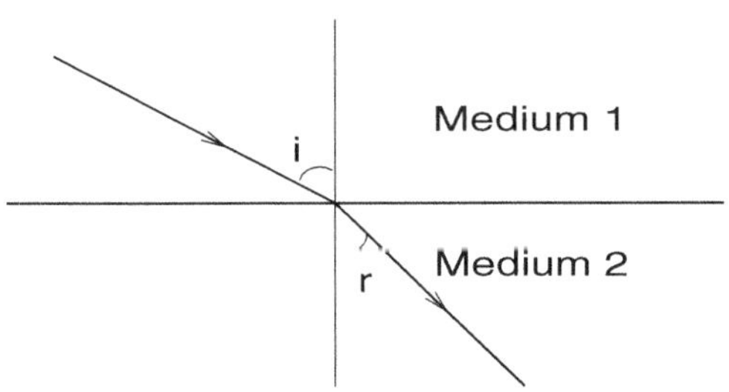

Figure 1.6 Schematic diagram of Snell's law: i and r are the incident and refraction angles respectively.

1.4 Verify that if I_ν is independent of φ, the azimuthal angle, the x- and y-components of the flux F_x, F_y vanish and that in a spherically symmetric medium F_r is non-zero and is given by

$$F_\nu(r,t) = 2\pi \int_{-1}^{+1} I(r,\mu,t)\mu\, d\mu.$$

1.5 If $I = \sum_{n=0}^{\infty} I_0 \mu^n$, where I_0 is a constant, show that only odd powers of μ will contribute to the flux and only even powers will contribute to the mean intensity.

1.6(a) If R is the radius of a star at a distance D from an observer ($D \gg R$) and if no radiation falls on the star from outside ($I(R,-\mu,\nu) = 0$), show that the flux from the star received by the observer is

$$2\pi \left(\frac{R}{D}\right)^2 \int_0^1 I(R,\mu,\nu)\mu\, d\mu.$$

(b) If I is independent of μ, write the expression for J, H and K in terms of (R/D) and show that as $(D/R) \to \infty$, $J = H = K \to 0$.

1.7 Show that $B(T) = \int_0^\infty B_\nu(T)\, d\nu = \sigma T^4$, where $B_\nu(T)$ is the Planck function, $\sigma = 2\pi^5 k^4/15c^2h^3$ and σ is called the Stefan–Boltzmann constant and is equal to 5.67×10^{-5} erg cm^{-2} s^{-1} deg^{-4}. (Hint: use the series $\sum_{n=1}^{\infty} 1/n^4 = \pi^4/90$. This is a Riemann zeta-function.)

1.8 Write down all the components of the stress tensor. From these relations show that the mean pressure is given by $\vec{p} = \frac{1}{3}[p_r(xx) + p_r(yy) + p_r(zz)]$.

1.9 Calculate the value of f, the Eddington factor: (a) when $I(\mu) = I_0 + \sum_n^\infty I_n \mu^n$, where the summation includes only odd powers of n and (b) when I is different, say a_1 and a_2, in the two ranges $(0 \le \mu \le 1)$ and $(-1 \le \mu \le 0)$.

1.10 Calculate J, H or F and K if $I(\mu) = |\mu|$.

1.11 Show that the angle-averaged R_I and R_{II} functions with dipole scattering are given respectively by

$$\begin{aligned} R_{I-A(dipole)}(x,x') &= \frac{3}{8}\pi^{-\frac{1}{2}} \int_{|\bar{x}|}^{\infty} \exp(-u^2) \\ &\quad \times \left[3 - \left(\frac{x}{u}\right)^2 - \left(\frac{x'}{u}\right)^2 + 3\left(\frac{x}{u}\right)^2 \left(\frac{x'}{u}\right)^2\right] du \\ &= \frac{3}{8}\left\{\frac{1}{2}\operatorname{erfc}(|\bar{x}|)\left[3 + 2(x^2 + x'^2) + 4x^2x'^2\right]\right. \\ &\quad \left. - \pi^{\frac{1}{2}} e^{-|\bar{x}|^2} |\bar{x}|(2|\underline{x}|^2 + 1)\right\} \end{aligned}$$

and

$$R_{II-A(dipole)}(x, x') = \frac{3}{8}\pi^{-\frac{3}{2}}\sigma \int_{\frac{1}{2}|\bar{x}-\underline{x}|}^{\infty} \exp(-u^2)$$
$$\times \int_{\bar{x}-u}^{\underline{x}+u} \left[3 - \left(\frac{x-t}{u}\right)^2 - \left(\frac{x'-t}{u}\right)^2\right.$$
$$\left.+3\left(\frac{x-t}{u}\right)^2 \left(\frac{x'-t}{u}\right)^2\right] \frac{dt\,du}{t^2+\sigma^2}.$$

1.12 Using expression (1.13.28), calculate the width for $T = 10\,000$ K, 30 000 K and $\lambda = 4000$ Å, 6000 Å for $\gamma = \pi/2$.

REFERENCES

Chandrasekhar, S., 1938, *An Introduction to the Study of Stellar Structure.* Dover, New York.

Chandrasekhar, S., 1960, *Radiative Transfer*, Dover, New York.

Collins, II, G.W., 1973, *A&A*, **26**, 315.

Domke, H., Hubený, I., 1988, *ApJ*, **334**, 527.

Fernández, J.E., Hubbel, J.H., Hanson, A.L., Spencer, L.V., 1993, *Radiat. Phys. Chem.*, **41**, 579.

Grant, I.P., Peraiah, A., 1972, *MNRAS*, **160**, 239.

Heinzel, P., 1981, *JQSRT*, **25**, 483.

Heinzel, P., 1982, *JQSRT*, **27**, 1.

Heinzel, P., Hubený, I., 1983, *JQSRT*, **30**, 77.

Henyey, L.G., 1941, *Proc. Nat. Acad. Sci.*, **26**, 50.

Hubený, I., 1982, *JQSRT*, **27**, 593.

Hubený, I., Oxenius, J., Simonneau, E., 1983, *JQSRT*, **29**, 477.

Hummer, D.G., 1962, *MNRAS*, **125**, 21.

Kandel, A.S., 1973, *A&A*, **22**, 155.

Kikoin, R.S., 1976, *A&A*, **22**, 155, L56.

Mihalas, D., 1978, *Stellar Atmospheres*, Freeman and Company, San Francisco.

Mihalas, D., Mihalas, B.W., 1984, *Foundation of Radiation Hydrodynamics*, Oxford University Press, New York.

Milne, E.A., 1930, *Handbuch der Astrophysik*, Vol. 3, Part I, Springer, Berlin.

Mohan Rao, D., Rangarajan, K.E., Peraiah, A., 1984, *J. Astrophys. Astr.*, **5**, 169.

Nagendra, K.N., 1994, *ApJ*, **432**, 274.

Nagirner, S.I., Poutanen, J., 1994, in *Astrophysics and Space Physics Reviews*, Vol. 9, ed. Sunyaev, R.A., Harwood, New York, page 1.

Poutanen, J., Nagendra, K.N., Svensson, R., 1990, *MNRAS*, **283**, 892.

Rangarajan, K.E., Mohan Rao, D., Peraiah, A., 1990, *A&A*, **235**, 305.

Rangarajan, K.E., Mohan Rao, D., Peraiah, A., 1991, *MNRAS*, **250**, 633.

Rybicki, G.B., 1969, *Spectrum Formation in Stars with Steady State Extended Atmospheres*, eds. Groth and Wellman, IAU Coll. No. 2, NBS Special Publication, Vol. 332, page 96.

Sen, K.K., Wilson, S.J., 1998, *Radiative Transfer in Moving Media*, Springer, Singapore.

Sobolev, V.V., 1955, Vestnik Leningradskogo Gosudarst vennogo Universitata, No. 5, 85.

Sobolev, V.V., 1963, *A Treatise on Radiative Transfer* (translated by S.I. Gaposchkin), Van Nostrand Company Inc., New York.

Streater, A., Cooper, J., Rees, D.E., 1988, *ApJ*, **335**, 503.

Thomas, R.N., 1947, *ApJ*, **125**, 260.

Unno, W., 1952, *Publ. Astron. Soc. Japan*, **4**, 100.

Chapter 2

The equation of radiative transfer

2.1 General derivation of the radiative transfer equation

The equation of radiative transfer is the mathematical expression of the conservation of radiant energy. We assume that radiation with intensity $I_\nu(\mathbf{r}, \mathbf{\Omega}, t)$ in the frequency interval $d\nu$, passes in time dt through an element of length ds and cross section $d\sigma$ normal to the direction of the ray $\mathbf{\Omega}$ into the solid angle $d\omega$ (see figure 2.1). Let the intensity of the radiation emerging at $\mathbf{r} + \Delta\mathbf{r}$ at the end of time $t + \Delta t$ be $I_\nu(\mathbf{r} + \Delta\mathbf{r}, \mathbf{\Omega}, t + \Delta t)$. This energy is the difference between the energy absorbed and that emitted in the volume element, therefore,

$$\begin{aligned}[I_\nu(\mathbf{r} + \Delta r, \mathbf{\Omega}, t + \Delta t) - I_\nu(\mathbf{r}, \mathbf{\Omega}, t)]\, d\sigma\, d\omega\, d\nu\, dt \\ = [j_\nu(\mathbf{r}, \mathbf{\Omega}, t) - \kappa_\nu(\mathbf{r}, \mathbf{\Omega}, t) I_\nu(\mathbf{r}, \mathbf{\Omega}, t)]\, ds\, d\sigma\, d\omega\, d\nu\, dt, \qquad (2.1.1)\end{aligned}$$

where j_ν and κ_ν are the emission and absorption coefficients respectively. Let s be the path length traced by the ray in passing through this volume element, then $\Delta t = \Delta s / c$ and

$$I_\nu(\mathbf{r} + \Delta\mathbf{r}, \mathbf{\Omega}, t + \Delta t) - I_\nu(\mathbf{r}, \mathbf{\Omega}, t) = \left(\frac{1}{c}\frac{\partial I_\nu}{\partial t} + \frac{\partial I_\nu}{\partial s}\right) ds, \qquad (2.1.2)$$

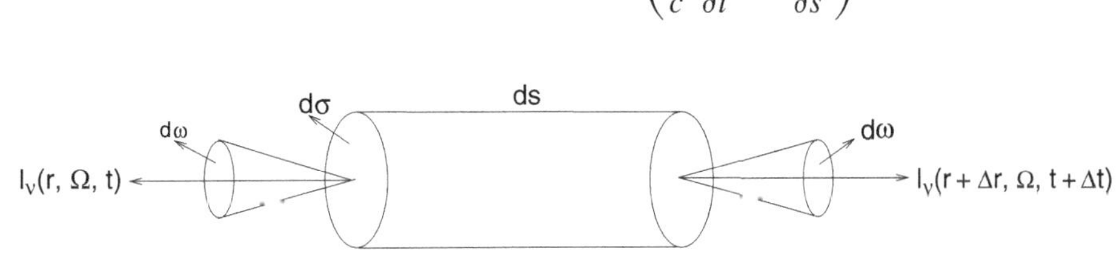

Figure 2.1 Schematic diagram of transfer of radiation.

c being the velocity of light. From equations (2.1.1) and (2.1.2) we obtain the transfer equation:

$$\left(\frac{1}{c}\frac{\partial}{\partial t}+\frac{\partial}{\partial s}\right)I_\nu(\mathbf{r},\mathbf{\Omega},t)=j_\nu(\mathbf{r},\Omega,t)-\kappa_\nu(\mathbf{r},\mathbf{\Omega},t)I_\nu(\mathbf{r},\mathbf{\Omega},t). \tag{2.1.3}$$

In the Cartesian coordinate system, the derivative $\partial/\partial s$ can be written as

$$\begin{aligned}\frac{\partial I_\nu}{\partial s}&=\left(\frac{\partial x}{\partial s}\right)\left(\frac{\partial I_\nu}{\partial x}\right)+\left(\frac{\partial y}{\partial s}\right)\left(\frac{\partial I_\nu}{\partial y}\right)+\left(\frac{\partial z}{\partial s}\right)\left(\frac{\partial I_\nu}{\partial z}\right)\\&=l\left(\frac{\partial I\nu}{\partial x}\right)+m\left(\frac{\partial I\nu}{\partial y}\right)+n\left(\frac{\partial I_\nu}{\partial z}\right),\end{aligned} \tag{2.1.4}$$

where l, m and n are the direction cosines of the ray, $\mathbf{\Omega}$. Equation (2.1.3) can also be written in divergent form as

$$\frac{1}{c}\frac{\partial I_\nu}{\partial t}+\operatorname{div}(\mathbf{\Omega}I_\nu)=j_\nu-\kappa_\nu I_\nu. \tag{2.1.5}$$

Generally, we study the time independent transfer equation, in which case equation (2.1.3) becomes

$$\frac{\partial I_\nu(\mathbf{r},\mathbf{\Omega})}{\partial s}=j_\nu(\mathbf{r},\mathbf{\Omega})-\kappa_\nu(r,\mathbf{\Omega})I_\nu(\mathbf{r},\mathbf{\Omega}). \tag{2.1.6}$$

In one-dimensional geometry when the medium is divided into plane parallel layers and if the angle made by the ray with the normal (z-axis) to the layers is $\theta(\mu=\cos\theta)$, equation (2.1.4) is written as

$$\mu\frac{dI_\nu}{dz}=j_\nu-\kappa_\nu I_\nu. \tag{2.1.7}$$

This is the usual equation of radiative transfer in plane parallel atmospheres.

2.2 The time-independent transfer equation in spherical symmetry

In this case the intensity changes with the radius vector $\mathbf{r}$ and the angle made by the ray with the radius vector. In the plane parallel case the angle made by the ray with the normal to the layers is constant. The transfer equation is

$$\frac{\partial I_\nu}{\partial s}=j_\nu-\kappa_\nu I_\nu. \tag{2.2.1}$$

Now the derivative $\partial I_\nu/\partial s$ can be written

$$\frac{\partial I_\nu}{\partial s}=\frac{\partial I_\nu}{\partial r}\frac{\partial r}{\partial s}+\frac{\partial I_\nu}{\partial\theta}\frac{\partial\theta}{\partial s}. \tag{2.2.2}$$

The ray path along s in the spherically symmetric case is illustrated in figure 2.2. ds is the length element in the direction of θ at r, so we have

$$dr = \cos\theta\, ds \quad \text{and} \quad r\, d\theta = -\sin\theta\, ds, \tag{2.2.3}$$

in which case, equation (2.2.2) becomes

$$\cos\theta \frac{\partial I_\nu}{\partial r} - \frac{\sin\theta}{r}\frac{\partial I_\nu}{\partial \theta} = j_\nu - \kappa_\nu I_\nu. \tag{2.2.4}$$

Or if we put $\mu = \cos\theta$, equation (2.2.4) becomes

$$\mu \frac{\partial I_\nu}{\partial r} + \frac{1-\mu^2}{r}\frac{\partial I_\nu}{\partial \mu} = j_\nu - \kappa_\nu I_\nu. \tag{2.2.5}$$

Alternatively the above equation can be derived as follows: from figure 2.2,

$$p^2 + s^2 = r^2 \quad \text{or} \quad \frac{\partial r}{\partial s} = \cos\theta = \frac{s}{r}, \tag{2.2.6}$$

where p is the impact parameter and we obtain

$$-r\sin\theta \frac{\partial \theta}{\partial s} + \cos\theta \frac{\partial r}{\partial s} = 1$$

or

$$-r\sin\theta \frac{\partial \theta}{\partial s} + \cos^2\theta = 1;$$

therefore

$$\frac{\partial \theta}{\partial s} = -\frac{1-\cos^2\theta}{r\sin\theta}. \tag{2.2.7}$$

Substituting equations (2.2.6) and (2.2.7) into equations (2.2.1) and (2.2.2) we obtain

$$\cos\theta \frac{\partial I_\nu}{\partial r} - \frac{1-\cos^2\theta}{r\sin\theta}\frac{\partial I_\nu}{\partial \theta} = j_\nu - \kappa_\nu I_\nu. \tag{2.2.8}$$

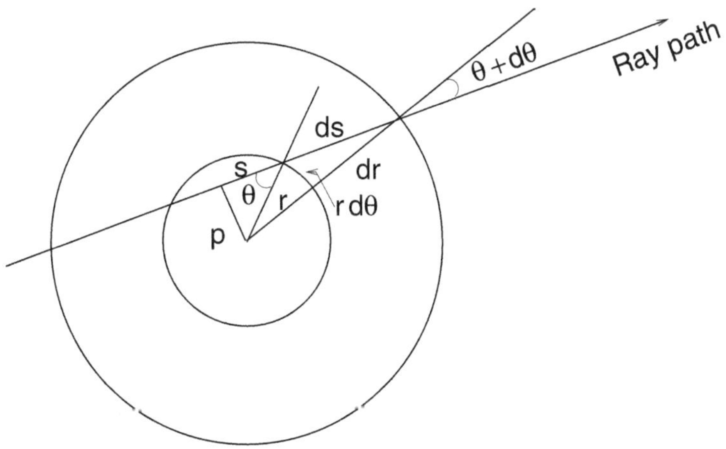

Figure 2.2 Ray path in spherical symmetry.

Using the fact that $\partial\mu = -\sin\theta\partial\theta$, equation (2.2.8) can be written as

$$\mu\frac{\partial I_\nu}{\partial r} + \frac{1-\mu^2}{r}\frac{\partial I_\nu}{\partial\mu} = j_\nu - \kappa_\nu I_\nu. \tag{2.2.9}$$

Equation (2.1.7) for plane parallel layers differs from equation (2.2.9) for spherical symmetry by the term

$$\frac{1-\mu^2}{r}\frac{\partial I_\nu}{\partial\mu}.$$

The geometrical meaning is illustrated in figures 2.3 and 2.4.

In plane parallel geometry the ray makes a constant angle θ with the normal while in the spherically symmetric shells the angle made by the radius vector and the ray direction changes constantly. This change is represented by the term

$$\frac{1-\mu^2}{r}\frac{\partial I_\nu}{\partial\mu}$$

in equation (2.2.5). The ray appears to peak with the radius vector towards the outer boundary of the spherical symmetry.

2.3 Cylindrical symmetry

We consider a cylindrical system in the z-direction in which r is the perpendicular distance from the z-direction. The direction of the ray $\mathbf{\Omega}$ requires two angles to be specified. One is the angle θ between the z-axis and $\mathbf{\Omega}$ and the other is the

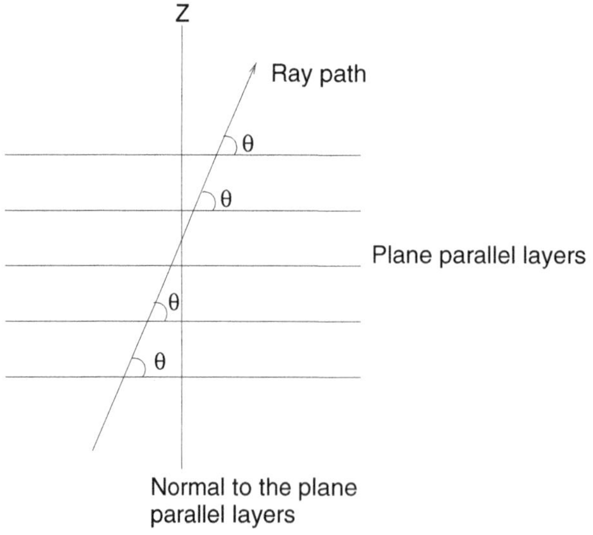

Figure 2.3 Ray path in plane parallel layers. Notice that the ray path does not change its angle with the normal to the plane parallel layers.

corresponding azimuthal angle φ, which is the angle between the projection of $\mathbf{\Omega}$ on the x–y plane and the coordinate r. The spatial derivative $\partial I_\nu/\partial s$ then becomes

$$\frac{\partial I_\nu}{\partial s} = \frac{\partial I_\nu}{\partial r}\left(\frac{dr}{ds}\right) + \frac{\partial I_\nu}{\partial \theta}\left(\frac{d\theta}{ds}\right) + \frac{\partial I_\nu}{\partial \varphi}\left(\frac{d\varphi}{ds}\right). \tag{2.3.1}$$

This translates into

$$\sin\theta\cos\varphi\frac{\partial I_\nu}{\partial r} - \frac{\sin\theta\sin\varphi}{r}\frac{\partial I_\nu}{\partial \varphi} = j_\nu - \kappa_\nu I_\nu. \tag{2.3.2}$$

2.4 The transfer equation in three-dimensional geometries

We will not go into the details of the derivation of the transfer equation in other geometries but will sketch them briefly. No symmetry is assumed (see Pomraning (1973)).

(a) Cartesian system

The transfer equation in the Cartesian system is given by

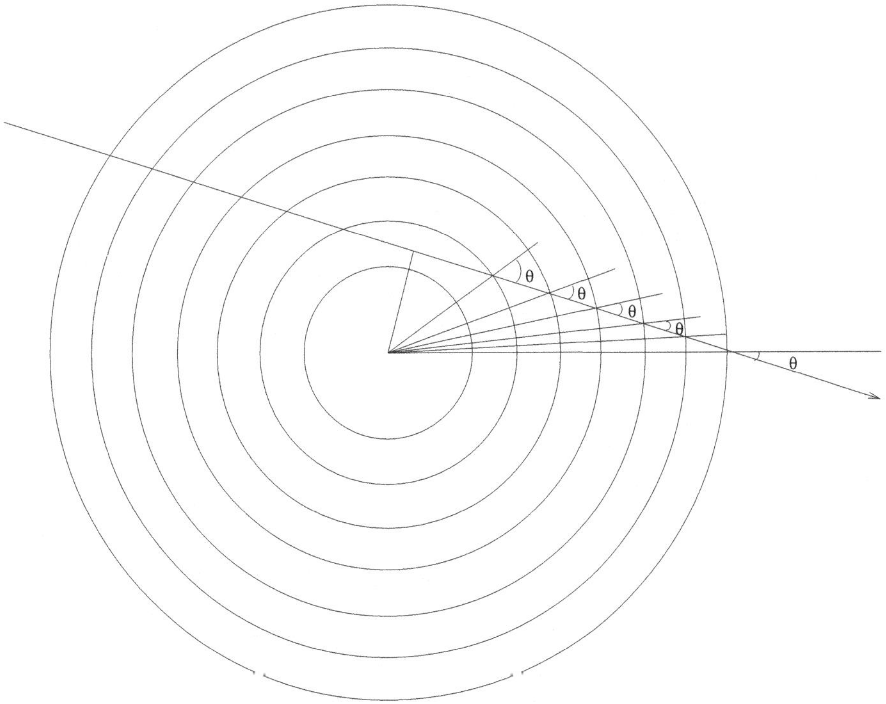

Figure 2.4 Ray path in spherically symmetric shells. Notice how the angle made by the ray with the radius vector changes.

$$\mu \frac{\partial I_\nu}{\partial x} + \eta \frac{\partial I_\nu}{\partial y} + \xi \frac{\partial I_\nu}{\partial z} = j_\nu - \kappa_\nu I_\nu, \tag{2.4.1}$$

where μ, η and ξ are the direction cosines of the direction $\mathbf{\Omega}$ with respect to the x-, y- and z-axes respectively. These direction cosines are related to each other by

$$\mu^2 + \eta^2 + \xi^2 = 1. \tag{2.4.2}$$

If any two of the three direction cosines are fixed then the third one is automatically known.

(b) General spherical geometry

Let Θ be the polar angle between the radius vector and the x-axis and Φ be the azimuthal angle between the x-axis and the projection of $\mathbf{r}$ in the x–y plane. The direction of $\mathbf{\Omega}$ is described by the polar angle θ between $\mathbf{r}$ and $\mathbf{\Omega}$ and the azimuthal angle φ between $\mathbf{r}$ and the projection of $\mathbf{\Omega}$ in the x–y plane. Then the transfer equation is

$$\mu \frac{\partial I_\nu}{\partial r} + \frac{\eta}{r} \frac{\partial I_\nu}{\partial \Theta} + \frac{\xi}{r \sin\Theta} \frac{\partial I_\nu}{\partial \Phi} + \frac{1-\mu^2}{r} \frac{\partial I_\nu}{\partial \mu} - \frac{\xi \cot\Theta}{r} \frac{\partial I_\nu}{\partial \varphi} = j_\nu - \kappa_\nu I_\nu, \tag{2.4.3}$$

where

$$I_\nu = I_\nu(r, \Theta, \Phi, \mu, \varphi) \tag{2.4.4}$$

and

$$\left.\begin{aligned} \mu &= \cos\theta, \\ \eta &= \sin\theta \cos\varphi, \\ \xi &= \sin\theta \sin\varphi. \end{aligned}\right\} \tag{2.4.5}$$

Equation (2.4.5) satisfies the relation (2.4.2).

(c) General cylindrical coordinates

The coordinates are r the perpendicular distance from the z-axis, the z coordinate (from x–y plane) and the angle Θ between r and the x-axis. The direction $\mathbf{\Omega}$ is set by the polar angle θ between the z-axis and $\mathbf{\Omega}$ and the azimuthal angle φ between r and the projection of $\mathbf{\Omega}$ in the x–y plane. Then the transfer equation is

$$\mu \frac{\partial I_\nu}{\partial r} + \frac{\eta}{r} \frac{\partial I_\nu}{d\Theta} + \xi \frac{\partial I_\nu}{\partial z} - \frac{\eta}{r} \frac{\partial I_\nu}{\partial \varphi} = j_\nu - \kappa_\nu I_\nu, \tag{2.4.6}$$

where

$$I_\nu = I_\nu(r, \Theta, z, \theta, \varphi) \tag{2.4.7}$$

and

$$\left.\begin{aligned} \xi &= \cos\theta, \\ \mu &= \sin\theta\cos\varphi, \\ \eta &= \sin\theta\sin\varphi, \end{aligned}\right\} \tag{2.4.8}$$

which follow the relation

$$\xi^2 + \mu^2 + \eta^2 = 1. \tag{2.4.9}$$

(d) Boltzmann equation for photons

The Boltzmann equation describes particle transport in kinetic theory. For a detailed description see Mihalas (1978) and Mihalas and Mihalas (1984). When the particles considered are photons, the Boltzmann equation becomes the transfer equation.

Let us have a system of particles with a distribution function $f(\mathbf{r}, \mathbf{p}, t)$, which gives the number density of particles in the phase space volume element $(\mathbf{r}, \mathbf{r} + d\mathbf{r}, \mathbf{p}, \mathbf{p}+d\mathbf{p})$, where $\mathbf{p}$ is the momentum. If we allow the function f to evolve within a particular phase space element during time interval dt during which $\mathbf{r} \to \mathbf{r} + \mathbf{v}\,dt$ and $\mathbf{p} \to \mathbf{p} + \mathbf{F}\,dt$, where $\mathbf{v}$ and $\mathbf{F}$ are the velocity and external force acting on the particle respectively, then the phase space element evolves from

$$(d^3r)_0(d^3p)_0 \to (d^3r)(d^3p) = J\left[(d^3r)_0(d^3p)_0\right], \tag{2.4.10}$$

where J is the Jacobian of the transformation. To the first order in dt, the Jacobian of the transformation of phase volume element is $J = 1$, which means that the phase volume remains unchanged although the phase space element is deformed. In the presence of a continuous external force $\mathbf{F}$, the deformation of the phase space element is continuous and all particles which were originally within the volume remain there and the volume and hence the particle density remain unchanged. If collisions occur, there will be a reshuffling from one element of phase space to another 'discontinuously', meaning that their neighbourhood remains unaffected during the same time interval. This leads us to the fact that the number density in a phase space element must equal the net number density introduced into the element by collisions, or

$$\begin{aligned} \frac{\partial f}{\partial t} &+ \left(\frac{\partial x}{\partial t}\right)\left(\frac{\partial f}{\partial x}\right) + \left(\frac{\partial y}{\partial t}\right)\left(\frac{\partial f}{\partial y}\right) + \left(\frac{\partial z}{\partial t}\right)\left(\frac{\partial f}{\partial z}\right) \\ &+ F_x\left(\frac{\partial f}{\partial p_x}\right) + F_y\left(\frac{\partial f}{\partial p_y}\right) + F_z\left(\frac{\partial f}{\partial p_z}\right) = \left(\frac{Df}{Dt}\right)_{coll}. \end{aligned} \tag{2.4.11}$$

This can also be written as,

$$\frac{\partial f}{\partial t} + (\mathbf{v}\cdot\boldsymbol{\nabla})f + (\mathbf{F}\cdot\boldsymbol{\nabla}_p)f = \left(\frac{Df}{Dt}\right)_{coll}. \tag{2.4.12}$$

For photons (rest mass = 0) and when no general relativistic effects are present, $\mathbf{F} = 0$, and photon propagation will be in straight lines with $\mathbf{v} = c\mathbf{n}$ and the frequency remaining constant.

The photon redistribution function is defined in such a way that $f(\mathbf{r}, \boldsymbol{\Omega}, \nu, t)\, d\omega\, d\nu$ is the number of photons per unit volume at point $\mathbf{r}$ and time t, with frequencies in the range $(\nu, \nu + d\nu)$, propagating with velocity c in the direction $\boldsymbol{\Omega}$ into a solid angle $d\omega$. Each photon has an energy $h\nu$. The number of photons crossing an element $d\sigma$ in time dt is $f(c \cdot dt)(\boldsymbol{\Omega} \cdot d\sigma)\, d\omega\, d\nu$ so that the energy transferred is

$$dE = ch\nu f \cos\theta\, d\omega\, d\nu\, dt. \tag{2.4.13}$$

In terms of specific intensity, dE is given by

$$dE = I(\mathbf{r}, \boldsymbol{\Omega}, \nu, t)\, d\sigma \cos\theta\, d\nu\, dt, \tag{2.4.14}$$

where θ is the angle between the direction of the beam and the normal to the surface or $d\sigma \cos\theta = \boldsymbol{\Omega} \cdot \mathbf{d\sigma}$. Therefore the distribution function can be written in terms of the specific intensity through equations (2.4.13) and (2.4.14). The photon interactions with the material and the net number of photons introduced into the volume will be the energy emitted minus the energy absorbed divided by the energy of each photon. Therefore for photons equation (2.4.12) can be written as

$$\frac{1}{ch\nu}\left[\frac{\partial I_\nu}{\partial t} + c\,(\boldsymbol{\Omega} \cdot \boldsymbol{\nabla})\, I_\nu\right] = \frac{j_\nu - \kappa_\nu I_\nu}{h\nu}, \tag{2.4.15}$$

which is same as equation (2.1.5). It is now understood that the transfer equation is a Boltzmann equation for a fluid which is not subject to external forces but which suffers strong collisional effects.

(e) The transfer equation in various curvilinear coordinate systems

For the sake of completeness we write a general formula for the differential operation in equation (2.1.5) in an arbitrary orthogonal coordinate system. Let (a_1, a_2, a_3) be a system of orthogonal curvilinear coordinates connected to the Cartesian coordinates (x, y, z) by the relations (see Grant (1968))

$$x = x(a_1, a_2, a_3), \quad y = y(a_1, a_2, a_3), \quad z = z(a_1, a_2, a_3). \tag{2.4.16}$$

The square of an element of arc length is given by

$$ds^2 = h_1^2\, da_1^2 + h_2^2\, da_2^2 + h_3^2\, da_3^2, \tag{2.4.17}$$

where

$$h_i^2 = \left(\frac{\partial x}{\partial a_i}\right)^2 + \left(\frac{\partial y}{\partial a_i}\right)^2 + \left(\frac{\partial z}{\partial a_i}\right)^2, \quad i = 1, 2, 3. \tag{2.4.18}$$

If f is any scalar function, then the components of (grad f) with respect to unit vectors along the orthogonal curvilinear directions are given by

$$(\text{grad } f)_{a_i} = \frac{1}{h_i}\frac{\partial f}{\partial a_i}, \quad i = 1, 2, 3 \tag{2.4.19}$$

and

$$\text{div}\,\mathbf{F} = \frac{1}{h_1 h_2 h_3}\left[\frac{\partial}{\partial a_1}(h_2 h_3 F_{a_1}) + \frac{\partial}{\partial a_2}(h_3 h_1 F_{a_2}) + \frac{\partial}{\partial a_3}(h_1 h_2 F_{a_3})\right]. \tag{2.4.20}$$

The volume element dV in orthogonal curvilinear coordinates is given by

$$dV = h_1 h_2 h_3 \, da_1 \, da_2 \, da_3. \tag{2.4.21}$$

Integration of div $\mathbf{F}$ over a volume given by $m_i \leq a_i \leq n_i, i = 1, 2, 3$ is then,

$$\int \text{div}\,\mathbf{F}\, dV = \left[\iint h_2 h_3 F_{a_1} \, da_2 \, da_3\right]_{m_1}^{n_1} + \left[\iint h_3 h_1 F_{a_2} \, da_3 \, da_1\right]_{m_2}^{n_2} + \left[\iint h_1 h_2 F_{a_3} \, da_1 \, da_2\right]_{m_3}^{n_3}. \tag{2.4.22}$$

We give a few special cases below:

1. Rectangular coordinates:

$$h_1 = h_2 = 1, \quad da_1 = dx, \quad da_2 = dy, \quad da_3 = dz. \tag{2.4.23}$$

2. Spherical coordinates:

$$\left.\begin{aligned} &x = r \sin\theta \cos\phi, \quad y = r \sin\theta \sin\phi, \quad z = r \cos\theta, \\ &h_1 = 1, \quad h_2 = r, \quad h_3 = r \sin\theta, \\ &da_1 = dr, \quad da_2 = d\theta, \quad da_3 = d\phi. \end{aligned}\right\} \tag{2.4.24}$$

3. Cylindrical coordinates:

$$\left.\begin{aligned} &x = r \cos\theta, \quad y = r \sin\theta, \quad z, \\ &h_1 = 1 \quad h_2 = r, \quad h_3 = 1, \\ &da_1 = dr, \quad da_2 = d\theta, \quad da_3 = dz. \end{aligned}\right\} \tag{2.4.25}$$

If we have $\mathbf{F} = \boldsymbol{\Omega} I$, then div$(\boldsymbol{\Omega} I)$ is written as

$$\begin{aligned} \text{div}(\boldsymbol{\Omega} I) &= \frac{1}{h_1 h_2 h_3}\left[\frac{\partial}{\partial a_1}\left(h_2 h_3 \Omega_{a_1} I\right) + \frac{\partial}{\partial a_2}(h_3 h_1 \Omega_{a_2} I) + \frac{\partial}{\partial a_3}(h_1 h_2 \Omega_{a_3} I)\right] \\ &= \frac{\Omega_{a_1}}{h_1}\frac{\partial I}{\partial a_1} + \frac{\Omega_{a_2}}{h_2}\frac{\partial I}{\partial a_2} + \frac{\Omega_{a_3}}{h_3}\frac{\partial I}{\partial a_3} = \boldsymbol{\Omega} \cdot \nabla I. \end{aligned} \tag{2.4.26}$$

Equation (2.4.26) is satisfied because of the following relation:

$$
\begin{aligned}
0 &= \operatorname{div} \boldsymbol{\Omega} \\
&= \frac{1}{h_1 h_2 h_3}\left[\frac{\partial}{\partial a_1}\left(h_2 h_3 \Omega_{a_1}\right) + \frac{\partial}{\partial a_2}(h_3 h_1 \Omega_{a_2}) + \frac{\partial}{\partial a_3}(h_1 h_2 \Omega_{a_3})\right].
\end{aligned}
\tag{2.4.27}
$$

Now one can derive the standard expressions for the differential operator of radiative transfer by substituting equations (2.4.23)–(2.4.25) into equation (2.4.27). We note that

$$
\Omega_{a_1} = \mu, \quad \Omega_{a_2} = \eta, \quad \Omega_{a_3} = \xi, \tag{2.4.28}
$$

where

$$
\mu^2 + \eta^2 + \xi^2 = 1. \tag{2.4.29}
$$

One can choose μ to be an independent quantity and then write $\eta^2 = (1-\mu^2)\sin^2\omega$, $\xi^2 = (1-\mu^2)\cos\omega$, although there is freedom in choosing this kind of representation. For example, in spherical geometry,

$$
\boldsymbol{\Omega} \cdot \nabla I = \mu \frac{\partial I}{\partial r} + \frac{(1-\mu^2)^{\frac{1}{2}} \sin\omega}{r} \frac{\partial I}{\partial \theta} + \frac{(1-\mu^2)^{\frac{1}{2}} \cos\omega}{r \sin\theta} \frac{\partial I}{\partial \phi}. \tag{2.4.30}
$$

If there is spherical symmetry, the axis can always be rotated so that $\boldsymbol{\Omega}$ lies in a plane $\phi =$ constant. Then $\omega = 3\pi/2$ and if Ω is parallel to the axis $\theta = 0$, $\mu = \cos\theta$ and we get the familiar equation for $\boldsymbol{\Omega} \cdot \nabla I$:

$$
\boldsymbol{\Omega} \cdot \nabla I = \mu \frac{\partial I}{\partial r} + \frac{1-\mu^2}{r} \frac{\partial I}{\partial \mu}, \tag{2.4.31}
$$

where μ is the cosine of the angle between $\boldsymbol{\Omega}$ and the radius vector.

2.5 Optical depth

We shall restrict ourselves to the time independent plane parallel transfer equation (2.1.7). The optical depth is an important quantity that is used in transfer theory. It is defined as

$$
\tau(z, \nu) = -\int_{z_1}^{z_2} \kappa_\nu(z', \nu)\, dz'. \tag{2.5.1}
$$

The negative sign appears because we shall adopt the convention that z and τ_ν run in opposite directions. $\tau_\nu(z)$ gives the integrated absorption of radiation along the z-direction in the segment $(z_2 - z_1)$. In the context of a stellar atmosphere τ_ν increases from the outside into the star while z (or r in the case of the spherical atmosphere) increases from the inside towards outside. That is,

$$\left.\begin{aligned} &\text{at } z = z_{max}, \quad \tau_\nu = 0, \\ &\text{at } z = 0, \quad \tau_\nu = \tau_{\nu,max}. \end{aligned}\right\} \tag{2.5.2}$$

The optical depth gives an estimate of how much one can 'see' through the medium. We have seen earlier that κ_ν^{-1} is the photon mean free path. This means that τ_ν gives the number of photon mean free paths.

2.6 Source function in the transfer equation

Another quantity that plays an important role in the study of radiative transfer is the source function (see chapter 1). This is defined as the ratio of total emission to the total absorption, or

$$S_\nu = j_\nu / \kappa_\nu. \tag{2.6.1}$$

The time independent transfer equation in plane parallel atmospheres (2.1.7) is then written as

$$\mu \frac{\partial I_\nu}{\partial \tau_\nu} = I_\nu - S_\nu. \tag{2.6.2}$$

The absorption coefficient κ_ν consists of contributions from pure or true absorption and scattering processes. Thus

$$\kappa_\nu = K_\nu + \sigma_\nu, \tag{2.6.3}$$

where K_ν is due to true absorption (see chapter 1) and the scattering term σ_ν consists of coherent, non-coherent, isotropic continuum scattering terms such as Thomson scattering or Rayleigh scattering. The emission consists of thermal emission and scattering and is given by

$$j_\nu = j_\nu^t + j_\nu^s = K_\nu B_\nu + \sigma_\nu J_\nu. \tag{2.6.4}$$

Therefore, the source function obtained from equation (2.6.1) is

$$S_\nu = (K_\nu B_\nu + \sigma_\nu J_\nu)/(K_\nu + \sigma_\nu). \tag{2.6.5}$$

If we have local thermodynamic equilibrium, then

$$S_\nu = B_\nu(T), \tag{2.6.6}$$

where $B_\nu(T)$ is the Planck function. The equation of transfer (2.6.2) is then

$$\mu \frac{\partial I_\nu}{\partial \tau} = I_\nu - B_\nu(T). \tag{2.6.7}$$

In the case of spectral lines, we need to consider the absorption from continuum and line:

$$\kappa_\nu = \kappa_c + \kappa_l(\nu) = \kappa_c + \kappa_L \phi_\nu, \tag{2.6.8}$$

where κ_c and κ_l are continuum and line absorption coefficients respectively. ϕ_ν is the profile function which is normalized such that

$$\int_{-\infty}^{+\infty} \phi(\nu)\, d\nu = 1. \tag{2.6.9}$$

The line source function comprises contributions from: (1) the continuum thermal emission κ_c, (2) a fraction ϵ (where ϵ is the probability per scatter that a photon is destroyed by collisional de-excitation) of the emission $\kappa_L \phi \epsilon B_\nu$ that comes from the thermal processes, and (3) the redistributed (complete or partial) photons. Thus the source function is given by (see chapter 1),

$$S_\nu = \kappa_L \phi(\nu)(1-\epsilon) \int_{-\infty}^{+\infty} J(\nu)\phi(\nu)\, d\nu + \kappa_L \phi_\nu \epsilon B_\nu + \kappa_c(\nu) B_\nu. \tag{2.6.10}$$

This is one of the more widely studied source functions for a two-level atom model.

2.7 Boundary conditions

We need to specify the boundary values of the specific intensities at the boundaries shown in figure 2.4. In stellar atmospheres we have two types of media: (1) finite media and (2) semi-infinite media. A finite medium is of finite optical depth with open boundaries on both sides which need the incident radiation on both sides of the medium, while a semi-infinite medium has an open boundary on one side but the other side is so thick that it is assumed that it extends to infinity.

We prescribe for example in stellar atmospheres

$$I_\nu^- = I(\tau = 0, \mu, \nu) \qquad (-1 \le \mu < 0) \tag{2.7.1}$$

and

$$I_\nu^+ = I(\tau = \tau_{max}, \mu, \nu) \qquad (0 \le \mu \ge 1) \tag{2.7.2}$$

(see figure 2.5).

In stellar atmospheres, we set $I_\nu^- = 0$ and I_ν^+ is given a finite value. In planetary work and binary stars, $I_\nu^- \neq 0$ and this plays the pivotal role particularly in planetary atmospheres. In the case of the lower boundary we set $\tau \to \infty$, and the diffusion approximation may be applied here.

2.8 Media with only either absorption or emission

When the medium neither absorbs (and/or scatters) nor emits radiation (vacuum) we have $j_\nu = \kappa_\nu = 0$, and the equation

$$\mu \frac{\partial I_\nu}{\partial z} = j_\nu - \kappa_\nu I_\nu \tag{2.8.1}$$

becomes

$$\mu \frac{\partial I_\nu}{\partial z} = 0, \tag{2.8.2}$$

or I_ν = constant, which is consistent with the idea that the specific intensity is invariant in the absence of sources and sinks. When the medium only absorbs and there is no emission from the medium, then

$$\mu \frac{\partial I_\nu}{\partial \tau_\nu} = -I_\nu. \tag{2.8.3}$$

Integrating the above equation, we get

$$I_\nu(\tau, \mu) = I_0(\nu) \exp[-(\tau_\nu - \tau_{\nu 0})/\mu]. \tag{2.8.4}$$

The incident intensity $I_0(\nu)$ is reduced by a factor of $\exp[-(\tau_\nu - \tau_{\nu 0})/\mu]$ when it passes through an absorbing medium with an optical depth of $(\tau_\nu - \tau_{\nu 0})$. If the

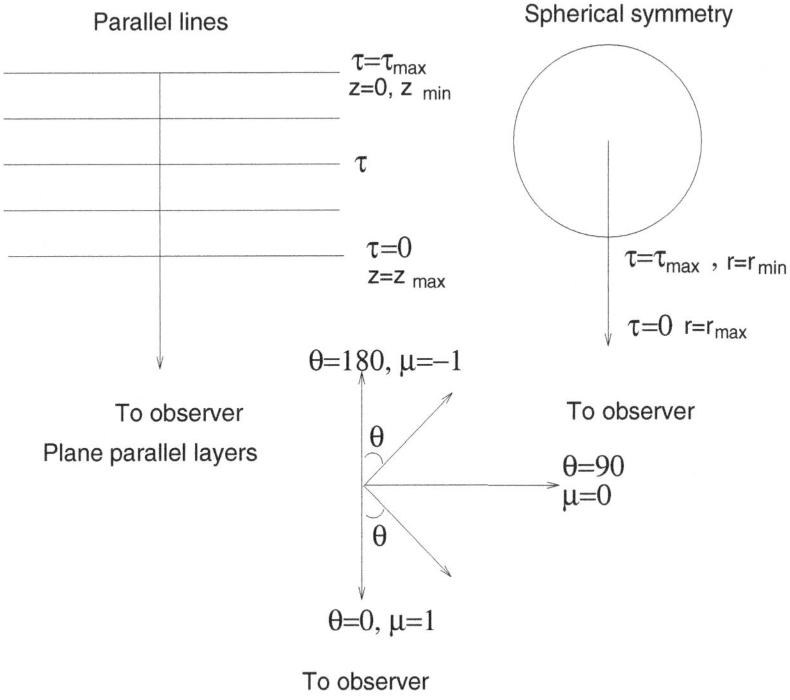

Figure 2.5 Boundary condition in plane parallel and spherical symmetries.

medium only emits and no absorption takes place, then the transfer equation is written as

$$\mu \frac{\partial I_\nu(z,\mu)}{\partial z} = j_\nu(z) \tag{2.8.5}$$

or

$$I_\nu(\mu, z) = \mu^{-1} \int_0^z j_\nu(z)\, dz + I_0(0, \nu, \mu), \tag{2.8.6}$$

which occurs in scattering media, such as planetary nebulae where optically forbidden lines are formed. Atoms that are excited to metastable levels due to collisions remain unperturbed in this state for longer times, due to low nebular densities and therefore little chance of collisions. When a large number of these atoms accumulate, some of them decay through 'forbidden' transitions with a very small (non-zero) transition probability of emitting photons. As these photons come through forbidden lines their probability of reabsorption is very small and they escape the nebulae.

2.9 Formal solution of the transfer equation

The general transfer equation in general is given by (from equation (2.1.6))

$$-\frac{\partial I_\nu}{\kappa_\nu \partial s} = I_\nu - S_\nu. \tag{2.9.1}$$

The optical depth is defined as (see figure 2.6)

$$\tau_\nu(s, s_1) = \int_{s_1}^{s} \kappa_\nu\, ds. \tag{2.9.2}$$

Equation (2.9.1) is a first order linear differential equation with a constant coefficient and therefore it will have the integration factor $\exp(-\tau_\nu)$. Using this we can immediately write the formal solution as (we leave out the subscript ν for simplicity)

$$I_\nu(s) = I_\nu(0) \exp[-\tau_\nu(s, 0)] + \int_0^s S_\nu(s') \exp[-\tau_\nu(s, s')]\kappa_\nu\, ds'. \tag{2.9.3}$$

The intensity $I_\nu(s)$ at a point s is given by equation (2.9.3). The first term on the RHS represents the intensity at $s = 0$ reduced by the factor $\exp[-\tau_\nu(s, 0)]$ and the

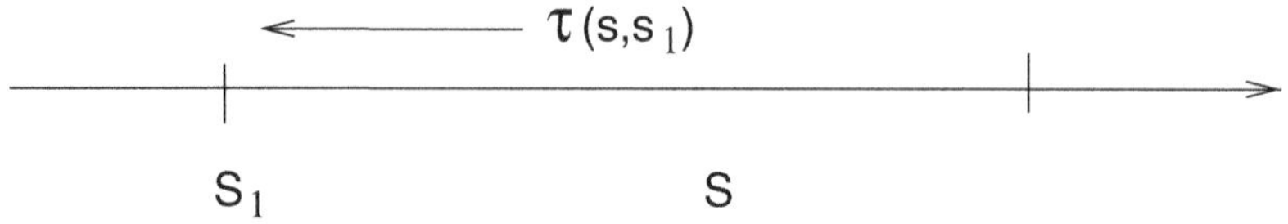

Figure 2.6 Direction of the optical depth.

second term represents the intensity at any point and in a given direction resulting from the emission at all the interior points s' reduced by the factor $\exp[-\tau_\nu(s, s')]$ to allow for absorption by the intervening matter. If the matter extends to infinity then the $I_\nu(0)\exp[-\tau(s, 0)]$ term vanishes and equation (2.9.3) reduces to

$$I_\nu(s) = \int_{-\infty}^{s} S_\nu(s')\exp[-\tau_\nu(s, s')]\kappa_\nu\, ds'. \tag{2.9.4}$$

If the source function S_ν contains the intensity then equations (2.9.3) and (2.9.4) become integral equations.

We shall consider the solution in a plane parallel atmosphere. We shall omit the subscript ν on τ for the sake of convenience. There are two types of atmospheres (see section 2.7): (1) the finite atmosphere which is bounded by the optical depths on the two sides $\tau = 0$ and $\tau = \tau_1$ (say), and (2) the semi-infinite atmosphere which is bounded by the optical depth on one side $\tau = 0$ and by $\tau \to \infty$ on the other side. The intensities at any point in a finite atmosphere are (see Chandrasekhar (1960))

$$I_\nu(\tau, +\mu) = I_\nu(\tau_1, \mu)\exp[-(\tau_1 - \tau)] + \int_{\tau}^{\tau_1} S_\nu(t, \mu)\exp[-(t - \tau)/\mu]\frac{dt}{\mu}$$
$$(1 \geq \mu > 0) \tag{2.9.5}$$

and

$$I_\nu(\tau, -\mu) = I(0, -\mu)\exp(-\tau/\mu) + \int_{0}^{\tau} S_\nu(t, -\mu)\exp[-(\tau - t)/\mu]\frac{dt}{\mu}$$
$$(1 \geq \mu > 0). \tag{2.9.6}$$

Equations (2.9.5) and (2.9.6) are the outward and inward intensities respectively (see figure 2.5), at each level. The emergent intensities are obtained by putting $\tau = 0$ in equation (2.9.5) and by putting $\tau = \tau_1$ in equation (2.9.6). Thus,

$$I_\nu(0, \mu) = I_\nu(\tau_1, \mu)\exp(-\tau_1/\mu) + \int_{0}^{\tau_1} S_\nu(t, \mu)\exp(-t/\mu)\frac{dt}{\mu} \tag{2.9.7}$$

and

$$I_\nu(\tau, -\mu) = I_\nu(0, -\mu)\exp(-\tau_1/\mu) + \int_{0}^{\tau_1} S_\nu(t, -\mu)\exp[-(\tau_1 - t)/\mu]\frac{dt}{\mu}. \tag{2.9.8}$$

Now we shall write similar equations for the semi-infinite atmosphere. These are

$$I(\tau, +\mu) = \int_{\tau}^{\infty} S_\nu(t, +\mu)\exp[-(t - \tau)/\mu]\frac{dt}{\mu}, \tag{2.9.9}$$

$$I_\nu(\tau, -\mu) = I(0, -\mu)\exp(-\tau/\mu) + \int_{0}^{\tau} S_\nu(t, -\mu)\exp[-(\tau - t)/\mu]\frac{dt}{\mu}, \tag{2.9.10}$$

and

$$I_\nu(0, +\mu) = \int_0^\infty S_\nu(t, +\mu)\exp(-t/\mu)\frac{dt}{\mu}. \tag{2.9.11}$$

The RHS of equation (2.9.11) is the Laplace transform of $S_\nu(t, \mu)$.

2.10 Scattering atmospheres

The source function is defined in equations (2.6.1) and (2.6.5) as the ratio of emission and absorption coefficients. The absorption consists of true (or pure) absorption and scattering (see equation (2.6.3)). We need to specify qualitatively the concept of scattering and the angular distribution of the scattered radiation. This is done by the phase function $p(\cos\Theta)$ so that the quantity (see Chandrasekhar (1960))

$$\sigma_\nu I_\nu p(\cos\Theta)\frac{d\omega'}{4\pi}\,dm\,d\nu\,d\omega, \tag{2.10.1}$$

gives the rate at which the energy is scattered into an element of solid angle $d\omega'$ in the direction Θ to the direction of the incident radiation on an element of mass dm, with σ_ν the mass scattering coefficient. From expression (2.10.1), we obtain the rate of loss of energy from the incident pencil due to scattering in all directions as

$$\sigma_\nu I_\nu\,dm\,d\nu\,d\omega\int p(\cos\Theta)\frac{d\omega'}{4\pi}. \tag{2.10.2}$$

This must be equal to

$$\sigma_\nu I_\nu\,dm\,d\nu\,d\omega, \tag{2.10.3}$$

which is the energy scattered from the pencil of radiation incident on the element of mass dm with cross section $d\sigma$, height ds and density ρ or $dm = \rho\cos\theta\,d\sigma\,ds$, or

$$\sigma_\nu I_\nu\,dm\,d\nu\,d\omega = \sigma_\nu I_\nu\,dm\,d\nu\,d\omega\int p(\cos\Theta)\frac{d\omega'}{4\pi}, \tag{2.10.4}$$

or

$$\int p(\cos\Theta)\frac{d\omega'}{4\pi} = 1, \tag{2.10.5}$$

which means that the phase function is normalized to unity. The scattering of a pencil of radiation from the direction (θ', φ') contributes energy to a pencil in the direction $(\theta, \varphi)(\varphi'$ and φ are the azimuthal angles) at a rate of

$$\sigma_\nu\,dm\,d\nu\,d\omega\,p(\theta, \varphi; \theta', \varphi')I_\nu(\theta', \varphi')\frac{\sin\theta'\,d\theta'\,d\varphi'}{4\pi}, \tag{2.10.6}$$

where $p(\theta, \varphi; \theta', \varphi')$ is the phase function for the angle between the directions (θ, φ) and (θ', φ'). Therefore the emission coefficient for scattering, j_ν^s, is

$$j_\nu^{(s)}(\theta,\varphi) = \frac{\sigma_\nu}{4\pi}\int_0^\pi\int_0^{2\pi} p(\theta,\varphi;\theta',\varphi')I_\nu(\theta',\varphi')\sin\theta'\,d\theta'\,d\varphi'. \tag{2.10.7}$$

Generally, the emission coefficients will have contributions from processes other than scattering. If these other processes are absent, we say that we have a scattering atmosphere. The alternative to a scattering atmosphere is an atmosphere in LTE which we have seen earlier. Therefore the source function in a scattering atmosphere is

$$S_\nu^{(s)}(\theta,\varphi) = \frac{1}{4\pi}\int_0^\pi\int_0^{2\pi} p(\theta,\varphi;\theta',\varphi')I_\nu(\theta',\varphi')\sin\theta'\,d\theta'\,d\varphi', \tag{2.10.8}$$

while for an atmosphere in LTE

$$S_\nu(LTE) = B_\nu(T). \tag{2.10.9}$$

We shall show that the flux is constant in a purely scattering atmosphere. Let $\mathbf{n}$ be a unit vector of a certain direction through a point $\mathbf{r}$. Then the source function (2.10.8) can be written as

$$S_\nu(\mathbf{r},\mathbf{n}) = \frac{1}{4\pi}\int p(\mathbf{n},\mathbf{n}')I(\mathbf{r},\mathbf{n}')\,d\omega(n'). \tag{2.10.10}$$

Then in this case the transfer equation can be written:

$$-\frac{1}{\kappa_\nu}(\mathbf{n}\cdot\mathrm{grad})I(\mathbf{r},\mathbf{n}) = I(\mathbf{r},\mathbf{n}) - \frac{1}{4\pi}\int p(\mathbf{n},\mathbf{n}')I(\mathbf{r},\mathbf{n}')\,d\omega(n'). \tag{2.10.11}$$

Integrating over all directions, equation (2.10.11) becomes

$$\begin{aligned}-\frac{1}{\kappa_\nu}(\mathbf{n}\cdot\mathrm{grad})I(\mathbf{r},\mathbf{n})\,d\omega(n) = &\int I(\mathbf{r},\mathbf{n})\,d\omega(n)\\ &-\frac{1}{4\pi}\iint p(\mathbf{n},\mathbf{n}')I(\mathbf{r},\mathbf{n}')\,d\omega(n)\,d\omega(n').\end{aligned} \tag{2.10.12}$$

The quantity on the LHS is the divergence of the vector $\pi\mathbf{F}$ with components parallel to x-, y- and z-axes. The first term on the RHS is $4\pi J$ and the second term is also $4\pi J$ if we use relation (2.10.5), Therefore equation (2.10.12) reduces to

$$\mathrm{div}\,F = 0. \tag{2.10.13}$$

This is the flux integral for a conservative problem. Therefore in a purely scattering atmosphere the flux is constant and in a plane parallel atmosphere it is equal to

$$\frac{dF}{dz} = 0 \quad \text{or} \quad F_z = \text{constant}. \tag{2.10.14}$$

In spherical symmetry, the flux integral becomes

$$F_r = \frac{F_0}{r^2}, \tag{2.10.15}$$

where F_0 is a constant.

2.11 The K-integral

The transfer equation in a plane parallel scattering atmosphere can be written as (see equation (2.10.7))

$$\mu \frac{dI(\tau, \mu, \varphi)}{d\tau} = I(\tau, \mu, \varphi) - \frac{1}{4\pi} \int_{-1}^{+1} \int_{0}^{2\pi} p(\mu, \varphi; \mu', \varphi') I(\tau, \mu', \varphi')\, d\mu'\, d\varphi'. \tag{2.11.1}$$

In the conservative case the albedo for single scattering ω equals 1, and the flux F is constant, where πF represents the flux of radiation normal to the plane of stratification. If we multiply equation (2.11.1) by μ and integrate over all solid angles, we obtain

$$\frac{d}{d\tau} \int_{-1}^{+1} \int_{0}^{2\pi} I(\tau, \mu, \varphi) \mu^2\, d\mu\, d\varphi = \pi F - \frac{1}{4\pi} \int_{-1}^{+1} \int_{0}^{2\pi} d\mu'\, d\varphi'\, I(\tau, \mu', \varphi') \times \int_{-1}^{+1} \int_{0}^{2\pi} d\mu\, d\varphi\, \mu p(\mu, \varphi; \mu', \varphi'). \tag{2.11.2}$$

The phase function $p(\mu, \varphi; \mu', \varphi')$ can be expanded in a series of Legendre polynomials as

$$p(\mu, \varphi; \mu', \varphi') = \sum \omega_l P_l \left[\mu\mu' + \left(1 - \mu^2\right)^{1/2} \left(1 - \mu'^2\right)^{1/2} \cos(\varphi - \varphi') \right], \tag{2.11.3}$$

where $\omega_0 = 1$, which is the albedo for single scattering. By applying the addition theorem of spherical harmonics, we find that

$$\frac{1}{4\pi} \int_{-1}^{+1} \int_{0}^{2\pi} p(\mu, \varphi; \mu', \varphi') \mu\, d\mu\, d\varphi = \frac{1}{2} \omega_1 \mu' \int_{-1}^{+1} \mu^2\, d\mu = \frac{1}{3} \omega_1 \mu'. \tag{2.11.4}$$

Now equation (2.11.2) becomes

$$\frac{1}{4\pi} \frac{d}{d\tau} \int_{-1}^{+1} \int_{0}^{2\pi} I(\tau, \mu, \varphi) \mu^2\, d\mu\, d\varphi = \frac{1}{4} \left(1 - \frac{1}{3} \omega_1\right) F, \tag{2.11.5}$$

or

$$\frac{dK}{d\tau} = \frac{1}{4} \left(1 - \frac{1}{3} \omega_1\right) F, \tag{2.11.6}$$

where

$$\frac{1}{4\pi}\int_{-1}^{+1}\int_{0}^{2\pi} I(\tau,\mu,\varphi)\mu^2\,d\mu\,d\varphi = K(\tau). \tag{2.11.7}$$

As F is constant, equation (2.11.6) can be written as

$$K = \frac{1}{4}F\left[\left(1-\frac{1}{3}\omega_1\right)\tau + Q\right], \tag{2.11.8}$$

where Q is a constant. This is called the K-integral.

2.12 Schwarzschild–Milne equations and Λ, Φ, X operators

We now consider a semi-infinite atmosphere. In a scattering medium, the transfer equation in plane parallel layers with an independent azimuthal angle is

$$\mu\frac{dI_\nu(\tau,\mu)}{d\tau} = I_\nu(\tau,\mu) - \frac{1}{2}\int_{-1}^{+1} p(\mu,\mu')I_\nu(\tau,\mu')\,d\mu', \tag{2.12.1}$$

where the phase function $p(\mu,\mu')$ is given by

$$p(\mu,\mu') = \frac{1}{2\pi}\int_{0}^{2\pi} p(\mu,\varphi;\mu,\varphi')\,d\varphi'. \tag{2.12.2}$$

In the case of isotropic scattering, $p(\mu,\mu') \equiv 1$ and equation (2.12.1) becomes

$$\mu\frac{dI_\nu(\tau,\mu)}{d\tau} = I_\nu(\tau,\mu) - J_\nu(\tau), \tag{2.12.3}$$

where

$$J_\nu(\tau) = \frac{1}{2}\int_{-1}^{+1} I_\nu(\tau,\mu)\,d\mu. \tag{2.12.4}$$

Equation (2.12.3) is the simplest of the transfer equations with scattering to have been studied extensively. If we give no incident radiation at $\tau = 0$, the formal solutions from equations (2.9.9) and (2.9.10) are written as

$$I_\nu(\tau,\mu) = \int_{\tau}^{\infty} \exp[-(t-\tau)/\mu]J_\nu(t)\frac{dt}{\mu} \qquad (0<\mu\le 1), \tag{2.12.5}$$

$$I_\nu(\tau,-\mu) = \int_{0}^{\tau} \exp[-(\tau-t)/\mu]J_\nu(t)\frac{dt}{\mu} \qquad (0<\mu\le 1). \tag{2.12.6}$$

The above solutions can be expressed in terms of $J_\nu(\tau)$. We shall omit the subscript ν for convenience. Let us write that

$$I_n(\tau) = \int_{-1}^{+1} I(\tau,\mu)\mu^n\,d\mu. \tag{2.12.7}$$

We substitute equations (2.12.5) and (2.12.6) for $I(\tau, \mu)$ into equation (2.12.7) and obtain

$$I_n(\tau) = \int_\tau^\infty dt\, J(t) \int_0^1 d\mu\, \mu^{n-1} \exp[-(t-\tau)/\mu] \\ + (-1)^n \int_0^\tau dt\, J(t) \int_0^1 d\mu\, \mu^{n-1} \exp[-(\tau-t)/\mu]. \tag{2.12.8}$$

If we substitute $1/x$ for μ, equation (2.12.8) becomes

$$I_n(\tau) = \int_\tau^\infty dt\, J(t) \int_1^\infty \frac{dx}{x^{n+1}} \exp[-x(t-\tau)] \\ + (-1)^n \int_0^\tau dt\, J(t) \int_1^\infty \frac{dx}{x^{n+1}} \exp[-x(\tau-t)]. \tag{2.12.9}$$

We can write this in terms of the exponential integral:

$$E_n(y) = \int_1^\infty \frac{dx}{x^n} \exp(-xy). \tag{2.12.10}$$

Using the above integral, equation (2.12.9) is written as (see Chandrasekhar (1960))

$$I_n(\tau) = \int_\tau^\infty J(t) E_{n+1}(t-\tau)\, dt + (-1)^n \int_0^\tau J(t) E_{n+1}(\tau-t)\, dt. \tag{2.12.11}$$

Using equation (2.12.4) we can write $J(\tau)$ in terms of the exponential integral E_n as follows:

$$J(\tau) = \frac{1}{2} \int_0^\infty J(t) E_1(|t-\tau|)\, dt. \tag{2.12.12}$$

This is called the Schwarzschild–Milne equation for the mean intensity. Furthermore, F and K are given by

$$F(\tau) = \frac{1}{2} \int_{-1}^{+1} I(\tau, \mu) \mu\, d\mu, \tag{2.12.13}$$

which is called the Milne equation for flux, and

$$K(\tau) = \frac{1}{2} \int_{-1}^{+1} I(\tau, \mu) \mu^2\, d\mu, \tag{2.12.14}$$

and can be written in terms of the exponential integral given in equation (2.12.10) as follows:

$$F(\tau) = 2 \int_\tau^\infty J(t) E_2(t-\tau)\, dt - 2 \int_0^\tau J(t) E_2(\tau-t)\, dt \tag{2.12.15}$$

and

$$K(\tau) = \frac{1}{2}\int_0^\infty J(\tau)E_3|t-\tau|\,dt. \tag{2.12.16}$$

Solving equation (2.12.12) is equivalent of solving transfer equation (2.12.3). The exponential integrals follow the recursion formulae:

$$nE_{n+1}(x) = \exp(-x) - xE_n(x) \quad (n \geq 1), \tag{2.12.17}$$

$$E'_{n+1}(x) = -E_n(x) \quad (n \geq 1) \tag{2.12.18}$$

and

$$E'_1(x) = -\exp(-x)/x. \tag{2.12.19}$$

The first of these relations (equation (2.12.17)) follows from

$$nE_{n+1}(x) = -\int_1^\infty \exp(-xt)\frac{d}{dt}t^{-n}\,dt, \tag{2.12.20}$$

after integration by parts. The second (equation (2.12.18)) follows from direct differentiation of

$$E_n(x) = \int_1^\infty \exp(-xt)\frac{dt}{t^n} = \int_0^1 \exp(-x/\mu)\mu^{n-1}\frac{d\mu}{\mu}. \tag{2.12.21}$$

Furthermore

$$E_n(0) = \int_0^1 \frac{dt}{t^n} = \frac{1}{n-1} \qquad (n \geq 2). \tag{2.12.22}$$

Equations (2.12.12)–(2.12.14) play a most important role in radiative transfer theory and can be written in operator notation. Equation (2.12.12) is written in operator notation as (see Kourganoff (1963))

$$\Lambda_\tau[f(t)] = \frac{1}{2}\int_0^\infty f(t)E_1|t-\tau|\,dt. \tag{2.12.23}$$

Similarly $F(\tau)$ and $K(\tau)$ written in operator notation are

$$\Phi_\tau[f(t)] = 2\int_\tau^\infty f(t)E_2(t-\tau)\,dt - 2\int_0^\tau f(t)E_2(\tau-t)\,dt \tag{2.12.24}$$

and

$$X_\iota[f(t)] - 2\int_0^\infty f(t)E_3|t-\tau|\,dt. \tag{2.12.25}$$

Several properties of these operators are given in Kourganoff (1963), Chapter 2. Some of them are (as quoted in Rutten (1999) also):

$$\left.\begin{aligned} \mathbf{\Lambda}_\tau[1] &= 1 - \frac{1}{2}E_2(\tau), \\ \mathbf{\Lambda}_\tau[t] &= \tau + \frac{1}{2}E_3(\tau), \\ \mathbf{\Lambda}_\tau[t^2] &= \frac{2}{3} + \tau^2 - E_4(\tau), \\ \mathbf{\Lambda}_\tau[t^p] &= \frac{1}{2}p!\left[\sum_{k=0}^{p}\frac{\tau^k}{k!}\delta_\alpha + (-1)^{p+1}E_{p+2}(\tau)\right], \end{aligned}\right\} \tag{2.12.26}$$

where $\delta_\alpha = 0$ for even $\alpha \equiv p + 1 - k$ and $\delta_\alpha = 2/\alpha$ for odd α. The Schwarzschild–Milne equation in terms of $\mathbf{\Lambda}$ the operator is

$$J(\tau) = \mathbf{\Lambda}_\tau[J(t)] = \frac{1}{2}\int_0^\infty J(t)E_1\,|t-\tau|\,dt, \tag{2.12.27}$$

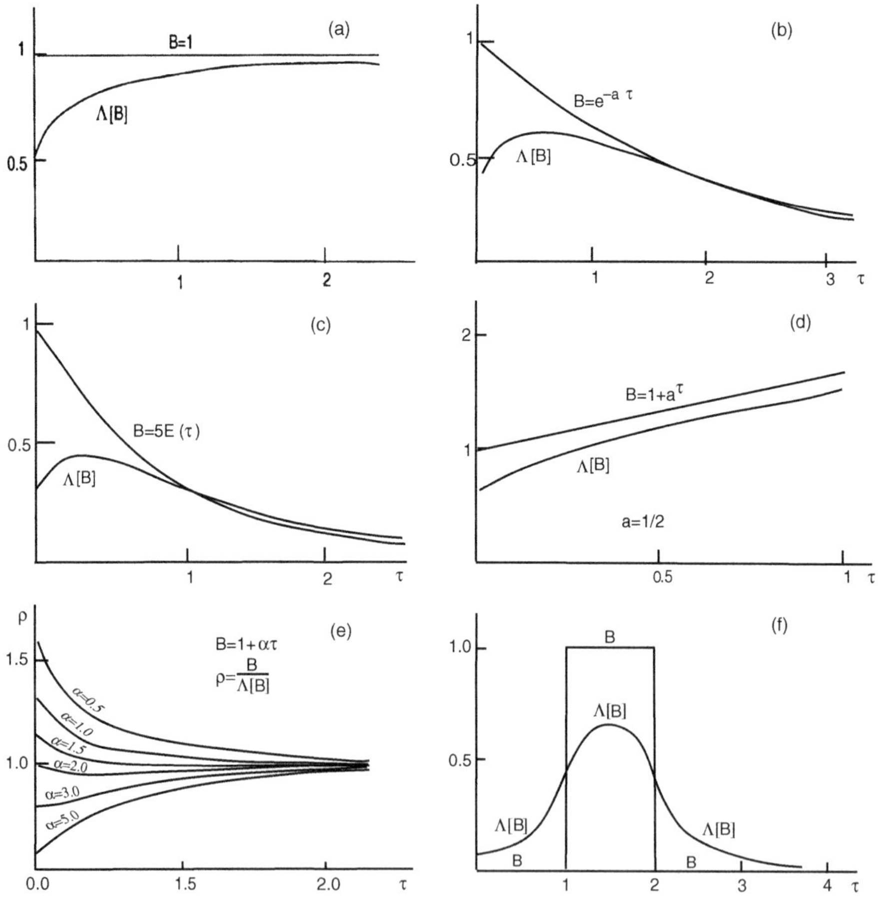

Figure 2.7 (a) B and $\Lambda(B)$ with $B = 1$, (b) B and $\Lambda(B)$ with $B = \exp(-a\tau)$ and $a = \frac{1}{2}$, (c) B and $\Lambda(B)$ with $B = (n-1)E_6(\tau)$, (d) B and $\Lambda(B)$ with $B = 1 + a\tau$ and $a = \frac{1}{2}$, (e) $\rho = B/\Lambda(B)$, $B = 1 + a\tau$, $\frac{1}{2} \leq a \leq 5$, (f) B and $\Lambda(B)$, when B is a 'pulse' function (see Kourganoff (1963)).

or from equations (2.9.9) and (2.9.10) for $I(0, -\mu) = 0$ we can write $J(\tau)$ as

$$J(\tau) = \frac{1}{2}\int_0^\infty S(\tau)E_1(|t-\tau|)\,dt = \mathbf{\Lambda}_\tau[S(t)]. \tag{2.12.28}$$

The operator $\mathbf{\Phi}_\tau$ is

$$\mathbf{\Phi}_\tau[S(\tau)] = 2\int_\tau^\infty S(t)E_2(t-\tau)\,dt - 2\int_0^\tau S(t)E_2(\tau-t)\,dt = F(\tau), \tag{2.12.29}$$

where $F(\tau)$ is the astrophysical flux. The operator $\mathbf{X}$ is

$$\mathbf{X}_\tau[S(t)] = 2\int_0^\infty S(t)E_3(|t-\tau|)\,dt = 4K(\tau). \tag{2.12.30}$$

Figures 2.7(a)–(f) and 2.8(a)–(d) illustrate the results of operators $\mathbf{\Lambda}$ and $\mathbf{\Phi}$ respectively with the source function $S = B$ for a given depth dependence.

2.13 Eddington–Barbier relation

The emergent intensity is (in a semi-infinite plane parallel atmosphere)

$$I(0,\mu) = \int_0^\infty S(t)\exp(-t/\mu)\frac{dt}{\mu} \tag{2.13.1}$$

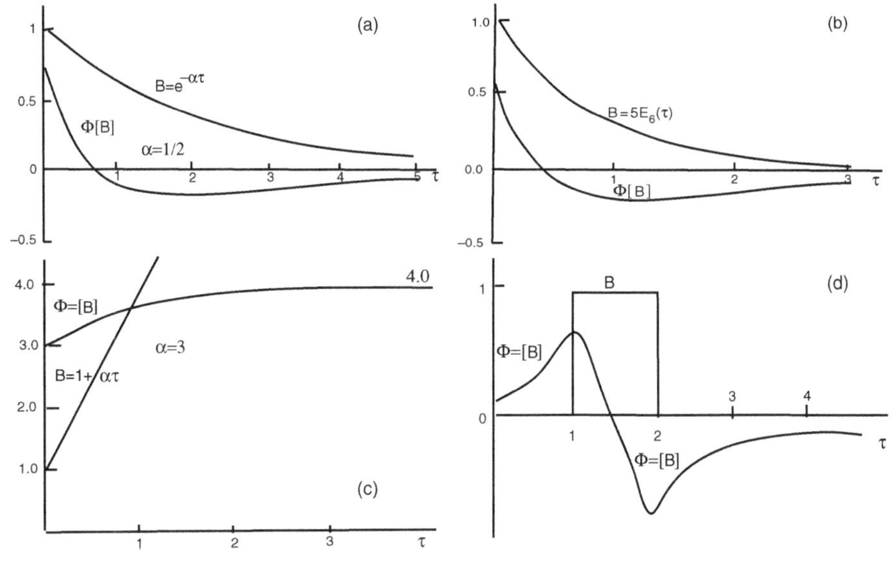

Figure 2.8 (a) B and $\Phi(B)$ with $b = \exp(-a\tau)$, $a = \frac{1}{2}$, (b) B and $\Phi(B)$, $B = 5E_6(\tau)$, (c) B and $\Phi(B)$ with $B = 1 + a\tau$ and $a = 3$, (d) B and $\Phi(B)$ when B is a 'pulse' function (see Kourganoff (1963)).

(see equation (2.9.11)). The emergent intensity is a weighted average of the source function along the line of sight. The weighting function is the fraction of the energy emitted that emerges at the surface from each point along the ray whose optical depth is modified by (τ/μ). The specific intensity is the Laplace transform of the source function. This is useful in solving several problems in radiative transfer theory.

If we assume that the source function varies linearly with the optical depth as

$$S(t) = S_0 + S_1 t, \tag{2.13.2}$$

and substitute this into equation (2.13.1), we obtain

$$I(0, \mu) = S_0 + S_1 \mu = S(\tau = \mu). \tag{2.13.3}$$

Relation (2.13.3) is known as the Eddington–Barbier relation. It implies that the emergent intensity is characteristic of the source function at about an optical depth of unity along the line of sight. This relation has been extensively used in stellar and solar work and has helped in the interpretation of several observations.

2.14 Moments of the transfer equation

In section 1.5 we defined the moments of the radiation field. We now apply these moments to the transfer equation. If we apply the zeroth moment to the transfer equation in plane parallel geometry we obtain

$$\int \mu \frac{dI_\nu}{d\tau} d\mu = \int I_\nu \, d\mu - \int S_\nu \, d\mu. \tag{2.14.1}$$

This reduces to

$$\frac{dH_\nu}{d\tau} = J_\nu - S_\nu. \tag{2.14.2}$$

In spherical geometry the zeroth moment of the transfer equation is

$$\frac{1}{r^2} \frac{\partial (r^2 H_\nu)}{\partial r} = j_\nu - \kappa_\nu J_\nu. \tag{2.14.3}$$

The first order moment gives in plane parallel geometry

$$\frac{\partial K_\nu}{\partial \tau_\nu} = H_\nu \tag{2.14.4}$$

and

$$\frac{\partial K_\nu}{\partial r} + r^{-1}(3K_\nu - J_\nu) = -\kappa_\nu H_\nu. \tag{2.14.5}$$

2.15 Condition of radiative equilibrium

In stellar atmospheres, energy is transported by radiation and convection. Transport by conduction is very ineffective except in the solar corona in which the temperature reaches the order of million degrees. When all the energy is transported by radiation, it is said that the atmosphere is in *radiative equilibrium* (see the discussion given in Mihalas (1978)).

Let us consider a static medium with a time independent radiation field. The condition of radiative equilibrium is that the total energy absorbed by a given volume of material should be equal to the energy emitted. Thus the energy removed from the beam by the volume element is

$$4\pi \int_0^\infty \kappa_\nu J_\nu \, d\nu \tag{2.15.1}$$

and the energy emitted by the volume element is

$$4\pi \int_0^\infty j_\nu \, d\nu = 4\pi \int_0^\infty \kappa_\nu S_\nu \, d\nu. \tag{2.15.2}$$

Therefore at each point in the atmosphere, using (2.15.1) and (2.15.2), we obtain

$$4\pi \int_0^\infty \kappa_\nu [S_\nu - J_\nu] \, d\nu = 0 \tag{2.15.3}$$

or

$$\int_0^\infty \kappa_\nu S_\nu \, d\nu = \int \kappa_\nu J_\nu \, d\nu. \tag{2.15.4}$$

Thus in planar geometry the condition of radiative equilibrium is equivalent to the condition that the derivative of the flux is zero or the flux is constant (see section 2.10). In spherical geometry we have

$$r^2 F = \text{constant} = L/4\pi, \tag{2.15.5}$$

where L is the luminosity.

2.16 The diffusion approximations

If we consider a semi-infinite atmosphere (with boundaries $\tau \to \infty$ at one side and $\tau = 0$ at the other), at large optical depths, the solution of the transfer equation becomes simple. At these depths the photon mean free path is quite small and the radiation is trapped inside the medium becomes isotropic and reaches thermal equilibrium, that is, the source function S_ν approaches the Planck function B_ν. In this case, when $\tau_\nu \gg 1$ one can expand the source function as follows (see Mihalas (1978)):

$$S_\nu(t_\nu) = \sum_{n=0}^{\infty} [d^n B_\nu / d\tau_\nu^n][t_\nu - \tau_\nu]^n / n!. \tag{2.16.1}$$

If we substitute this into equation (2.9.9), we obtain (in the range $0 < \mu \leq 1$)

$$I_\nu(\tau_\nu, \mu) = \sum_{n=0}^{\infty} \mu^n \frac{d^n B_\nu}{d\tau_\nu^n} = B_\nu(\tau_\nu) + \mu \frac{dB_\nu}{d\tau_\nu} + \mu^2 \frac{d^2 B_\nu}{d\tau_\nu^2} + \cdots. \tag{2.16.2}$$

Substitution of equation (2.16.1) into equation (2.9.10) with $I(0, -\mu) = 0$ gives us a similar equation in the range $-1 \leq \mu < 0$. Similarly,

$$J_\nu(\tau_\nu) = \sum_{n=0}^{\infty} (2n+1)^{-1} \frac{d^{2n} B_\nu}{d\tau_\nu^{2n}} = B_\nu(\tau_\nu) + \frac{1}{3} \frac{d^2 B_\nu}{d\tau_\nu^2} + \cdots \quad , \tag{2.16.3}$$

$$H_\nu(\tau_\nu) = \sum_{n=0}^{\infty} (2n+3)^{-1} \frac{d^{2n+1} B_\nu}{d\tau_\nu^{2n+1}} = \frac{1}{3} \frac{dB_\nu}{d\tau_\nu} + \cdots, \tag{2.16.4}$$

$$K_\nu(\tau_\nu) = \sum_{n=0}^{\infty} (2n+3)^{-1} \frac{d^{2n} B_\nu}{d\tau_\nu^{2n}} = \frac{1}{3} B_\nu(\tau_\nu) + \frac{1}{5} \frac{d^2 B_\nu}{d\tau_\nu^2} + \cdots. \tag{2.16.5}$$

J_ν and K_ν contain only even order derivatives, while F_ν contains only odd order derivatives. Equations (2.16.2)–(2.16.5) converge rapidly. For example, if we write $d^2 B_\nu / d\tau_n^n \sim B_\nu / \tau_\nu^n$, then the ratio of successive terms is $1/\tau^2$ or of the order $1/\langle\kappa_\nu\rangle^2 \Delta z^2$, where $\langle\kappa_\nu\rangle$ is the average opacity along the path length Δz. The convergence is quite rapid. In the limit of large optical depths, one can write equations (2.16.2)–(2.16.5) as:

$$I_\nu(\tau_\nu, \mu) \approx B_\nu(\tau_\nu) + \mu \frac{dB_\nu}{d\tau_\nu}, \tag{2.16.6}$$

$$J_\nu(\tau_\nu) \approx B_\nu(\tau_\nu), \tag{2.16.7}$$

$$H_\nu(\tau_\nu) \approx \frac{1}{3} \frac{dB_\nu}{d\tau_\nu} \tag{2.16.8}$$

and

$$K_\nu(\tau_\nu) \approx \frac{1}{3} B_\nu(\tau_\nu). \tag{2.16.9}$$

Note that

$$K_\nu(\tau_\nu)/J_\nu(\tau_\nu) = \frac{1}{3}, \tag{2.16.10}$$

which is true for isotropic radiation.

2.17 The grey approximation

Although the radiation in stellar atmospheres and other objects is highly frequency dependent, the approximation of frequency independent radiative transfer gives some useful results which provide us with a good understanding of the atmospheres, particularly when mean opacities are used. We write $\kappa_\nu = \kappa$ and then

$$I = \int_0^\infty I_\nu \, d\nu. \tag{2.17.1}$$

Similarly J_ν, S_ν, B_ν etc. become J, S, B etc. respectively. Then the transfer equation in plane parallel layers becomes

$$\mu \frac{dI}{d\tau} = I - S. \tag{2.17.2}$$

When radiative equilibrium exists

$$\int_0^\infty \kappa_\nu J_\nu \, d\nu = \int_0^\infty \kappa_\nu S_\nu \, d\nu \tag{2.17.3}$$

or

$$J = S. \tag{2.17.4}$$

Equation (2.17.2) then becomes

$$\mu \frac{dI}{d\tau} = I - J, \tag{2.17.5}$$

the solution of which is

$$J(\tau) = \frac{1}{2} \int_0^\infty J(t) E_1 |t - \tau| \, dt. \tag{2.17.6}$$

This integral equation is called Milne's equation for the grey problem or Milne's problem. The solution of equation (2.17.6) simultaneously satisfies the transfer equation and the condition of radiative equilibrium.

When LTE exists

$$J(\tau) = S(\tau) = B[(T(\tau)] = \frac{\sigma_R T^4}{\pi}. \tag{2.17.7}$$

Taking the zeroth order moment of equation (2.17.5), we obtain

$$\frac{dH}{d\tau} = J - S = J - J = 0, \tag{2.17.8}$$

which means that the flux is constant. The first order moment of equation (2.17.5) gives us

$$\frac{dK}{d\tau} = H, \tag{2.17.9}$$

which leads to

$$K(\tau) = H\tau + C = \frac{1}{4}F\tau + C. \tag{2.17.10}$$

At great depths the specific intensity is represented to a good accuracy by

$$I(\mu) = I_0 + \mu I_1, \tag{2.17.11}$$

and, at $\tau \gg 1$, that $K(\tau) = \frac{1}{3}J(\tau)$ and $K(\tau) \to \frac{1}{4}F\tau$ give us

$$J(\tau) \to \frac{3}{4}F\tau \qquad (\tau \gg 1), \tag{2.17.12}$$

which means that the mean intensity varies linearly as the optical depth. However, at the surface of the star, this variation need not apply. Therefore we use the expression

$$J(\tau) = \frac{3}{4}F[\tau + q(\tau)] = \frac{3}{4\pi}(\sigma_R T_{eff}^4)[\tau + q(\tau)], \tag{2.17.13}$$

where $q(\tau)$ is called *Hopf function*. We need to determine this function.

From equation (2.17.6), we can write

$$\tau + q(\tau) = \frac{1}{2}\int_0^\infty [t + q(t)]E_1|t - \tau|\,dt. \tag{2.17.14}$$

In the limit of $\tau \to \infty$, we obtain

$$\lim_{\tau\to\infty}\left[\frac{1}{3}J(\tau) - K(\tau)\right] = \frac{1}{4}F\lim_{\tau\to\infty}[\tau + q(\tau) - \tau - C] = 0. \tag{2.17.15}$$

Thus we have $C = q(\infty)$. Therefore the equation (2.17.10) can be written as

$$K(\tau) = \frac{1}{4}F[\tau + q(\infty)]. \tag{2.17.16}$$

2.18 Eddington's approximation

Eddington made the assumption that $J = 3K$ everywhere. Then from equation (2.17.10) we have

$$J_E(\tau) = \frac{3}{4}F\tau + C_1. \tag{2.18.1}$$

The constant C_1 can be calculated from the emergent flux. Therefore from equation (2.12.15) we obtain,

$$F(0) = 2\int_0^\infty \left(\frac{3}{4}F\tau + C_1\right)E_2(\tau)\,d\tau = 2c'E_3(0) + \frac{3}{4}F\left[\frac{4}{3} - 2E_4(0)\right]. \tag{2.18.2}$$

We have $E_n(0) = 1/(n-1)$ and letting $F = F(0)$ we obtain

$$C_1 = \frac{1}{2}F \tag{2.18.3}$$

or

$$J_E(\tau) = \frac{3}{4}F\left(\tau + \frac{2}{3}\right). \tag{2.18.4}$$

Therefore in Eddington's approximation $q(\tau) = \frac{2}{3}$. In the case of LTE and radiative equilibrium we have

$$T^4 \approx \frac{3}{4}T_{eff}^4\left(\tau + \frac{2}{3}\right), \tag{2.18.5}$$

where T_{eff} is the effective temperature of the star. (The effective temperature of the star is equivalent to the black body temperature emitting the same flux or it is the kinetic temperature at the optical depth of unity where continuum radiation originates.)

Equation (2.18.5) predicts the boundary temperature in terms of the effective temperature at $z = 0$. This is

$$\frac{T_0}{T_{eff}} = \left(\frac{1}{2}\right)^{1/4} = 0.841, \tag{2.18.6}$$

the exact value being given by

$$\frac{T_0}{T_{eff}} = \left(\frac{3^{1/2}}{4}\right)^{1/4} = 0.8114.$$

The angle dependence of the emergent radiation field is obtained from equation (2.18.2) and the equation

$$I(0, \mu) = \int_0^\infty S(t)\exp(-t/\mu)\frac{dt}{\mu},$$

and is given by

$$I(0, \mu) = \frac{2}{3}F\left(\mu + \frac{2}{3}\right). \tag{2.18.7}$$

The *limb darkening* which is defined as $I(0, \mu)/I(0, 1)$ is given (in Eddington's approximation) by

$$I(0, \mu)/I(0, 1) = \frac{3}{5}\left(\mu + \frac{2}{3}\right), \tag{2.18.8}$$

which is the intensity at an angle $\cos^{-1}\mu$ relative to that of the disc centre.

Exercises

2.1(a) The radiative transfer equation in spherical symmetry is sometimes written as

$$\mu\frac{\partial I_\nu}{\partial r}+\frac{\partial}{\partial\mu}\left[(1-\mu^2)\frac{\partial I_\nu}{\partial\mu}\right]=j_\nu-\kappa_\nu I_\nu.$$

Compare this with equation (2.2.9). How do you interpret I_ν?

(b) Integration of equation (2.8.2) leads to the invariance of the specific intensity in the absence of absorption and emission in the medium. In such a medium how do you interpret the equation in spherical symmetry

$$\mu\frac{\partial I}{\partial r}+\frac{1-\mu^2}{r}\frac{\partial I}{\partial\mu}=0\ ?$$

2.2 Show that in conservative cases the transfer equation in plane parallel atmospheres admits a solution of the form

$$I(\tau,\mu)=C\left[\tau+\mu\left(1-\frac{1}{3}\omega_1\right)^{-1}\right].$$

2.3 If we have

$$\frac{dI^+}{d\tau}+I^+=S^+$$

and

$$\frac{dI^-}{d\tau}+I^-=S^-,$$

show that

$$I^+(\tau)=I_1\exp(-\tau)+\int_0^\tau\exp[-(\tau-t)]S^+(t)\,dt$$

and

$$I^-(\tau)=I_2\exp[-(T-\tau)]+\int_\tau^T\exp[-(t-\tau)]S^-(t)\,dt,$$

with $I_1=I^+(0)$, $I_2=I^-(\tau)$ and T the total optical depth.

2.4 Consider a spherical planetary nebula (see figure 2.9) with the radial optical depth $K'r$, where r is the radius. Let the ray in the line of sight PQ make an angle θ with the radius at Q and the total optical depth along this ray be $2\tau\cos\theta$ and let t and $K'l$ be the running length and optical depth along the ray. J is the constant emission coefficient (cm^{-3} ster^{-1} s^{-1}). Show that the emergent intensity at Q is (no incident radiation at P).

$$I(\theta) = \frac{J}{K'}\left[1 - \exp(-2\tau\cos\theta)\right]$$

and that the outward flux πF is given by

$$\pi F = \frac{\pi J}{K'}\left[1 - \frac{1}{2\tau^2} + \frac{\exp(-2\tau)}{2\tau^2}(1 + 2\tau)\right].$$

If $J = J_0 t$, where J_0 is constant, show that the emergent intensity is

$$I(\theta) = \frac{J_0}{K'}\left\{(1 + 2\tau\cos\theta)\left[1 - \exp(-2\tau\cos\theta)\right]\right\},$$

and the flux πF is

$$\pi F = \frac{2\pi J_0}{K'}\left\{\frac{1}{2} + \frac{2\tau}{3} - \frac{1}{4\tau^2}\left[3 - \exp(-2\tau)\left(4\tau^2 + 6\tau + 3\right)\right]\right\}.$$

2.5 Derive equations (2.4.3) and (2.4.6).

2.6 Milne's planetary nebula boundary condition is as follows: at the inner boundary of a spherical shell with radius r_i, which envelopes a spherical vacuum, the boundary condition is $I(r_i, +\mu) = I(r_i, -\mu)$. Prove this. Furthermore, if there is a point source of intensity I_0 at the centre of the void sphere with radius r_i then show that $I(r_i, +\mu) = I_0\delta(\mu - 1)$.

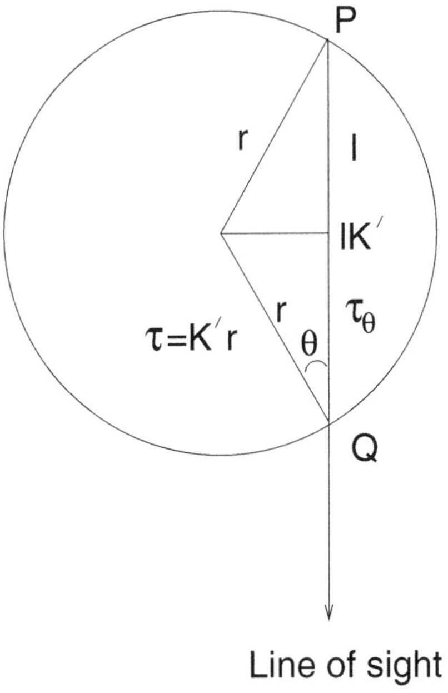

Figure 2.9 Schematic diagram of a planetary nebula.

2.7 If the source function S is expanded by the following series around the optical depth $\tau_0 = 1$

$$S(\tau) = S(\tau_0) + S'(\tau_0)(\tau - \tau_0) + \frac{1}{2}S''(\tau_0)(\tau - \tau_0)^2,$$

find the emergent intensity.

2.8 Show that the normally emergent intensity from a finite slab of optical thickness T in which the source function S is constant and with no incident radiation is given by

$$I(0, 1) = S[1 - \exp(-T)].$$

2.9 Derive equations (2.12.17)–(2.12.19).

2.10 Derive equation (2.14.5).

2.11 If the source function is given by

$$S_\nu = (K_\nu B_\nu + \sigma_\nu J_\nu)/(K_\nu + \sigma_\nu),$$

where K_ν is the absorption coefficient, σ_ν is the scattering coefficient, B_ν is the Planck function and J_ν is the mean intensity, show that in radiative equilibrium

$$\int_0^\infty K_\nu B_\nu \, d\nu = \int_0^\infty K_\nu J_\nu \, d\nu.$$

2.12 In the case of perfect scattering ($\omega = 1$) show that

$$\operatorname{div} \mathbf{F} = 0.$$

Show that F_z = constant in planar geometry, $F_r = F_0/r^2$ in spherical geometry, where F_0 is constant, and $F_r = F_0/r$ in cylindrical symmetry.

2.13 Assuming that

$$\frac{dI_\nu(\tau_\nu)}{d\tau_\nu} = -I_\nu(\tau_\nu).$$

Show that the magnitude change Δm is given by

$$\Delta m = 1.086\tau_\nu$$

given that

$$\Delta m = -2.5 \log_{10}\left[\frac{I_\nu}{I_\nu(0)}\right].$$

2.14 Using equation (2.9.11) with $S(t) = \sum_{n=0}^{\infty} a_n t^n$ and $\int_0^{\infty} p^n \exp(-p)\,dp = n!$, derive the Eddington–Barbier approximation. Show that the Laplace transform of $S(t)$ given above will give the relation (Rutten 1999)

$$I(0,+\mu) = L_{(\frac{1}{\mu})}[S(t)] = \sum_{n=0}^{\infty} n! a_n \mu^n.$$

Also show that

$$J(\tau) = \Lambda[S(t)] = a_0 + a\tau + a_2\tau^2 + \frac{2}{3}a_2 - \frac{1}{2}a_0 E_2(\tau) + \frac{1}{2}a_1 E_3(\tau) - a_2 E_4(\tau),$$

$$F(\tau) = \Phi(\tau)[S(t)] = 2a_0 E_3(\tau) + a_1\left[\frac{4}{3} - 2E_4(\tau)\right] + \cdots$$

and

$$J_\nu(0) = \frac{1}{2}a_0 + \frac{1}{4}a_1 + \frac{1}{3}a_2 = \frac{1}{2}S \quad \left(\tau = \frac{1}{2}\right)$$

$$F(0) = S \quad \left(\tau = \frac{2}{3}\right),$$

2.15 Show that

$$\Lambda_\tau(a+bt) = a + b\tau + \frac{1}{2}[bE_3(\tau) - aE_2(\tau)],$$

$$\Phi_\tau(a+bt) = \frac{4}{3}b + 2[aE_3(\tau) - bE_4(\tau)],$$

$$X_\tau(a+b\tau) = \frac{4}{3}(a+b\tau) + 2[bE_5(\tau) - aE_4(\tau)],$$

2.16 Assuming the source function is given by

$$S_\nu = \frac{\kappa_\nu B_\nu + \sigma_\nu J_\nu}{\kappa_\nu + \sigma_\nu},$$

show that

$$J_\nu(\tau_\nu) = \Lambda_{\tau_\nu}[\rho_\nu(t_\nu)J_\nu(t_\nu)] + \Lambda_{\tau_\nu}\{[1-\rho_\nu(t_\nu)]\,B_\nu(t_\nu)\},$$

where $\rho_\nu = \sigma_\nu/(\kappa_\nu + \sigma_\nu)$.

2.17 In spherical geometry, show that

$$\frac{\partial K_\nu}{\partial r} + \frac{1}{r}(3K_\nu - J_\nu) = -\kappa_\nu H_\nu,$$

where j_ν, H_ν, K_ν are the zeroth, first and second moments of I_ν, the specific intensity.

2.18 If S_ν is given as in exercise 2.16, show that the condition of radiative equilibrium gives

$$\int_0^\infty \kappa_\nu B_\nu \, d\nu = \int_0^\infty \kappa_\nu J_\nu \, d\nu.$$

2.19 Show that the equations

$$I(\tau, \mu) = \int_0^\infty S(t) \exp\left[-\frac{(t-\tau)}{\mu}\right] \frac{dt}{\mu} \qquad (0 \le \mu \le 1)$$

(for outgoing radiation) and

$$I(\tau, \mu) = \int^\tau (t) \exp\left[-\frac{(\tau - t)}{(-\mu)}\right] \frac{dt}{(-\mu)} \qquad (-1 \le \mu \le 0)$$

(for incoming radiation) are equivalent to

$$I(\mathbf{r}, \mathbf{n}) = \int_0^{s_{max}} j(r') \exp[-\tau(\mathbf{r}, \mathbf{r}')] \, d|\mathbf{r}' - \mathbf{r}|,$$

where

$$\mathbf{r}'(s) \equiv \mathbf{r} - s\mathbf{n}, \quad \tau(\mathbf{r}', \mathbf{r}) = \int_0^{|\mathbf{r}'-\mathbf{r}|} \kappa(s)\mathbf{r}'(s) \, ds$$

and s_{max} is the distance along the ray to any boundary surface in the direction $(-\mathbf{n})$, $s_{max} = \infty$ for outward directed rays in a semi-infinite medium. Substitute the above result into the definition of $J(\mathbf{r})$ to get

$$J(\mathbf{r}) = \frac{1}{4\pi} \int_V \{jr') \exp[-\tau(r', r)]/|\mathbf{r}' - \mathbf{r}|^2\} \, d^3 r',$$

where V represents the volume of the material. This is Peierl's equation.

2.20 Consider an axially symmetric but not spherically symmetric atmosphere (for example a rotationally flattened star). (a) Show that $I = I(r, \Theta, \theta, \phi)$. (b) For the general case where $I = I(r, \Theta, \Phi, \theta, \phi)$ (for example, a rotationally flattened star illuminated by a companion), show that the transfer equation in spherical coordinates, accounting for all the spatial coordinates, is given by

$$\frac{1}{c}\frac{\partial I}{\partial t} + \mu \frac{\partial I}{\partial r} + \frac{\gamma}{r}\frac{\partial I}{\partial \Theta} + \frac{\sigma'}{r} \sin\Theta \frac{\partial I}{\partial \Phi} + \frac{1-\mu^2}{r}\frac{\partial I}{\partial \mu} - \sigma \cot\Theta \frac{\partial I}{\partial \phi}$$
$$= j - \kappa I,$$

where $\gamma = \cos\phi \sin\theta$, $\sigma' = \sin\phi \sin\theta$ and θ is the polar angle between the direction of the pencil of radiation and the normal to atmospheric layers, Θ is the polar angle of the point on a spherical surface, ϕ is the azimuthal angle of a pencil of radiation around the normal to the atmospheric layers and Φ is the azimuthal angle of a point on a spherical surface.

REFERENCES

Chandrasekhar, S., 1960, *Radiative Transfer*, Dover, New York.

Grant, I.P., 1968, New Methods in Radiative Transfer: Lectures Notes, unpublished, Oxford.

Kourganoff, V., 1963, *Basic Methods in Radiative Transfer*, Dover, New York.

Menzel, D.H., ed., 1966, *Selected papers on the Transfer of Radiation*, Dover, New York.

Mihalas, D., 1978, *Stellar Atmospheres*. Freeman and Company, San Francisco.

Mihalas, D., Mihalas, B.W., 1984, *Foundation of Radiation Hydrodynamics*, Oxford University Press, Oxford.

Osterbrock, D.E., 1974, *Astrophysics of Gaseous Nebulae*, W.H. Freeman and Company, San Francisco.

Pomraning, G.C., 1973, *The Equations of Radiative Hydrodynamics*, Pergamon Press, Oxford.

Rutten, R.J., 1999, Radiative Transfer in Stellar Atmospheres, Lecture Notes: http://www.astro.un.nl/ rutten.

Sobolev, V.V., 1963, *A Treatise on Radiative Transfer*, (translated by S.I. Gaposchkin), Van Nostrand Company Inc., New York.

Chapter 3

Methods of solution of the transfer equation

We shall now describe some methods of obtaining the solution of the transfer equation in a plane parallel medium.

3.1 Chandrasekhar's solution

The radiation field is divided into streams as was done by Joule in the context of the kinetic theory of gases. The molecules in a box are presumed to be moving in three equal pairs of streams – parallel to the length, breadth and depth of the box in which the gas is situated. Each pair of streams is presumed to be moving in opposite directions. Schuster and Schwarzschild wrote similar equations in transfer theory. The transfer equation in plane parallel stratification (see equation (2.12.3)),

$$\mu \frac{dI(\tau, \mu)}{d\tau} = I(\tau, \mu) - \frac{1}{2} \int_{-1}^{+1} I(\tau, \mu')\, d\mu', \tag{3.1.1}$$

is replaced by a pair of equations for I_+ and I_-, the outward and inward intensities, thus

$$\left.\begin{aligned} +\frac{1}{2}\frac{dI_+}{d\tau} &= I_+ - \frac{1}{2}(I_+ + I_-), \\ -\frac{1}{2}\frac{dI_-}{d\tau} &= I_- - \frac{1}{2}(I_+ + I_-). \end{aligned}\right\} \tag{3.1.2}$$

The factor $\frac{1}{2}$ on the LHS is chosen arbitrarily as in the kinetic theory of gases.

The division of the radiation field into only two streams does not represent reality. However, by increasing the number of rays one can improve the accuracy and can solve many problems involving, for example, polarization, anisotropic scattering.

By following the concept of Schuster–Schwarzschild, we retain the principal advantage and increase the number of streams. Wick (1943) suggested that the integral in the transfer equation (2.12.3) can be replaced by the Gaussian sums for numerical quadrature. This can be written as

$$\int_{-1}^{+1} I(\tau, \mu)\, d\mu \approx \sum_{j=-n}^{j=+n} a_j I(\tau, \mu_j), \tag{3.1.3}$$

where the μ_js are the $2n$ roots of the Legendre polynomials $P_{2n}(\mu)$ of order $2n$ denoted by $\mu_n, \mu_{n-1}, \ldots, \mu_1, \mu_2 \ldots \mu_{n-1}, \mu_n$ and the a_js are the weight factors (see figure 3.1). Furthermore (see Abramowitz and Stegun (1964)),

$$\left.\begin{aligned} \mu_{-i} &= -\mu_i, \\ a_j &= a_{-j}. \end{aligned}\right\} \tag{3.1.4}$$

It is well known that for a given number of subdivisions of the interval (−1, +1), Gauss's choice of μ_js and a_js gives the best value of the integral, for example

$$\sum_{j=1}^{n} a_j \mu_j^m = \int_0^1 \mu^m\, d\mu = \frac{1}{m+1}. \tag{3.1.5}$$

Chandrasekhar (1944) used Gauss's formulae to solve the transfer equation in a plane parallel, semi-infinite, isotropically scattering atmosphere with constant net flux (see equation (2.12.3)):

$$\mu \frac{dI(\tau, \mu)}{d\tau} = I(\tau, \mu) - \frac{1}{2} \int_{-1}^{+1} I(\tau, \mu')\, d\mu'. \tag{3.1.6}$$

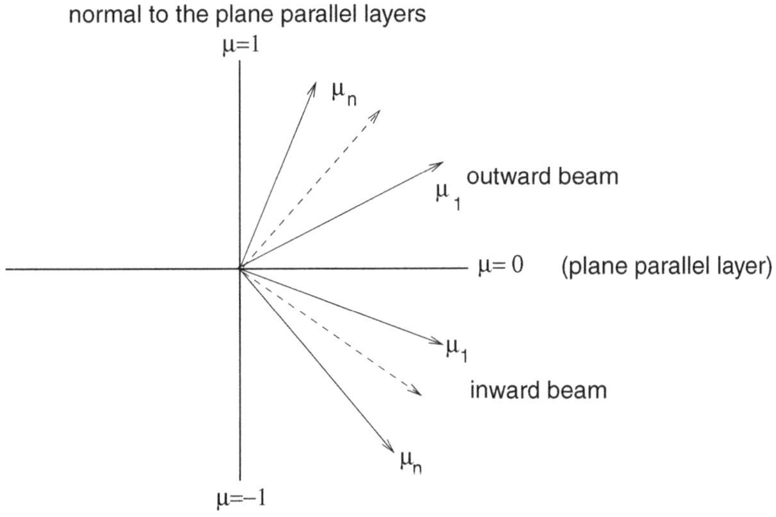

Figure 3.1 Gaussian points, μs.

Using equation (3.1.3), equation (3.1.6) is written for $2n$ linear equations as

$$\mu_i \frac{dI_i}{d\tau} = I_i - \frac{1}{2}\sum_j a_j I_j \quad (i, j = \pm 1, \ldots, \pm n), \tag{3.1.7}$$

where the μ_is are zeros of the Legendre polynomials $P_{2n}(\mu)$, the a_js are the corresponding Gaussian weights and

$$I_i = I(\tau, \mu_i). \tag{3.1.8}$$

We now describe Chandrasekhar's solution for the equation (3.1.7) (see Chandrasekhar (1960)). We seek solutions of the form

$$I_i = g_i \exp(-k\tau) \quad (i = \pm 1, \pm 2, \ldots, \pm n), \tag{3.1.9}$$

where the g_is and k are the unspecified $2n+2$ constants. Introducing solution (3.1.9) into equation (3.1.7), we obtain

$$g_i(1 + \mu_i k) = \frac{1}{2}\sum_j a_j g_j \tag{3.1.10}$$

or

$$g_i = \frac{\frac{1}{2}\sum a_j g_j}{1 + \mu_i k} = \frac{C}{1 + \mu_i k}, \tag{3.1.11}$$

where C is a constant that is independent of i. Substituting equation (3.1.11) into equation (3.1.10), we obtain the characteristic equation

$$\sum_j \frac{a_j}{1 + \mu_j k} = 2. \tag{3.1.12}$$

Using equation (3.1.4), the above equation can be written as

$$\sum_{j=1}^{n} \frac{a_j}{1 - \mu_j^2 k^2} = 1. \tag{3.1.13}$$

This is an algebraic equation of order n which has $k^2 = 0$ as a root as we have

$$\sum_{j=1}^{n} a_j = 1. \tag{3.1.14}$$

Consequently, the characteristic equation (3.1.13) admits $2n - 2$ distinct, non-zero roots in the form of pairs:

$$\pm k_\alpha \quad (\alpha = 1, \ldots, n-1). \tag{3.1.15}$$

It is understood that equation (3.1.7) will have $2n - 2$ independent solutions. Equation (3.1.7) admits another solution of the form

$$I_i = b(\tau + q_i) \quad (i = \pm 1, \ldots, \pm n), \tag{3.1.16}$$

where b is a constant. Substituting this into equation (3.1.7), we get

$$\mu_i = q_i - \frac{1}{2}\sum_j a_j q_i \tag{3.1.17}$$

or

$$q_i = Q + \mu_i \quad (i = \pm 1, \pm 2, \ldots, \pm n), \tag{3.1.18}$$

where Q is a constant and equation (3.1.18) will satisfy equation (3.1.17). The general solution of the system of equations (3.1.9) can be written as

$$I_i = b\left\{\sum_{\alpha=1}^{n-1} \frac{L_\alpha \exp(-k_\alpha \tau)}{1 - \mu_i k_\alpha} + \sum_{\alpha=1}^{n-1} \frac{L_{-\alpha} \exp(k_\alpha \tau)}{1 - \mu_i k_\alpha} + \tau + \mu_i + Q\right\}$$
$$(i = \pm 1, \ldots, \pm n), \tag{3.1.19}$$

where b, $L_{\pm\alpha}$ $(\alpha = 1, \ldots, n-1)$ and Q are the $2n$ constants of integration. We now apply the following boundary conditions:

$$I(0, -\mu) = 0 \quad (0 < \mu \leq 1). \tag{3.1.20}$$

Furthermore, as $\tau \rightarrow \infty$ all the integrals that occur in the formal solution (see equations (2.9.5)–(2.9.11)) converge. The convergence of these integrals over the source function requires that

$$S(\tau, \mu)\exp(-\tau) \rightarrow 0 \quad \text{as} \quad \tau \rightarrow \infty. \tag{3.1.21}$$

None of the terms in I_i should increase more rapidly than $\exp(\tau)$ as $\tau \rightarrow \infty$. This requires that in the solution (3.1.19), we omit all terms containing $\exp(+k_\alpha \tau)$, which gives us

$$I_i = b\left\{\sum_{\alpha=1}^{n-1} \frac{L_\alpha \exp(-k_\alpha \tau)}{1 + \mu_i k_\alpha} + \tau + \mu_i + Q\right\} \quad (i = \pm 1, \ldots, \pm n). \tag{3.1.22}$$

The boundary condition (3.1.20) implies that

$$L_{-i} = 0 \text{ at } \tau = 0, \quad i = 1, \ldots, n. \tag{3.1.23}$$

Therefore from equation (3.1.22), we have

$$\sum_{\alpha=1}^{n-1} \frac{L_\alpha}{1 - \mu_i k_\alpha} - \mu_i + Q = 0 \quad (i = 1, \ldots, n). \tag{3.1.24}$$

From the above set of equations, one can determine the n constants of integration, L_α $(\alpha = 1, \ldots, n-1)$ and Q.

We shall now calculate the flux given by

$$F = 2\int_{-1}^{+1} I\mu\, d\mu. \tag{3.1.25}$$

By replacing the integral by the sums of $I_i\mu_i$s and using equation (3.1.22), the flux can be written as

$$F = 2b\left[\sum_{\alpha=1}^{n-1} L_\alpha \exp(-k_\alpha\tau)\sum_i \frac{a_i\mu_i}{1+\mu_i k_\alpha} + \sum_i a_i\mu_i^2 + (Q+\tau)\sum_i a_i\mu_i\right]. \tag{3.1.26}$$

By using equations (3.1.4) and (3.1.5) we obtain

$$\sum a_i\mu_i^2 = \frac{2}{3} \quad \text{and} \quad \sum a_i\mu_i = 0. \tag{3.1.27}$$

From the identity (3.1.12) we have

$$\begin{aligned}\sum \frac{a_i\mu_i}{1+\mu_i k_\alpha} &= \frac{1}{k_\alpha}\sum_i a_i\left(1 - \frac{1}{1+\mu_i k_\alpha}\right)\\ &= \frac{1}{k_\alpha}\left(2 - \sum \frac{a_i}{1+\mu_i k_\alpha}\right) = 0.\end{aligned} \tag{3.1.28}$$

Substituting the results of equations (3.1.27) and (3.1.28) into equation (3.1.22), we obtain

$$F = \frac{4}{3}b = \text{constant}. \tag{3.1.29}$$

We can write I_i in terms of F, using equation (3.1.22), as

$$I_i = \frac{3}{4}F\left\{\sum_{\alpha=1}^{n-1} \frac{L_\alpha \exp(-k_\alpha\tau)}{1+\mu_i k_\alpha} + \tau + \mu_i + Q\right\} \quad (i = \pm 1, \ldots, \pm n). \tag{3.1.30}$$

The mean intensity J is given by

$$J = \frac{1}{2}\int_{-1}^{+1} I\, d\mu = \frac{1}{2}\sum_i a_i I_i. \tag{3.1.31}$$

Using equations (3.1.5), (3.1.13), (3.1.22), (3.1.27) and (3.1.28), we obtain for the mean intensity

$$J = \frac{3}{4}F\left[\tau + Q + \sum_{\alpha=1}^{n-1} L_\alpha \exp(-k_\alpha\tau)\right]. \tag{3.1.32}$$

If we set

$$q(\tau) = Q + \sum_{\alpha=1}^{n-1} L_\alpha \exp(-k_\alpha \tau), \tag{3.1.33}$$

then

$$J = \frac{3}{4} F(\tau + q(\tau)). \tag{3.1.34}$$

If there is no incident radiation on the surface at $\tau = 0$, then from equations (2.9.9) and (2.9.10) the formal solution becomes

$$I(\tau, \mu) = \int_\tau^\infty \exp[-(t-\tau)/\mu] S(t, \mu) \frac{dt}{\mu} \tag{3.1.35}$$

and

$$I(\tau, -\mu) = \int_0^\tau \exp[-(\tau - t)/\mu] S(t, \mu) \frac{dt}{\mu} \quad (0 < \mu < 1). \tag{3.1.36}$$

From equations (3.1.32), (3.1.35) and (3.1.36), we get

$$I(\tau, +\mu) = \frac{3}{4} F \left\{ \sum_{\alpha=1}^{n-1} \frac{L_\alpha \exp(-k_\alpha \tau)}{1 + k_\alpha \mu} + \tau + \mu + Q \right\} \tag{3.1.37}$$

and

$$I(\tau, -\mu) = \frac{3}{4} F \left\{ \sum_{\alpha=1}^{n-1} \frac{L_\alpha}{1 - k_\alpha \mu} [\exp(-k_\alpha \tau) - \exp(-\tau/\mu)] \right. \\ \left. + \tau + (Q - \mu)[1 - \exp(-\tau/\mu)] \right\}. \tag{3.1.38}$$

The law of limb darkening or the angular distribution of the emergent radiation is obtained by putting $\tau = 0$ in equation (3.1.37), thus,

$$I(0, \mu) = \frac{3}{4} F \left\{ \sum_{\alpha+1}^{n-1} \frac{L_\alpha}{1 + k_\alpha \mu} + \mu + Q \right\}. \tag{3.1.39}$$

We now evaluate the K-integral. By virtue of the relation (Chandrasekhar 1960),

$$\sum_{j=\pm 1}^{\pm n} a_j \mu_j^l = \frac{2\delta_{l,e}}{l+1} \quad (l < 4n - 1), \tag{3.1.40}$$

where

$$\delta_{l,e} = \begin{cases} 1 & \text{if } l \text{ is even} \\ 0 & \text{if } l \text{ is odd,} \end{cases}$$

we have

$$\sum_i a_i \mu_i^3 = 0. \tag{3.1.41}$$

Using equation (3.1.29), equation (3.1.22) can be written as

$$I_i = \frac{3}{4}F\left\{\sum_{\alpha=1}^{n-1} \frac{L_\alpha \exp(-k_\alpha \tau)}{1+\mu_i k_\alpha} + \tau + \mu_i + Q\right\} \quad (i = \pm 1, \ldots, \pm n). \tag{3.1.42}$$

The K-integral is now written using equation (3.1.42) as

$$K = \frac{3}{8}F\left\{\sum_{\alpha=1}^{n-1} L_\alpha \exp(-k_\alpha \tau) \sum_i \frac{a_i \mu_i^2}{1+\mu_i k_\alpha} + (\tau + Q)\sum_i a_i \mu_i^2 + \sum_i a_i \mu_i^3\right\}$$
$$(i = \pm 1, \ldots, \pm n). \tag{3.1.43}$$

By using equation (3.1.40), the K-integral becomes

$$K = \frac{1}{4}F(\tau + Q). \tag{3.1.44}$$

3.2 The H-function

The characteristic roots and the zeros of the Legendre polynomial are connected by the relationship (see exercise 3.6)

$$k_1 \ldots k_{n-1}, \ \mu_1 \ldots \mu_n = \frac{1}{3^{1/2}}. \tag{3.2.1}$$

Let

$$S(\mu) = \sum_{\alpha=1}^{n-1} \frac{L_\alpha}{1 - k_\alpha \mu} - \mu + Q. \tag{3.2.2}$$

The boundary condition requires that (see equation (3.1.24))

$$S(\mu_i) = 0 \quad (i = 1, \ldots, n). \tag{3.2.3}$$

Therefore the angular distribution of the emergent radiation given by equation (3.1.39) can be written as

$$I(0, \mu) = \frac{3}{4}FS(-\mu). \tag{3.2.4}$$

$S(\mu)$ can be calculated without solving for the constants L_α and Q.

We define

$$R(\mu) = \prod_{\alpha=1}^{n-1}(1 - k_\alpha \mu). \qquad (3.2.5)$$

The product $S(\mu)R(\mu)$ is a polynomial of degree n in μ which vanishes for $\mu = \mu_i$, $i = 1, \ldots, n$. Therefore the product $S(\mu)R(\mu)$ and the polynomial

$$P(\mu) = \prod_{i=1}^{n}(\mu - \mu_i), \qquad (3.2.6)$$

differ only by a constant factor. The constant of proportionality can be obtained by comparing the coefficients of the highest powers of μ (namely μ^n) in $P(\mu)$ and $S(\mu)R(\mu)$ and is given by

$$(-1)^n k_1 k_2 \cdots k_{n-1}. \qquad (3.2.7)$$

Therefore, the required equation for $S(\mu)$ is

$$S(\mu) = (-1)^n k_1 k_2 \cdots k_{n-1} \frac{P(\mu)}{R(\mu)} \qquad (3.2.8)$$

or

$$S(-\mu) = k_1 \cdots k_{n-1} \frac{\prod_{i=1}^{n}(\mu + \mu_i)}{\prod_{\alpha=1}^{n-1}(1 + k_\alpha \mu)}. \qquad (3.2.9)$$

By using relationship (3.2.1), we get

$$S(-\mu) = 3^{-1/2} H(\mu), \qquad (3.2.10)$$

where

$$H(\mu) = \frac{1}{\mu_1 \cdots \mu_n} \frac{\prod_{i=1}^{n}(\mu + \mu_i)}{\prod_{\alpha=1}^{n-1}(1 + k_\alpha \mu)}. \qquad (3.2.11)$$

In terms of H-functions, the angular distribution of the emergent radiation is (see equation (3.2.4))

$$I(0, \mu) = \frac{3^{1/2}}{4} F H(\mu). \qquad (3.2.12)$$

From equations (3.1.32), (3.1.39) and (3.2.12), we can write

$$I(0, 0) = J(0) = \frac{3}{4} F \left(\sum_{\alpha=1}^{n-1} L_\alpha + Q \right) = \frac{3^{1/2}}{4} F H(0). \qquad (3.2.13)$$

But from equation (3.2.11)

$$H(0) = 1. \tag{3.2.14}$$

Therefore

$$J(0) = \frac{\sqrt{3}}{4}F. \tag{3.2.15}$$

This result was first derived by Hopf and Bronstein (see Hopf (1934)).

The constants of integration are given by (see Chandrasekhar (1960))

$$L_\alpha = (-1)^n k_n \cdots k_{n-1} \frac{P(1/k_\alpha)}{R_\alpha(1/k_\alpha)} \quad (\alpha = 1, \ldots, n-1), \tag{3.2.16}$$

where

$$R_\alpha(x) = \prod_{\beta \neq \alpha} (1 - k_\beta x), \tag{3.2.17}$$

and

$$Q = \sum_{i=1}^{n} \mu_i - \sum_{\alpha=1}^{n-1} \frac{1}{k_\alpha}. \tag{3.2.18}$$

These will give the solution in the nth approximation.

Now we shall illustrate the solutions in the first and second approximations. The values of as and μs can be found in Abramowitz and Stegun (1964).

3.2.1 The first approximation

This is obtained for $n = 1$. We have

$$a_1 = a_{-1} = 1 \quad \text{and} \quad \mu_1 = -\mu_1 = \frac{1}{\sqrt{3}}. \tag{3.2.19}$$

From relation (3.1.13), namely

$$\sum_{j=1}^{n} \frac{a_j}{1 - \mu_j^2 k^2} = 1, \tag{3.2.20}$$

we get $k = 0$, from equation (3.1.24) we get

$$Q = \mu_1 = \frac{1}{\sqrt{3}}, \tag{3.2.21}$$

and from equation (3.1.33) we get

$$q(\tau) = \frac{1}{\sqrt{3}}. \tag{3.2.22}$$

Then we can find J from equation (3.1.34):

$$J = \frac{3}{4}F\left(\tau + \frac{1}{\sqrt{3}}\right) \tag{3.2.23}$$

and from equation (3.1.39) we get the angular distribution of the emergent radiation or the law of limb darkening $I(0, \mu)$:

$$I(0, \mu) = \frac{3}{4}F\left(\mu + \frac{1}{\sqrt{3}}\right). \tag{3.2.24}$$

This predicts a boundary value for $q(\tau)$ which is in exact agreement with the Hopf–Bronstein value.

The inward and outward beams respectively obey the equations

$$\frac{1}{\sqrt{3}}\frac{dI_1}{d\tau} = I_1 - \frac{1}{2}(I_1 + I_{-1}), \tag{3.2.25}$$

$$-\frac{1}{\sqrt{3}}\frac{dI_{-1}}{d\tau} = I_{-1} - \frac{1}{2}(I_1 + I_{-1}). \tag{3.2.26}$$

These are essentially the equations that Schuster and Schwarzschild derived several years ago (see equations (3.1.1)) making use of the ideas of kinetic theory of gases. The difference is that on the LHS of the foregoing equations, we now have the factor $1/\sqrt{3}$ instead of the factor $1/2$ in equations (3.1.2). It appears that if Schwarzschild had used the present Gaussian formalism, he might have discovered the Hopf–Bronstein relation some 25 years earlier.

3.2.2 The second approximation

In this approximation,

$$\left.\begin{array}{ll} a_1 = a_{-1} = 0.652\,145; & \mu_1 = -\mu_{-1} = 0.339\,981; \\ a_2 = a_{-2} = 0.347\,855; & \mu_2 = -\mu_{-2} = 0.861\,136. \end{array}\right\} \tag{3.2.27}$$

Using equation (3.2.20) we get

$$k = 0, \quad \mu_1^2\mu_2^2k^2 = a_1\mu_1^2 + a_2\mu_2^2 = \frac{1}{3} \tag{3.2.28}$$

or

$$k_1 = \frac{1}{\sqrt{3}\mu_1\mu_2} = 1.972\,027. \tag{3.2.29}$$

Solving for Q and L_1, we obtain

$$Q = 0.694\,025, \; L_1 = -0.116\,675. \tag{3.2.30}$$

The quantity $q(\tau)$ is given by

$$q(\tau) = 0.694\,025 - 0.116\,675\exp(-1.972\,027\tau). \tag{3.2.31}$$

The solution in the second approximation is therefore

$$I(0,\mu) = \frac{3}{4}F\left(\mu + 0.694\,025 - \frac{0.116\,675}{1 + 1.972\,027\mu}\right). \tag{3.2.32}$$

Third, fourth and higher approximations can be calculated similarly.

In the spherical harmonic method (SHM), the specific intensity is expanded into a series of Legendre polynomials $p_l(\mu)$ (to the desired accuracy), which forms a complete orthogonal set within the range (−1,1). Chandrasekhar stated that the SHM and the discrete ordinate method are equivalent in all details. However, Kourganoff drew attention to certain serious limitations of SHM (see Kourganoff (1963), Peraiah and Grant (1973)). These two methods have difficulty in the analytical representation of the boundary condition at the free surface where the radiation is discontinuous (exercises 3.2, 3.13 and 6.11).

3.3 Radiative equilibrium of a planetary nebula

The transfer equation for the 'ultraviolet' radiation (see Ambarzumian (1931), (1932), Chandrasekhar (1944)) (i.e. radiation beyond the head of Lyman series) consistent with Zanstra's theory is

$$\mu\frac{dI}{d\tau} = I - \frac{1}{2}p\int_{-1}^{+1} I\,d\mu - \frac{1}{4}pS\exp[-(\tau_1 - \tau)], \tag{3.3.1}$$

where τ_1 is the optical thickness of the nebula in the ultraviolet radiation incident on the inner surface of the nebula at $\tau = \tau_1$ and p is a quantity less than unity. The boundary conditions (see exercise 2.6) for the planetary nebula due to Milne (1930a) are

$$I_i = I_{-i} \quad \text{at } \tau = \tau_1 \quad \text{for } i = 1, \ldots, n \tag{3.3.2}$$

and

$$I_{-i} = 0 \quad \text{at } \tau = 0 \quad \text{for } i = 1, \ldots, n \tag{3.3.3}$$

Equation (3.3.1) can be written, using the Gaussian quadratures in the nth approximation, as

$$\mu_i = \frac{dI_i}{d\tau} = I_i - \frac{1}{2}p\sum a_j I_j - \frac{1}{4}pS\exp[-(\tau_1 - \tau)] \quad (i = \pm 1, \ldots, \pm n). \tag{3.3.4}$$

One can obtain its general solution by substituting

$$I_i = S\{g_i\exp(-k\tau) + h_i\exp[-(\tau_1 - \tau)]\} \quad (i = \pm 1, \ldots, \pm n) \tag{3.3.5}$$

(with g_i, h_i, and k constants) into equation (3.3.4), giving

$$g_i(1 + k\mu_i) = \frac{1}{2} p \sum a_j g_j \tag{3.3.6}$$

and

$$h_i(1 - \mu_i) = \frac{1}{2} p \sum a_j h_j + \frac{1}{4} p. \tag{3.3.7}$$

Applying Chandrasekhar's method, we get the general solution as

$$I_i = S \left\{ \sum_{\alpha=1}^{n} \frac{L_\alpha \exp(-k_\alpha \tau)}{1 + \mu_i k_\alpha} + \frac{L_{-\alpha} \exp(+k_\alpha \tau)}{1 - \mu_i k_\alpha} + \frac{p \exp[-(\tau_1 - \tau)]}{4(1 - \mu_i)} \left(1 - p \sum_{j=1}^{n} \frac{a_j}{1 - \mu_j^2} \right) \right\}, \tag{3.3.8}$$

where $L_{\pm\alpha}, \alpha = 1, \ldots, n$ are $2n$ constants of integration.

3.4 Incident radiation from an outside source

If a plane parallel layer receives a parallel beam of radiation in the direction shown in figure 3.2 with a flux of πF per unit area normal to itself on the medium with optical depth τ_1 (see Milne (1930a), Chandrasekhar (1960)) the transfer equation for isotropic scattering becomes

$$\mu \frac{dI(\tau, \mu)}{d\tau} = I(\tau, \mu) - \frac{1}{2} \varpi \int_{-1}^{+1} I(\tau, \mu') \, d\mu' - \frac{1}{4} \omega F \exp(-\tau/\mu_0), \tag{3.4.1}$$

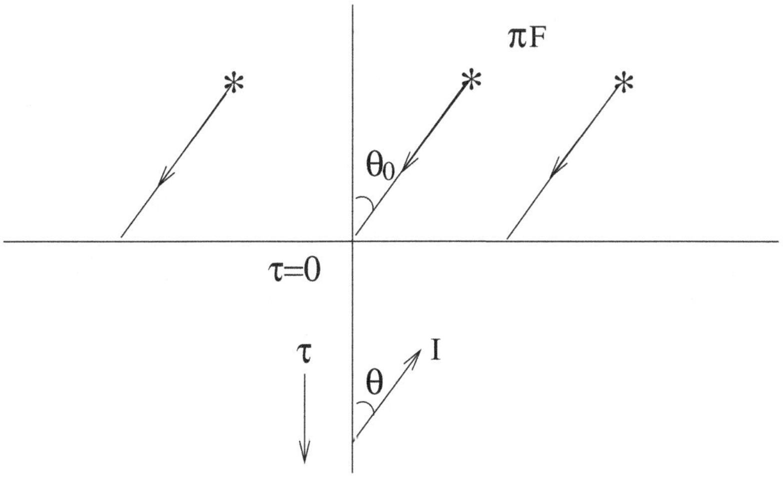

Figure 3.2 Incidence of a parallel beam of radiation.

where $\mu_0 = \cos\theta_0$ and ω is the albedo for single scattering. The most fundamental problem of radiative transfer is the study of the diffuse radiation field. This field is generated by one or more scatterings of radiation by matter. In deriving the quantities $I(0, \mu)$ and $I(\tau_1, -\mu)$ we obtain the transmitted and reflected radiation from a plane parallel layer of optical thickness τ_1 with the boundary conditions $I(0, -\mu) = 0$ and $I(\tau_1, -\mu) = I_0$. When radiation with a net flux of πF is incident on the layer at $\tau = 0$ as shown in figure 3.2, the reflected and transmitted fractions of this radiation emerge at $\tau = 0$ and $\tau = \tau_1$ respectively. This is in addition to the radiation represented by $I(0, +\mu)$ and $I(\tau_1, -\mu)$. The radiation with net flux πF generates diffuse radiation and the emergent intensities at $\tau = 0$ and $\tau = \tau_1$ are associated with what are called the diffuse reflection and transmission of the incident parallel beam with net flux πF.

The solution of equation (3.4.1) can be obtained by applying the method described in the preceding sections. The equivalent of this equation in the nth approximation can be written using Gaussian divisions. Thus

$$\mu_i \frac{dI_i}{d\tau} = I_i - \frac{1}{2}\omega \sum_j a_j I_j - \frac{1}{4}\omega F \exp(-\tau/\mu_0) \quad (i = \pm 1, \ldots, \pm n). \tag{3.4.2}$$

The last term on the RHS of the above equation requires a particular integral which can be written as

$$I_i = \frac{1}{4}\varpi F h_i \exp(-\tau/\mu_0) \quad (i = \pm 1, \ldots, \pm n), \tag{3.4.3}$$

where the h_is are constants which follow the relation

$$h_i \left(1 + \frac{\mu_i}{\mu_0}\right) = \frac{1}{2}\varpi \sum_j a_j h_j + 1. \tag{3.4.4}$$

Therefore the particular integral is written as

$$I_i = \frac{1}{4}\varpi F \left(1 + \frac{\mu_i}{\mu_0}\right)^{-1} \gamma \exp(-\tau/\mu_0) \quad (i = \pm 1, \ldots, \pm n), \tag{3.4.5}$$

where

$$\gamma = \left[1 - \varpi \sum_{j=1}^{n} a_j \left(1 - \frac{\mu_j^2}{\mu_0^2}\right)^{-1}\right]^{-1}. \tag{3.4.6}$$

Therefore the solution in the nth approximation is

$$I_i = \frac{1}{4}\omega F \left[\sum_{\alpha=1}^{n} L_\alpha \exp(-k_\alpha \tau)(1 + \mu_i k_\alpha)^{-1} + \gamma \exp(-\tau/\mu_0)\left(1 + \frac{\mu_i}{\mu_0}\right)^{-1}\right]$$
$$(i = \pm 1, \ldots, \pm n), \tag{3.4.7}$$

where the constants $L_\alpha (\alpha = 1, \ldots, n)$ can be found from the boundary conditions

$$I_i = 0 \quad \text{at } \tau = 0 \quad (i = 1, \ldots, n) \tag{3.4.8}$$

and from the equations

$$\sum_{\alpha=1}^{n} L_\alpha (1 - \mu_i k_\alpha)^{-1} + \gamma \left(1 - \frac{\mu_i}{\mu_0}\right)^{-1} = 0 \quad (i = 1, \ldots, n). \tag{3.4.9}$$

The source function $S(\tau)$ can be written as

$$S(\tau) = \frac{1}{4}\omega F \left[\sum_{\alpha=1}^{n} L_\alpha \exp(-k_\alpha \tau) + \gamma \exp(-\tau/\mu_0)\right]. \tag{3.4.10}$$

The radiation field at $(\tau, \pm\mu)$ can be written (see equations (2.12.5) and (2.12.6)) using the above source function as

$$\begin{aligned} I(\tau, +\mu) = \frac{1}{4}\varpi F \Bigg[& \sum_{\alpha=1}^{n} L_\alpha \exp(-k_\alpha \tau)(1 + \mu k_\alpha)^{-1} \\ & + \gamma \exp(-\tau/\mu_0)\left(1 + \frac{\mu}{\mu_0}\right)^{-1}\Bigg], \end{aligned} \tag{3.4.11}$$

and

$$\begin{aligned} I(\tau, -\mu) = \frac{1}{4}\varpi F \Bigg\{ & \sum_{\alpha=1}^{n} L_\alpha \exp(-k_\alpha \tau)(1 - \mu k_\alpha)^{-1} \\ & \times \left[\exp(-k_\alpha \tau) - \exp(-\tau/\mu)\right] \\ & + \gamma \left(1 - \frac{\mu}{\mu_0}\right)^{-1} \left[\exp(-\tau/\mu_0) - \exp(-\tau/\mu)\right]\Bigg\}. \end{aligned} \tag{3.4.12}$$

The angular distribution of the reflected radiation is

$$I(0, \mu) = \frac{1}{4}\omega F \left\{\sum_{\alpha=1}^{n} L_\alpha (1 + k_\alpha \mu)^{-1} + \gamma \left(1 + \frac{\mu}{\mu_0}\right)^{-1}\right\}. \tag{3.4.13}$$

Now, we can express the angular distribution $I(0, \mu)$ in terms of the H-functions. Consider equation (3.4.5) and write the function

$$N(p) = 1 - \frac{1}{2}\omega p \sum_j a_j (p + \mu_j)^{-1}, \tag{3.4.14}$$

or

$$N(p) = 1 - \omega p^2 \sum_{j=1}^{n} a_j (p^2 - \mu_j^2)^{-1}. \tag{3.4.15}$$

Comparing this with the characteristic equations (3.1.12), we get

$$N(p) = 0 \quad \text{for} \quad p = \pm k_\alpha^{-1} (\alpha = 1, \ldots, n). \tag{3.4.16}$$

By comparing the two quantities $\prod_{j=1}^{n}(p^2 - \mu_j^2)N(p)$ and $\prod_{\alpha=1}^{n}(1 - k_\alpha^2 p^2)$, we find that these two quantities are related through a constant factor $(-1)^n \mu_1^2 \cdots \mu_n^2$. Thus we can write

$$N(p) \prod_{j=1}^{n}(p^n - \mu_j^2) = (-1)^n \mu_1^2 \ldots \mu_n^2 \prod_{\alpha=1}^{n}(1 - k_\alpha^2 p^2). \tag{3.4.17}$$

Using equation (3.2.11), we can write equation (3.4.17) in terms of the H-functions as

$$1 - \omega p^2 \sum_{j=1}^{n} a_j (p^2 - \mu_j^2)^{-1} = \frac{1}{H(p)H(-p)}. \tag{3.4.18}$$

Therefore the quantity γ from equation (3.4.5) is

$$\gamma = \frac{1}{N(\mu_0)} = H(\mu_0)H(-\mu_0). \tag{3.4.19}$$

The angular distribution of the emergent radiation, which is also the law of diffuse reflection, is written in terms of H-functions (see equation (3.4.13)) as

$$I(0, \mu) = \frac{1}{4}\omega F \frac{\mu_0}{\mu + \mu_0} H(\mu)H(\mu_0). \tag{3.4.20}$$

3.5 Diffuse reflection when $\omega = 1$ (conservative case)

In this case equation (3.4.10) can be written, using equation (3.4.19), as

$$S(\tau) = \frac{1}{4}F\left[\sum_{\alpha=1}^{n-1} L_\alpha \exp(-k_\alpha \tau) + L_n + H(\mu_0)H(-\mu_0)\exp(-\tau/\mu_0)\right], \tag{3.5.1}$$

and the angular distribution of the emergent radiation is

$$I(0, \mu) = \frac{1}{4}F\left[\sum_{\alpha=1}^{n-1} L_\alpha (1 + k_\alpha \mu)^{-1} + L_n + H(\mu_0)H(-\mu_0)\left(1 + \frac{\mu}{\mu_0}\right)^{-1}\right]. \tag{3.5.2}$$

Once the constants are eliminated the law of diffuse reflection is written in terms of the H-functions:

$$I(0, \mu) = \frac{1}{4}F \frac{\mu_0}{\mu + \mu_0} H(\mu)H(\mu_0). \tag{3.5.3}$$

If $\tau \to \infty$, equation (3.5.1) becomes

$$S(\infty) = \frac{1}{4} F L_n. \tag{3.5.4}$$

Furthermore, at $\tau = 0$, from equations (3.5.1) and (3.5.2) we get

$$S(0) = I(0, 0) = \frac{1}{4} F H(\mu_0). \tag{3.5.5}$$

Multiplying equation (3.5.2) by

$$\left(1 + \frac{\mu}{\mu_0}\right) \prod_{\alpha=1}^{n-1} (1 + k_\alpha \mu)$$

and comparing the coefficients of μ^n on either side we get

$$k_1 \cdots k_{n-1} L_n \mu_1 \cdots \mu_n = \mu_0 H(\mu_0). \tag{3.5.6}$$

Therefore

$$L_n = \frac{\mu_0 H(\mu_0)}{k_1 \cdots k_{n-1} \mu_1 \cdots \mu_n}. \tag{3.5.7}$$

By making use of the relation (3.2.1), we get

$$L_n = 3^{1/2} \mu_0 H(\mu_0). \tag{3.5.8}$$

From equations (3.5.4), (3.5.5) and (3.5.8) we get

$$\frac{S(\infty)}{S(0)} = 3^{1/2} \mu_0. \tag{3.5.9}$$

which is independent of the order of the approximation. The remarkable fact is that the H-function describes not only the distribution of the emergent radiation with constant net flux but also the problem of diffuse reflection of radiation. More of this will be seen in chapter 5 which describes the invariance principles.

3.6 Iteration of the integral equation

Consider the monochromatic transfer equation in a plane parallel medium

$$\mu \frac{dI_\nu}{d\tau_\nu} = I_\nu - S_\nu, \tag{3.6.1}$$

where the source function S_ν is given by

$$S_\nu = (1 - \varpi_\nu) B_\nu + \varpi_\nu J_\nu, \tag{3.6.2}$$

and

$$\varpi_\nu = \frac{\sigma_\nu}{\kappa_\nu + \sigma_\nu}. \tag{3.6.3}$$

From equation (2.12.28) we have

$$J_\nu(\tau_\nu) = \Lambda_{\tau_\nu}[S_\nu(t_\nu)] = \Lambda_{\tau_\nu}[B_\nu] + \Lambda_{\tau_\nu}[\varpi_\nu(J_\nu - B_\nu)]. \tag{3.6.4}$$

The quantity $J_\nu(\tau_\nu)$ is the solution of an integral equation. One can solve the above equation by iteration. In LTE $J_\nu \to B_\nu$ and $\tau_\nu \to \infty$, we can write (Mihalas 1970)

$$(J_\nu - B_\nu) = \bar{B}_\nu - B_\nu + \Lambda_{\tau_\nu}[\varpi_\nu(J_\nu - B_\nu)], \tag{3.6.5}$$

where

$$\bar{B}_\nu = \Lambda_{\tau_\nu}[B_\nu]. \tag{3.6.6}$$

If $\varpi_\nu = 0$, then $J_\nu - B_\nu$ will be equal to $\bar{B}_\nu - B_\nu$. But if $\varpi_\nu \neq 0$, we can consider equation (3.6.5) as a first approximation for $J_\nu - B_\nu$ and calculate improved estimates by iteration of

$$(J_\nu - B_\nu)^{(n)} = (\bar{B}_\nu - B_\nu) + \sum_{j=1}^{n} Q^{(j)}, \tag{3.6.7}$$

where

$$Q^{(j)} = \Lambda_{\tau_\nu}[\varpi_\nu Q^{(j-1)}] \tag{3.6.8}$$

and

$$Q^{(1)} = \Lambda_{\tau_\nu}[\varpi_\nu(\bar{B}_\nu - B_\nu)]. \tag{3.6.9}$$

The criterion for convergence is

$$\left| \frac{2Q^{(j)}}{(J_\nu - B_\nu)^{(j)} + (J_\nu - B_\nu)^{(j-1)}} \right| \leq \epsilon \tag{3.6.10}$$

at all optical depth ranges. The quantity ϵ is chosen to be as small as possible depending on the required accuracy and the available computing time. Normally it is chosen to be between 10^{-3} and 10^{-4}. The convergence is very slow and one may need as many as $1/\epsilon$ iterations. Let us consider a different analysis. We assume that the Planck function is expressed as

$$B_\nu(\tau_\nu) = a_\nu + b_\nu \tau_\nu, \tag{3.6.11}$$

and that ϖ_ν is constant with depth. If we take the zeroth order moment of the transfer equation we get

$$\frac{dH_\nu}{d\tau_\nu} = J_\nu - S_\nu = (1 - \varpi_\nu)(J_\nu - B_\nu) = \lambda_\nu(J_\nu - B_\nu), \tag{3.6.12}$$

where S_ν is given by equation (3.6.2) and $H_\nu = \int I_\nu(\mu)\mu\, d\mu$. The first order moment gives us

$$\frac{dK_\nu}{d\tau_\nu} = H_\nu. \tag{3.6.13}$$

Using the Eddington approximation that $J_\nu = 3K_\nu$ and substituting equation (3.6.13) into equation (3.6.12), we get

$$\frac{1}{3}\frac{d^2 J_\nu}{d\tau_\nu^2} = \lambda_\nu (J_\nu - B_\nu). \tag{3.6.14}$$

From equation (3.6.11), $d^2 B_\nu / d\tau_\nu^2 = 0$ and in view of this we can write equation (3.6.14) as

$$\frac{1}{3}\frac{d^2(J_\nu - B_\nu)}{d\tau_\nu^2} = \lambda_\nu (J_\nu - B_\nu), \tag{3.6.15}$$

which has a solution

$$J_\nu - B_\nu = p_\nu \exp[-(3\lambda_\nu)^{1/2}\tau_\nu] + q_\nu \exp[(3\lambda_\nu)^{1/2}\tau_\nu]. \tag{3.6.16}$$

In LTE, we have $J_\nu \to B_\nu$ as $\tau_\nu \to \infty$, then $q_\nu = 0$ and hence

$$J_\nu(\tau_\nu) = a_\nu + b_\nu \tau_\nu + p_\nu \exp[-(3\lambda_\nu)^{1/2}\tau_\nu]. \tag{3.6.17}$$

Using the grey solution that $J_\nu(0) = 3^{1/2} H_\nu(0)$ we can evaluate p_ν. From equation (3.6.13), we get

$$H_\nu = \frac{1}{3}\frac{dJ_\nu}{d\tau_\nu} = \frac{1}{3}\left\{b_\nu - p_\nu (3\lambda_\nu)^{1/2} \exp[-(3\lambda_\nu)^{1/2}]\right\}. \tag{3.6.18}$$

Applying the boundary condition, we obtain

$$J_\nu(0) = a_\nu + p_\nu = 3^{1/2} H_\nu(0) = \frac{1}{3^{1/2}}[b_\nu - p_\nu (2\lambda_\nu)^{1/2}], \tag{3.6.19}$$

which gives

$$p_\nu = \frac{b_\nu - 3^{1/2} a_\nu}{3^{1/2}}(1 + \lambda_\nu^{1/2}). \tag{3.6.20}$$

We now get

$$J_\nu(\tau_\nu) = a_\nu + b_\nu \tau_\nu + \frac{b_\nu - 3^{1/2} a_\nu}{3^{1/2}(1 + \lambda_\nu^{1/2})} \exp[-(3\lambda_\nu)^{1/2}\tau_\nu]. \tag{3.6.21}$$

In a predominantly scattering atmosphere the mean intensity will depart considerably from the thermal source term at large optical depths. If, in equation (3.6.11), $b_\nu = 0$, then $B_\nu(\tau_\nu) = a_\nu$ and equation (3.6.21) becomes

$$J_\nu(0) = \frac{\lambda_\nu^{1/2}}{1 + \lambda_\nu^{1/2}} a_\nu = \frac{\lambda^{1/2}}{1 + \lambda_\nu^{1/2}} B_\nu. \tag{3.6.22}$$

When λ_ν is small, $J_\nu(0)$ will be much smaller than B_ν and then $\tau \to \infty$.

In early type stars, the main source of opacity is electron scattering and the scattering term dominates over the thermal terms. The solution of equation (3.6.4) by iteration will fail if we start iteration with the initial estimate of $J_\nu \approx B_\nu$ because each Λ-iteration propagates information about the departure of J_ν from B_ν only over a depth of $\Delta\tau \approx 1$. This needs $1/\lambda_\nu^{1/2}$ iterations to obtain the correct solution which is prohibitive particularly when $\lambda_\nu \ll 1$, which is the case in a predominantly scattering atmosphere.

3.7 Integral equation method. Solution by linear equations

Alternatively, the source function can be approximated by an interpolation formula which is valid between a discrete set of depth points. This means the integral operator will be replaced by a sum and a set of linear algebraic equations will replace the integral equations. Several authors used this method (see Gebbie (1967)). We shall give Kurucz's (1969) description of the method.

Assume that the integration in the Λ-operator is the sum of integrals over N discrete intervals and write (omitting the subscript ν for convenience)

$$J_l = J(\tau_l) = \frac{1}{2}\sum_{j=1}^{N}\int_{\tau_j}^{\tau_{j+1}} S(t)E_1\left|t-\tau_1\right| dt. \tag{3.7.1}$$

The source function $S(t)$ can be represented by a quadratic interpolation formula of the form

$$S(t) = \sum_{k=1}^{3} t^{k-1}\sum_{i=1}^{N} C_{jki}S_i, \tag{3.7.2}$$

where the C_{jki}s are the interpolation coefficients (see Kurucz (1969)). Inserting equation (3.7.2) into equation (3.7.1), the jth terms of the sum is given by

$$J_{lj} = \frac{1}{2}\int_{\tau_j}^{\tau_{j+1}} dt\, E_1\left|t-\tau_l\right|\sum_{k=1}^{3} t^{k-1}\sum_{i=1}^{N} C_{jki}S_i = \sum_{k=1}^{3}\eta_{ljk}\sum_{i=1}^{N} C_{jki}S_i, \tag{3.7.3}$$

where

$$\eta_{ljk} = \frac{1}{2}\int_{\tau_j}^{\tau_{j+1}} t^{k-1}E_1\left|t-\tau_l\right| dt. \tag{3.7.4}$$

By defining Λ_{li} as

$$\Lambda_{li} = \sum_{j=1}^{N}\sum_{k=1}^{3}\eta_{ljk}C_{jki} \tag{3.7.5}$$

we can write J_l in equation (3.7.1) in terms of Λ_{li} as

$$J_l = \sum_{i=1}^{N} \Lambda_{li} S_i. \tag{3.7.6}$$

Then we can write the source function as

$$S_l = (1 - \varpi_l)B_l + \varpi_l J_l = (1 - \varpi_l)B_l + \varpi_l \sum_{i=1}^{N} \Lambda_{li} S_i. \tag{3.7.7}$$

In matrix notation,

$$\mathbf{S} = (\mathbf{E} - \boldsymbol{\varpi})\mathbf{B} + \boldsymbol{\varpi}\boldsymbol{\Lambda}\mathbf{S}, \tag{3.7.8}$$

where $\mathbf{S}$ and $\mathbf{B}$ are N-component vectors given by $\mathbf{S} = (S_1, S_2, \ldots, S_N)$ and $\mathbf{B} = (B_1, B_2, \ldots, B_N)$, $\boldsymbol{\varpi}$ is the diagonal matrix with elements $(\boldsymbol{\rho})_{ii} = \rho_i$, $\mathbf{E}$ is the unit matrix and $\boldsymbol{\Lambda}$ is the matrix defined in equation (3.7.5). We can get $\mathbf{S}$ from

$$\mathbf{S} = (\mathbf{E} - \boldsymbol{\varpi}\Lambda)^{-1}(\mathbf{E} - \boldsymbol{\varpi})\mathbf{B}. \tag{3.7.9}$$

For a given set τ_i the $\boldsymbol{\Lambda}$ matrix may be computed once and for all. After obtaining $\mathbf{S}$ from (3.7.9), we can get $\mathbf{J}$:

$$\mathbf{J} = \boldsymbol{\Lambda}\mathbf{S}. \tag{3.7.10}$$

Unlike the iteration method, this gives a direct solution of the integral equation and is therefore free of the difficulties faced in the iteration procedure.

Other methods which use the two-point boundary character will be described in chapter 4.

Exercises

3.1 If $J = \frac{1}{2}(I_+ + I_-)$ and $F = I_+ - I_-$, show using equation (3.1.2) that

$$J = \frac{1}{2}F(2\tau + 1),$$

where τ is the optical depth and $I_- = 0$ at $\tau = 0$.

3.2 If $I(\tau, \mu) = \sum_{l=0}^{\infty} I_l(\tau)P_l(\mu)$, where the P_ls are the Legendre polynomials, show that:

(a)

$$\mu\frac{dI}{dt} = I - J, \quad \left(J = \frac{1}{2}\int_{-1}^{+1} I(\tau, \mu)\, d\mu\right)$$

will become

$$\mu\frac{dI}{d\tau} = I - I_0;$$

(b)

$$\frac{l}{2l-1}\frac{dI_{l-1}}{d\tau} + \frac{l+1}{2l+3}\frac{dI_{l+1}}{d\tau} = I_l \quad (l = 1, 2, \ldots)$$

and

$$\frac{1}{3}\frac{dI_1}{d\tau} = 0 \quad (l = 0).$$

3.3 Show that the relation

$$\sum_{j=-n}^{+n} \frac{a_j}{1+\mu_j k} = 2$$

becomes

$$\sum_{j=1}^{n} \frac{a_j}{1-\mu_j^2 k^2} = 1,$$

where the μ_js and a_js are the roots and weights of the Gauss–Legendre quadrature formula over the interval $(+1, -1)$ of μ.

3.4 Show that

$$\sum_{i=-n}^{+n} \frac{a_i \mu_i}{1+\mu_i k} = 0.$$

3.5 Show that the solution $I_i = b(\tau + q_i)$, $(i = \pm 1, \ldots, \pm n)$ when substituted into the equation

$$\mu_i \frac{dI_i}{d\tau} = I_i - \frac{1}{2}\sum_i a_j I_j \quad (i, j = \pm 1, \ldots, \pm n)$$

will give

$$\mu_i = q_i - \frac{1}{2}\sum a_j q_j.$$

3.6 Show that the characteristic roots and zeros of the Legendre polynomials satisfy the relation

$$\prod_{\alpha=1}^{n-1} k_\alpha \prod_{\alpha=1}^{n} \mu_\alpha = 3^{-\frac{1}{2}}.$$

3.7 Show that the third and fourth approximation of Chandrasekhar's solution gives the laws of darkening respectively as

$$I(0, \mu) = \frac{3}{4}F\left(\mu + 0.703\,899 - \frac{0.101\,245}{1+3.202\,95\mu} - \frac{0.025\,30}{1+1.225\,21\mu}\right)$$

and

$$I(0, \mu) = \frac{3}{4}F\left(\mu + 0.706\,92 - \frac{0.083\,92}{1+4.458\,08\mu} - \frac{0.036\,19}{1+1.591\,78\mu} - \frac{0.009\,46}{1+1.103\,194\mu}\right).$$

3.8 Write the mean intensity J, the net flux F and K in terms of the Gaussian sums in the first approximation of Chandrasekhar's solution.

3.9 Obtain the first two approximations of the solution of the planetary nebula problem given in the equation (3.3.8).

3.10 If the transfer equation in plane parallel layers with the Rayleigh phase function $\gamma(\theta', \varphi', \theta)$ is given by

$$\cos\theta \frac{dI(\tau,\theta)}{d\tau} = I(\tau,\theta) - \frac{1}{4\pi}\int_0^{\pi}\int_0^{2\pi} I(\tau,\theta')\gamma(\theta',\varphi';\theta)\sin\theta'\,d\theta'\,d\varphi',$$

where

$$\gamma(\theta',\varphi',\theta) = \frac{3}{4}\left(1+\cos^2\Theta\right)d\omega'$$

is the Rayleigh phase function and

$$\cos\Theta = \mu\mu' + (1-\mu^2)^{\frac{1}{2}}\left(1-\mu'^2\right)^{\frac{1}{2}}\cos\varphi',$$

$\mu = \cos\theta,\ \mu' = \cos\theta'$:

(a) Show the above transfer equation can be written (after performing integration over φ') as

$$\mu\frac{dI}{d\tau} = I - \frac{3}{8}\left[(3-\mu^2)J + (3\mu^2-1)K\right].$$

(b) Furthermore, if the boundary condition is $I_{-i} = 0$ for $\tau = 0$ and none of the I_is increases as $\tau \to \infty$, show that the angular distribution of the emergent radiation is given (in the nth approximation) by

$$I(0,\mu) = \frac{3}{4}F\left[\mu + Q + (3-\mu^2)\sum_{\alpha=1}^{n-1} L_\alpha(1+k_\alpha\mu)^{-1}\right],$$

where the L_αs and Q are the constants of integration.

(c) Show that in the first and second approximations the solutions are given by respectively

$$I(0,\mu) = \frac{3}{4}F(\mu + 3^{\frac{1}{2}})$$

and

$$I(0,\mu) = \frac{3}{4}F\left[\mu + 0.695\,39 - (3-\mu^2)\frac{0.044\,845}{1+1.870\,83\mu}\right].$$

(d) Calculate the third and fourth approximations.

3.11 Derive the analytical expression η_{ljk} in equation (3.7.4).

3.12 Assuming radiative equilibrium (equation (2.15.4)) in LTE, then if the temperature changes from T to $T_0 + \Delta T$, show that to the first order

$$\Delta T = \int_0^{\infty} \frac{\kappa_\nu\left[J_\nu - B_\nu(T_0)\right]d\nu}{\displaystyle\int_0^{\infty}\kappa_\nu \frac{\partial B_\nu}{\partial T}d\nu}.$$

3.13 The spherical term

$$\frac{1-\mu^2}{r}\frac{\partial I}{\partial \mu}$$

in the equation

$$\mu\frac{\partial I}{\partial r}+\frac{1-\mu^2}{r}\frac{\partial I}{\partial \mu}=-\kappa I+\frac{1}{2}\kappa\int_{-1}^{+1} I(\tau,\mu)\,d\mu$$

is replaced by the Gaussian division points as $[\partial I/\partial\mu]_{\mu=\mu_i}, i=\pm 1,\ldots,\pm n$. Define a polynomial

$$P_n(\mu)=-\frac{dQ_n}{d\mu}$$

with $Q_m=0$ for $|\mu|=1$ $(m=1,\ldots,2n)$ and $Q_m(\mu)=(P_{m-1}-P_{m+1})/(2\mu+1)$ + constant. Define $\eta(\mu)$ such that

$$Q_m(\mu)=\eta(\mu)(1-\mu^2)$$

and

$$\int_{-1}^{+1} Q_m(\mu)\frac{\partial I}{\partial\mu}\,d\mu=-\int_{-1}^{+1} I\frac{dQ_m}{d\mu}\,d\mu=\int_{-1}^{+1} I\,P_m(\mu)\,d\mu.$$

Using the Gaussian sums in the nth approximation find the values of $P_m(\mu)$, $Q_m(\mu)$, $\eta(\mu)$ for $m=1,\ldots,5$.

REFERENCES

Abramowitz, M., Stegun, I., 1964, *Handbook of Mathematical Functions*, Washington, DC, US Department of Commerce.

Ambarzumian, V.A., 1931, *Pulkova Obs. Bull.* No. 13.

Ambarzumian, V.A., 1932, *MNRAS*, **93**, 50.

Chandrasekhar, S., 1935, *Z. f. Ap.*, **9**, 266.

Chandrasekhar, S., 1944, *ApJ*, **100**, 76.

Chandrasekhar, S., 1960, *Radiative Transfer*, Dover, New York.

Gebbie, K., 1967, *MNRAS*, **135**, 181.

Hopf, E.A., 1934, *Mathematical Problems of Radiative Equilibrium*, Cambridge Mathematical Tract, No. 311, Cambridge University Press, Cambridge.

Kourganoff, V., 1963, *Basic Methods in Radiative Transfer*, Dover, New York.

Kurucz, R., 1969, *ApJ*, **156**, 235.

Mihalas, D., 1970, *Stellar Atmospheres*, 1st Edition, W.H. Freeman and Company, San Francisco.

Milne, E.A., 1930a, *Z. f. Ap.*, **1**, 90.

Milne, E.A., 1930a, *Handbuch der Astrophysik*, **3**, Part 1, Springer, Berlin, pp. 65–173.

Peraiah, A., Grant, I.P., 1973, *J. Inst. Math. Applics.*, **12**, 75.

Wick, G.C., 1943, *Z. f. Phys.*, **120**, 702.

Chapter 4

Two-point boundary problems

4.1 Boundary conditions

The solution of the radiative transfer equation is complete only when the boundary conditions are specified. We have seen that there are two types of problems in astrophysics: (1) those in a finite medium and (2) those in a semi-infinite medium. A finite medium is specified by a given geometrical thickness and the total optical thickness τ changes from 0 to T corresponding to the geometrical thickness $z = z_{max}$ to 0. In a semi-infinite medium at $z = z_{max}$, $\tau = 0$ while at $z = 0$, $\tau \to \infty$ and we need to specify the incident radiation on both sides. Therefore we specify these conditions as

$$I_-(\tau = 0, z = z_{max}) = \alpha \tag{4.1.1}$$

and

$$I_+(\tau = \tau_{max}, z = 0) = \beta. \tag{4.1.2}$$

We need to specify the quantities α and β. We shall give an example of how important it is that one must specify the boundary condition correctly, particularly in a semi-infinite atmosphere.

Consider a radiation field divided into an outgoing (I_+) and an incoming (I_-) radiation field (Mihalas 1970). Let $\mu_\pm = \pm\frac{1}{2}$, then

$$\frac{1}{2}\frac{dI_+}{d\tau} = I_+ - B \tag{4.1.3}$$

and

$$-\frac{1}{2}\frac{dI_-}{d\tau} = I_- - B, \tag{4.1.4}$$

where B is the LTE source function which is assumed to be constant. By subtracting equation (4.1.4) from equation (4.1.3) and then adding, we get

$$\frac{1}{2}\frac{d(I_+ + I_-)}{d\tau} = I_+ - I_- \tag{4.1.5}$$

and

$$\frac{1}{2}\frac{d(I_+ - I_-)}{d\tau} = I_+ + I_- - 2B. \tag{4.1.6}$$

By putting

$$J = \frac{1}{2}(I_+ + I_-) \tag{4.1.7}$$

and substituting equations (4.1.5) and (4.1.7) into equation (4.1.6) we get

$$\frac{1}{4}\frac{d^2 J}{d\tau^2} = J - B, \tag{4.1.8}$$

the solution of which is given by

$$J = a\exp(2\tau) + b\exp(-2\tau) + B. \tag{4.1.9}$$

From equations (4.1.9) and (4.1.5) we get

$$\frac{dJ}{d\tau} = 2a\exp(2\tau) - 2b\exp(-2\tau) = I_+ - I_- \tag{4.1.10}$$

or

$$I_+ = 2a\exp(2\tau) + B, \tag{4.1.11}$$

and

$$I_- = 2be^{-2\tau} + B. \tag{4.1.12}$$

We now specify the boundary conditions

$$I_+(\tau_{max}) = B \tag{4.1.13}$$

and

$$I_-(0) = 0. \tag{4.1.14}$$

Then using these boundary conditions in equations (4.1.11) and (4.1.12), we get

$$a = 0 \quad \text{and} \quad b = -\frac{B}{2}. \tag{4.1.15}$$

Therefore the solutions are

$$I_+(\tau) = B, \tag{4.1.16}$$

$$I_-(\tau) = B[1 - \exp(-2\tau)] \tag{4.1.17}$$

and

$$J(\tau) = B\left[1 - \frac{1}{2}\exp(-2\tau)\right]. \tag{4.1.18}$$

We may guess a value for I_- at the lower boundary $\tau = \tau_{max}$ and integrate towards the surface and at large optical depths this could be

$$I_-(\tau_{max}) = B. \tag{4.1.19}$$

This differs from condition (4.1.17) by ϵ', where

$$\epsilon' = B\exp(2\tau_{max}). \tag{4.1.20}$$

If τ_{max} is large then this error is small. However, it will propagate to the surface and we get an erroneous solution. One needs to give correct boundary conditions to obtain correct solutions.

4.2 Differential equation method. Riccati transformation

This method of solving the transfer equation by Riccati transformation was introduced by Rybicki (1965). The transfer equation in differential equation form is

$$\mu\frac{dI}{d\tau} = I - \varpi J - (1-\varpi)B. \tag{4.2.1}$$

Let us assume that $I^1_\alpha(\alpha = 1, \ldots, n)$ denotes the outgoing streams and I^2_α the incoming streams. Let ϖ_α be the quadratic weights chosen so that

$$J = \frac{1}{2}\int_{-1}^{+1} I(\mu)\,d\mu \approx \sum_\alpha w_\alpha(I^1_\alpha + I^2_\alpha). \tag{4.2.2}$$

In view of equation (4.2.2), equation (4.2.1) can be written as

$$\mu_\alpha\frac{dI_\alpha}{d\tau} = I^1_\alpha - \varpi\sum_\alpha w_\alpha(I^1_\alpha + I^2_\alpha) - (1-\varpi)B \tag{4.2.3}$$

and

$$-\mu_\alpha\frac{dI^2_\alpha}{d\tau} = I^2_\alpha - \varpi\sum_\alpha w_\alpha(I^1_\alpha + I^2_\alpha) - (1-\varpi)B. \tag{4.2.4}$$

If the outgoing radiation field is represented by a set of numbers $\mathbf{f}^1$ related to I^1 and the incoming radiation field by the set of numbers $\mathbf{f}^2$ related to I^2, we can write equations (4.2.3) and (4.2.4) in the vector form:

$$\frac{d\mathbf{f}^1}{d\tau} = \mathbf{\Gamma}^{11}\mathbf{f}^1 + \mathbf{\Gamma}^{12}\mathbf{f}^2 + \mathbf{h}^1 \tag{4.2.5}$$

and

$$-\frac{d\mathbf{f}^2}{d\tau} = \mathbf{\Gamma}^{21}\mathbf{f}^1 + \mathbf{\Gamma}^{22}\mathbf{f}^2 + \mathbf{h}^2, \tag{4.2.6}$$

where the $\mathbf{\Gamma}$s represent various matrices coupling various components and the **h**s represent the inhomogeneous thermal terms. The boundary conditions can be written as

$$\mathbf{f}^1(T) = \mathbf{E}^1 \tag{4.2.7}$$

and

$$\mathbf{f}^2(0) = \mathbf{E}^2. \tag{4.2.8}$$

We shall rewrite the equations so that we are dealing with only those quantities whose boundary conditions are known at say $\tau = T$. For this purpose Rybicki suggested that the quantity $\mathbf{f}^2$ be eliminated by the following transformation

$$\mathbf{f}^1 = \boldsymbol{\psi} + \mathbf{R}\mathbf{f}^2, \tag{4.2.9}$$

where $\boldsymbol{\psi}$ is a vector and $\mathbf{R}$ is a matrix. Substitution of this equation into equations (4.2.5) and (4.2.6) gives us

$$\frac{d\boldsymbol{\psi}}{d\tau} + \mathbf{R}\frac{d\mathbf{f}^2}{d\tau} + \mathbf{f}^2\frac{d\mathbf{R}}{d\tau} = \mathbf{\Gamma}^{11}\boldsymbol{\psi} + \mathbf{\Gamma}^{11}\mathbf{R}\mathbf{f}^2 + \mathbf{\Gamma}^{12}\mathbf{f}^2 + \mathbf{h}^1 \tag{4.2.10}$$

and

$$-\mathbf{R}\frac{d\mathbf{f}^2}{d\tau} = \mathbf{R}\mathbf{\Gamma}^{21}\boldsymbol{\psi} + \mathbf{R}\mathbf{\Gamma}^{21}\mathbf{R}\mathbf{f}^2 + \mathbf{R}\mathbf{\Gamma}^{22}\mathbf{f}^2 + \mathbf{R}\mathbf{h}^2. \tag{4.2.11}$$

Addition of equations (4.2.10) and (4.2.11) gives us

$$\begin{aligned} \frac{d\boldsymbol{\psi}}{\tau} &= (\mathbf{\Gamma}^{11} + \mathbf{R}\mathbf{\Gamma}^{22})\boldsymbol{\psi} + (\mathbf{h}^1 + \mathbf{R}\mathbf{h}^2) \\ &\quad + \left(\mathbf{\Gamma}^{11}\mathbf{R} + \mathbf{R}\mathbf{\Gamma}^{21}\mathbf{R} + \mathbf{R}\mathbf{\Gamma}^{22} + \mathbf{\Gamma}^{12} - \frac{d\mathbf{R}}{d\tau}\right)\mathbf{f}^2. \end{aligned} \tag{4.2.12}$$

If we want the third term on the RHS containing $\mathbf{f}^2$ to vanish, the quantity in the large brackets must equal zero or

$$\frac{d\mathbf{R}}{d\tau} = \mathbf{\Gamma}^{11}\mathbf{R} + \mathbf{R}\mathbf{\Gamma}^{21}\mathbf{R} + \mathbf{R}\mathbf{\Gamma}^{22} + \mathbf{\Gamma}^{12}. \tag{4.2.13}$$

If we substitute equation (4.2.13) into equation (4.2.12), we obtain

$$\frac{d\boldsymbol{\psi}}{d\tau} = (\mathbf{\Gamma}^{11} + \mathbf{R}\mathbf{\Gamma}^{22})\boldsymbol{\psi} + (\mathbf{h}^1 + \mathbf{R}\mathbf{h}^2). \tag{4.2.14}$$

The above two equations are then solved simultaneously with the starting values:

$$\boldsymbol{\psi}(T) = \mathbf{E} \tag{4.2.15}$$

and

$$\mathbf{R}(T) = 0. \tag{4.2.16}$$

We thus have the values of $\mathbf{R}$ and $\boldsymbol{\psi}$ in the range $0 \leq \tau \leq T$ after the integration is completed. Now we integrate equation (4.2.11) in the form

$$-\frac{d\mathbf{f}^2}{d\tau} = \left(\boldsymbol{\Gamma}^{21}\mathbf{R} + \boldsymbol{\Gamma}^{22}\right)\mathbf{f}^2 + \left(\boldsymbol{\Gamma}^{21}\boldsymbol{\psi} + \mathbf{h}^2\right), \tag{4.2.17}$$

from 0 to T, with the starting value

$$\mathbf{f}^2(0) = \mathbf{E}^2. \tag{4.2.18}$$

The final solution is obtained from equation (4.2.9). This method reduces the two-point boundary value problem to two simple initial value problems. One can give a simple physical interpretation. The quantity $\boldsymbol{\psi}$ can be regarded as part of the outgoing radiation field due to sources in deeper layers and the term $\mathbf{R}\mathbf{f}^2$ as the reflection matrix acting on the outgoing radiation field.

This method can give accurate results where the iteration technique fails and can treat more general cases (see Rybicki (1967)).

4.3 Feautrier method for plane parallel and stationary media

The transfer equation can be written as a two-point boundary value problem in the form of a second order differential equation. We now describe a method due to Feautrier (1964) for obtaining the solution of the transfer equation in stationary plane parallel media . The specific intensity is changed into mean-intensity-like and flux-like variables. The transfer equation is written in terms of these two variables as a finite difference equation with two boundary conditions. The discretization is effected in depth, angle and frequency. These equations with the boundary conditions are written in block tri-diagonal form which is solved by the Gaussian elimination scheme. This has been applied in plane parallel, spherically symmetric and moving media (see Mihalas (1978) and Sen and Wilson (1998)).

We shall first consider the transfer equation in a plane parallel, stationary medium. This is written (see chapter 2) for two oppositely directed pencils $\pm\mu$ as

$$\pm\frac{\partial I(z, \pm\mu, \nu)}{\partial z} = \kappa(z, \nu)\,[S(z, \nu) - I(z, \pm\mu, \nu)] \quad 0 \leq \mu \leq 1, \tag{4.3.1}$$

where the symbols have their usual meanings. We define

$$u(z, \mu, \nu) = \frac{1}{2}\,[I(z, \mu, \nu) + I(z, -\mu, \nu)], \tag{4.3.2}$$

$$v(z, \mu, \nu) = \frac{1}{2}\left[I(z, \mu, \nu) - I(z, -\mu, \nu)\right], \tag{4.3.3}$$

where u and v are the mean-intensity-like and flux-like variables. Adding the two equations in (4.3.1) and using equations (4.3.2) and (4.3.3), we get

$$\mu \frac{\partial v(z, \mu, \nu)}{\partial z} = \kappa(z, \nu)\left[S(z, \nu) - u(z, \mu, \nu)\right] \tag{4.3.4}$$

and

$$\mu \frac{\partial u(z, \mu, \nu)}{\partial z} = -\kappa(z, \nu) v(z, \mu, \nu). \tag{4.3.5}$$

Eliminating v between equations (4.3.4) and (4.3.5) gives us a second order differential equation in u. Thus, we have

$$\mu^2 \frac{\partial^2 u(\tau_\nu, \mu, \nu)}{\partial \tau_\nu^2} = u(\tau_\nu, \mu, \nu) - S(\tau_\nu, \nu), \tag{4.3.6}$$

where $d\tau_\nu = -\kappa(\nu)\, dz$.

The source function can be independent of μ if it is of the form

$$S(\nu) = \alpha_\nu \int \phi_{\nu'} J_{\nu'}\, d\nu' + \beta_\nu \tag{4.3.7}$$

or

$$S(\nu) = \alpha_\nu \int R(\nu', \nu) J_{\nu'}\, d\nu' + \beta_\nu. \tag{4.3.8}$$

However, if the redistribution function is angle dependent then we need to estimate $S(\mu, \nu)$ instead of $S(\nu)$. In equations (4.3.7) and (4.3.8) α stands for the scattering part of the radiation and β represents the thermal terms. If we consider the radiative equilibrium, then the source function needs to contain radiation of all frequencies.

4.4 Boundary conditions

The solution of equation (4.3.6) can be found when the boundary conditions at $\tau = 0$ (upper boundary) and at $\tau = \tau_{max}$ (lower boundary) are specified. At the upper boundary $\tau = 0$, if we set $I^-(0, \mu, \nu) = 0$, then we get $v(0, \mu, \nu) = u(0, \mu, \nu)$ from equations (4.3.2) and (4.3.3). This gives us

$$\mu \left. \frac{\partial u(\tau_\nu, \mu, \nu))}{\partial \tau_\nu} \right|_{\tau_\nu = 0} = u(0, \mu, \nu). \tag{4.4.1}$$

At the lower boundary, $\tau_\nu = \tau_{max}$, $I(\tau_{max}, \mu, \nu) = I^+(\tau_{max}, \mu, \nu)$ and this implies that

$$v(\tau_{max}, \mu, \nu) = I^+(\tau_{max}, \mu, \nu) - u(\tau_{max}, \mu, \nu). \tag{4.4.2}$$

By using relation (4.3.5), we get

$$\mu \frac{\partial u(\tau_\nu, \mu, \nu)}{\partial \tau_\nu}\bigg|_{\tau_{max}} = I^+(\tau_{max}, \mu, \nu) - u(\tau_{max}, \mu, \nu). \qquad (4.4.3)$$

Equations (4.4.1) and (4.4.2) are the upper and lower boundary conditions.

In a semi-infinite atmosphere at large depths in the atmosphere, the radiation is effectively trapped and the radiation becomes isotropic and $S_\nu \rightarrow B_\nu$. In such situations we have what is called the diffusion approximation (equations (2.16.1) and (2.16.2)) at $\tau = \tau_{max}$. Then we have

$$I(\tau_{max}, \mu, \nu) = B_\nu(\tau_{max}) + \mu \frac{\partial B_\nu}{\partial \tau_\nu}\bigg|_{\tau_{max}}. \qquad (4.4.4)$$

This gives us

$$u(\tau_{max}, \mu, \nu) = B_\nu(\tau_{max}) \qquad (4.4.5)$$

and

$$v(\tau_{max}, \mu, \nu) = \mu \left| \frac{\partial B_\nu}{\partial \tau_\nu} \right|_{\tau_{max}}. \qquad (4.4.6)$$

4.5 The difference equation

We now write the second order differential equation (4.3.6) in its equivalent difference equation form. The variables τ, μ and ν are discretized at conveniently chosen discrete points. The depth points are discretized as $\{\tau_d\}$, $(d = 1, 2, \ldots, D)$, $\tau_1 < \tau_2 < \tau_3 < \cdots < \tau_D$ (see figure 4.1). Similarly the angle points are discretized as $\{\mu_m\}$, $(m = 1, 2, \ldots, M)$, $0 < \mu_1 < \mu_2 < \cdots < \mu_M \leq 1$ and the frequency points as $\{\nu_n\}$, $(n = 1, 2, \ldots, N)$ $0 < \nu_\infty$. In what follows any variable with a subscript such as P_{dmn} denotes $P_{dmn} = P(z_d, \mu_m, \nu_n)$. The integrals are replaced by quadrature sums. We write a composite set of discrete points combining the angle and frequency discrete points. This is given by

$$i = m + (n-1)M, \quad I = NM. \qquad (4.5.1)$$

The derivatives are written in the form of difference equations: for example,

$$\frac{dP}{d\tau}\bigg|_{d+\frac{1}{2}} = \frac{\Delta P_{d+\frac{1}{2}}}{\Delta \tau_{d+\frac{1}{2}}} = \frac{P_{d+1} - P_d}{\tau_{d+1} - \tau_d} \qquad (4.5.2)$$

and

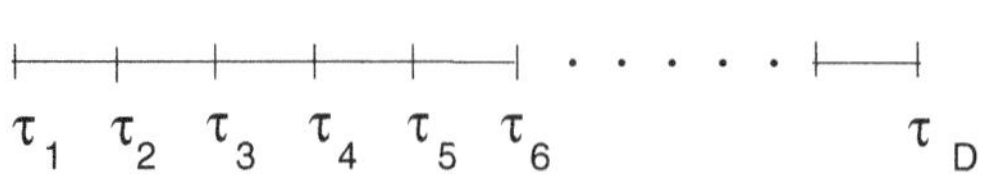

Figure 4.1 Discretization of the optical depth.

$$\left.\frac{d^2P}{d\tau^2}\right|_d = \frac{\left(\left.\frac{dP}{d\tau}\right|_{d+\frac{1}{2}} - \left.\frac{dP}{d\tau}\right|_d\right)}{\frac{1}{2}\left(\Delta\tau_{d+\frac{1}{2}} + \Delta\tau_{d-\frac{1}{2}}\right)}, \tag{4.5.3}$$

where

$$\Delta\tau_{d\pm\frac{1}{2},i} = \frac{1}{2}\left(\kappa_{d\pm1,i} + \kappa_{d,i}\right)|z_{d\pm1} - z_d| \tag{4.5.4}$$

and

$$\Delta\tau_{d,i} = \frac{1}{2}\left(\Delta\tau_{d-\frac{1}{2},i} + \Delta\tau_{d+\frac{1}{2},i}\right). \tag{4.5.5}$$

In view of the above, equation (4.3.6) can be written as

$$\begin{aligned}&\left(\frac{\mu_i^2}{\Delta\tau_{d-\frac{1}{2},i}\,\Delta_{d,i}}\right)u_{d-1,i} - \frac{\mu_i^2}{\Delta\tau_{d,i}}\left(\frac{1}{\Delta\tau_{d-\frac{1}{2},i}} + \frac{1}{\Delta\tau_{d+\frac{1}{2},i}}\right)u_{d,i}\\ &\quad + \left(\frac{\mu_i^2}{\Delta\tau_{d,i}\Delta\tau_{d+\frac{1}{2},i}}\right)u_{d+1,i} = u_{d,i} - S_{d,i},\\ &\quad i = 1,\ldots,I(=NM), \quad d = 2,\ldots,D-1.\end{aligned} \tag{4.5.6}$$

In equation (4.5.6), there are i equations at $D-2$ depth points. The source function S in (4.3.7) and (4.3.8) are written in terms of quadrature sums in the discretized form as

$$S_{di} = \alpha_{di}\sum_{i'=1}^{I}\omega_{i'}\phi_{d_{i'}}u_{d_{i'}} + \beta_{di} \quad (i = 1,\ldots,I), \tag{4.5.7}$$

and

$$S_{di} = \alpha_{di}\sum_{i'=1}^{I}R_{d,i',i}u_{d_{i'}} + \beta_{di} \quad (i = 1,\ldots,I). \tag{4.5.8}$$

At each of the $D-2$ depth points, there is one equation for each value of i. Therefore we have $\mathbf{u}_d$ of dimension I (which contains the angle and frequency discrete points) as $(u_d)_i$, and the difference equation (4.5.6) can be written in matrix form as

$$-\mathbf{A}_d\mathbf{U}_{d-1} + \mathbf{B}_d\mathbf{U}_d - \mathbf{C}_d\mathbf{U}_{d+1} = \mathbf{L}_d. \tag{4.5.9}$$

Matrices $\mathbf{A}_d$ and $\mathbf{C}_d$ are the $(I \times I)$ diagonal matrices and represent the finite-difference operator. Matrix $\mathbf{B}_d$ is a full matrix containing the diagonal and off-diagonal terms due to the scattering terms in equations (4.5.7) and (4.5.8). The vector $\mathbf{L}_d$ represents the thermal source terms. We now use the boundary conditions.

At $\tau = 0$ (the upper boundary) we write the discretized form of equation (4.4.1) to the first order as

$$\frac{\mu_i\left(u_{2,i} - u_{1,i}\right)}{\Delta\tau_{\frac{3}{2},i}} = u_{1,i}. \tag{4.5.10}$$

The second order accuracy can be obtained by expanding u_2 by Taylor's expansion. Thus

$$u_2 = u_1 + \Delta\tau_{\frac{3}{2}}\left(\frac{du}{d\tau}\right)_1 + \frac{1}{2}\Delta\tau_{\frac{3}{2}}^2\left(\frac{d^2u}{d\tau^2}\right)_1. \tag{4.5.11}$$

Using equations (4.3.6) for $\left(d^2u/d\tau^2\right)$ and equation (4.4.1), equation (4.5.11) gives

$$\frac{\mu_i\left(u_{2,i} - u_{1,i}\right)}{\Delta\tau_{\frac{3}{2},i}} = u_{1,i} + \left(\frac{1}{2}\frac{\Delta\tau_{\frac{3}{2},i}}{\mu_i}\right)(u_{1i} - S_{1i}) \tag{4.5.12}$$

or

$$\mathbf{B}_1\mathbf{U}_1 - \mathbf{C}_1\mathbf{U}_2 = \mathbf{L}_1. \tag{4.5.13}$$

Similarly the lower boundary given in equation (4.4.3) gives

$$\frac{\mu_i\left(u_{Di} - u_{D-1,i}\right)}{\Delta\tau_{D-\frac{1}{2},i}} = I_{Di}^{+} - U_{Di} - \left(\frac{\frac{1}{2}\Delta\tau_{D-\frac{1}{2},i}}{\mu_i}\right)(U_{Di} - S_{Di}) \tag{4.5.14}$$

or

$$-\mathbf{A}_D\mathbf{U}_{D-1} + \mathbf{B}_D\mathbf{U}_D = \mathbf{L}_D. \tag{4.5.15}$$

Here $\mathbf{A}_1 \equiv 0$ and $\mathbf{C}_D \equiv 0$.

The three equations (4.5.9), (4.5.13) and (4.5.15) can be written in matrix form:

$$\begin{pmatrix} \mathbf{B}_1 & -\mathbf{C}_1 & & & & & \\ -\mathbf{A}_2 & \mathbf{B}_2 & -\mathbf{C}_2 & & & & \\ & -\mathbf{A}_3 & \mathbf{B}_3 & -\mathbf{C}_3 & & & \\ & & & \ddots & \ddots & & \\ & & & & -\mathbf{A}_{D-1} & \mathbf{B}_{D-1} & -\mathbf{C}_{D-1} \\ & & & & & -\mathbf{A}_D & \mathbf{B}_D \end{pmatrix} \begin{pmatrix} \mathbf{U}_1 \\ \mathbf{U}_2 \\ \vdots \\ \\ \mathbf{U}_{D-1} \\ \mathbf{U}_D \end{pmatrix}$$

$$= \begin{pmatrix} \mathbf{L}_1 \\ \mathbf{L}_2 \\ \vdots \\ \\ \mathbf{L}_{D-1} \\ \mathbf{L}_D \end{pmatrix}. \tag{4.5.16}$$

The solution is obtained as follows. Each of elements of the matrix on the LHS are of dimension $I \times I$, while the vectors $\mathbf{U}_1, \ldots, -\mathbf{U}_D$ and $\mathbf{L}_1, \ldots, -\mathbf{L}_D$ are of dimension I. Equation (4.5.16), which contains a block tri-diagonal matrix, can be solved by the process of forward elimination and backward substitution – the Gaussian elimination scheme. The vector $\mathbf{U}_d$ is expressed in terms of $\mathbf{U}_{d+1}$ and substituted into the next equation. We obtain $\mathbf{U}_1$ from equation (4.5.13)

$$\mathbf{U}_1 = \mathbf{B}_1^{-1}\mathbf{C}_1\mathbf{U}_1 + \mathbf{B}_1^{-1}\mathbf{L}_1 = \mathbf{D}_1\mathbf{U}_2 + \mathbf{V}_1. \tag{4.5.17}$$

Substituting equation (4.5.17) into equation (4.5.9) for $d = 2$ we get

$$\mathbf{U}_2 = \mathbf{D}_2\mathbf{U}_3 + \mathbf{V}_2, \tag{4.5.18}$$

where

$$\mathbf{D}_2 = \mathbf{D}_{21}\mathbf{C}_2 \quad \text{and} \quad \mathbf{V}_2 = \mathbf{D}_{21}\left(\mathbf{L}_2 + \mathbf{A}_2\mathbf{V}_1\right) \tag{4.5.19}$$

and

$$\mathbf{D}_{21} = \left(\mathbf{B}_2 - \mathbf{A}_2\mathbf{D}_1\right)^{-1}. \tag{4.5.20}$$

The above equations can be generalized and we write the scheme as follows

$$\mathbf{U}_d = \mathbf{D}_d\mathbf{U}_{d+1} + \mathbf{V}_d, \tag{4.5.21}$$

where

$$\mathbf{D}_d = \mathbf{D}_{d,d-1}\mathbf{C}_d, \tag{4.5.22}$$

$$\mathbf{V}_d = \mathbf{D}_{d,d-1}\left(\mathbf{L}_d + \mathbf{A}_d\mathbf{V}_{d-1}\right) \tag{4.5.23}$$

and

$$\mathbf{D}_{d,d-1} = \left(\mathbf{B}_d - \mathbf{A}_d\mathbf{D}_{d-1}\right)^{-1}, \tag{4.5.24}$$

with $d = 1, 2, \ldots, D$

Successive values of $\mathbf{D}_d$ and $\mathbf{V}_d$ are computed from $d = 1$ to $d = D - 1$. At $d = D$, $\mathbf{C}_D = 0$ and therefore $D_D = 0$ and $\mathbf{U}_d = \mathbf{U}_D$. When $\mathbf{U}_D$ is obtained, successive back substitutions ($d = D - 1, \ldots, 2, 1$) into equation (4.5.23) will give the vector $\mathbf{U}_d$ which will give us $\mathbf{U}_{dmn}$. We can evaluate the mean intensity from the relation

$$J_{dn} = \sum_{m=1}^{M} b_m U_{dmn}, \tag{4.5.25}$$

where the bs are the angle quadrature weights. The source function is obtained as

$$S_{dn} = \alpha_{dn} \sum_{n'} \omega_{n'}\phi_{dn'} J_{dn'} + \beta_{dn}; \tag{4.5.26}$$

see equations (4.5.7) and (4.5.8).

The forward and backward sweep in the Feautrier method takes care of the scattering integrals. This method is stable, compact and can be easily implemented in terms of extension to more complex problems. However, it does not give a physical understanding of the diffusion properties of the radiation field such as diffuse reflection or transmission (see chapters 5 and 6).

One interesting property is that at depth, the system becomes more diagonal $(1/\Delta\tau^2 \to 0)$ and the mean intensity $J_d \to S_d$ or $J \to S + \mu^2 d^2 S/d\tau^2$, which is equivalent to the diffusion approximation. The discretization in depth is done at equally spaced points in $\log \tau$. This is an advantage in the widely differing optical depths that are characteristic in a line.

The computing time goes as cDM^3N^3, where D is the number of depth points, M is the number of angle points and N is the number of frequency points. Certain problems do not require the full set of angle–frequency grid points in which case one can economize on computer time. In the case of problems which involve coherent scattering $N = 1$ and M can be made as small as possible. In such cases, the Feautrier method becomes quite efficient. But in problems which involve radiative equilibrium, the number of frequencies can be large. However, the angular information is not essential as it is J_ν that is needed (and not $U_{\mu\nu}$) in these calculations. J_ν can be obtained through the Eddington factors (Auer and Mihalas 1970)

$$f_\nu = \frac{K_\nu}{J_\nu}. \tag{4.5.27}$$

Integration of equation (4.3.6) over μ yields

$$\frac{\partial^2 (f_2 J_\nu)}{\partial \tau_\nu^2} = J_\nu - S_\nu, \tag{4.5.28}$$

and the boundary conditions give us

$$\left. \frac{\partial (f_\nu J_\nu)}{\partial \tau_\nu} \right|_{\tau=0} = h_\nu J_\nu(0) \tag{4.5.29}$$

and

$$\left. \frac{\partial (f_\nu J_\nu)}{\partial \tau_\nu} \right|_{\tau=\tau_{max}} = \frac{1}{3} \left(\frac{1}{\kappa_\nu} \left| \frac{\partial B_\nu}{\partial z} \right| \right)_{\tau=\tau_{max}}, \tag{4.5.30}$$

where

$$h_\nu \equiv \frac{H_\nu(0)}{J_\nu(0)}. \tag{4.5.31}$$

Equations (4.5.28)–(4.5.30) can be solved through difference equations as the angle dependent equations and the time required goes only as CDN^3. We need to know the depth variation of f_ν to obtain the solution for each frequency. We obtain the solution by following the steps given below:

1. Starting with an assumed S_ν (such as $S_\nu = B_\nu$), we solve equation (4.3.6) and obtain $U_{\mu\nu}$ one angle and frequency at a time. This solution will have a matrix form given by

$$\mathbf{T}_i \mathbf{U}_i = \mathbf{S}_i, \tag{4.5.32}$$

   where $\mathbf{T}_i$ is diagonal and $\mathbf{U}_i$ and $\mathbf{S}_i$ represent U_{di} and S_{di} respectively. The time required to obtain the full angle dependent radiation field for a given S_ν goes as *CDMN*.

2. Once U_{dmn} is obtained, we can get the depth variation of the fs:

$$f_{dn} = \sum_m b_m \mu_m^2 U_{dmn} / \sum_m b_m U_{dmn} \tag{4.5.33}$$

   and

$$h_n = \sum_m b_m \mu_m U_{1dn} / \sum_m b_m U_{1mn}. \tag{4.5.34}$$

3. Once the Eddington factor fs are determined we solve equations (4.5.28)–(4.5.30) for S_ν in terms of J_ν. Then S_ν is reevaluated using the new values of J_ν.

4. If S_ν obtained in step (3) differs from that used in step (1), we iterate steps (1)–(3) to convergence. Normally convergence takes place in about three or four iterations and the saving on the computation time is about a factor of 10.

4.6 Rybicki method

Rybicki (1971) developed a technique that can save a considerable amount of computer time if we consider complete redistribution of photons in the source function. In the Feautrier method described above, we can obtain a full frequency dependent source function with partial frequency redistribution. The calculation requires a computation proportional to the cube of the number of frequency points.

In several problems of line formation, complete redistribution is sufficient and in this case information regarding several frequencies is redundant. The source function then requires the computation of $\bar{J} = \int \phi_\nu J_\nu \, d\nu$. Rybicki developed a method of great power and generality that can compare with that of Feautrier and which requires much less computing time.

Rybicki described the radiation field in terms of the depth variation vectors at a given frequency instead of the frequency variation vectors at a given depth. These are defined as

$$\mathbf{U}_i = (U_{1i}, U_{2i}, U_{3i}, \ldots, U_{Di})^T. \tag{4.6.1}$$

We similarly write

$$\bar{\mathbf{J}} = \left(\bar{J}_1, \bar{J}_2, \ldots, \bar{J}_D\right)^T. \tag{4.6.2}$$

Now, for each angle–frequency point i, equations (4.5.6), (4.5.12) and (4.5.14) can be compactly written as

$$\mathbf{T}_i\mathbf{U}_i + \mathbf{U}_i\bar{\mathbf{J}} = \mathbf{K}_i \quad (i = 1, \ldots, I), \tag{4.6.3}$$

where $\mathbf{T}_i$ is a tri-diagonal $(D \times D)$ matrix and represents the differential operator at frequency i, $\mathbf{U}_i$ is a diagonal matrix containing the depth variation of the scattering coefficients α_{di} in equations (4.5.7) and (4.5.8) and $\mathbf{K}_i$ is a vector that contains the depth distribution of the thermal terms at the angle–frequency point i. Furthermore, we have D equations that define $\bar{\mathbf{J}}_d$:

$$\sum_{i'=i}^{I} \omega_{i'}\phi_{di'}U_{di'} - \bar{J}_d = 0 \quad (d = 1, \ldots, D), \tag{4.6.4}$$

or

$$\bar{J}_d = \sum_{i'=1}^{I} \mathbf{V}_{i'}\mathbf{U}_{i'}. \tag{4.6.5}$$

Combining equations (4.6.1)–(4.6.5) , we get

$$\begin{pmatrix} \mathbf{T}_1 & & & & \mathbf{U}_1 \\ & \mathbf{T}_2 & & & \mathbf{U}_2 \\ & & \ddots & & \vdots \\ & & & \mathbf{T}_I & \mathbf{U}_I \\ \mathbf{V}_1 & \mathbf{V}_2 & \ldots & \mathbf{V}_I & \mathbf{E} \end{pmatrix} \begin{pmatrix} \mathbf{u}_1 \\ \mathbf{u}_2 \\ \vdots \\ \mathbf{u}_I \\ \mathbf{E} \end{pmatrix} = \begin{pmatrix} \mathbf{K}_1 \\ \mathbf{K}_2 \\ \vdots \\ \mathbf{K}_I \\ \mathbf{P} \end{pmatrix}, \tag{4.6.6}$$

where the $\mathbf{V}$s are the $(D \times D)$ diagonal matrices which contain the depth variation of the profile function multiplied by the quadrature weights given in (4.6.4). $\mathbf{E}$ is the negative identity matrix and $\mathbf{P}$ is the void for equation (4.6.4). However, this will become useful during the computation of LTE model atmospheres. The solution of equation (4.6.6) can be found by writing

$$\mathbf{u}_i = \left(\mathbf{T}_i^{-1}\mathbf{K}_i\right) - \left(\mathbf{T}_i^{-1}\mathbf{U}_i\right)\bar{\mathbf{J}}, \quad (i = 1, \ldots I). \tag{4.6.7}$$

Then substituting (4.6.7) in (4.6.5), we get

$$\mathbf{W}\bar{\mathbf{J}} = \mathbf{Q}, \tag{4.6.8}$$

where

$$\mathbf{W} = \mathbf{E} - \sum_{i=1}^{I} \mathbf{V}_i\mathbf{T}_i^{-1}\mathbf{U}_i \tag{4.6.9}$$

and

$$\mathbf{Q} = \mathbf{P} - \sum_{i=1}^{I} \mathbf{V}_i \mathbf{T}_i^{-1} \mathbf{K}_i, \tag{4.6.10}$$

where $\mathbf{W}$ is a full $(D \times D)$ matrix and $\mathbf{Q}$ is a vector. We thus solve for $\mathbf{J}$ which is used to obtain the source function S_d using equation (4.5.28). We can also solve the full angle–frequency variation of the radiation field from equation (4.6.7).

The computing time for Rybicki's method is favourable when compared to that of the Feautrier method as the former is linear with MN while the latter goes as M^3N^3. Rybicki's method is more economical when variable Eddington factors and a large number of frequency points are needed. However, Rybicki's method is good only when the source function is written in terms of $\bar{J}$ in the scattering integral while the Feautrier method works for general scattering functions. One can combine Rybicki's method with variable Eddington factors but one needs to iterate for the solution which may not be of any advantage.

Rybicki's method is equivalent to the integral equation method where one can write

$$\mathbf{u}_i = \Lambda_i \bar{J} + M_i, \tag{4.6.11}$$

where Λ_i is the matrix of the kernel function that is equal to $\mathbf{T}_i^{-1}$ representing $\bar{J}$ and the inversion of $\mathbf{T}_i$ which is less costly than any other way for generating Λ. Rybicki's method has been employed by Rogers (1984) and Rogers and Martin (1984, 1986).

4.7 Solution in spherically symmetric media

We shall now see how Feautrier's method can be applied to a spherically symmetric system (Hummer and Rybicki 1971, Mihalas 1978, Sen and Wilson 1998).

We use the Feautrier solution of the moment equation through the variable Eddington factor f_ν. In this way we avoid involving large matrices. We shall consider the moment equations. The transfer equation in stationary spherically symmetric media in the steady state condition is written (see chapter 2) as

$$\mu \frac{\partial I(r, \mu, \nu)}{\partial r} + \frac{1 - \mu^2}{r} \frac{\partial I(r, \mu, \nu)}{\partial \mu} = \kappa_\nu [I(r, \mu, \nu) - S(r, \mu)], \tag{4.7.1}$$

with the radial optical depth

$$d\tau_\nu = -\kappa_\nu \, dr. \tag{4.7.2}$$

We can write the first two angular moments of equation (4.7.1) as

$$\frac{\partial [r^2 H_\nu(r)]}{\partial \tau_\nu} = r^2 [J_\nu(r) - S_\nu(r)] \tag{4.7.3}$$

and

$$\frac{\partial K_\nu(r)}{\partial \tau_\nu} - \frac{1}{\kappa_\nu r}\,[3K_\nu(r) - J_\nu(r)] = H_\nu(r). \tag{4.7.4}$$

The variable Eddington factor $f_\nu(r)$ is given by (see chapter 1)

$$f_\nu(r) = \frac{K_\nu(r)}{J_\nu(r)}. \tag{4.7.5}$$

Replacing $K_\nu(r)$ in equation (4.7.4) from equation (4.7.5) we obtain

$$\frac{\partial[f_\nu(r)J_\nu(r)]}{\partial \tau_\nu} - \frac{(3f_\nu(r) - 1)J_\nu(r)}{\kappa_\nu(r)r} = H_\nu(r). \tag{4.7.6}$$

Here $S_\nu(r)$ in equation (4.7.3) will have the general form

$$S_\nu(r) = \alpha_\nu(r)\int R(r;\nu',\nu)J_\nu(r)\,d\nu' + \beta_\nu(r). \tag{4.7.7}$$

We can solve the two equations (4.7.3) and (4.7.6) for H_ν simultaneously. However, there are difficulties in doing so: (1) we end up with a second and a first order derivative which will be more difficult to solve and (2) the term $(\kappa_\nu(r)r)^{-1}$ in equation (4.7.6) tends to diverge towards the surface (opacity per unit volume) as the particle density changes over several orders of magnitude. This term has a destabilizing effect on the system. These difficulties can be avoided by making the transformation (Auer 1971) through what is called the sphericity factor q_ν, defined as

$$\ln(r^2 q_\nu) = \int_{r_c}^{r}\left[\frac{(3f_\nu - 1)}{r' f_\nu}\right]dr' + \ln r_c^2, \tag{4.7.8}$$

where r_c is the 'core radius', which corresponds to the deepest point in the atmosphere. q_ν is known if f_ν is known. Using equation (4.7.8) we can write equation (4.7.6) as

$$\frac{\partial(f_\nu q_\nu r^2 J_\nu)}{\partial \tau_\nu} = q_\nu r^2 H_\nu. \tag{4.7.9}$$

Substituting equation (4.7.9) into equation (4.7.3), we obtain

$$\frac{\partial}{\partial \tau_\nu}\left[\frac{1}{q_\nu}\frac{\partial(f_\nu q_\nu r^2 J_\nu)}{\partial \tau_\nu}\right] = r^2\,(J_\nu - S_\nu). \tag{4.7.10}$$

This is the combined moment equation. If we introduce a new variable,

$$dX_\nu = -q_\nu \kappa_\nu\,dr = q_\nu\,d\tau_\nu, \tag{4.7.11}$$

then equation (4.7.10) becomes

$$\frac{\partial^2(f_\nu q_\nu r^2 J_\nu)}{\partial X_\nu^2} = q_\nu^{-1} r^2\,(J_\nu - S_\nu). \tag{4.7.12}$$

The upper boundary condition is obtained at $r = R$ (from equation (4.7.9)) as

$$\left.\frac{\partial(f_\nu q_\nu r^2 J_\nu)}{\partial X_\nu}\right|_{r=R} = h_\nu (r^2 J_\nu)\Big|_{r=R}, \tag{4.7.13}$$

where

$$h_\nu = \int_0^1 I(R,\mu,\nu)\mu\, d\mu \Big/ \int_0^1 I(R,\mu,\nu)\, d\mu. \tag{4.7.14}$$

The lower boundary condition is obtained from the planar diffusion approximation at the core of the star $r = r_c$, as given by

$$H_\nu(r_c) = \frac{1}{3}\left[\frac{1}{\kappa_\nu}\left|\frac{\partial B_\nu}{\partial r}\right|\right]_{r=r_c}, \tag{4.7.15}$$

so that the gradient is set to satisfy the condition that the integrated $H_\nu(r_c)$ over all frequencies equals the flux:

$$H_c = \frac{L}{16\pi^2 r_c^2}, \tag{4.7.16}$$

L being the luminosity. Then the lower boundary condition becomes

$$\left.\frac{\partial(f_\nu q_\nu r^2 J_\nu)}{\partial X_\nu}\right|_{r=r_c} = r_c^2 H_c \left[\kappa_\nu^{-1} \frac{\dfrac{\partial B_\nu}{\partial t}}{\displaystyle\int_0^\infty \kappa_\nu^{-1}\left(\frac{\partial B_\nu}{\partial T}\right) d\nu}\right]_{r=r_c}. \tag{4.7.17}$$

The diffusion approximation is valid deep in the atmosphere where the photon mean free path $\kappa_\nu^{-1} \ll \epsilon R$, with ϵ a small number. The lower boundary condition should be modified according to the physical conditions of the objects under study such as nebulae etc.

A discrete radial mesh $\{r_d\}$ $(d = 1, \ldots, D)$, where $R = r_1 > r_2 > \cdots > r_D = r_c$ and a discrete frequency mesh ν_n $(n = 1, \ldots, N)$ are introduced to replace the derivatives with differences in the equations (4.7.12), (4.7.13) and (4.7.17). Splines (Kunasz and Hummer 1974, Mihalas 1974) or Hermite polynomials (Auer 1976) can also replace them. The frequency integrals which occur in the source function can be replaced by the quadrature formulae. The above three equations then become of the tri-diagonal form suitable for a Feautrier type solution. To obtain the solution we now need to know the variable Eddington factors $f_\nu(r)$, which can be found if we know the angular dependence of the radiation field at each radial point for each frequency or the ratio of $\kappa_\nu / J_\nu(r)$ and the angular moments at the boundaries. A ray-by-ray solution of the transfer equation for each frequency is obtained for a set of impact parameters $\{p_i\}$, which are chosen to be tangents to the discrete radial shells, in addition to another set of C impact parameters, which are chosen to intersect the core including the central ray (see figure 4.2). The impact parameters are p_i $(i = 1, \ldots, I$ and $I = D + C$, where D is the number of radial and C is the number of impact parameters in the core). P_1 represents the central ray, p_c is the last ray inside the core ($p_c < r_c$) and $p_{c+1} = r_c$ and $P_I = R$. Each ray p_i intersects all shells with $r_d \gg p_i$. These intersections define another mesh of discrete z-points $\{z_{di}; d = 1, \ldots, D_i,\ D_i = D + C - i$, for $i > C,\ D_i = D$ for $i \leq C\}$. Therefore,

$$z_{di} = \left(r_d^2 - p_i^2\right)^{\frac{1}{2}}. \tag{4.7.18}$$

Also the ray p_i intersects the radial shell r_d at an angle whose cosine μ_{di} is given by

$$\mu_{di} = \mu(r_d, p_i) = \frac{(r_d^2 - p_d^2)^{\frac{1}{2}}}{r_d} = \frac{z_{di}}{r_d}. \tag{4.7.19}$$

If we compute the solution along the rays p_i at a particular r_d, then knowledge of $I_\nu(z_{di}, p_i)$ $(i = 1, \ldots, I_d)$ is equivalent to knowledge of the variation of $I_\nu(r_d, \mu)$ on the mesh μ_{di} $(i = 1, \ldots, I_d,\ I_d = I + 1 - d)$ over the interval $1 \geq \mu \geq 0$. In this way one can construct the Eddington factors.

The (r, μ) coordinate system is changed to the (z, p) coordinate system along the rays; z is measured positive in the direction of the observer and negative in the opposite direction. p is the impact parameter, the radius r is given by

$$r = r(z, p) = (z^2 + p^2)^{\frac{1}{2}} \tag{4.7.20}$$

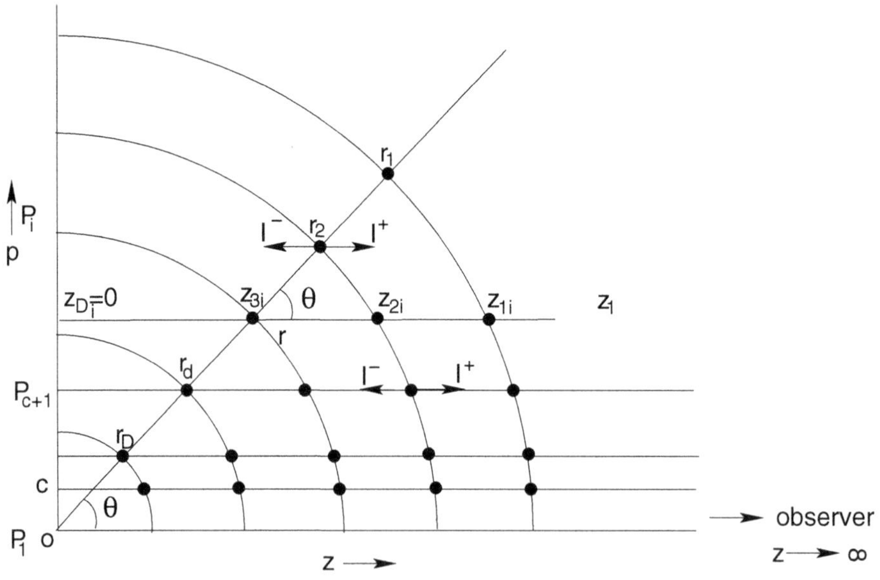

Figure 4.2 Schematic diagram of discrete (p, z) mesh. The p_is are the impact parameters which are chosen parallel to the central ray directed towards the observer and tangent to the spherical shell chosen to describe the depth variation of the physical properties of the envelope. The intersections of the rays with radial shells define a z coordinate along each ray (see Mihalas (1978), page 253 for a similar picture).

and the differential operator in z is given by

$$\frac{\partial}{\partial z} = \mu \frac{\partial}{\partial r} + \frac{1-\mu^2}{r}\frac{\partial}{\partial \mu}. \tag{4.7.21}$$

In terms of equation (4.7.21), the spherical transfer equation of the rays travelling along $\pm z$ for a ray with impact parameter p_i is written as

$$\pm \frac{\partial I^{\pm}(z, p_i, \nu)}{\partial z} = j(r, \nu) - \kappa(r, \nu) I^{\pm}(z, p_i, \nu). \tag{4.7.22}$$

The quantities j and κ have been written as functions of r in the sense of equation (4.7.20). We define the optical depth $d\tau$ as

$$d\tau(z, p_i, \nu) = -\kappa(r, \nu)\, dz. \tag{4.7.23}$$

The source function $S(r, \nu) = j(r, \nu)/\kappa(r, \nu)$ is assumed to be known at (r, ν).

We shall introduce the mean-intensity-like and flux-like variables:

$$u(z, p_i, \nu) = \frac{I^{+}(z, p_i, \nu) + I^{-}(z, p_i, \nu)}{2} \tag{4.7.24}$$

and

$$v(z, p_i, \nu) = \frac{I^{+}(z, p_i, \nu) - I^{-}(z, p_i, \nu)}{2}. \tag{4.7.25}$$

Adding and subtracting the equations in (4.7.22) and using equations (4.7.23)–(4.7.25), we obtain

$$\frac{\partial v(z, p_i, \nu)}{\partial \tau(z, p_i, \nu)} = u(z, p_i, \nu) - S(z, p_i, \nu) \tag{4.7.26}$$

and

$$\frac{\partial u(z, p_i, \nu)}{\partial \tau(z, p_i, \nu)} = v(z, p_i, \nu). \tag{4.7.27}$$

Combination of these two equations will give us the second order equation

$$\frac{\partial^2 u(z, p_i, \nu)}{\partial \tau^2(z, p_i, \nu)} = u(z, p_i, \nu) - S(z, p_i, \nu). \tag{4.7.28}$$

We now need to specify the boundary conditions. (a) The upper boundary condition at $r = R$ is

$$\frac{\partial u(z, p_i, \nu)}{\partial \tau(z, p_i, \nu)} = u(z_{max}, p_i, \nu), \tag{4.7.29}$$

where $z_{max} = (R^2 - p_i^2)^{1/2}$. (b) There are two cases in the lower boundary: (1) When the ray intersects the core $r_{Di} = r_c$, we give the diffusion approximation in equation (4.7.17) as the boundary condition; (2) When the ray intersects the plane $z = 0$, symmetry considerations give us that

$$v(0, p_i, \nu) = 0. \tag{4.7.30}$$

This gives us the boundary condition that

$$\left.\frac{\partial u(z, p_i, \nu)}{\partial \tau(z, p_i, \nu)}\right|_{z=0} = 0. \tag{4.7.31}$$

Equations (4.7.28), (4.7.29) and (4.7.31) are written as finite-difference equations to give a combined single tri-diagonal system, which is solved by the Feautrier method. It is assumed that the source function is known.

After computing $u_{din} = u(z_d, p_i, \nu_n)$ the moments of the radiation field are determined as

$$J_{dn} = \sum_{i=1}^{I_d} W_{di}^{(0)} u_{din}, \tag{4.7.32}$$

$$K_{dn} = \sum_{i=1}^{I_d} W_{di}^{(2)} u_{din}. \tag{4.7.33}$$

Here the Ws are the appropriate quadrature weights. Now the Eddington factors f_n can be calculated from $f_{dn} = K_{dn}/J_{dn}$. Using these factors, the moment equations are solved again and the process is iterated to convergence.

If the source function consists of a single scattering integral involving J or $\bar{J}$ in a line instead of a frequency dependent scattering integral with partial frequency redistribution, the iteration between the ray equations and the moment equations can be avoided and a direct solution can be obtained by a Rybicki type solution (Mihalas 1974).

4.8 Ray-by-ray treatment of Schmid-Burgk

Schmid-Burgk (1975) applied the integral equation method to compute the I^+ and I^- along each ray (see figure 4.2) in the spherical system. The solutions for I^- and I^+ are (see chapter 2)

$$I^-(p, z) = \int_x^y \kappa(p_\zeta) S(p_\zeta) \exp[-\tau(p; \zeta, z_{min})]\, d\zeta, \tag{4.8.1}$$

$$I^+(p, z) = (AI^-(p, z_0) + I_s(p) \exp[-\tau(p; z, z_0)] + \int_{z_0}^{z} \kappa(p_\zeta) S(p_\zeta) \exp[-\tau(p; z, \zeta)]\, d\zeta, \tag{4.8.2}$$

where

$$p_\zeta = (p^2 + \zeta^2)^{\frac{1}{2}}. \tag{4.8.3}$$

Here

$$\tau(p; a, b) = \int_b^a \kappa\left[(p^2 + z^2)^{\frac{1}{2}}\right] dz \tag{4.8.4}$$

and

$$z_{min} = \max(z, z_0). \tag{4.8.5}$$

The spherical medium is divided into N shells, r_N being the outer radius of the outermost shell. The outer boundary condition is given such that $I(r_N = r_{max}, \mu) = 0$, $\mu \in (-1, 0)$ which means there is no radiation incident from the outside.

In the case of the inner boundary conditions as seen in section 4.7 two cases are considered: (1) For shells $z_0 = (r_0^2 - p^2)^{1/2}$ for $p \leq r_0$ (= the radius of the inner shell surface), $z_0 = 0$ for $p > r_0$, $A = 0$ for $p \leq r_s$, $A = 1$ for $p > r_s$, where r_s is the radius of the central source. This is assumed to be totally opaque and to emit intensity $I_s(p)$ from its surface. For spheres on the other hand, $z_0 = 0$, $A = 1$ and $I_s(p) = 0$, for all ps. To calculate the intensities in the equation (4.8.1) and (4.8.2), the medium is divided into N subshells, each with an outer radius r_N and an inner radius r_{n-1}, $n = 1, 2, 3, \ldots, N$ and $r_0 = 0$. In each of the subshells, the source function is expanded in powers of the subshell's radial optical depth variable $\tau_n(r) = \tau(o; r_n, r)$:

$$S_n(r) = \frac{1}{r^2}\sum_{m=0}^{M} a_{nm}\tau_n^m(r) \quad \text{for } r_{n-1} \leq r \leq r_n, \tag{4.8.6}$$

where a_{nm} are constants.

The intensity at a point on the outer surface of the jth subshell of the ray p with $p \leq r_j$ (that is, at $z_j = (r_j^2 - p^2)^{1/2}$) is given (using equations (4.8.1) and (4.8.2)) by

$$\begin{aligned} I^-(p, z_j) &= \sum_{n=j+1}^{N}\sum_{m=0}^{M} a_{nm}\int_{z_{n-1}}^{z_n} \frac{\kappa\left(p^2+\zeta^2\right)^{1/2}}{p^2+\zeta^2}\tau_n^m\left(p^2+\zeta^2\right)^{1/2} \\ &\quad \times \tau_n^m\left(p^2+\xi^2\right)^{1/2}\exp\left[-\int_{z_j}^{\zeta}\kappa\left(p^2+t^2\right)^{1/2}dt\right]d\zeta \\ &\equiv \sum_{n=j+1}^{N}\sum_{m=0}^{M} a_{nm}\iota_{nm}(p)E_{nj}(p), \end{aligned} \tag{4.8.7}$$

for $0 \leq j < N$. Furthermore,

$$I^-(p, z_N) = 0 \tag{4.8.8}$$

and

$$E_{nj}(p) = \prod_{n=j+1}^{\prime n-1} \eta_h(p), \tag{4.8.9}$$

$$\eta_h(p) = \exp(-\tau(p; z_h, z_{h-1}), \tag{4.8.10}$$

where the prime on the product represents the fact that the product is equal to 1 for $n = j + 1$. A similar expression can be derived for $I^+(p, z_j)$. These quantities can be used in calculating the moment

$$M_i(r) = \frac{1}{2}\int_{-1}^{1} I(r, \mu)\mu^i \, d\mu \tag{4.8.11}$$

which in terms of a_{nm} is given by

$$M_i(r_j) = \sum_{n=1}^{N}\sum_{m=0}^{M} a_{nm}\mathcal{M}_{i,nm_j}, \quad \text{for } 0 \le j \le N. \tag{4.8.12}$$

The term with $n = 0$ stands for the contribution from a central source, with $a_{0m} = 0$ for $m = 0$. The condition of luminosity conservation is used to determine the coefficients a_{nm}:

$$\frac{L}{16\pi^2 r_j^2} = \sum_{n=0}^{N}\sum_{m=0}^{M} a_{nm}\mathcal{M}_{1,nm_j}. \tag{4.8.13}$$

Third degree spline functions are used to obtain the source function ($M = 3$) as

$$\left.\begin{aligned} &a_{n0} + a_{n1}\Theta_n + a_{n2}\Theta_n^2 + a_{n3}\Theta_n^3 = a_{n-1,0},\\ &a_{n1} + a_{n2}\Theta_n + 3a_{n3}\Theta_n^2 = a_{n-1,1},\\ &a_{n,2} + 3a_{n3} = a_{n-1,2} \end{aligned}\right\} \tag{4.8.14}$$

for $2 \le n \le N$ and $\Theta_n = \tau_n(r_{n-1})$.

The radiation field at the observer can be computed using the ray-by-ray procedure provided we have the prior knowledge of the source functions (see Peraiah (1980a,b)). The source function is obtained by solving the transfer equation in the comoving frame (in spherical symmetry) in the (r, μ) coordinate system.

Once we know the source function we can use the formal solution to obtain the intensities I^+ and I^- along a set of rays parallel to the line of sight. Radial optical depths are used in solving the transfer equation in spherical symmetry while optical depths along the rays are used to obtain the radiation field in the observer's frame. This is described in chapter 7.

4.9 Discrete space representation

In the ray-by-ray procedure, one needs to know the source function. Following Mihalas and Mihalas (1984) we present an alternative procedure.

In a grey and isotropically scattering spherically symmetric medium, the transfer equation can be written as

$$3\mu\frac{\partial(r^2 I)}{\partial(r^3)} + \frac{1}{r}\frac{\partial}{\partial\mu}\left[\left(1-\mu^2\right)I\right] = -\kappa\,[I - S]. \tag{4.9.1}$$

Equation (4.9.1) can be written for the oppositely directed beams and then the sum and differences taken to get

$$3\mu\frac{\partial(r^2 v)}{\partial(r^3)} + \frac{1}{r}\frac{\partial}{\partial\mu}\left[\left(1-\mu^2\right)v\right] = \kappa\,[u - S] \tag{4.9.2}$$

and

$$3\mu\frac{\partial(r^2 u)}{\partial(r^3)} + \frac{1}{r}\frac{\partial}{\partial\mu}\left[\left(1-\mu^2\right)u\right] = -\kappa v, \tag{4.9.3}$$

where u and v are defined in equations (4.3.2) and (4.3.3). The μ-integration is done on the mesh $[\mu_{j-\frac{1}{2}}, \mu_{j+\frac{1}{2}}]$ (see chapter 6), where

$$\mu_{j+\frac{1}{2}} = \sum_{k=1}^{j} c_k, \quad j = 1, 2, \ldots, J, \tag{4.9.4}$$

the cs being the quadrature weights. Furthermore, $\mu_{-\frac{1}{2}} = \mu_{J+\frac{1}{2}} = 0$.

Now using equation (4.9.4), equations (4.9.2) and (4.9.3) can be written as

$$3c_j\mu_j\left[\frac{\partial(r^2 v_j)}{\partial(r^3)}\right] + \frac{1}{r}\left\{\left(1-\mu^2_{j+\frac{1}{2}}\right)v_{j+\frac{1}{2}} - \left(1-\mu^2_{j-1}\right)v_{j-\frac{1}{2}}\right\}$$
$$= c_j\kappa(S - U_j) \tag{4.9.5}$$

and

$$3c_j\mu_j^2\left[\frac{\partial(r^2 u_j)}{\partial(r^3)}\right] + \frac{c_j(\mu_j^2 - 1)}{r}U_j \frac{1}{r}\left\{\mu_{j+\frac{1}{2}}\left[1-\mu^2_{j+\frac{1}{2}}\right]U_{j+\frac{1}{2}}\right.$$
$$\left. - \mu_{j-\frac{1}{2}}\left[1-\mu^2_{j-\frac{1}{2}}\right]\right\}u_{j-\frac{1}{2}} = -\kappa(U_j\mu_j v_j). \tag{4.9.6}$$

The quantities $u_{j\pm\frac{1}{2}}$ and $v_{j+\frac{1}{2}}$ are replaced by a linear spline approximation (see Peraiah and Grant (1973)). The resulting differential equations together with the appropriate boundary conditions are discretized to obtain a tri-diagonal system (Mihalas and Mihalas 1984, page 385). This scheme avoids the formal solution of the transfer equation and needs little computer time.

Exercises

4.1 In the case of reflection nebulae or components of close binary stars, write the boundary condition at $\tau = 0$.

4.2 Derive equations (4.5.12) and (4.5.14) and using the diffusion approximation of the lower boundary condition derive the equation equivalent to equation (4.5.14).

4.3 Derive equation (4.7.10) using equation (4.7.8).

4.4 Write the difference approximations of equations (4.7.28), (4.7.29) and (4.7.31).

4.5 Using the symmetry of the u_{din} about the plane $z = 0$ and the difference equations of the second order differential equation (4.7.26), derive a second order lower boundary condition beyond the central $z = 0$.

4.6 Write down the moment equations in the (p, z) coordinate system for the spherical geometry.

4.7 Write a computer program for comparing the computer times taken by the Feautrier method and the Rybicki method in a finite atmosphere with a total optical depth of 10^3 (at the line centre) assuming a Doppler profile, 21 frequency points and four angle points.

REFERENCES

Auer, L., 1971, *JQSRT*, **11**, 573.

Auer, L., 1976, *JQSRT*, **16**, 931.

Auer, L., Mihalas, D., 1970, *MNRAS*, **149**, 60.

Feautrier, P., 1964, *C. R. Acad Sc. Paris*, **258**, 3189.

Hummer, D.G., Rybicki, G., 1971, *MNRAS*, **152**, 1.

Kunasz, P., Hummer, D., 1974, *MNRAS*, **166**, 19.

Mihalas, D. 1970, *Stellar Atmospheres*, 1st Edition, W.H. Freeman and Co., San Francisco.

Mihalas, D. 1974, *ApJ Suppl.*, No. 265, **28**, 343.

Mihalas, D. 1978, *Stellar Atmospheres*, Freeman and Company, San Francisco.

Mihalas, D. and Mihalas, B.W., 1984, *Foundations of Radiation Hydrodynamics*, Oxford University Press, New York.

Peraiah, A., 1980a, *J. Astrophys. Astr.*, **1**, 3.

Peraiah, A., 1980b, *J. Astrophys. Astr.*, **1**, 17.

Peraiah, A., Grant, I.P., 1973, *J. Inst. Maths. Applics.* **12**, 75.

Rogers, C., 1984, *ApJ*, **286**, 659.

Rogers, C., Martin, P.G., 1984, *ApJ*, **284**, 327.

Rogers, C., Martin, P.G., 1986, *ApJ*, **311**, 800.

Rybicki, G.B., 1965, in *Harvard-Smithsonian Conference on Stellar Atmospheres, Proceedings of the Second Conference*, SAO Special Report No. 174, Smithsonian Astrophysical Observatory, Cambridge, MA.

Rybicki, G.B., 1967, *ApJ*, **150**, 607.

Rybicki, G., 1971, *JQSRT*, **11**, 589.

Schmid-Burgk, T., 1975, *A&A*, **40**, 249.

Sen, K.K., Wilson, S.J., 1998, *Radiative Transfer in Moving Media*, Springer, Singapore.

Chapter 5

Principle of invariance

5.1 Glass plates theory

The most fundamental characteristic of the radiation field in dispersive media such as stellar atmospheres, planetary atmospheres, planetary nebulae is the diffuse radiation which arises from multiple scattering of radiation by the media. This has been studied through an approach called the principle of invariance, or invariant imbedding, due to Ambarzumian (see books by Chandrasekhar (1960), Sobolev (1963), Kourganoff (1963), Wing (1962), Preisendorfer (1965)). Bellman and his collaborators have published several papers on this subject (see the bibliography at the end of the chapter). Before we study this principle, we shall see how the concept was developed by Sir George Stokes in his glass plate theory (1852, 1862). In remarkably simple papers, he derived the transmission and reflection factors when a ray of light passes through a system of glass plates. We shall see below how he obtained the principle of invariance of reflectance when several glass plates are arranged parallel to each other, one on top of the other.

He obtained difference equations for the reflection of radiation by a pile of identical glass plates and derived certain commutation relations for sets of glass plates. It is remarkable that he was able to obtain transmission and reflection factors which look similar to those obtained in more complicated media such as stellar atmospheres. We shall derive the transmission and reflection factors for the set of glass plates following the treatment given in Hottel and Sarofim (1967).

Consider a single glass plate with a reflectivity r at each face (see figure 5.1) and transmittivity t for one travel of the ray between the inside surfaces excluding the loss due to boundary reflections. If χ is the angle made by the refracting ray with the surface normal, L is the plate thickness, and λ is the wavelength, then

$$t = \exp(-\mathcal{K}L/\cos\chi) = \exp[-4\pi nKL/(\lambda\cos\chi)], \tag{5.1.1}$$

where $\mathcal{K}$ is the absorption index, K is absorption coefficient and n is the index of refraction.

In figure 5.1, a ray with unit intensity incident at the surface A of the glass plate AB is traced through the glass plate. The ray undergoes reflections and transmissions as shown in the figure. Let R and T be defined as the ratios of the reflected and transmitted fluxes to that of the incident flux, then these quantities are given by

$$\begin{aligned} R &= r + t^2(1-r)^2 r\left(1 + r^2t^2 + r^4t^4 + r^6t^6 + \cdots\right) \\ &= r\left\{1 + \left[\frac{t^2(1-r)^2}{1-r^2t^2}\right]\right\} \end{aligned} \tag{5.1.2}$$

and

$$T = (1-r)^2 t\left(1 + r^2t^2 + r^4t^4 + \cdots\right) = \frac{(1-r)^2 t}{1-r^2t^2}. \tag{5.1.3}$$

Therefore the absorbed part, A, is given by

$$A = 1 - (T + R) = \frac{1 + rt - (r+T)}{1 - rt}. \tag{5.1.4}$$

If we combine a system of n plates with transmittance T_n and reflectance R_n with another system of m plates with transmittance T_m and the reflectance R_m

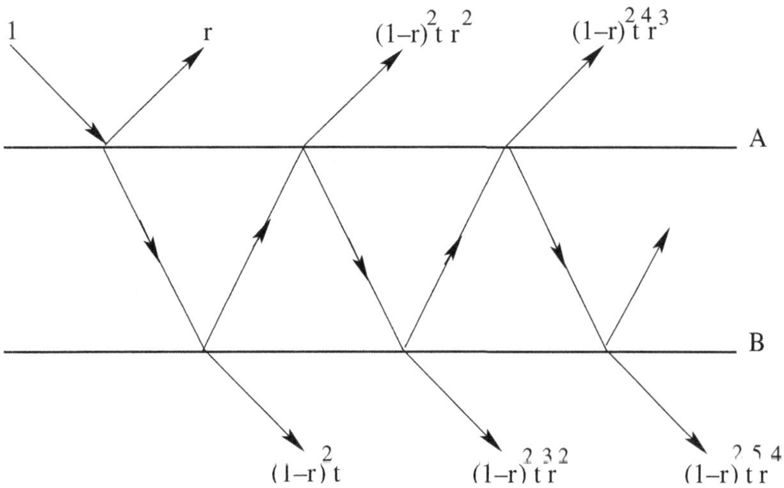

Figure 5.1 Schematic diagram of a ray traced through a single glass plate (see Hottel and Sarofim (1967) for a similar picture).

respectively (see figure 5.2), the composite transmittance and reflectance are given by

$$T_{m+n} = T_n T_m \left(1 + R_n R_m + R_n^2 R_m^2 + \cdots\right) = \frac{T_n T_m}{1 - R_n R_m} \tag{5.1.5}$$

and

$$R_{n+m} = R_n + R_m T_n^2 \left(1 + R_n R_m + \cdots\right) = R_n + \frac{R_m T_n^2}{1 - R_n R_m}. \tag{5.1.6}$$

These equations were first derived by Sir George Stokes (1852, 1862); see also Rayleigh (1920). These equations look remarkably similar to those derived in discrete space theory (see chapter 6). The above equations can be used to produce the properties of a multiple system starting from a single plate by adding plates or by the successive doubling of the number of plates.

In this process we can add an infinite number of plates (we shall see the similarity to Ambarzumian's principle of invariance). In figure 5.3 we show how the reflectance remains invariant after the addition of certain number of plates. The t and r factors for a single plate are 0.99 and 0.004, respectively.

It is seen that as more and more plates are added the reflectance reaches an asymptote: in other words, the reflection property of the system of plates remains

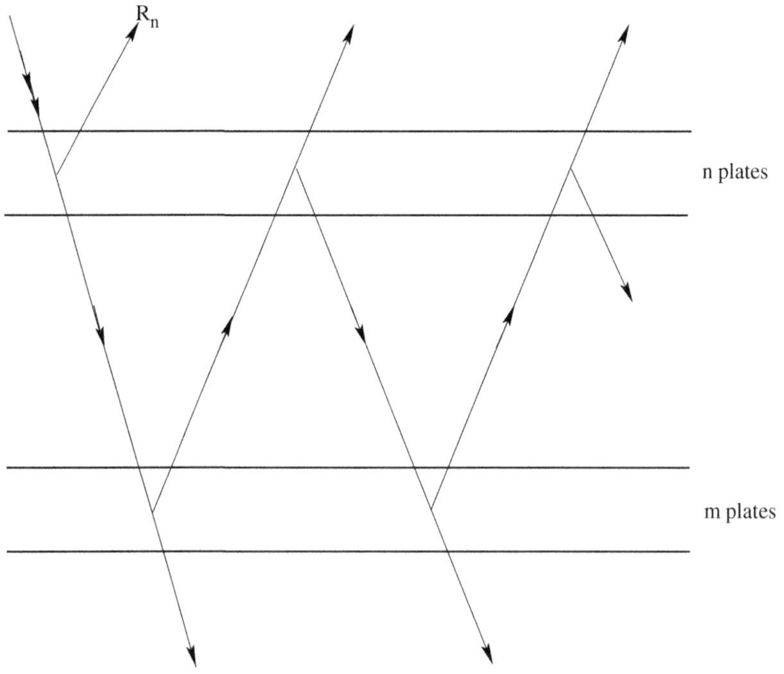

Figure 5.2 Schematic diagram of a ray traced through composite of m and n plates (see Hottel and Sarofim (1967) for a similar picture).

invariant to the addition of an extra plate. We can equate the reflectivity of $m+1$ plates to that of m plates if m is large. If we set $n=1$ and $m=\infty$, equation (5.1.6) becomes

$$R_\infty = R + R_\infty T^2 (1 - R_\infty R)^{-1}. \tag{5.1.7}$$

This gives us

$$R_\infty = \left\{1 + R^2 - T^2 \pm [(1 + R^2 - T^2)^2 - 4R^2]^{1/2}\right\}. \tag{5.1.8}$$

The addition of glass plates is similar to the addition of layers in dispersive and continuous media. McClelland (1906) studied the continuous problem and calculated the internal fluxes by adding a thin layer to a homogeneous slab. Schmidt (1907) modified Stokes's original approach by adding an arbitrarily thin layer and obtained the equations

$$\frac{dr}{ds} = kt^2 \tag{5.1.9}$$

and

$$\frac{dt}{ds} = [kr - (k + l)]\,t \tag{5.1.10}$$

where s is the thickness of the slab and the quantities k, l are constants.

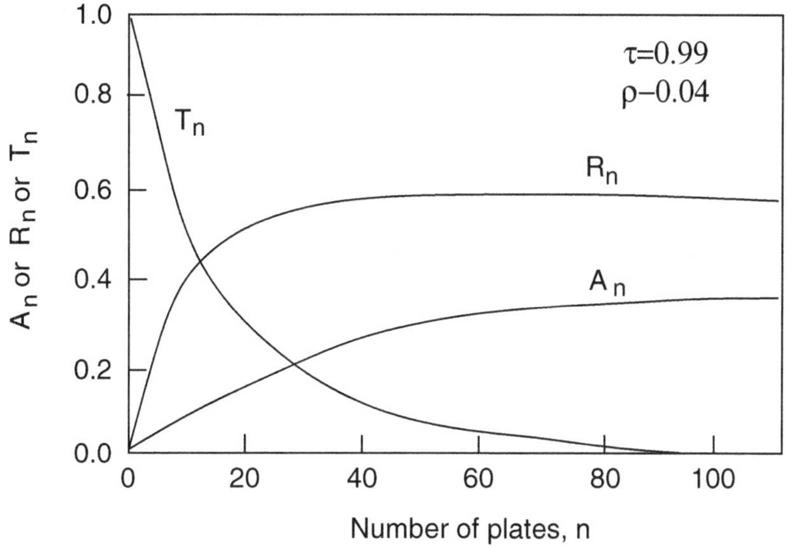

Figure 5.3 The reflectance R_n, transmittance T_n and absorbance A_n of n parallel plates are shown above. The transmittivity and reflectivity for a single plate are 0.99 and 0.004 respectively, at the plate interface. Notice that from plate 50 onwards the reflectance R_n remains invariant when an additional plate is added to the system (from Peraiah (1999), with permission; see also Hottel and Sarofim (1967)).

5.2 The principle of invariance

The glass plate examples show clearly how the reflection and transmission parameters become invariant (see figure 5.3). Ambarzumian (1942, 1943, 1944) formulated a similar principle in dispersive media. He considered a layer of infinite thickness with another layer of infinitesimally small thickness $\Delta\tau$ added to it. This results in the appearance of additional components of the radiation field in four ways: (1) the additional layer $\Delta\tau$ scatters part of the direct radiation that crosses it; (2) another part of the direct radiation scattered by the layer $\Delta\tau$ is directed towards the boundary and partly reflected by it; (3) the reflected radiation from the boundary of the medium is also scattered by $\Delta\tau$ layer; and (4) part of the radiation reflected from the medium is scattered by the layer $\Delta\tau$ back into this surface and partly reflected again from it.

Fundamental to these ideas is that the intensities at any point τ inside a medium are the combination of several other reflected and transmitted intensities. We shall explain this point using figure 5.4 as an example. Let us divide the semi-infinite medium at τ into two adjacent parts – the upper part and the lower part with $\tau = T^+$ and $\tau = T^-$ and $T^+ = T^-$ (Kourganoff 1963). The intensities in the upper part are $I^+(T^+)$ and $I^-(T^+)$ in the upward ($\tau \to 0$) and downward ($\tau \to \infty$) directions. Similarly the intensities in the lower part are $I^+(T^-)$ and $I^-(T^-)$ in the upward and downward direction. The intensity $I^+(T^+)$ is not necessarily equal to the intensity $I^+(T^-)$. The intensity $I^+(T^+)$ is a combination of $I^+(T^-)$ from the lower part and the reflected part of $I^-(T^+)$ from the upper part. Thus

$$I^+(T^+) = I^+(T^-) + R_\mu I^-(T^+), \tag{5.2.1}$$

where R_μ is the reflection parameter. This is a fundamental concept in Ambarzumian's analysis.

Ambarzumian presented his arguments in the form of the principle of invariance which expresses the invariance of the intensity emergent from a semi-infinite

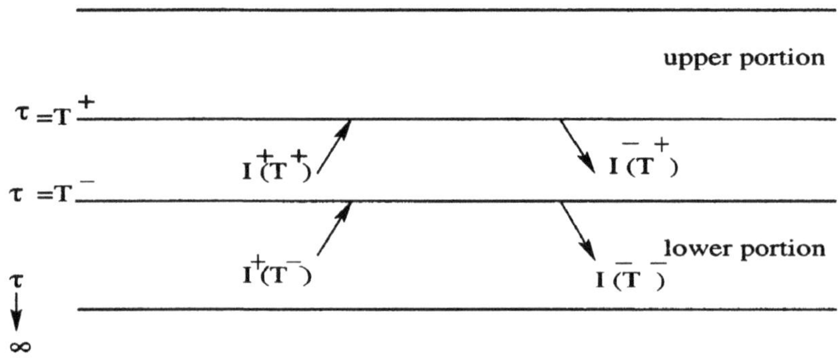

Figure 5.4 Reflected intensities.

atmosphere to the addition or subtraction of a layer of small thickness (see the similarity with the Stokes's glass plate theory in figure 5.3). This principle has been extensively used by Chandrasekhar (1960). The relationship between the two problems of constant net flux and the diffuse reflection was discussed in chapter 3. It is of considerable interest to see that these two problems are connected to the principle of invariance which can be formulated in the problem of diffuse reflection. Thus the law of diffuse reflection by a semi-infinite plane parallel medium must be invariant to the addition (or subtraction) of layers of arbitrary optical thickness to (or from) the atmosphere. This was first formulated by Ambarzumian (1942, 1943, 1944).

5.3 Diffuse reflection and transmission

The following treatment of diffuse radiation follows closely Chandrasekhar (1960). If a parallel beam of radiation of net flux πF per unit area normal to itself is incident on a plane parallel atmosphere of optical thickness τ (see figure 5.5) in the direction $(-\mu_0, \varphi_0)$, then we need to find the diffusely reflected radiation at $\tau = 0$ and the diffusely transmitted radiation at $\tau = \tau_1$. If we assign two functions called the scattering function $S(\tau_1; \mu, \varphi; \mu_0, \varphi_0)$ and the transmission function $T(\tau_1; \mu, \varphi; \mu_0, \varphi_0)$, then it is easy to express the reflected and transmitted intensities respectively as follows:

$$I(0, +\mu, \varphi) = \frac{F}{4\mu} S(\tau_1; \mu, \varphi; \mu_0, \varphi_0) \tag{5.3.1}$$

and

$$I(\tau_1, -\mu, \varphi) = \frac{F}{4\mu} T(\tau_1; \mu, \varphi; \mu_0, \varphi_0) \quad (0 \leq \mu \leq 1). \tag{5.3.2}$$

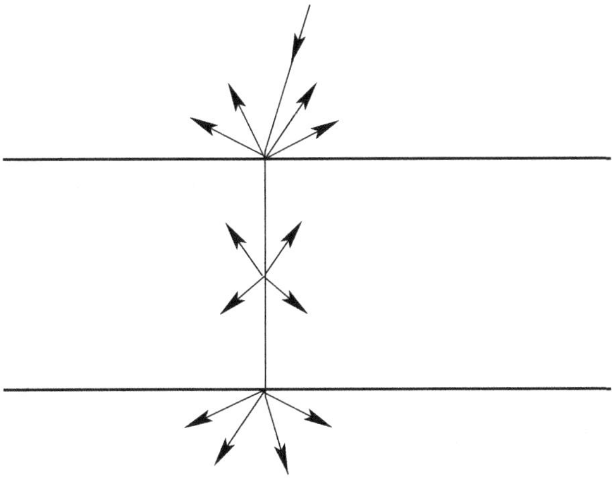

Figure 5.5 Incidence of a parallel beam of radiation on a plane parallel atmosphere.

It is important to note that the intensities in equations (5.3.1) and (5.3.2) refer only to the diffuse radiation or scattered radiation. For example, the intensity $I(\tau, -\mu, \varphi)$ does not include the directly transmitted flux $\frac{1}{4}F\exp(-\tau_1/\mu_0)$ in the direction $(-\mu_0, \varphi)$. Similarly the intensity $I(0, \mu, \varphi)$ does not include the diffusely reflected part of the directly transmitted flux $\frac{1}{4}F\exp(-\tau_1/\mu_0)$.

If there is arbitrary incident radiation with the same angular distribution at all points on the surface at $\tau = 0$, then the solution can be expressed in terms of S and T. If $I_{inc}(\mu', \varphi')$ is the incident radiation on $\tau = 0$ in the direction $(-\mu', \varphi')$, the angular distribution of the reflected and transmitted intensities are

$$I(0, \mu, \varphi) = \frac{1}{4\pi\mu}\int_0^1\int_0^{2\pi} S(\tau_1; \mu, \varphi; \mu', \varphi')I_{inc}(\mu', \varphi')\,d\mu'\,d\varphi' \quad (5.3.3)$$

and

$$I(\tau_1, -\mu, \varphi) = \frac{1}{4\pi\mu}\int_0^1\int_0^{2\pi} T(\tau_1; \mu, \varphi; \mu', \varphi')I_{inc}(\mu', \varphi')\,d\mu'\,d\varphi'. \quad (5.3.4)$$

The factor $1/\mu$ in equations (5.3.1) and (5.3.2) ensures the symmetry of S and T in the two sets of (μ, φ) and (μ_0, φ_0) which is the basis in the Helmholtz principle of reciprocity, which means

$$S(\tau_1; \mu, \varphi; \mu_0, \varphi_0) = S(\tau_1; \mu_0, \varphi_0; \mu, \varphi) \quad (5.3.5)$$

and

$$T(\tau_1; \mu, \varphi; \mu_0, \varphi_0) = T(\tau_1; \mu_0, \varphi_0; \mu, \varphi) \quad (5.3.6)$$

A plane parallel beam of radiation incident in the direction $(-\mu_0, \varphi_0)$ will have an intensity in terms of Dirac's δ-function of

$$I_{inc}(\mu', \varphi') = \pi F\delta(\mu' - \mu_0)\delta(\varphi' - \varphi_0). \quad (5.3.7)$$

In a semi-infinite medium the law of diffuse reflection is therefore written as

$$I(0, \mu, \varphi) = \frac{F}{4\mu}S(\mu, \varphi; \mu_0, \varphi_0). \quad (5.3.8)$$

Here, we need to distinguish between two types of radiation: (1) the reduced incident radiation $\pi F\exp(-\tau/\mu_0)$ which travels to the depth τ without suffering any scattering or absorption process; and (2) the diffuse radiation field which is generated by the multiple scattering. If we represent the latter by the intensity

$I(\tau, \mu, \varphi)$, the transfer equation is written

$$\mu \frac{dI(\tau, \mu, \varphi)}{d\tau} = I(\tau, \mu, \varphi) - \frac{1}{4\pi} \int_{-1}^{1} \int_{0}^{2\pi} p(\mu, \varphi; \mu', \varphi') I(\tau, \mu', \varphi') \, d\mu' \, d\varphi' - \frac{1}{4} F \exp(-\tau/\mu_0) p(\mu, \varphi; -\mu_0, \varphi_0), \quad (5.3.9)$$

where $p(\mu, \varphi; \mu', \varphi')$ is the phase function. The boundary conditions are

$$\text{at } \tau = 0, \quad I(0, -\mu, \varphi) = 0 \quad (0 < \mu \leq 1) \quad (5.3.10)$$

and

$$\text{at } \tau = \tau_1, \quad I(\tau_1, -\mu, \varphi) = 0 \quad (0 < \mu \leq 1). \quad (5.3.11)$$

It is assumed that at $\tau = \tau_1$, there is complete absorption or vacuum. In the case of a semi-infinite atmosphere which emits diffuse radiation, the boundary condition is $\tau_1 \to \infty$.

For a radiation field with axial symmetry and isotropic scattering ($P(\mu, \varphi; \mu', \varphi') = 1$) with albedo for single scattering ϖ, equation (5.3.9) reduces to

$$\mu \frac{dI(\tau, \mu)}{d\tau} = I(\tau, \mu) - \frac{1}{2} \varpi \int_{-1}^{1} I(\tau, \mu') \, d\mu' - \frac{1}{4} \varepsilon F \exp(-\tau/\mu_0) \quad (5.3.12)$$

5.4 The invariance of the law of diffuse reflection

If a parallel beam of radiation of net flux πF per unit area normal to itself is incident on a semi-infinite plane parallel atmosphere in the direction $(-\mu_0, \varphi_0)$, the diffusely reflected intensity $I(0, \mu, \varphi)$ expressed through scattering function $S(\mu, \varphi; \mu_0, \varphi_0)$ is given by

$$I(0, \mu, \varphi) = \frac{F}{4\mu} S(\mu, \varphi; \mu_0, \varphi_0). \quad (5.4.1)$$

The outward and inward intensities are $I(\tau, +\mu, \varphi)$ $(0 \leq \mu \leq 1)$ and $I(\tau, -\mu, \varphi)$ $(0 \leq \mu \leq 1)$. The semi-infinite medium will diffusely reflect the reduced incident flux and the inward directed radiation $I(\tau, -\mu, \varphi)$ and these will contribute to the outwardly directed intensity, $I(\tau, +\mu, \varphi)$, which is given by,

$$I(\tau, +\mu, \varphi) = \frac{F}{4\mu} \exp(-\tau/\mu_0) S(\mu, \varphi; \mu_0, \varphi_0) + \frac{1}{4\pi\mu} \int_0^1 \int_0^{2\pi} S(\mu, \varphi; \mu', \varphi') I(\tau, -\mu', \varphi') \, d\mu' \, d\varphi' \quad (5.4.2)$$

This is the invariance of $S(\mu, \varphi; \mu_0, \varphi_0)$ to the addition or subtraction of layers.

Let us consider an axially symmetric radiation field in a semi-infinite medium with constant net flux πF. Let $I(0, \mu)$ be the emergent radiation and $I(\tau, \mu)$ $(0 \leq \mu \leq 1)$ be the outward directed radiation at τ (see figure 5.6). The difference between these two intensities is the diffusely reflected radiation $I(\tau, -\mu)$ at τ. When the medium above τ is removed, the intensity $I(\tau, \mu)$ is restored to $I(0, \mu)$. Therefore the invariance of the emergent radiation $I(0, \mu)$ to the addition and subtraction of layers of arbitrary optical thickness follows. The removal of the medium above τ should restore $I(\tau, +\mu)$ to $I(0, \mu)$. This leads to

$$I(\tau, +\mu) = I(0, \mu) + \frac{1}{4\pi\mu} \int_0^1 \int_0^{2\pi} S(\mu, \varphi; \mu', \varphi') I(\tau, -\mu')\, d\mu'\, d\varphi'. \tag{5.4.3}$$

In the case of axial symmetry, we have

$$I(\tau, +\mu) = I(0, \mu) + \frac{1}{2\mu} \int_0^1 S_0(\mu, \mu') I(\tau, -\mu')\, d\mu', \tag{5.4.4}$$

where

$$S_0(\mu, \mu') = \frac{1}{2\pi} \int_0^1 S(\mu, \varphi; \mu', \varphi')\, d\varphi'. \tag{5.4.5}$$

5.5 Evaluation of the scattering function

So far we have not specified the scattering function S. We shall evaluate this function from the principles of the law of diffuse reflection and the law of darkening described in the last section. We shall derive certain non-linear integral equations which are nothing but the expressions of the invariance principles.

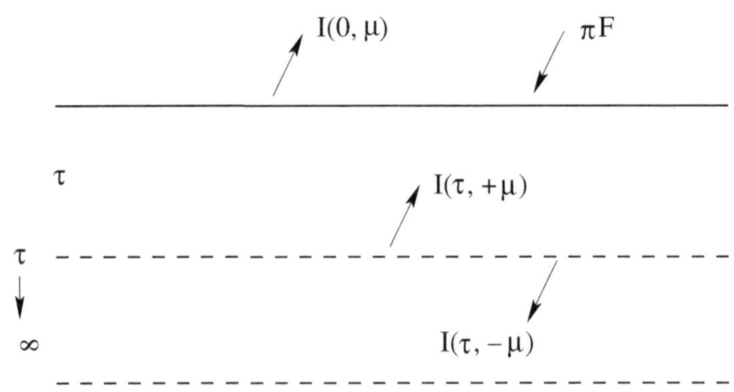

Figure 5.6 Diffuse reflection of radiation from a semi-infinite medium.

The first step is to differentiate equation (5.4.2) which involves the law of diffuse reflection and then pass onto $\tau \to 0$. Differentiating equation (5.4.2) and then setting the resulting equation to $\tau = 0$, we get

$$\left[\frac{dI(\tau, +\mu, \varphi)}{d\tau}\right]_{\tau=0} = -\frac{F}{4\mu\mu_0} S(\mu, \varphi; \mu_0, \varphi_0)$$
$$+ \frac{1}{4\pi\mu} \int_0^1 \int_0^{2\pi} S(\mu, \varphi; \mu', \varphi') \left[\frac{dI(\tau, -\mu', \varphi')}{d\tau}\right]_{\tau=0} d\mu' \, d\varphi'. \quad (5.5.1)$$

The derivatives in the square brackets, namely $dI(\tau, +\mu, \varphi)/d\tau$ and $dI(\tau, -\mu', \varphi')/d\tau$ can be taken from equation (5.3.9),

$$\mu \frac{dI(\tau, \mu, \varphi)}{d\tau} = I(\tau, \mu, \varphi) - B(\tau, \mu, \varphi), \quad (5.5.2)$$

where

$$B(\tau, \mu, \varphi) = \frac{1}{4} \exp(-\tau/\mu_0) F p(\mu, \varphi; -\mu_0, \varphi_0)$$
$$+ \frac{1}{4\pi} \int_{-1}^1 \int_0^{2\pi} p(\mu, \varphi; \mu'', \varphi'') I(\tau, \mu'', \varphi'') \, d\mu'' d\varphi'', \quad (5.5.3)$$

p being the phase function. Equation (5.5.2) gives

$$\left[\frac{dI(\tau, +\mu, \varphi)}{d\tau}\right]_{\tau=0} = \frac{1}{+\mu} [I(0, +\mu, \varphi) - B(0, +\mu, \varphi)] \quad (5.5.4)$$

and

$$\left[\frac{dI(\tau, -\mu', \varphi')}{d\tau}\right]_{\tau=0} = \frac{1}{+\mu'} B(0, -\mu', \varphi'). \quad (5.5.5)$$

We have made use of the boundary condition that

$$I(0, -\mu, \varphi) = 0 \quad (0 \le \mu \le 1) \quad (5.5.6)$$

in deriving equation (5.5.5). Eliminating $dI(\tau, +\mu, \varphi)/d\tau$ and $dI(\tau, -\mu', \varphi')/d\tau$ from equations (5.5.4), (5.5.5) and (5.5.1), we obtain

$$I(0, +\mu, \varphi) - B(0, +\mu, \varphi) = -\frac{F}{4\mu_0} S(\mu, \varphi; \mu_0, \varphi_0)$$
$$+ \frac{1}{4\pi} \int_0^1 \int_0^{2\pi} S(\mu, \varphi; \mu', \varphi') B(0, -\mu', \varphi') \frac{d\mu'}{\mu'} d\varphi'. \quad (5.5.7)$$

By using equation (5.4.1) to replace $I(0, +\mu, \varphi)$ in the above equation, we get

$$\frac{1}{4} F \left(\frac{1}{\mu} + \frac{1}{\mu_0}\right) S(\mu, \varphi; \mu_0, \varphi_0) - B(0, +\mu, \psi)$$
$$+ \frac{1}{4\pi} \int_0^1 \int_0^{2\pi} S(\mu, \varphi; \mu', \varphi') B(0, -\mu', \varphi') \frac{d\mu'}{\mu'} d\varphi'. \quad (5.5.8)$$

According to equations (5.4.1), (5.5.3) and (5.5.6), we have

$$B(0, +\mu, \varphi) = \frac{1}{4} F p(\mu, \varphi; -\mu_0, \varphi_0) + \frac{F}{16\pi} \int_0^1 \int_0^{2\pi} p(\mu, \varphi; \mu''\varphi'') S(\mu''\varphi''; \mu_0, \varphi_0) \frac{d\mu''}{\mu''} d\varphi'' \quad (-1 \leq \mu \leq 1). \tag{5.5.9}$$

Substituting equation (5.5.9) into equation (5.5.8) we get

$$\begin{aligned} \left(\frac{1}{\mu} + \frac{1}{\mu_0}\right) S(\mu, \varphi; \mu_0, \varphi_0) &= p(\mu, \varphi; -\mu_0, \varphi_0) \\ &+ \frac{1}{4\pi} \int_0^1 \int_0^{2\pi} p(\mu, \varphi; \mu''\varphi'') S(\mu'', \varphi''; \mu_0, \varphi_0) \frac{d\mu''}{\mu''} d\varphi'' \\ &+ \frac{1}{4\pi} \int_0^1 \int_0^{2\pi} S(\mu, \varphi; \mu'\varphi') p(-\mu', \varphi'; -\mu_0, \varphi_0) \frac{d\mu'}{\mu'} d\varphi' \\ &+ \frac{1}{16\pi^2} \int_0^1 \int_0^{2\pi} \int_0^1 \int_0^{2\pi} S(\mu, \varphi; \mu'\varphi') p(-\mu', \varphi'; \mu'', \varphi'') \\ &\times \mu S(\mu'', \varphi''; \mu_0 \varphi_0) \frac{d\mu'}{\mu'} d\varphi' \frac{d\mu''}{\mu''} d\varphi''. \end{aligned} \tag{5.5.10}$$

This is the required integral equation for S.

If there is axial symmetry in the radiation field, equation (5.5.10) becomes

$$\begin{aligned} &\left(\frac{1}{\mu} + \frac{1}{\mu_0}\right) S(\mu, \mu_0) \\ &= \varpi_0 \left[1 + \frac{1}{2} \int_0^1 S(\mu, \mu') \frac{d\mu'}{\mu'} + \frac{1}{2} \int_0^1 S(\mu'', \mu_0) \frac{d\mu''}{\mu''} \right. \\ &\left. + \frac{1}{4} \int_0^1 \int_0^1 S(\mu, \mu') S(\mu'', \mu_0) \frac{d\mu'}{\mu'} \frac{d\mu''}{\mu''} \right]. \end{aligned} \tag{5.5.11}$$

The quantities $S(\mu', \mu_0)$ and $S(\mu'', \mu_0)$ are separable in the variable, μ and μ_0 and therefore equation (5.5.11) can be rewritten as

$$\begin{aligned} \left(\frac{1}{\mu} + \frac{1}{\mu_0}\right) S(\mu, \mu_0) = \varpi_0 &\left[1 + \frac{1}{2} \int_0^1 S(\mu, \mu') \frac{d\mu'}{\mu'}\right] \\ &\times \left[1 + \frac{1}{2} \int_0^1 S(\mu'', \mu_0) \frac{d\mu''}{\mu''}\right]. \end{aligned} \tag{5.5.12}$$

Since $S(\mu, \mu')$ is symmetrical in μ and μ' the two factors on the RHS of the above equation should be values for μ and μ_0 of the same function. Therefore, we can express the scattering function as

$$\left(\frac{1}{\mu} + \frac{1}{\mu_0}\right) S(\mu, \mu_0) = \varpi H(\mu) H(\mu_0), \tag{5.5.13}$$

where

$$\begin{aligned} H(\mu) &= 1 + \frac{1}{2}\int_0^1 S(\mu, \mu')\frac{d\mu'}{\mu'} \\ &= 1 + \frac{1}{2}\int_0^1 S(\mu', \mu)\frac{d\mu'}{\mu'}. \end{aligned} \tag{5.5.14}$$

If we substitute the value of $H(\mu)$ into equation (5.5.13), we obtain the non-linear integral equation for $H(\mu)$:

$$H(\mu) = 1 + \frac{1}{2}\varpi \mu H(\mu) \int_0^1 \frac{H(\mu')}{\mu + \mu'} d\mu'. \tag{5.5.15}$$

From equation (3.4.20) we have

$$I(0, \mu) = \frac{1}{4}\varpi F \frac{\mu}{\mu + \mu_0} H(\mu) H(\mu_0). \tag{5.5.16}$$

If the reflected intensity $I(0, \mu)$ is expressed in terms of the scattering function given in equation (5.4.1), we get

$$\left(\frac{1}{\mu} + \frac{1}{\mu_0}\right) S(\mu, \mu_0) = \varpi H(\mu) H(\mu_0). \tag{5.5.17}$$

If we compare the results (5.5.13), (5.5.16) and (5.5.17) which are derived from the two approaches, it is confirmed that these solutions are exact. The H-function derived in equation (3.2.11) is in the limit as $n \to \infty$. This is the solution of equation (5.5.15)

5.6 An equation connecting $I(0, \mu)$ and $S_0(\mu, \mu')$

If we differentiate equation (5.4.4) and put $\tau = 0$, we obtain

$$\left[\frac{dI(\tau, +\mu)}{d\tau}\right]_{\tau=0} = \frac{1}{2\mu}\int_0^1 S_0(\mu, \mu') \left[\frac{dI(\tau, -\mu')}{d\tau}\right]_{\tau=0} d\mu'. \tag{5.6.1}$$

And from equations (5.5.4) and (5.5.5) we get

$$\left[\frac{dI(\tau, +\mu)}{d\tau}\right]_{\tau=0} = \frac{1}{\mu}\left[I(0, \mu) - B(0, +\mu)\right] \tag{5.6.2}$$

and

$$\left[\frac{dI(\tau, -\mu')}{d\tau}\right]_{\tau=0} = \frac{1}{\mu'} B(0, -\mu'), \tag{5.6.3}$$

where

$$B(0,\mu) = \frac{1}{2}\int_0^1 p^{(0)}(\mu,\mu'')I(0,\mu'')\,d\mu'' \quad (-1 \le \mu \le +1) \tag{5.6.4}$$

(see equations (2.12.1) and (2.12.2)). Substituting equations (5.6.2) and (5.6.3) into equation (5.6.1), we get

$$I(0,\mu) = B(0,\mu) + \frac{1}{2}\int_0^1 S_0(\mu,\mu')B(0,-\mu')\frac{d\mu'}{\mu'} \tag{5.6.5}$$

or

$$\begin{aligned} I(0,\mu) = {} & \frac{1}{2}\int_0^1 p^{(0)}(\mu,\mu'')I(0,\mu'') \\ & + \frac{1}{4}\int_0^1\int_0^1 S_0(\mu,\mu')p^{(0)}(-\mu',\mu'')I(0,\mu'')\frac{d\mu'}{\mu'}d\mu''. \end{aligned} \tag{5.6.6}$$

This is the relation between $I(0,\mu)$ and $S_0(\mu,\mu')$.

For conservative isotropic scattering with constant net flux, equation (5.6.6) becomes

$$I(0,\mu) = \frac{1}{2}\int_0^1 I(0,\mu'')\,d\mu'' + \frac{1}{4}\int_0^1\int_0^1 S(\mu,\mu')I(0,\mu'')\frac{d\mu'}{\mu'}d\mu'', \tag{5.6.7}$$

which can be written as,

$$I(0,\mu) = J(0)\left[1 + \frac{1}{2}\int_0^1 S(\mu,\mu')\frac{d\mu'}{\mu'}\right], \tag{5.6.8}$$

with

$$J(0) = \frac{1}{2}\int_0^1 I(0,\mu)\,d\mu. \tag{5.6.9}$$

From equation (5.5.13), we substitute the value of $S(\mu,\mu')$ and obtain

$$I(0,\mu) = J(0)\left[1 + \frac{1}{2}\mu H(\mu)\int_0^1 \frac{H(\mu)}{\mu+\mu'}d\mu'\right]. \tag{5.6.10}$$

Using the integral equation (5.5.15) for $H(\mu)$ with $\varpi = 1$, we can write equation (5.6.10) as

$$I(0,\mu) = J(0)H(\mu). \tag{5.6.11}$$

This relation can also be derived directly from the principle of invariance (see exercise 6.2).

5.7 The integral for S with $p(\cos\Theta) = \varpi(1 + x\cos\Theta)$

The phase function p is given by

$$p(\mu, \varphi; \mu', \varphi') = \varpi\left[1 + x\mu\mu' + x\left(1-\mu^2\right)^{\frac{1}{2}}\left(1-\mu'^2\right)^{\frac{1}{2}}\cos(\varphi' - \varphi)\right]. \tag{5.7.1}$$

The scattering function $S(\mu, \varphi; \mu_0, \varphi_0)$ can be written in the form

$$S(\mu, \varphi; \mu_0, \varphi_0) = \varpi\Big[S_0(\mu, \mu_0) + x\left(1-\mu^2\right)^{\frac{1}{2}}\left(1-\mu_0^2\right)^{\frac{1}{2}} S_1(\mu, \mu_0)\cos(\varphi_0 - \varphi)\Big]. \tag{5.7.2}$$

Substituting equations (5.7.1) and (5.7.2) into the integral equation (5.5.10) for S, we obtain

$$\begin{aligned} S_0(\mu, \mu_0) = {} & \left(\frac{1}{\mu} + \frac{1}{\mu_0}\right)^{-1} \\ & \times \Bigg[1 - x\mu\mu_0 + \frac{1}{2}\varpi\int_0^1 (1 + x\mu\mu'')S_0(\mu'', \mu_0)\frac{d\mu}{\mu''} \\ & + \frac{1}{2}\varpi\int_0^1 S_0(\mu, \mu')\left(1 + x\mu'\mu_0\frac{d\mu'}{\mu'}\right) \\ & + \frac{1}{4}\varpi^2\int_0^1\int_0^1 S_0(\mu, \mu')(1 - x\mu'\mu'')S_0(\mu'', \mu_0)\frac{d\mu'}{\mu'}\frac{d\mu''}{\mu''}\Bigg] \end{aligned} \tag{5.7.3}$$

and

$$\begin{aligned} S_1(\mu, \mu_0) = {} & \left(\frac{1}{\mu} + \frac{1}{\mu_0}\right)^{-1}\Bigg\{1 + \frac{1}{4}x\varpi\int_0^1 \frac{d\mu''}{\mu''}(1 - \mu''^2)S_1(\mu, \mu_0) \\ & \times\left[1 + \frac{1}{4}x\varpi\int_0^1 \frac{d\mu'}{\mu'}(1 - \mu'^2)S_1(\mu, \mu')\right]\Bigg\}. \end{aligned} \tag{5.7.4}$$

The equation for S_0 can be reduced to

$$\left(\frac{1}{\mu} + \frac{1}{\mu_0}\right)S_0(\mu, \mu_0) = \psi(\mu)\psi(\mu_0) - x\phi(\mu)\phi(\mu_0), \tag{5.7.5}$$

where

$$\psi(\mu) = 1 + \frac{1}{2}\varpi\int_0^1 S_0(\mu, \mu')\frac{d\mu'}{\mu'} \tag{5.7.6}$$

and

$$\phi(\mu) = \mu - \frac{1}{2}\varpi \int_0^1 S(\mu, \mu')\, d\mu'. \tag{5.7.7}$$

If equation (5.7.5) is substituted for S_0 in equations (5.7.6) and (5.7.7), we obtain

$$\psi(\mu) = 1 + \frac{1}{2}\varpi\mu\psi(\mu)\int_0^1 \frac{\psi(\mu')}{\mu+\mu'}d\mu' - \frac{1}{2}x\varpi\mu\phi(\mu)\int_0^1 \frac{\phi(\mu')}{\mu+\mu'}d\mu' \tag{5.7.8}$$

and

$$\phi(\mu) = \mu - \frac{1}{2}\varpi\psi(\mu)\int_0^1 \frac{\psi(\mu')}{\mu+\mu'}\mu'\, d\mu' + \frac{1}{2}x\varpi\mu\phi(\mu)\int_0^1 \frac{\phi(\mu')}{\mu+\mu'}\mu'\, d\mu'. \tag{5.7.9}$$

The function S can be expressed in terms of the H-functions as follows:

$$\left(\frac{1}{\mu} + \frac{1}{\mu_0}\right) S_1(\mu, \mu_0) = H_1(\mu)H_1(\mu_0), \tag{5.7.10}$$

where

$$H_1(\mu) = 1 + \frac{1}{4}x\mu_0 \int_0^1 (1-\mu'^2)S_1(\mu, \mu')\frac{d\mu'}{\mu'}. \tag{5.7.11}$$

From the above two equations we obtain the non-linear integral equations for $H(\mu)$:

$$H_1(\mu) = 1 + \frac{1}{4}x\varpi\mu H_1(\mu)\int_0^1 \frac{(1-\mu'^2)}{\mu+\mu'}H_1(\mu')\, d\mu'. \tag{5.7.12}$$

The properties of the function $H(\mu)$ and its numerical evaluation are given in Chandrasekhar (1960).

5.8 The principle of invariance in a finite medium

So far, we have considered the principle of invariance in a semi-infinite medium. This principle is not restricted to this medium only. There are many problems which require the solution of the radiative transfer in finite media. These principles can be formulated in finite media. This technique has been used in many classes of problems successfully.

Let us consider a plane parallel medium bounded by $\tau = 0$ and $\tau = \tau_1$. Let a parallel beam of radiation with net flux πF per unit area normal to itself be incident at $\tau = 0$ in the direction $(-\mu_0, \varphi_0)$. We need to consider two intensities: (1) the diffusely reflected intensity $I(0, \mu, \varphi)(0 \le \mu \le 1)$ in the direction (μ, φ); and (2) the diffusely transmitted intensity $I(\tau_1, -\mu, \varphi)$ at $\tau = \tau_1$ in the direction $(-\mu_1, \varphi)$.

In general $I(\tau, +\mu, \varphi)$ and $I(\tau, -\mu, \varphi)(0 \leq \mu \leq 1)$ are the outward (that is towards $\tau \to 0$) and inward (that is $\tau \to \tau_1$) intensities.

The diffusely reflected and transmitted intensities $I(0, \mu, \varphi)$ and $I(\tau_1, -\mu, \varphi)$ are expressed in terms of the scattering and transmission functions $S(\tau_1; \mu_1, \varphi; \mu_0, \varphi_0)$ and $T(\tau_1; \mu_1, \varphi; \mu_0, \varphi_0)$ (Chandrasekhar 1960, Peraiah 1999). Thus,

$$I(0, \mu, \varphi) = \frac{F}{4\mu} S(\tau_1; \mu_1, \varphi; \mu_0, \varphi_0) \tag{5.8.1}$$

and

$$I(\tau, -\mu, \varphi) = \frac{F}{4\mu} T(\tau_1; \mu_1, \varphi; \mu_0, \varphi_0). \tag{5.8.2}$$

In addition to the diffuse radiation field in the medium ($\tau = 0$ to $\tau = \tau_1$), we have the reduced incident flux in the direction $(-\mu_0, \varphi_0)$ given by

$$F_R = \pi F \exp(-\tau_1/\mu_0) \tag{5.8.3}$$

Our aim is to find the radiation field at any point τ inside the medium. Essentially we need to find:

1. $I(\tau, \mu, \varphi) =$ the intensity in the outward direction at any level τ;
2. $I(\tau, -\mu, \varphi) =$ the intensity in the downward direction at any level τ;
3. $(F/4\mu)S(\tau_1; \mu, \varphi; \mu_0, \varphi_0) =$ the diffuse reflection of the incident radiant flux by the whole medium from $\tau = 0$ to $\tau = \tau_1$; and
4. $(F/4\mu)T(\tau_1; \mu, \varphi; \mu_0, \varphi_0) =$ the diffuse transmission of the incident light by the whole medium from $\tau = 0$ to $\tau = \tau_1$.

We shall see below how these are estimated.

1. The upward intensity $I(\tau, +\mu, \varphi)$ (see figure 5.7) is the combination of: (a) the reduced incident flux $\pi F \exp(-\tau/\mu_0)$ at τ reflected by the medium $\tau_1 - \tau$; and (b) the inward radiation $(I(\tau, -\mu, \varphi)$ reflected by the medium

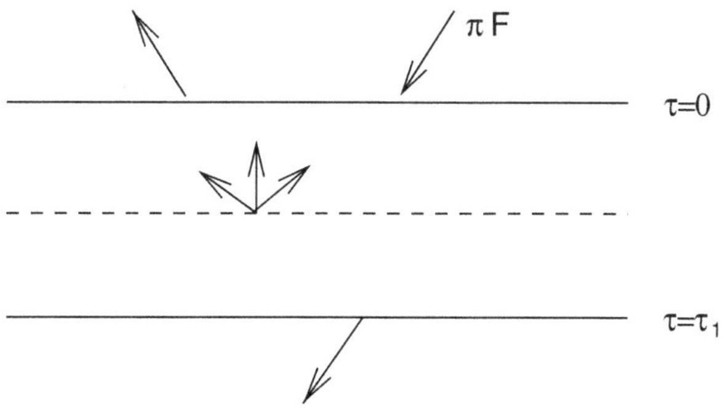

Figure 5.7 The upward intensity $I(\tau, +\mu, \varpi)$.

$\tau_1 - \tau$. If $S(\tau_1 - \tau; \mu, \varphi; \mu_0, \varphi_0)$ is the scattering function corresponding to the medium $(\tau_1 - \tau)$, then the intensity $I(\tau, +\mu, \varphi)$ is the combination of (a) and (b) and is given by

$$I(\tau, +\mu, \varphi) = \frac{F}{4\mu} \exp(-\tau/\mu_0) S(\tau_1, -\tau; \mu, \varphi; \mu_0, \varphi_0) + \frac{1}{4\pi\mu} \int_0^1 \int_0^{2\pi} S(\tau_1 - \tau; \mu, \varphi; \mu', \varphi') I(\tau, -\mu', \varphi')\, d\mu'\, d\varphi'. \tag{5.8.4}$$

2. The inward directed intensity $I(\tau - \mu, \varphi)$ at any τ (see figure 5.8) is the combination of: (a) the transmission of the incident flux πF by the medium above τ; and (b) the diffusely reflected radiation of the upward directed intensity $I(\tau, +\mu, \varphi)$ by the medium above τ. Therefore,

$$I(\tau, -\mu, \varphi) = \frac{F}{4\pi} T(\tau; \mu, \varphi; \mu_0) + \frac{1}{4\pi\mu} \int_0^1 \int_0^{2\pi} S(\tau; \mu, \varphi; \mu', \varphi') I(\tau, +\mu', \varphi')\, d\mu'\, d\varphi'. \tag{5.8.5}$$

3. The diffuse reflection of the incident light by the whole medium is the combination of three components: (a) the reflection of the incident flux by the part of the medium between $\tau = 0$ and τ; (b) the direct transmission of the diffuse intensity $I(\tau, +\mu, \varphi)$; and (c) the diffuse transmission of the upward directed intensity incident on the surface at τ from the medium below τ. Combining (a), (b) and (c), we get

$$\frac{F}{4\mu} S(\tau; \mu, \varphi; \mu_0, \varphi_0) = \frac{F}{4\mu} S(\tau; \mu, \varphi; \mu_0, \varphi_0) + \exp(-\tau/\mu) I(\tau, +\mu, \varphi) + \frac{1}{4\pi\mu} \int_0^1 \int_0^{2\pi} T(\tau; \mu, \varphi; \mu', \varphi') I(\tau, +\mu', \varphi')\, d\mu'\, d\varphi'. \tag{5.8.6}$$

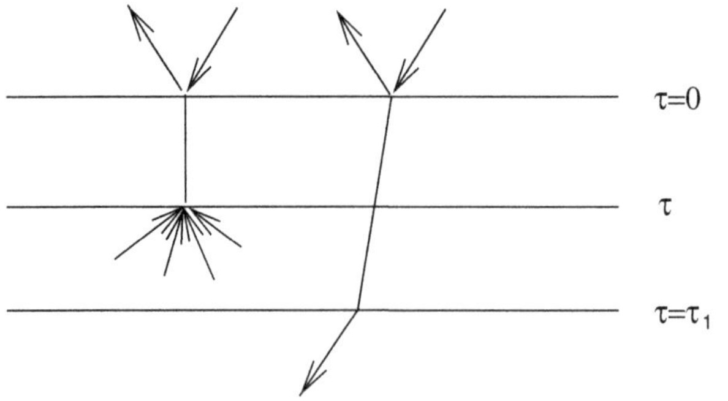

Figure 5.8 The inward intensity $I(c, -\mu, \varpi)$.

4. The diffuse transmission of the incident light by the medium $\tau = 0$ to $\tau = \tau_1$ consists of two parts: (a) the transmission of the reduced incident flux $\pi F \exp(-\tau/\mu_0)$; and (b) diffuse radiation $I(\tau, -\mu', \varphi')(0 < \mu' \leq 1)$ τ by the medium of optical thickness $(\tau, -\tau)$ incident on the surface below τ (see figure 5.9). Therefore we have

$$\frac{F}{4\mu} T(\tau_1; \mu, u; \mu_0, \varphi_0) = \frac{F}{4\mu} \exp(-\tau/\mu_0) T(\tau_1, -\tau; \mu, \varphi; \mu_0, \varphi_0) + \exp[-(\tau_1 - \tau)/\mu] I(\tau, -\mu, \varphi) + \frac{1}{4\pi\mu} \int_0^1 \int_0^{2\pi} T(\tau_1 - \tau; \mu, \varphi; \mu', \varphi') I(\tau, -\mu', \varphi') \, d\mu' \, d\varphi'. \tag{5.8.7}$$

The first term on the RHS represents the transmitted intensity by the reduced incident flux $\pi F \exp(-\tau/\mu_0)$ transmitted through the medium $(\tau_1 - \tau)$, the second term represents the direct transmission of the diffuse intensity $I(\tau, -\mu, \varphi)$ and the third term represents the diffuse transmission of the radiation field $I(\tau, -\mu, \varphi)$ by the medium below τ or by the medium with optical thickness $\tau_1 - \tau$.

Equations (5.8.4)–(5.8.7) determine the radiation field in a plane parallel finite medium with the incident flux πF in terms of the scattering and transmission functions. If the radiation field is axially symmetric then we can write the solution as (see figure 5.10)

$$I(T, +\mu) = I(0, \mu) + \frac{1}{2\mu} \int_0^1 S_0(\infty, \mu, \mu') I(\tau, -\mu') \, d\mu', \tag{5.8.8}$$

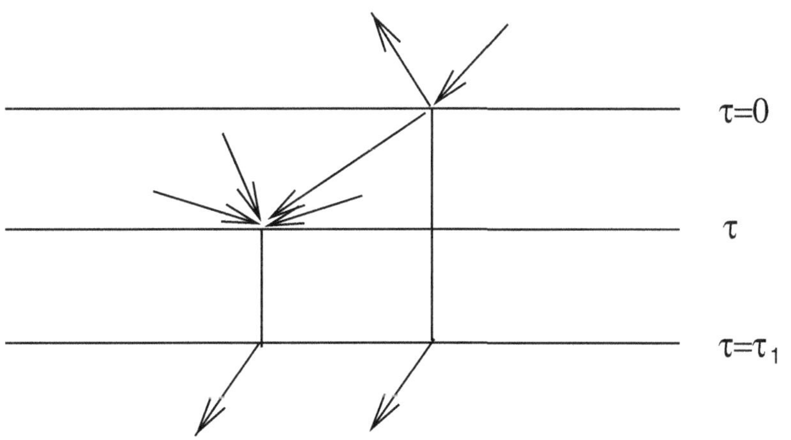

Figure 5.9 Diffuse transmission of the incident light.

$$I(0, +\mu) = \exp(-\tau/\mu) I(\tau, +\mu) + \frac{1}{2\mu} \int_0^1 T_0(\tau; \mu, \mu') I(\tau, +\mu')\, d\mu'. \tag{5.8.9}$$

and

$$I(\tau, -\mu) = \frac{1}{2\mu} \int_0^1 S_0(\tau; \mu, \mu') I(\tau, +\mu')\, d\mu', \tag{5.8.10}$$

where S_0 and T_0 are the azimuth independent scattering and transmission functions respectively. The above three equations enunciate the principle of invariance of the emergent radiation to the addition (or subtraction) of layers of arbitrary optical thickness to (or from) a semi-infinite plane parallel medium with a constant net flux. Furthermore, the emergent radiation is the transmitted incident flux on the surface τ, from below and the inward directed radiation at any τ is the reflection of the outward directed radiation by the layers overlying τ.

5.9 Integral equations for the scattering and transmission functions

The radiation field described in equations (5.8.4)–(5.8.7) contains the unknown scattering and transmission functions. To determine the radiation field we have to calculate these functions. We derive a set of four non-linear and inhomogeneous integral equations from equations (5.8.4)–(5.8.7) with the boundary conditions

$$\left.\begin{aligned} I(0, -\mu, \varphi) &= 0 \quad (0 < \mu \le 1), \\ I(\tau_1, +\mu, \varphi) &= 0 \quad (0 < \mu \le 1). \end{aligned}\right\} \tag{5.9.1}$$

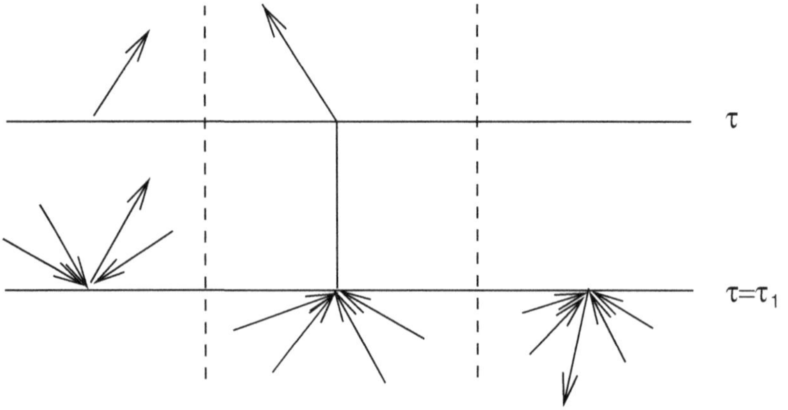

Figure 5.10 Schematic diagram of transfer of radiation.

Equations (5.8.4)–(5.8.7) are differentiated first and then passing on to the limit of $\tau = 0$ or $\tau = \tau_1$ (equations (5.8.5) and (5.8.6)), we get

$$\left[\frac{dI(\tau, +\mu, \varphi)}{d\tau}\right]_{\tau=0} = \frac{F}{4\mu}\left[-\frac{1}{\mu_0}S(\tau; \mu, \varphi; \mu_0, \varphi_0) - \frac{\partial S(\tau_1; \mu, \varphi; \mu_0, \varphi_0)}{\partial \tau_1}\right]$$
$$+ \frac{1}{4\pi\mu}\int_0^1\int_0^{2\pi} S(\tau_1; \mu, \varphi; \mu', \varphi')\left[\frac{dI(\tau, -\mu', \varphi'}{d\tau}\right]_{\tau=0} d\mu'\, d\varphi', \tag{5.9.2}$$

$$\left[\frac{dI(\tau, -\mu, \varphi)}{d\tau}\right]_{\tau=\tau_1} = \frac{F}{4\mu}\frac{\partial T(\tau_1; \mu, \varphi; \mu', \varphi')}{\partial \tau_1}$$
$$+ \frac{1}{4\pi\mu}\int_0^1\int_0^{2\pi} S(\tau_1; \mu, \varphi; \mu', \varphi')\left[\frac{dI(\tau, +\mu', \varphi')}{d\tau}\right]_{\tau=\tau_1} d\mu'\, d\varphi', \tag{5.9.3}$$

$$0 = \frac{F}{4\mu}\frac{\partial S(\tau_1; \mu, \varphi; \mu', \varphi')}{\partial \tau_1} + \exp(-\tau/\mu)\left[\frac{dI(\tau, \mu, \varphi)}{d\tau}\right]_{\tau=\tau_1}$$
$$+ \frac{1}{4\pi\mu}\int_0^1\int_0^{2\pi} T(\tau_1; \mu, \varphi; \mu', \varphi')\left[\frac{dI(\tau, +\mu', \varphi')}{d\tau}\right]_{\tau=\tau_1} d\mu'\, d\varphi' \tag{5.9.4}$$

and

$$0 = \frac{F}{4\mu}\left[-\frac{1}{\mu_0}T(\tau_1; \mu, \varphi; \mu_0, \varphi_0) - \frac{\partial T(\tau_1; \mu, \varphi; \mu_0, \mu_0)}{\partial \tau_1}\right]$$
$$+ \exp(-\tau_1/\mu)\left[\frac{dI(\tau_1 - \mu_1\varphi)}{d\tau}\right]_{\tau=0}$$
$$+ \frac{1}{4\pi\mu}\int_0^1\int_0^{2\pi} T(\tau_1; \mu, \varphi; \mu', \varphi')\left[\frac{dI(\tau_1 - \mu', \varphi')}{d\tau}\right]_{\tau=0} d\mu'\, d\varphi'. \tag{5.9.5}$$

The derivatives in the above equations are taken from the transfer equation (5.5.2). By making use of equations (5.8.2) and (5.9.1) and the boundary conditions in equation (5.9.1), we can write these derivatives as

$$\left[\frac{dI(\tau, +\mu, \varphi)}{d\tau}\right]_{\tau=0} = +\frac{1}{\mu}\left[\frac{F}{4\mu}S(\tau_1; \mu, \varphi; \mu_0, \varphi_0) - B(0, \mu, \varphi)\right], \tag{5.9.6}$$

$$\left[\frac{dI(\tau, -\mu, \varphi)}{d\tau}\right]_{\tau=0} = +\frac{1}{\mu}B(0, -\mu, \varphi), \tag{5.9.7}$$

$$\left[\frac{dI(\tau,+\mu,\varphi)}{d\tau}\right]_{\tau=\tau_1} = -\frac{1}{\mu}B(\tau_1,+\mu,\varphi), \tag{5.9.8}$$

$$\left[\frac{dI(\tau,-\mu,\varphi)}{d\tau}\right]_{\tau=\tau_1} = -\frac{1}{\mu}\left[\frac{F}{4\mu}T(\tau_1;\mu,\varphi;\mu_0,\varphi_0) - B(\tau_1,-\mu,\varphi)\right], \tag{5.9.9}$$

where the quantity B is defined in equation (5.5.3). Substituting equations (5.9.6)–(5.9.9) into the four integral equations which describe the problem of diffuse reflection and transmission by finite plane parallel medium, we obtain

$$\begin{aligned}
&\left(\frac{1}{\mu}+\frac{1}{\mu_0}\right)S(\tau_1,\mu,\varphi;\mu_0,\varphi_0) + \frac{\partial S_1(\tau_1;\mu_1,\varphi;\mu_0,\varphi_0)}{\partial\tau_1}\\
&\quad = p(\mu,\varphi;-\mu_0,\varphi_0)\\
&\quad + \frac{1}{4\pi}\int_0^1\int_0^{2\pi} p(\mu,\varphi;\mu'',\varphi'')S(\tau,\mu'',\varphi'';\mu_0,\varphi_0)\,d\mu'\,d\varphi'\\
&\quad + \frac{1}{4\pi}\int_0^1\int_0^{2\pi} S(\tau_1,\mu,\varphi,\mu',\varphi')p(-\mu',\varphi';-\mu_0,\varphi_0)\frac{d\mu'}{\mu'}d\varphi'\\
&\quad + \frac{1}{16\pi^2}\int_0^1\int_0^{2\pi}\int_0^1\int_0^{2\pi} S(\tau_1;\mu,\varphi;\mu',\varphi')p(-\mu',\varphi';\mu'',\varphi'')\\
&\quad \times S(\tau_1;\mu'',\varphi'',\mu_0,\varphi_0)\frac{d\mu'}{\mu'}\varphi'\frac{d\mu''}{\mu''}d\varphi'',
\end{aligned} \tag{5.9.10}$$

$$\begin{aligned}
&\frac{\partial S(\tau_1;\mu_1,\varphi;\mu_0,\varphi_0)}{\partial\tau_1} = p(\mu,\varphi;-\mu_0,\varphi_0)\exp\left[-\tau_1\left(\frac{1}{\mu}+\frac{1}{\mu_0}\right)\right]\\
&\quad + \frac{1}{4\pi}\exp(-\tau_1/\mu)\\
&\quad \times \int_0^1\int_0^{2\pi} p(\mu,\varphi;-\mu'',\varphi'')T(\tau_1,\mu'',\varphi'';\mu_0,\varphi_0)\frac{d\mu''}{\mu''}d\varphi''\\
&\quad + \frac{1}{4\pi}\exp(-\tau_1/\mu)\\
&\quad \times \int_0^1\int_0^{2\pi} T(\tau_1;\mu,\varphi;\mu'\varphi')p(\mu',\varphi';-\mu_0,\varphi_0)\frac{d\mu'}{\mu'}d\varphi'\\
&\quad + \frac{1}{16\pi^2}\int_0^1\int_0^{2\pi}\int_0^1\int_0^{2\pi} T(\tau_1,\mu,\varphi;\mu'\varphi')p(\mu';-\mu'',\varphi'')\\
&\quad \times T(\tau_1;\mu'',\varphi'';\mu_0,\varphi_0)\frac{d\mu'}{\mu'}d\varphi'\frac{d\mu''}{\mu''}d\varphi'',
\end{aligned} \tag{5.9.11}$$

$$\begin{aligned}
&\frac{1}{\mu}T(\tau_1;\mu,\varphi;\mu_0,\varphi_0) + \frac{\partial T(\tau_1,\mu,\varphi;\mu_0,\varphi_0)}{\partial\tau_1}\\
&\quad = \exp(-\tau_1/\mu_0)p(-\mu,\varphi;-\mu_0,\varphi_0)
\end{aligned}$$

$$
\begin{aligned}
&+\frac{1}{4\pi}\int_0^1\int_0^{2\pi} p(-\mu,\varphi;-\mu'',\varphi'')T(\tau_1,\mu'',\varphi'';\mu_0,\varphi_0)\frac{d\mu''}{\mu''}d\varphi''\\
&+\frac{\exp(-\tau_1/\mu_0)}{4\pi}\int_0^1\int_0^{2\pi} S(\tau_1,\mu,\varphi;\mu',\varphi')p(\mu',\varphi';-\mu_0,\varphi_0)\frac{d\mu'}{\mu'}d\varphi'\\
&+\frac{1}{16\pi^2}\int_0^1\int_0^{2\pi}\int_0^1\int_0^{2\pi} S(\tau_1,\mu,\varphi;\mu',\varphi')p(\mu',\varphi';-\mu'',\varphi'')\\
&\times T(\tau_1;\mu'',\varphi'';\mu_0,\varphi_0)\frac{d\mu'}{\mu'}\varphi\frac{d\mu''}{\mu''}d\varphi'',
\end{aligned}
\tag{5.9.12}
$$

$$
\begin{aligned}
&\frac{1}{\mu_0}T(\tau_1;\mu,\varphi;\mu_0,\varphi_0)+\frac{\partial T(\tau_1,\mu,\varphi;\mu_0,\varphi_0)}{\partial\tau_1}\\
&\quad=\exp(-\tau_1/\mu)p(-\mu,\varphi;-\mu_0,\varphi_0)\\
&\quad+\frac{\exp(\tau_1/\mu)}{4\pi}\int_0^1\int_0^{2\pi} p(-\mu,\varphi;\mu'',\varphi'')S(\tau_1,\mu'',\varphi'';\mu_0,\varphi_0)\frac{d\mu''}{\mu''}d\varphi''\\
&\quad+\frac{1}{4\pi}\int_0^1\int_0^{2\pi} T(\tau_1;\mu,\varphi;\mu',\varphi')p(-\mu',\varphi';-\mu_0,\varphi_0)\frac{\mu'}{\mu'}\\
&\quad+\frac{1}{16\pi^2}\int_0^1\int_0^{2\pi}\int_0^1\int_0^{2\pi} T(\tau_1,\mu,\varphi;\mu',\varphi')p(-\mu',\varphi';\mu'',\varphi'')\\
&\quad\times S(\tau_1,\mu'',\varphi'',\mu_0,\varphi_0)\frac{d\mu'}{\mu'}d\varphi'\frac{d\mu''}{\mu''}d\varphi''.
\end{aligned}
\tag{5.9.13}
$$

The term $\partial S/\partial\tau_1$ can be eliminated from equations (5.9.10) and (5.9.11), while $\partial T/\partial\tau_1$ can be eliminated from equations (5.9.12) and (5.9.13). The resulting two equations express the invariance laws of diffuse reflection and transmission to the addition (or removal) of layers of arbitrary optical thickness to (or from) the medium at the top and simultaneous removal (or addition) of layers of equal optical thickness from (or to) the medium at the bottom.

5.10 The X- and the Y-functions

The scattering and transmission functions S and T can be expressed (see Chandrasekhar (1960)) in terms of certain functions called the X- and Y-functions. In the case of isotropic scattering we write these as follows:

$$
\left(\frac{1}{\mu_0}+\frac{1}{\mu}\right)S(\tau;\mu,\mu_0)=\varpi\,[X(\mu)X(\mu_0)-Y(\mu)Y(\mu_0)],
\tag{5.10.1}
$$

$$
\left(\frac{1}{\mu_0}-\frac{1}{\mu}\right)T(\tau_1;\mu_1,\mu_0)-\varpi\,[Y(\mu)X(\mu_0)-X(\mu)Y(\mu_0)],
\tag{5.10.2}
$$

$$
\frac{\partial S(\tau_1;\mu,\mu_0)}{\partial\tau_1}=\varpi\,Y(\mu)Y(\mu_0),
\tag{5.10.3}
$$

$$\left(\frac{1}{\mu_0}-\frac{1}{\mu}\right)\frac{\partial T(\tau_1;\mu,\mu_0)}{\partial\tau_1}=\varpi\left[\frac{1}{\mu_0}X(\mu)Y(\mu_0)-\frac{1}{\mu}Y(\mu)X(\mu_0)\right], \tag{5.10.4}$$

where

$$X(\mu)=1+\frac{1}{2}\varpi\mu\int_0^1\frac{d\mu}{\mu+\mu'}\left[X(\mu)X(\mu')-Y(\mu)Y(\mu')\right], \tag{5.10.5}$$

$$Y(\mu)=\exp(-\tau_1/\mu)+\frac{1}{2}\varpi\mu\int_0^1\frac{d\mu}{\mu-\mu'}\left[Y(\mu)X(\mu')-X(\mu)Y(\mu')\right]. \tag{5.10.6}$$

If we compare these equations with those of the semi-infinite atmosphere (see equations (5.5.13) and (5.5.15)), we can write the X- and Y-functions as

$$X(\mu)=1+\mu\int_0^1\frac{\Psi(\mu')}{\mu+\mu'}\left[X(\mu)X(\mu')-Y(\mu)Y(\mu')\right]d\mu' \tag{5.10.7}$$

and

$$Y(\mu)=\exp(-\tau_1/\mu)+\mu\int_0^1\frac{\Psi(\mu')}{\mu-\mu'}\left[Y(\mu)X(\mu')-X(\mu)Y(\mu')\right]d\mu', \tag{5.10.8}$$

where $\Psi(\mu)$ is a characteristic function. These equations play the same role in a finite plane parallel medium as the H-function, given by

$$H(\mu)=1+\mu H(\mu)\int_0^1\frac{\Psi(\mu')}{\mu+\mu'}H(\mu')\,d\mu', \tag{5.10.9}$$

plays in the semi-infinite atmosphere or medium. The characteristic function $\Psi(\mu)$ is generally an even polynomial in μ which satisfies the condition

$$\int_0^1\Psi(\mu)\,d\mu\le\frac{1}{2}. \tag{5.10.10}$$

The X- and Y-functions become

$$X(\mu)\to H(\mu)\quad\text{and}\quad Y(\mu)\to 0\quad\text{as }\tau_1\to\infty, \tag{5.10.11}$$

where $H(\mu)$ satisfies equation (5.10.9). Furthermore,

$$X(\mu)\to 1\quad\text{and}\quad Y(\mu)\to\exp(-\tau_1/\mu)\quad\text{as }\tau_1\to 0. \tag{5.10.12}$$

Combining equations (5.8.1), (5.8.2), (5.10.1) and (5.10.2), we can write the reflected and transmitted intensities in terms of the X- and Y-functions:

$$I(0,\mu)=\frac{1}{4}\varpi F\frac{\mu_0}{\mu+\mu_0}\left[X(\mu)X(\mu_0)-Y(\mu)Y(\mu_0)\right] \tag{5.10.13}$$

and

$$I(0, -\mu) = \frac{1}{4}\varpi F \frac{\mu_0}{\mu - \mu_0}[Y(\mu)X(\mu_0) - X(\mu)Y(\mu_0)]. \quad (5.10.14)$$

The H-, X- and Y-functions are solved by iterative methods (see Chandrasekhar (1960)).

The physical meaning of the X- and Y-functions is shown in figure 5.11. If the point source of unit brightness (with total flux of 4π being emitted) illuminates a plane parallel atmosphere from a distance, then this atmosphere transmits and reflects part of this radiation due to multiple scattering. The combination of the point source and the illuminated part of the atmosphere will again appear as a point source from a large distance. The brightness of this combined source is represented by X in the same direction as that of the point source and by Y in the opposite direction. In either case $1/\mu$ is always positive.

5.11 Non-uniqueness of the solution in the conservative case

In the conservative case, we have

$$\int_0^1 \Psi(\mu)\, d\mu = \frac{1}{2}, \quad (5.11.1)$$

and the solutions of equations (5.10.7) and (5.10.8) are not then unique. If $X(\mu)$ and $Y(\mu)$ are solutions,

$$X(\mu) + Q\mu\,[X(\mu) + Y(\mu)] \quad (5.11.2)$$

and

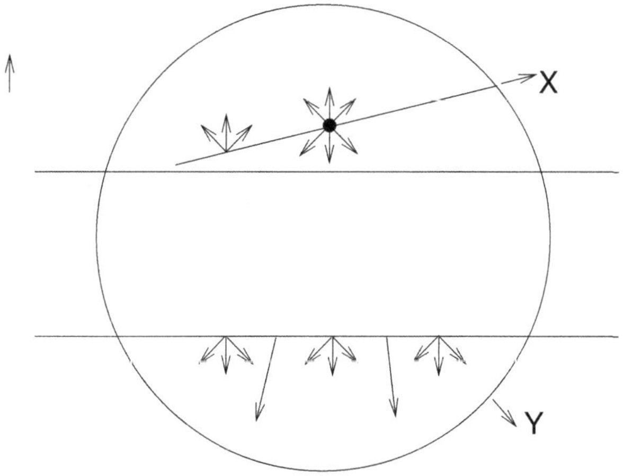

Figure 5.11 Physical meaning of the X- and Y-functions. The combined brightness of the light source and the illuminated atmosphere in a given direction is X and Y times the brightness of the source alone (from van de Hulst (1948), with permission).

$$Y(\mu) - Q\mu\,[X(\mu) + Y(\mu)] \tag{5.11.3}$$

are also solutions. Here Q is an arbitrary constant. In a finite atmosphere, unlike in the case of semi-infinite atmosphere, in all conservative cases of perfect scattering, the integral equations derived from the principle of invariance do not satisfy the condition of uniqueness (see Chandrasekhar (1960)). There is a multiplicity of solutions. These problems have been extensively studied by Anselone (1960, 1961), Mullikan (1964a,b,c) and Busbridge (1960).

Fymat and Abhyankar (1969a,b) applied the perturbation technique to the X- and Y-functions in homogeneous media. They considered a layer whose upper and lower boundaries are given by $\tau = \alpha$ and $\tau = \beta$ which has azimuthal independence with Chandrasekhar's $X(\alpha, \beta; \mu)$ and $Y(\alpha, \beta; \mu)$ for the radiation incident from above and $X^\star(\alpha, \beta; \mu)$ and $Y^\star(\alpha, \beta; \mu)$ for the radiation incident from below. They assumed that the albedo for single scattering $\Omega(\tau)$ differs from a constant value $\Omega_0 (0 \le \Omega_0 \le 1)$ by a small amount throughout the atmosphere. They expressed $\Omega(\tau)$ as

$$\Omega(\tau) = \Omega_0[1 + \omega(\tau)], \tag{5.11.4}$$

where $\omega(\tau) \ll 1$. Then $X(\alpha, \beta; \tau)$ and $X^\star(\alpha, \beta; \tau)$ can be expressed as

$$\left.\begin{aligned} X^\star(\alpha, \tau; \mu) &= X_0^\star(\alpha, \tau; \mu)\left[1 + x^\star(\alpha, \tau; \mu)\right], \\ X(\tau, \beta; \mu) &= X_0(\tau, \beta; \mu)\,[1 + x(\tau, \beta; \mu)] \end{aligned}\right\} \tag{5.11.5}$$

$$\begin{aligned} x^\star(\alpha, \beta; \mu) = {} & \frac{\Omega_0}{X_0^\star(\alpha, \beta; \mu)} \int_0^1 \int_\alpha^\beta \Big(X_0^\star(\alpha, \beta'; \mu) X_0^\star(\alpha, \beta'; \mu) \\ & \times \left\{\omega(\beta') + (1 + \omega(\beta')\left[x^\star(\alpha, \beta; \mu) + x^\star(\alpha, \beta'; \mu')\right]\right\} \\ & \times \exp\left[-(\beta - \beta')\left(\frac{1}{\mu} + \frac{1}{\mu'}\right)\right] d\beta' \Big) \psi(\mu') \frac{d\mu'}{\mu'}. \end{aligned} \tag{5.11.6}$$

The above equation was first derived by Sobolev (1956); Busbridge (1961) derived it independently. Abhyankar and Fymat (1970) studied the imperfect Rayleigh scattering in a semi-infinite medium of a planetary atmosphere. They computed Stokes's vectors and studied the Babinet and Brewster neutral points using the imperfect Rayleigh scattering. As the scattering process becomes more and more imperfect (or Ω decreases), the two neutral points approach the sun and for much smaller values of Ω, they coalesce which confirms the observational fact that the neutral points are closer to the sun.

X- and Y-functions have been tabulated for the finite medium with isotropic scattering by Chandrasekhar *et al.* (1952), Sobouti (1963), Carlstedt and Mullikan (1966).

Ambarzumian's mathematical equation was developed to solve Milne's integral equation. An auxiliary equation for this equation was developed and the solution sought in the form of a Neumann series whose existence and uniqueness are established. The scattering and transmission functions are defined in terms of N-solutions. The integro-differential equations of the scattering and transmission functions are obtained in terms of an auxiliary equation in a form suitable for numerical solution as an initial value problem (see Kourganoff (1963)).

5.12 Particle counting method

Bellman and his colleagues in a series of papers studied the radiative transfer problems using the principle of invariant imbedding (see bibliography). They used what is called the particle counting method. Their procedure consists of the addition of an infinitesimal layer, then the first order contributions to transmission and reflection are counted and the integro-differential equations for transmission and reflection coefficients derived in the limit as the layer thickness vanishes. The X- and Y-functions of Chandrasekhar are obtained in a similar way. They obtained a system of simultaneous non-linear ordinary differential equations by replacing the quadrature over angles by Gaussian sums. These are solved by standard numerical methods. We shall briefly describe this method for a plane parallel slab following Sen and Wilson (1990).

A layer of infinitesimal thickness $\Delta\tau$ is added to an inhomogeneous layer of thickness τ as shown in figure 5.12. A flux πF is incident on the surface at A making an angle θ_0 with the inward normal at A. The emergent intensity is assumed to make an angle θ with the outward normal. The intensity of light reflected from the whole medium in the θ-direction at a given point on the surface is

$$I(\theta, \mu) = FR(\mu, \mu_0; \tau) \tag{5.12.1}$$

($\mu = \cos\theta$, $\mu_0 = \cos\theta_0$). The reflection function $R(\mu, \mu_0; \tau)$ gives the fraction of the reflected energy from the whole medium. Bellman and Kalaba (1956) derived the following invariant imbedding relation:

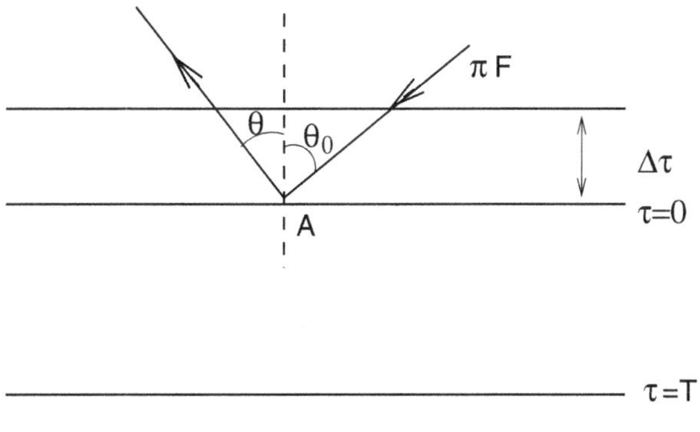

Figure 5.12 Particle counting method.

$$R(\mu, \mu_0; \tau + \Delta\tau) = R(\mu, \mu_0; \tau)\left(1 - \frac{\Delta\tau}{\mu} - \frac{\Delta\tau}{\mu_0}\right) + \frac{\varpi(\tau)\Delta\tau}{4\mu}$$
$$+ \frac{\varpi(\tau)}{2}\frac{\Delta\tau}{\mu}\int_0^1 R(\mu', \mu_0; \tau)\, d\mu' + \frac{\varpi(\tau)}{2}\Delta\tau \int_0^1 R(\mu, \mu''; \tau)\frac{d\mu''}{\mu''}$$
$$+ \varpi(\tau)\Delta\tau \int_0^1 R(\mu', \mu_0; \tau)\, d\mu' \int_0^1 R(\mu, \mu''; \tau)\frac{d\mu''}{\mu''}, \tag{5.12.2}$$

where ϖ is the albedo for single scattering. In the above equation, only the first order terms in $\Delta\tau$ have been retained. The terms on the RHS in equation (5.12.2) can be interpreted as follows. The first term represents the Lagrangian residue of intensity due to absorption losses in passing through $\Delta\tau$ on the way in and on the way out. The second term represents the part directly scattered from $\Delta\tau$, if the incident radiation on it is in the direction θ. The third term represents the contribution from the light that is scattered in a layer of thickness $\Delta\tau$ and reflected from the slab extending from 0 to τ. The fourth term represents the contribution from the light reflected from the layer 0 to τ and scattered in the layer $\Delta\tau$. The fifth term is the contribution from the reflection in $(0, \tau)$ followed by scattering in $(\tau, \tau + \Delta\tau)$ and then reflection in $(0, \tau)$ again. If $S(\mu, \mu_0; \tau)$ is the scattering function such that

$$R(\mu, \mu_0; \tau) = \frac{S(\mu, \mu_0; \tau)}{4\mu}, \tag{5.12.3}$$

then $I(0, \mu)$ in equation (5.12.1) becomes

$$I(0, \mu) = \frac{4}{4\mu} S(\mu, \mu_0; \tau). \tag{5.12.4}$$

If we let $\Delta\tau \to 0$ in equation (5.12.2), we get

$$\frac{\partial}{\partial\tau} S(\mu, \mu_0; \tau) + \left(\frac{1}{\mu} + \frac{1}{\mu_0}\right) S(\mu, \mu_0; \tau)$$
$$= \varpi(\tau)\left[1 + \frac{1}{2}\int_0^1 S(\mu, \mu''; \tau)\frac{d\mu''}{\mu''}\right]\left[1 + \frac{1}{2}\int_0^1 S(\mu', \mu_0; \tau)\frac{d\mu'}{\mu'}\right]. \tag{5.12.5}$$

If we write

$$\phi(\mu, \tau) = 1 + \frac{1}{2}\int_0^1 S(\mu, \mu''; \tau)\frac{d\mu''}{\mu''}, \tag{5.12.6}$$

then solving equation (5.12.5) we get

$$S(\mu, \mu_0; \tau) = \int_0^\tau \exp\left[-\left(\frac{1}{\mu} + \frac{1}{\mu_0}\right)(\tau - t)\right] \varpi(t)\phi(\mu_0, t)\phi(\mu, t)\, dt. \tag{5.12.7}$$

At $\tau = 0$, we assume that

$$S(\mu, \mu_0; 0) = 0. \tag{5.12.8}$$

Substituting equation (5.12.7) into equation (5.12.6), we obtain

$$\phi(\mu, \tau) = 1 + \frac{1}{2}\int_0^1 \left\{ \int_0^\tau \exp\left[-\left(\frac{1}{\mu} + \frac{1}{\mu_0}\right)(\tau - t)\right] \right.$$
$$\left. \times\ \varpi(t)\phi(\mu_0, t)\phi(\mu, t)\, dt \right\} \frac{d\mu_0}{\mu_0}. \tag{5.12.9}$$

Once we know $\phi(\mu, \tau)$ from the above equation we can compute $S(\mu, \mu_0; \tau)$ from equation (5.12.7) and $I(0, \mu)$ can be obtained from (5.12.4). We have described above the important steps that lead to the invariant imbedding technique of Bellman and Kalaba (1956) for solving the transfer problems in plane parallel media. Bellman *et al.* (1963) tabulated the diffuse reflection functions $R(\mu_i, \mu_j; \tau)$ for $\tau = 0$ to τ_{max} with intervals of 0.1 and $\varpi = 0, 0.1, 0.2, \ldots, 1.0$. Kagiwada and Kalaba (1967) estimated the local anisotropic function by using the principle of invariance, (see Adams and Kattawa (1970)).

5.13 The exit function

Hovenier (1978) introduced a function call the exit function by using the symmetry properties. This function can be used instead of two functions to describe the reflected and transmitted intensities of a homogeneous plane parallel atmosphere. It contains three double integrals replacing the traditional pair of simultaneous integral equations for the reflection and transmission functions containing sixteen double integrals. The integral equation of the exit function can be used for a rapid reduction of the problem to functions of only one variable.

We assume a purely scattering (no internal sources) plane parallel, homogeneous layer of optical thickness τ with ϖ as the albedo for single scattering and $P(\cos\Theta)$ as the phase function. We specify the azimuthal angle ϕ and an incident parallel beam in the direction (μ_0, ϕ_0) with a net flux per unit area normalized to π. The intensities of the diffuse radiation leaving the slab at the top and bottom are given in terms of the reflection functions $R(\mu, \mu_0, \phi - \phi_0)$ and the transmission function $T(\mu, \mu_0, \phi - \phi_0)$ respectively. These are $(4\mu\mu_0)^{-1}$ times the corresponding S- and T-functions of Chandrasekhar. The standard problem consists of finding the reflection and transmission functions for a given τ, ϖ and $P(\cos\Theta)$. The symmetry properties of the R- and T-functions depend on μ, μ_0 and $\phi - \phi_0$. The azimuthal difference occurs because of the rotational symmetry of the problem about the vertical. We have

$$R(\mu, \mu_0, \phi - \phi_0) = R(\mu, \mu_0, \phi_0 - \phi), \tag{5.13.1}$$

$$T(\mu, \mu_0, \phi - \phi_0) = T(\mu, \mu_0, \phi_0 - \phi). \tag{5.13.2}$$

The principle of reciprocity states that the scattering and transmission functions are unaltered when the directions of incidence and emergence are interchanged. The principle of reciprocity requires that

$$R(\mu, \mu_0, \phi - \phi_0) = R(\mu_0, \mu, \phi_0 - \phi). \tag{5.13.3}$$

By using equation (5.13.1), we get

$$R(\mu, \mu_0, \phi - \phi_0) = R(\mu_0, \mu, \phi - \phi_0). \tag{5.13.4}$$

Turning the layer and beams of incoming and outgoing light upside down, we find

$$T(\mu, \mu_0, \phi - \phi_0) = T(\mu_0, \mu, \phi - \phi_0). \tag{5.13.5}$$

Equations (5.13.4) and (5.13.5) imply that the reflection and transmission functions are symmetric in μ and μ_0. Using this property of symmetry, one can obtain a unified treatment for replacing the reflection and transmission functions by one function of the same variables, not necessarily symmetric in μ and μ_0. We may choose to define the function:

$$\begin{aligned} E(\mu, \mu_0, \phi - \phi_0) = &(\mu_0 + \mu) R(\mu, \mu_0, \phi - \phi_0) \\ &+ (\mu_0 - \mu) T(\mu, \mu_0, \phi - \phi_0). \end{aligned} \tag{5.13.6}$$

Interchanging μ and μ_0 gives

$$\begin{aligned} E(\mu_0, \mu, \phi - \phi_0) = &(\mu_0 + \mu) R(\mu, \mu_0, \phi - \phi_0) \\ &+ (\mu_0 - \mu) T(\mu, \mu_0, \phi - \phi_0). \end{aligned} \tag{5.13.7}$$

Adding and subtracting equations (5.13.6) and (5.13.7) gives us

$$R(\mu, \mu_0, \phi - \phi_0) = \frac{1}{2(\mu_0 + \mu)} [E(\mu, \mu_0, \phi - \phi_0) + E(\mu_0, \mu, \phi - \phi_0)], \tag{5.13.8}$$

$$T(\mu, \mu_0, \phi - \phi_0) = \frac{1}{2(\mu_0 - \mu)} [E(\mu, \mu_0, \phi - \phi_0) - E(\mu_0, \mu, \phi - \phi_0)]. \tag{5.13.9}$$

The function $E(\mu, \mu_0, \phi - \phi_0)$ which fully describes the intensity of radiation through the functions $R(\mu, \mu_0, \phi - \phi_0)$ and $T(\mu, \mu_0, \phi - \phi_0)$ at either the top or bottom is called the exit function. It is clear that knowledge of $E(\mu, \mu_0, \phi - \phi_0)$ ensures knowledge of $R(\mu, \mu_0, \phi - \phi_0)$ and $T(\mu, \mu_0, \phi - \phi_0)$. It can be seen from equation (5.13.9) that the function $T(\mu, \mu_0, \phi - \phi_0)$ cannot be obtained when $\mu = \mu_0$ in which case one should apply L'Hôpital's rule or interpolation.

From the symmetry properties,

$$E(\mu, \mu_0, \phi - \phi_0) = E(\mu, \mu_0, \phi_0 - \phi). \quad (5.13.10)$$

In a semi-infinite layer ($T \to 0$ as $\tau \to \infty$)

$$E_\infty(\mu, \mu_0, \phi - \phi_0) = (\mu + \mu_0)R_\infty(\mu, \mu_0, \phi - \phi_0). \quad (5.13.11)$$

Furthermore, if and only if $\tau = \infty$, then

$$E(\mu, \mu_0, \phi - \phi_0) = E(\mu_0, \mu, \phi - \phi_0). \quad (5.13.12)$$

In more general terms, one can write the R- and T-functions in terms of a function U such that

$$U(\mu, \mu_0, \phi - \phi_0) = f(\mu, \mu_0)R(\mu, \mu_0, \phi - \phi_0) + g(\mu, \mu_0)T(\mu, \mu_0, \phi - \phi_0), \quad (5.13.13)$$

where $f(\mu, \mu_0)$ and $g(\mu, \mu_0)$ are known functions. Interchanging μ and μ_0 gives us

$$U(\mu_0, \mu, \phi - \phi_0) = f(\mu_0, \mu)R(\mu, \mu_0, \phi - \phi_0) + g(\mu_0, \mu)T(\mu, \mu_0, \phi - \phi_0). \quad (5.13.14)$$

The R- and T-functions can be determined if $U(\mu, \mu_0, \phi - \phi_0)$ is completely known on the condition that the determinant

$$\Delta(\mu, \mu_0) = f(\mu, \mu_0)g(\mu_0, \mu) - f(\mu_0, \mu)g(\mu, \mu_0) \quad (5.13.15)$$

is non-vanishing everywhere.

We shall now derive the fundamental integral equation for the exit function. We symmetrize the problem by putting two sources of light with exactly the same physical properties one on each side of the slab. Both the sources send parallel beams of radiation into the slab in the direction (μ_0, ϕ_0). This follows from the symmetry property: the angular distributions of the outgoing radiation are the same at the top and at the bottom for any optical thickness of the atmosphere. This can be written as the sum for diffuse light:

$$R(\mu, \mu_0, \phi - \phi_0) + T(\mu, \mu_0, \phi - \phi_0). \quad (5.13.16)$$

The unscattered directly transmitted component of light is $\pi \exp(-\tau/\mu_0)$. Let us add a thin layer of optical thickness $\Delta\tau$ which has similar physical properties. $\Delta\tau$ is so small that only its direct transmission and single scattering are important.

Direct transmission through $\Delta\tau$ in the direction (μ, ϕ) with the attenuation factor $\exp(-\Delta\tau/\mu)$ to the first order, single scattering gives (see Chandrasekhar (1960))

$$\frac{\varpi\,\Delta\tau}{4\mu\mu_0} P\left[-\mu\mu_0 + (1-\mu^2)^{\frac{1}{2}}(1-\mu_0^2)^{\frac{1}{2}}\cos(\phi-\phi_0)\right]$$
$$= \frac{\varpi\,\Delta\tau}{4\mu\mu_0} P_r(\mu, \mu_0, \phi-\phi_0) \tag{5.13.17}$$

and

$$\frac{\varpi\,\Delta\tau}{4\mu\mu_0} P[\mu\mu_0 + (1-\mu^2)^{\frac{1}{2}}(1-\mu_0)^{\frac{1}{2}}\cos(\phi-\phi_0)$$
$$= \frac{\varpi\,\Delta\tau}{4\mu\mu_0} P_t(\mu, \mu_0, \phi-\phi_0). \tag{5.13.18}$$

Hovenier (1978) gives the exit function as

$$E(\mu, \mu_0, \phi-\phi_0) = \frac{\varpi}{4}\left[\exp(-\tau/\mu_0) - \exp(-\tau\mu)\right] P_t(\mu, \mu_0, \phi-\phi_0)$$
$$+ \frac{\varpi}{4}\left[1 - \exp(-\tau/\mu - \tau/\mu_0)\right] P_r(\mu, \mu_0, \phi-\phi_0)$$
$$+ \frac{\varpi}{4\pi}\int_0^1 ds \int_0^{2\pi} d\phi\, \frac{\mu E(\mu, s, \phi-\phi_0) - sE(s, \mu, \phi-\phi')}{\mu^2 - s^2}$$
$$\times \left[P_t(s, \mu_0, \phi'-\phi_0) - P_r(s, \mu_0, \phi'-\phi_0)\exp(-\tau/\mu_0)\right]$$
$$+ \frac{\varpi}{4\pi}\mu_0 \int_0^1 dt \int_0^{2\pi} d\phi' \Big[P_t(\mu, t, \phi-\phi') - \exp(-\tau/\mu) P_r(\mu, t, \phi-\phi'')$$
$$+ \frac{\mu}{\pi}\int_0^1 ds \int_0^{2\pi} d\phi'\, \mu E(\mu, s, \phi-\phi')$$
$$- sE(s, \mu, \phi-\phi')P_r(s, t, \phi-\phi'')\Big]$$
$$\times \frac{\mu_0 E(t, \mu_0, \phi''-\phi_0) - tE(\mu_0, t, \phi''-\phi_0)}{\mu_0^2 - t^2}. \tag{5.13.19}$$

This is the general fundamental integral equation for the exit function. This equation contains the same information as those derived from the principle of invariance of Ambarzumian and of Chandrasekhar (1960). For further developments of the exit function, see Hovenier (1978, 1980).

Nikoghossian (1997) tried to connect the Rybicki quadratic integrals (Rybicki 1977) with the principle of invariance. Ivanov (1978) generalized Rybicki's results in a plane parallel medium which scatters monochromatic radiation isotropically. The concept of two-point bilinear integrals was introduced to connect two different radiation fields. It is argued that the majority of non-linear equations in transfer theory are connected through the principle of invariance. Hubený (1987a,b) provided some physical understanding of the quadratic nature of the theory. The quadratic Q-relation is understood to have a relationship with the principle of invariance. Nikoghossian (1997) supplied general rigorous mathematical derivations of quadratic and bilinear relations on the basis of the principle of invariance. Kirkorian and Nikoghossian (1996) applied the variational principle and showed that a strong connection exists among the conservation laws, the invariance principle and the quadratic relations. Nikoghossian (1984) computed the mean number of scatterings using the principle of invariance in a medium with internal sources.

Exercises

5.1 Using the principle of invariance show that $I(0, \mu) = J(0)H(\mu)$.

5.2 Derive the Hopf–Bronstein relation from the principle of invariance.

5.3 Derive the integral equations for S (in a plane parallel semi-infinite medium) with the phase function $p(\cos\Theta) = \frac{3}{4}(1 + \cos^2\Theta)$.

5.4 Assuming that

$$p_n = \int_0^1 X(\mu)\Psi(\mu)\mu^n \, d\mu,$$

$$q_n = \int_0^1 Y(\mu)\Psi(\mu)\mu^n \, d\mu,$$

prove the following:

(a) $$\int_0^1 \frac{\mu'\Psi(\mu)}{\mu + \mu'} \left[X(\mu)X(\mu') - Y(\mu)Y(\mu')\right] d\mu = 1 - [(1 - p_0)X(\mu) + q_0 Y(\mu)],$$

(b) $$\int_0^1 \frac{\mu'\Psi(\mu)}{\mu - \mu'} \left[Y(\mu)X(\mu') - X(\mu)Y(\mu')\right] d\mu = \exp(-\tau_1/\mu) + [q_0 X(\mu) - (1 - p_0)Y(\mu)],$$

(c) $$\int_0^1 \frac{\mu'^2\Psi(\mu')}{\mu + \mu'} \left[X(\mu)X(\mu') - Y(\mu)Y(\mu')\right] d\mu' = p_1 X(\mu) - q_1 Y(\mu) - \mu + \mu\,[(1 - p_0)X(\mu) + p_0 Y(\mu)]$$

and

(d) $$\int_0^1 \frac{\mu'^2\Psi(\mu')}{\mu - \mu'} \left[Y(\mu)X(\mu') - X(\mu)Y(\mu')\right] d\mu' = q_1 X(\mu) - p_1 Y(\mu) - \mu\exp(-\tau_1/\mu) + \mu\,[q_0 X(\mu) + (1 - p_0)Y(\mu)].$$

5.5 Derive the exit function for a homogeneous semi-infinite medium.

5.6 Consider the incidence of a narrow beam of radiation, instead of a parallel beam, on a plane parallel slab (search light problem of a slab) and derive expressions for the diffuse reflection and transmission functions.

REFERENCES

Abhyankar, K.D., Fymat, A.L., 1970, *A&A*, **4**, 101.

Adams, C.N., Kattawa, G.W., 1970, *JQSRT*, **10**, 341.

Ambarzumian, V.A., 1942, *Russ. Astron. J.*, **19**, 1.

Ambarzumian, V.A., 1943, *C.R. Dokl. Acad. Sci. USSR*, **38**, 257.

Ambarzumian, V.A., 1944, *J. Phys. Acad. Sci. USSR*, **8**, 65.

Anselone, P.M., 1960, *MNRAS*, **120**, 498.

Anselone, P.M., 1961, *J. Math. Mech.*, **10**, 537.

Bellman, R., Kagiwada, H., Kakaba, R., Ueno, S., 1966, *J. Franklin Institute*, **282**, 330.

Bellman, R., Kalaba, R., 1956, *Proc. Nat. Acad. Sci.*, **42**, 629.

Bellman, R., Kalaba, R., Prestrud, M., 1963, *Invariant Imbedding and Radiative Transfer in Slabs of Finite Thickness*, Elsevier, New York.

Busbridge, I.W., 1960, *The Mathematics of Radiative Transfer*, Cambridge University Press, Cambridge.

Busbridge, I.W., 1961, *ApJ*, **133**, 198.

Busbridge, I.W., 1967, *ApJ*, **149**, 195.

Carlstedt, J.L., Mullikan, T.W., 1966, *ApJ Suppl.*, **12**, 449.

Chandrasekhar, S., 1960, *Radiative Transfer*, Dover, New York.

Chandrasekhar, S., Elbert, D., Franklin, A., 1952, *ApJ*, **115**, 244.

Fymat, A.L., Abhyankar, K.D., 1969a, *ApJ*, **158**, 315.

Fymat, A.L., Abhyankar, K.D., 1969b, *ApJ*, **158**, 325.

Hottel, H.C. and Sarofim, A.F., 1967, *Radiative Transfer*, McGraw-Hill, New York.

Hovenier, J.W., 1978, *A&A*, **68**, 239.

Hovenier, J.W., 1980, *A&A*, **82**, 61.

Hubený, I., 1987a, *A&A*, **185**, 332.

Hubený, I., 1987b, *A&A*, **185**, 336.

Hulst, van de, 1948, *Astrophys. J.*, **107**, 220.

Ivanov, V.V., 1978, *Astron. Zh.*, **22**, 612.

Kagiwada, H.K., Kalaba, R.E., 1967, *JQSRT*, **7**, 295.

Kourganoff, V., 1963, *Basic Methods in Transfer Problems*, Dover, New York.

Kirkorian, R.A., Nikoghossian, A.G., 1996, *JQSRT*, **56**, 465.

McClelland, J.A., 1906, *Roy. Dublin Soc. Sci. Trans.*(2), **9**, 9.

Mullikan, T.W., 1962, *Astrophys. J.*, **136**, 627.

Mullikan, T.W., 1964a, in P.M. Anselone (ed), *Nonlinear Equations of Radiative Transfer, Nonlinear Integral Equations*, University of Wisconsin Press, Wisconsin.

Mullikan, T.W., 1964b, *Astrophys. J.*, **139**, 379.

Mullikan, T.W., 1964c, *Astrophys. J.*, **139**, 1267.

Mullikan, T.W., 1964d, *Trans. Amer. Math. Soc.*, **113**, 316.

Nikoghossian, A.G., 1984, *Astrophysics*, **20**, 685.

Nikoghossian, A.G., 1997, *ApJ*, **483**, 849.

Peraiah, A., 1999, *Space Sci. Rev.*, **87**, 465.

Preisendorfer, R.W., 1965, *Radiative Transfer on Discrete Spaces*, Pergamon, Oxford.

Rayleigh, Lord, 1920, *Scientific Papers of Lord Rayleigh IV*, Cambridge, p. 482.

Rybicki, G.B., 1977, *ApJ*, **213**, 165.

Schmidt, H.W., 1907, *Ann. der Phys.*, **23**, 671.

Sen, K.K., Wilson, S.J., 1990, *Radiative Transfer in Curved Media*, World Scientific, Singapore.

Sobolev, V.V., 1956, *C.R. Dokl. Acad. Sci. USSR*, **111**, 1000.

Sobolev, V.V., 1963, *A Treatise on Radiative Transfer* (translated by S.I. Gaposchkin), Van Nostrand Company Inc., New York.

Sobouti, Y., 1963, *ApJ Suppl.*, **7**, 411.

Stokes, Sir George, 1852, *Trans. Cambridge Phil. Soc.*, **9**, 399.

Stokes, Sir George, 1862, *Proc. Roy. Soc.*, **11**, 545.

Wing, G.M., 1962, *An Introduction to Transport Theory*, John Wiley, New York.

Chapter 6

Discrete space theory

6.1 Introduction

We have studied homogeneous, plane parallel scattering atmospheres in chapter 5, using the principles of invariance in semi-infinite and finite media. These problems are solvable by the standard techniques of differential equations and expressible in standard functions. The X- and Y-functions of Chandrasekhar are solutions of certain integral equations. These cannot be used in a non-homogeneous media unless one sacrifices the physical characteristics of the medium. These solutions have been tabulated and it is difficult to use them in practical problems. One has to make serious physical approximations or resort to a numerical approximation. The principles of invariance are essentially the statement of the conservation of energy. Conservation of energy in a finite region can be expressed by what is called the 'interaction principle'. In the limit of vanishing thickness of the medium these principles lead to the integro-differential equations of radiative transfer. The principle of interaction (see Redheffer (1962), Preisendorfer (1965), Grant and Hunt (1969b)) generalizes the invariance principles particularly in a finite medium. The basic idea of the interaction principle is to specify the radiation field in terms of the transmitted and reflected radiation at any given point in the medium.

Carlson (1963) and Lathrop and Carlson (1967) used a numerical version of the discrete ordinate technique in neutron reactor calculations. By integrating the radiative transfer equation over a finite volume in space coordinates and using the mean value theorem for integrals, we can develop difference equations that conserve flux. These difference equations are of quite general use in non-uniform media and curvilinear coordinate systems. This system of equations is solved by iterative methods. As these equations are the expression of the conservation of flux, invariance principles can be expected to be deduced from them. One needs to study the errors and stability factors of any system of equations. Carlson's S_n methods did not have a well studied error and stability analysis. This can be overcome by rewriting equations in what is called 'invariant S_n' form. In this way, one can test the stability and estimate the errors due to truncation and round-off of the terms.

The reflection and transmission operators can be expressed in the form of matrices. The matrix structure allows us to perform the desired analysis and to obtain an explicit solution which essentially expresses the results in terms of the Green's function of the transport operator which is related to the probability of quantum exit as defined and exploited by Ueno (1965) (see also chapter 9). The matrix structure is the discrete equivalent of the equation of Rybicki and Usher (1966) and converges to it when we pass the limit of infinitesimally thin segments.

6.2 The rod model

In this, we develop a simple method of solving a one-dimensional monochromatic problem (see Wing (1962), Sobolev (1963) and Grant (1968a)). We shall describe this below (see figure 6.1).

We assume that all quantities depend on the length l in a steady state. The optical depth is defined as

$$\tau = \tau(l) = \int_0^l \sigma(l')\,dl' \tag{6.2.1}$$

or

$$\tau(L) = T. \tag{6.2.2}$$

We also assume monochromatic radiation and a source which will be increasing in the direction of τ. A fraction $p(\tau)$ of the beam is scattered in the direction of the beam and the remaining fraction $1 - p(\tau)$ of the beam is scattered in the opposite direction and we set $\varpi(\tau)$ as the albedo for single scattering. If $I^+(\tau)$ and $I^-(\tau)$ are the intensities in the increasing and decreasing directions of τ, then

$$\frac{dI^+}{d\tau} + I^+ = S^+, \tag{6.2.3}$$

$$-\frac{dI^-}{d\tau} + I^- = S^-, \tag{6.2.4}$$

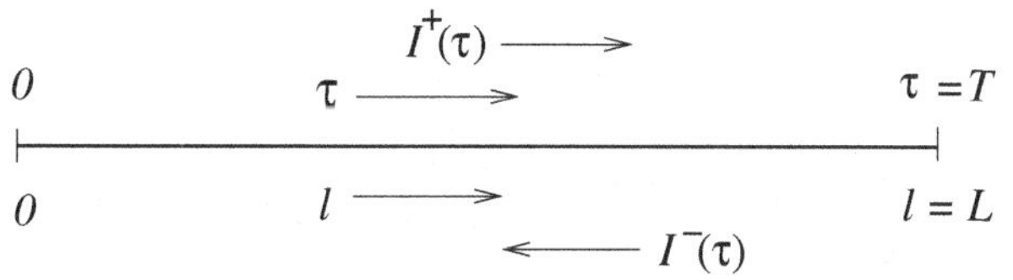

Figure 6.1 The rod model.

where the source functions S^+ and S^- are given by

$$S^+(\tau) = B^+(\tau) + \varpi(\tau)\left[p(\tau)I^+(\tau) + (1 - p(\tau))I^-(\tau)\right] \tag{6.2.5}$$

and

$$S^-(\tau) = B^-(\tau) + \varpi(\tau)\left[(1 - p(\tau))I^+(\tau) + p(\tau)I^-(\tau)\right]. \tag{6.2.6}$$

We need to specify the boundary conditions at $\tau = 0$ and $\tau = T$. These can be as follows:

$$I^+(0) = I_1 \tag{6.2.7}$$

and

$$I^-(T) = I_2. \tag{6.2.8}$$

There are two aspects of the radiation field: one is the diffusely scattered radiation field and the other is due to the incident radiation; the intensities I^+ and I^- refer to the total radiation field. The total source functions S_d^+ and S_d^- are written as

$$S_d^+(\tau) = S_d^+(\tau) + \varpi(\tau)\left\{p(\tau)I_1 \exp(-\tau) + (1 - p(\tau))I_2 \exp[-(T - \tau)]\right\} \tag{6.2.9}$$

and

$$S_d^-(\tau) = S_d^-(\tau) + \varpi(\tau)\left\{(1 - p(\tau))I_1 \exp(-\tau) + p(\tau)I_2 \exp[-(T - \tau)]\right\}, \tag{6.2.10}$$

and solutions (6.2.3) and (6.2.4) can be written as

$$I^+(\tau) = I_1 \exp(-\tau) + \int_0^\tau \exp[-(\tau - t)]S^+(t)\,dt \tag{6.2.11}$$

and

$$I^-(\tau) = I_2 \exp[-(T - \tau)] + \int_0^\tau \exp[-(t - \tau)]S^-(t)\,dt. \tag{6.2.12}$$

If the quantities S^+ and S^- are known, I^+ and I^- are calculated using the quadratures.

6.3 The interaction principle for the rod

Consider a rod with boundaries at 1 and 2 (see figure 6.2). Intensities I_1^+ and I_2^- are incident at points 1 and 2 while intensities I_1^- and I_2^+ are emergent at the points 1 and 2 respectively. We shall write the emergent intensities in terms of the incident intensities. The emergent intensity I_2^+ at point 2 is the combination of the transmitted intensity $t_{3/2}I_1^+$ and the reflected intensity $r'_{3/2}I_2^-$, together with the sources $\Sigma_{3/2}^+$ generated within the rod between points 1 and 2. The emergent intensity I^- is the combination of the reflected intensity $r_{3/2}I_1^+$ and the transmitted intensity $t'_{3/2}I_2^-$, together with the sources generated within the rod between points 1 and 2. Thus

$$I_2^+ = t_{3/2}I_1^+ + r'_{3/2}I_2^- + \Sigma_{3/2}^+ \tag{6.3.1}$$

and

$$I_1^- = r_{3/2}I_1^+ + t'_{3/2}I_2^- + \Sigma_{3/2}^-, \tag{6.3.2}$$

where the subscript 3/2 refers to the averages of the quantities subscripted between points 1 and 2 of the rod where the flux is measured and the ts and rs are the transmission and reflection coefficients. Equations (6.3.1) and (6.3.2) are termed the interaction principle for the rod.

We define the cell matrix as

$$S_{3/2} = \begin{pmatrix} t_{3/2} & r'_{3/2} \\ r_{3/2} & t'_{3/2} \end{pmatrix} = \begin{pmatrix} t(2,1) & r(1,2) \\ r(2,1) & t(1,2) \end{pmatrix} = S(1,2). \tag{6.3.3}$$

In terms of the cell matrix, equations (6.3.1) and (6.3.2) are written as

$$\begin{pmatrix} I_2^+ \\ I_1^- \end{pmatrix} = S_{3/2}\begin{pmatrix} I_1^+ \\ I_2^- \end{pmatrix} + \begin{pmatrix} \Sigma_{3/2}^+ \\ \Sigma_{3/2}^- \end{pmatrix}. \tag{6.3.4}$$

The transmission (t) and reflection (r) coefficients are the ordinary numbers less than unity in the rod model. They will be matrix operators in the case of beams of radiation in higher geometries. Therefore one should be careful about the order of products. We shall interpret the operators t and r in connection with equations (6.2.3) and (6.2.4).

Let us now examine

$$\lim_{\Delta\tau\to 0}\begin{pmatrix} I_2^+ - I_1^+ \\ I_1^- - I_2^- \end{pmatrix}(\Delta\tau)^{-1} = \begin{pmatrix} \dfrac{dI^+}{d\tau} \\ -\dfrac{dI^-}{d\tau} \end{pmatrix} \tag{6.3.5}$$

$$= \lim_{\Delta\tau\to 0}\left\{(\Delta\tau)^{-1}[S-E]\begin{pmatrix} I_1^+ \\ I_2^- \end{pmatrix} + (\Delta\tau)^{-1}\begin{pmatrix} \Sigma^+ \\ \Sigma^- \end{pmatrix}\right\}, \tag{6.3.6}$$

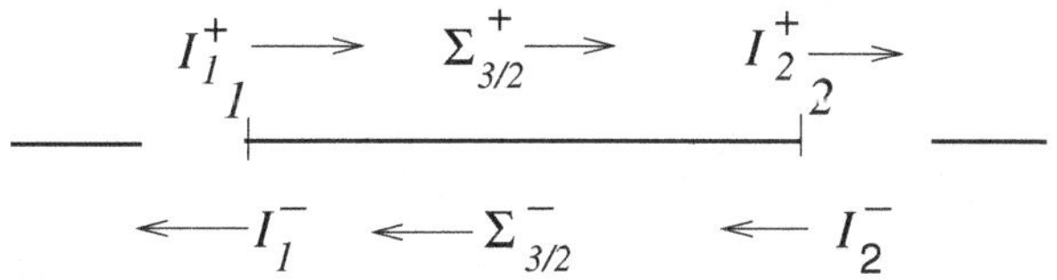

Figure 6.2 Interaction principle for the rod

where E is the unit matrix. Let us write that

$$\Sigma^+ = B^+\Delta\tau + O(\Delta\tau), \quad \Sigma^- = B^-\Delta\tau + O(\Delta\tau) \tag{6.3.7}$$

and

$$S - E = \begin{pmatrix} (1-\varpi p)\Delta\tau & -\varpi(1-p)\Delta\tau \\ -\varpi(1-p)\Delta\tau & (1-\varpi p)\Delta\tau \end{pmatrix} + O(\Delta\tau). \tag{6.3.8}$$

Equation (6.3.4) tends (in the limit) to

$$\frac{dI^+}{d\tau} = -I_1^+ + \varpi\left[pI_1^+ + (1-p)I_2^-\right] + B^+, \tag{6.3.9}$$

$$-\frac{dI^-}{d\tau} = -I_2^- + \varpi\left[(1-p)I_1^+ + pI_2^-\right] + B^-. \tag{6.3.10}$$

It appears that equations (6.2.3) and (6.2.4) are consistent with equations (6.3.9) and (6.3.10) in the limit $\Delta\tau \to 0$. This will be true if and only if at each point of the rod we have

$$t, t' = 1 - (1-\varpi p)\Delta\tau + O(\Delta\tau), \tag{6.3.11}$$

$$r, r' = \varpi(1-p)\Delta\tau + O(\Delta\tau), \tag{6.3.12}$$

$$\Sigma^+ = B^+\Delta x + O(\Delta\tau), \tag{6.3.13}$$

$$\Sigma^- = B^-\Delta x + O(\Delta\tau) \tag{6.3.14}$$

as $\Delta\tau \to 0$.

6.4 Multiple rods: star products

If we know the transmission and reflection properties, we can construct cell matrices in a given segment. This will give us the full description of the radiation field with prescribed boundary conditions. If we have two such cells and place them one after another as shown in figure 6.3, we would like to know the reflection and transmission

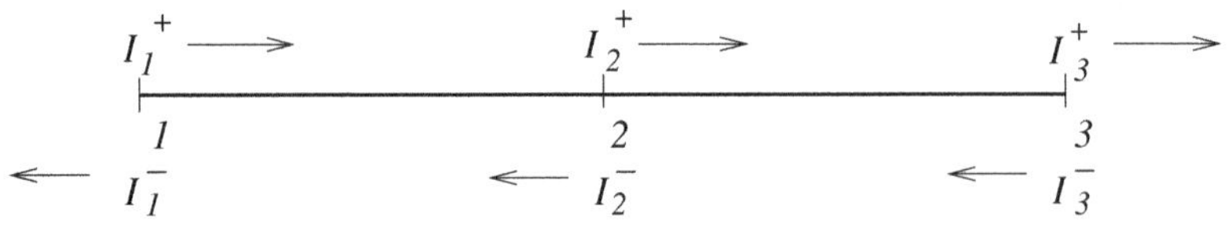

Figure 6.3 Composite rod.

properties of this composite cell. For the two cells, we can write from the interaction principle by using equation (6.3.4)

$$\begin{pmatrix} I_2^+ \\ I_1^- \end{pmatrix} = S(1,2)\begin{pmatrix} I_1^+ \\ I_2^- \end{pmatrix} + \Sigma(1,2) \tag{6.4.1}$$

and

$$\begin{pmatrix} I_3^+ \\ I_2^- \end{pmatrix} = S(2,3)\begin{pmatrix} I_2^+ \\ I_3^- \end{pmatrix} + \Sigma(2,3). \tag{6.4.2}$$

The emergent intensities for the composite cell are written as

$$\begin{pmatrix} I_3^+ \\ I_1^- \end{pmatrix} = S(1,3)\begin{pmatrix} I_1^+ \\ I_3^- \end{pmatrix} + \Sigma(1,3), \tag{6.4.3}$$

where $S(1,3)$ is given by what is called by Redheffer (1962) the 'star product'

$$S(1,3) = S(1,2) * S(2,3) \tag{6.4.4}$$

and

$$\Sigma(1,3) = \Sigma(1,2) * \Sigma(2,3). \tag{6.4.5}$$

The quantities $S(1,3)$ and $\Sigma(1,3)$ consist of the transmission and reflection factors of the two cells. These are given by

$$S(1,3) = \begin{pmatrix} t(3,1) & r(1,3) \\ r(3,1) & t(1,3) \end{pmatrix}, \tag{6.4.6}$$

where

$$t(3,1) = t(3,2)\,[1 - r(1,2)r(3,2)]^{-1}\,t(2,1), \tag{6.4.7}$$

$$r(3,1) = r(2,1) + t(1,2)r(3,2)\,[1 - r(1,2)r(3,2)]^{-1}\,t(2,1), \tag{6.4.8}$$

$$t(1,3) = t(1,2) + [1 - r(3,2)r(1,2)]^{-1}\,t(2,3), \tag{6.4.9}$$

$$r(1,3) = r(2,3) + t(3,2)r(1,2)\,[1 - r(3,2)r(1,2)]^{-1}\,t(2,3), \tag{6.4.10}$$

and the source terms are given by

$$\begin{aligned}\Sigma^+(3,1) = \Sigma^+(3,2) + t(3,2)\,&[1 - r(1,2)r(3,2)]^{-1} \\ &\times \left[\Sigma^-(2,1) + r(1,2)\Sigma^-(2,3)\right]\end{aligned} \tag{6.4.11}$$

and

$$\begin{aligned}\Sigma^-(1,3) = \Sigma^-(1,2) + t(1,2)\,&[1 - r(3,2)r(1,2)]^{-1} \\ &\times \left[\Sigma^-(2,3) + r(3,2)\Sigma^+(2,1)\right].\end{aligned} \tag{6.4.12}$$

Any number of composite rods can be added to get the radiation field for the whole system.

It is clear that the star operation is associative, which means that it does not matter how we divide the rod and this division does not have any bearing on the results. More of this aspect will be seen later. All the previous analysis is applicable to situations of higher dimensions. We need to divide the region of interest into a one-parameter set of cells.

6.5 The interaction principle for a slab

In this section we shall study the interaction principle in a slab following closely Grant and Hunt (1969a,b). Dividing the space into several cells, each immediately adjacent to the next one, is called by Preisendorfer (1965) 'the quotient space decomposition'. This kind of division holds true for other geometries, specially spherical symmetry. In the case of a slab, the radiation field depends on a single space coordinate x (or the optical depth) and the angle $\mu = \cos\theta$ (see chapter 1). Therefore at any X, we can write the intensities $I^{+}(X, \mu)$ and $I^{-}(X, \mu)$, where as usual, μ is the cosine of the angle made by the common normal to the stratification in the direction in which X increases (see figure 6.4). The intensities $I^{+}(X)$ and $I^{-}(X)$ are written as

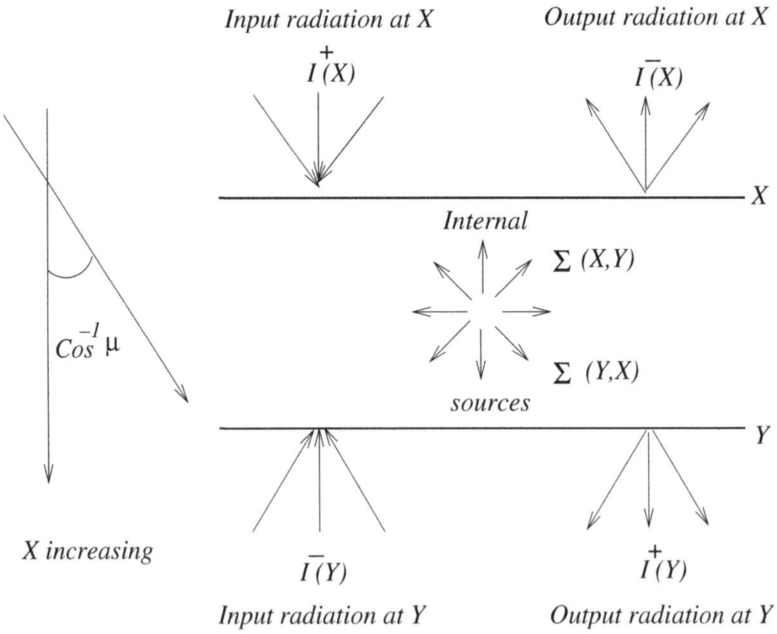

Figure 6.4 The interaction principle for a slab.

$$\left.\begin{aligned} I^+(X) &= I(X, +\mu): \quad 0 < \mu \le 1; a \le X \le b, \text{ say}, \\ I^-(X) &= I(X, -\mu): \quad 0 < \mu \le 1; a \le X \le b, \text{ say}, \end{aligned}\right\} \tag{6.5.1}$$

where $I(X, \mu)$ is the specific intensity in the direction of μ, while $I(X, -\mu)$ is the specific intensity in the opposite direction. We select a finite set of angles $\mu_1 < \mu_2 < \mu_3 < \cdots < \mu_j < \cdots < \mu_m \le 1$, and write the intensities as vectors in m-dimensional Euclidean space:

$$I^+(X) = \begin{pmatrix} I(X, \mu_1) \\ \vdots \\ I(X, \mu_m) \end{pmatrix}, \quad I^-(X) = \begin{pmatrix} I(X, -\mu_1) \\ \vdots \\ I(X, -\mu_m) \end{pmatrix}. \tag{6.5.2}$$

Let us consider a layer bounded by X and Y (see figure 6.4). We let the intensities $I^+(X)$ and $I^-(Y)$ be incident on this layer and the intensities $I^-(X)$ and $I^+(Y)$ emerge from the layer. The emergent intensity $I^+(Y)$ is the combination of the transmitted intensity $I^+(X)$, the reflected intensity $I^-(Y)$ and the sources $\Sigma^+(X)$. Similarly the emergent intensity $I^-(X)$ is the combination of the reflected intensity $I^+(X)$, the transmitted intensity $I^-(Y)$ and the sources $\Sigma^-(Y, X)$ in the layer. This is the principle of conservation of radiant energy, which can be written as

$$I^+(Y) = t(Y, X)I^+(X) + r(X, Y)I^-(Y) + \Sigma^+(Y, X) \tag{6.5.3}$$

and

$$I^-(X) = r(Y, X)I^+(X) + t(X, Y)I^-(Y) + \Sigma^-(X, Y), \tag{6.5.4}$$

where ts and rs are the diffuse transmission and diffuse reflection operators respectively. Caution must be exercised in writing the order of these operators (which are assumed linear) as the product of two matrices is non-commutative. Equations (6.5.3) and (6.5.4) constitute what is called the principle of interaction. Notice the similarity between equations (6.5.3) and (6.5.4) and the principle of invariance for a finite medium given in chapter 5.

Equations (6.5.3) and (6.5.4) are similar to those of Redheffer (1962) except for the source terms Σ^+ and Σ^- added by Grant and Hunt (1969a). This is an important addition particularly in the context of stellar and planetary atmospheres, planetary nebulae and other similar objects. Redheffer's (1962) work refers to the transmission line theory. Preisendorfer (1965) explains this principle in greater detail. The interaction principle is similar to van de Hulst's work (1965).

Equations (6.5.3) and (6.5.4) can be written as

$$\begin{pmatrix} I^+(Y) \\ I^-(X) \end{pmatrix} = S(X, Y) \begin{pmatrix} I^+(X) \\ I^-(Y) \end{pmatrix} + \Sigma(x, y), \tag{6.5.5}$$

where $S(X, Y)$ is called the S-matrix.

The operators ts and rs can be easily understood. For example the operator $r(X, Y)$ can be understood as

$$r(X, Y)I^-(Y) = \left\{\int_0^1 r(X, \mu'; Y, -\mu')I(Y, -\mu')\, d\mu' : \ 0 < \mu \le 1\right\}. \tag{6.5.6}$$

The discrete ordinate analogue of $r(X, Y)$ is a matrix:

$$r(X, Y) = \left[r(X, \mu_j; Y, -\mu_k)\right], \tag{6.5.7}$$

and this is multiplied by $I^-(Y)$ on its right, where $I^-(Y)$ is defined as in equation (6.5.2).

The interaction principle as stated above is extremely general as we have not included any restrictions (of physical type). The diffuse reflection and transmission operators and the source terms express the physical characteristics of any type of medium which we choose to study. Any inhomogeneities can be introduced into the medium. As the physical properties are specified point by point, we need to connect these with the global properties of the medium. That is to say that we need to derive the global radiation field from the local radiation field through the diffuse transmission and reflection operators.

We need to consider two problems: (1) the external response problems which means estimation of the fluxes emerging from the layer say (XY), and (2) the internal field of radiation which means we should be able to find the radiation field at a point inside the medium in any given direction, that is, $I^+(\mu)$ and $I^-(\mu)$. In the second case we need to divide the region of interest into several layers and apply the interaction principle (6.5.3) and (6.5.4). A numerical scheme to solve this system of simultaneous equations in the discrete ordinate approximations was given in Grant and Hunt (1969b).

6.6 The star product for the slab

We shall consider two adjacent layers bounded by X, Y and Z. Applying the interaction principle (6.5.5) to layers (X, Y), (Y, X) and the composite layer (X, Z), where $a \le X < Y < Z \le b$, we get

$$\begin{pmatrix} I^+(Y) \\ I^-(X) \end{pmatrix} = S(X, Y)\begin{pmatrix} I^+(X) \\ I^-(Y) \end{pmatrix} + \Sigma(X, Y), \tag{6.6.1}$$

$$\begin{pmatrix} I^+(Z) \\ I^-(Y) \end{pmatrix} = S(Y, Z)\begin{pmatrix} I^+(Y) \\ I^-(Z) \end{pmatrix} + \Sigma(Y, Z) \tag{6.6.2}$$

and

$$\begin{pmatrix} I^{+}(Z) \\ I^{-}(X) \end{pmatrix} = S(X, Z) \begin{pmatrix} I^{+}(X) \\ I^{-}(X) \end{pmatrix} + \Sigma(X, Z). \tag{6.6.3}$$

Redheffer (1962) defined the 'star product' as

$$S(X, Z) = S(X, Y) * S(Y, Z), \tag{6.6.4}$$

$$\Sigma(X, Z) = \Sigma(X, Y) * \Sigma(Y, Z), \tag{6.6.5}$$

where the S-matrix $S(x, z)$ is written as

$$S(X, Z) = \begin{pmatrix} t(Z, X) & r(X, Z) \\ r(Z, X) & t(X, Z) \end{pmatrix}, \tag{6.6.6}$$

$$\Sigma(X, Z) = \begin{pmatrix} \Sigma(Z, X) \\ \Sigma(X, Z) \end{pmatrix}. \tag{6.6.7}$$

The operators in the above S-matrix can be obtained by eliminating $I^{+}(Y)$ and $I^{-}(Y)$ from equations (6.6.2) and (6.6.3) and these are given by

$$t(Z, X) = t(Z, Y)\,[E - r(X, Y)r(Z, Y)]^{-1}\, t(Y, X), \tag{6.6.8}$$

$$t(X, Z) = t(X, Y)\,[E - r(Z, Y)r(X, Y)]^{-1}\, t(Y, Z), \tag{6.6.9}$$

$$r(X, Z) = r(Y, Z) + t(Z, Y)\,[E - r(X, Y)r(Z, Y)]^{-1}\, r(X, Y)t(Y, Z), \tag{6.6.10}$$

$$r(Z, X) = r(Y, X) + t(X, Y)\,[E - r(Z, Y)r(X, Y)]^{-1}\, r(Z, Y)t(Y, X), \tag{6.6.11}$$

where E is the identity matrix. The source terms are given by

$$\begin{aligned} \Sigma^{+}(Z, X) &= \Sigma^{+}(Z, Y) + t(Z, Y)\,[E - r(X, Y)r(Z, Y)]^{-1} \\ &\quad \times \left[r(X, Y)\Sigma^{-}(Y, Z) + \Sigma^{+}(Y, X)\right] \end{aligned} \tag{6.6.12}$$

$$\begin{aligned} \Sigma^{+}(X, Z) &= \Sigma^{-}(X, Y) + t(X, Y)\,[E - r(Z, Y)r(X, Y)]^{-1} \\ &\quad \times \left[r(Z, Y)\Sigma^{+}(Y, X) + \Sigma^{-}(Y, Z)\right]. \end{aligned} \tag{6.6.13}$$

The above transmission and reflection operators and the source vectors exist whenever the inverses $[E - r(X, Y)r(Z, Y)]^{-1}$ and $[E - r(Z, Y)r(X, Y)]^{-1}$ exist.

A revealing physical interpretation can be given of the operators in equations (6.6.8)–(6.6.11). For example, if we expand $t(Z, X)$ in the form

$$t(Z, X) = \sum_{k=0}^{\infty} t_k(Z, X), \tag{6.6.14}$$

where

$$t_k(Z, X) = t(Z, Y)\,[r(X, Y)r(Z, Y)]^k\,t(Y, X). \tag{6.6.15}$$

From equation (6.6.3) we notice that $t(Z, X)$ acts on $I^+(X)$ to give rise $I^+(Z)$. We consider different orders of scattering using equation (6.6.14). By letting $k = 0$ we get the zeroth order of scattering

$$t(Z, X) = t_0(Z, X) = t(Z, Y)t(Y, X), \tag{6.6.16}$$

that is the operator $t(Z, X)$ results from the transmission from layer X to layer Y which is $t(Y, X)$ and that from layer Y to layer Z or $t(Z, Y)$ only. If $k = 1$, first order scattering occurs:

$$\begin{aligned} t(Z, X) &= \sum_0^1 t_k(Z, X) \\ &= t(Z, Y)t(Y, X) + t(Z, Y)r(X, Y)r(Z, Y)t(Y, X). \end{aligned} \tag{6.6.17}$$

The meaning of equation (6.6.17) is as follows: the first order scattering consists of diffusely transmitted radiation through the first term on the RHS and diffusely reflected radiation of the transmitted radiation through the second term on the RHS. If we put $k = 2$ we obtain the second order scattering:

$$\begin{aligned} t(Z, X) = t(Z, Y)t(Y, X) &+ t(Z, Y)r(X, Y)r(Z, Y)t(Y, X) + t(Z, Y) \\ &\times [r(X, Y)r(Z, Y)]^2\,t(Y, X). \end{aligned} \tag{6.6.18}$$

The three terms on the RHS have the following physical meanings:

the first term is the diffusely transmitted radiation;

the second term is the once diffusely reflected radiation of the diffusely transmitted radiation of the first term;

the third term is the twice diffusely reflected radiation of the diffusely transmitted radiation of the first term or once diffusely reflected radiation in the radiation of the second term.

Similar expansions of equation (6.6.15) for $k = 3, 4, \ldots, \infty$ gives different orders of scattering of diffuse radiation. This gives us the infinitely transmitted and reflected radiation or the multiple scattering of radiation. Therefore the operators in equations (6.6.8)–(6.6.11) represent the diffuse radiation field provided $[E - r(X, Y)r(Z, Y)]^{-1}$ and $[E - r(Z, Y)r(X, Y)]^{-1}$ exist. These operators are easy to compute. The equivalents of these operators are the integral equations for S and T of chapter 5, whose kernels are too difficult to compute. Here one needs only the usual matrix operations and a simple matrix inversion gives the diffuse radiation field.

If we have two layers L and M and $S(L)$ and $S(M)$ are the S-matrices of these two layers, then

$$S(L * M) = S(L) * S(M). \tag{6.6.19}$$

$L * M$ signifies the fact that the two layers L and M are placed side by side with M to the right. If L and M are inhomogeneous, then

$$L * M \neq M * L. \tag{6.6.20}$$

In general star multiplication is non-commutative. If we have three layers L, M and N side by side, then we have

$$S\left[L * (M * N)\right] = S\left[(L * M) * N\right]. \tag{6.6.21}$$

Star multiplication is associative. If $S(Z)$ represents a layer of zero thickness, then

$$S(Z) = \begin{pmatrix} E & 0 \\ 0 & E \end{pmatrix} \tag{6.6.22}$$

and we have

$$S(Z) * S(L) = S(L) * S(Z) = S(L * Z) = S(L). \tag{6.6.23}$$

We can construct the star product series for any number of layers. If the system is homogeneous, relation (6.6.4) for the layers L and M can be written as

$$S(L + M) = S(L) * S(M) = S(M) * S(L), \tag{6.6.24}$$

as the order of the layers cannot affect the result. If

$$\lim_{M \to \infty} S(M) = S(\infty) \tag{6.6.25}$$

exists, then equation (6.6.24) becomes

$$S(\infty) = S(L) * S(\infty) = S(\infty) * S(L). \tag{6.6.26}$$

This is nothing but the principle of invariance of Ambarzumian (1943). Van de Hulst (1965) has given a fast scheme for obtaining S-operators for a thick homogeneous medium.

6.7 Emergent radiation

We divide the medium into several layers, say N (we will give detailed reasons why a medium should be so divided in later sections of this chapter) and assume that the r and t operators are known in advance. We need to find the intensities emerging from the medium, that is, I^+_{N+1} and I^-_1, and the source vectors $\Sigma^+(N+1, 1)$ and

$\Sigma^-(1, N+1)$. We can find these by using the interaction principle (6.5.5). This can be written with I_1^+ and I_{N+1}^- as the incident intensities as

$$\begin{pmatrix} I_{N+1}^+ \\ I_1^- \end{pmatrix} = S(1, N+1) \begin{pmatrix} I_1^+ \\ I_{N+1}^- \end{pmatrix} + \begin{pmatrix} \Sigma^+(N+1, 1) \\ \Sigma^-(1, N+1) \end{pmatrix}, \tag{6.7.1}$$

where

$$S(1, N+1) = \begin{pmatrix} t(N+1, 1) & r(1, N+1) \\ r(N+1, 1) & t(1, N+1) \end{pmatrix}. \tag{6.7.2}$$

The quantities $t(N+1, 1)$, $\Sigma^+(N+1, 1)$ and $\Sigma^-(1, N+1)$ can be computed using relations (6.6.8)–(6.6.11) repeatedly, which is the same as saying

$$S(1, N+1) = S(1, 2) * S(2, 3) * \cdots * S(N, N+1). \tag{6.7.3}$$

Equation (6.7.1) will give us only the emergent radiation field. If we need to know the radiation field at any point inside the medium, we have to follow a different algorithm.

6.8 The internal radiation field

As in the previous section, we divide the medium into N layers (with $N+1$ boundaries). The interaction principle is written for any layer n inside the medium as

$$I_{n+1}^+ = t(n+1, n)I_n^+ + r(n, n+1)I_{n+1}^- + \Sigma_{n+\frac{1}{2}}^+, \tag{6.8.1}$$

$$I_n^- = r(n+1, n)I_n^+ + t(n, n+1)I_{n+1}^- + \Sigma_{n+\frac{1}{2}}^-, \tag{6.8.2}$$

where $n = 1, 2, 3, \ldots, N$. We shall solve these equations with the boundary incident intensities I_1^+ and I_{N+1}^- at $n = 1$ and $N+1$ respectively. Our aim is to obtain the internal intensities I_n^+ and I_n^- at the boundary of each layer and I_1^- and I_{N+1}^+ at the boundary of the medium. We follow the procedure outlined in Grant and Hunt (1968) and Grant (1968a).

The system of equations (6.8.1) and (6.8.2) can be written for $n = 1$ as

$$\left.\begin{aligned} I_2^+ &= t(2, 1)I_1^+ + r(1, 2)I_2^- + \Sigma_{\frac{3}{2}}^+, \\ I_1^- &= r(2, 1)I_1^+ + t(1, 2)I_2^- + \Sigma_{\frac{3}{2}}^-, \end{aligned}\right\} \tag{6.8.3}$$

for $n = 2$ as

$$\left.\begin{aligned} I_3^+ &= t(3, 2)I_2^+ + r(2, 3)I_3^- + \Sigma_{\frac{5}{2}}^+ \\ I_2^- &= r(3, 2)I_2^+ + t(2, 3)I_3^- + \Sigma_{\frac{5}{2}}^- \end{aligned}\right\} \tag{6.8.4}$$

and so on for $n = 3, 4, \ldots, N$. This system of equations for $n = 1, \ldots, N$ will be solved to obtain the internal radiation field. Equations (6.8.3) and (6.8.4) can be written as

$$\mathbf{U}^+_{n+\frac{1}{2}} = \mathbf{T}(n+1, n)\mathbf{U}^+_{n-\frac{1}{2}} + \mathbf{R}(n, n+1)\mathbf{U}^-_{n+\frac{3}{2}} + \Sigma^+_{n+\frac{1}{2}}, \tag{6.8.5}$$

$$\mathbf{U}^-_{n+\frac{1}{2}} = \mathbf{R}(n+1, n)\mathbf{U}^+_{n-\frac{1}{2}} + \mathbf{T}(n, n+1)\mathbf{U}^-_{n+\frac{3}{2}} + \Sigma^-_{n+\frac{1}{2}}, \tag{6.8.6}$$

where

$$\left.\begin{aligned}
&\mathbf{U}^+_{n+\frac{1}{2}} = \begin{pmatrix} I^+_n \\ I^+_{n+1} \end{pmatrix}, \quad \mathbf{U}^-_{n+\frac{1}{2}} = \begin{pmatrix} I^-_{n+1} \\ I^-_n \end{pmatrix}, \\
&\mathbf{T}(n+1, n) = \begin{pmatrix} 0 & E \\ 0 & t(n+1, n) \end{pmatrix}, \quad \mathbf{T}(n, n+1) = \begin{pmatrix} 0 & E \\ 0 & t(n, n+1) \end{pmatrix}, \\
&\mathbf{R}(n+1, n) = \begin{pmatrix} 0 & 0 \\ 0 & r(n+1, n) \end{pmatrix}, \quad \mathbf{R}(n, n+1) = \begin{pmatrix} 0 & 0 \\ 0 & r(n, n+1) \end{pmatrix}, \\
&\mathbf{\Sigma}^+_{n+\frac{1}{2}} = \begin{pmatrix} 0 \\ \Sigma^+_{n+\frac{1}{2}} \end{pmatrix}, \quad \mathbf{\Sigma}^-_{n+\frac{1}{2}} = \begin{pmatrix} 0 \\ \Sigma^-_{n+\frac{1}{2}} \end{pmatrix}
\end{aligned}\right\} \tag{6.8.7}$$

and the subscript $n + \frac{1}{2}$ indicates the average of the subscripted quantity over the layers n and $n + 1$.

Equations (6.8.5) and (6.8.6) together with the equations (6.8.1) and (6.8.2) give us

$$-\mathbf{A}_{n+\frac{1}{2}}\mathbf{U}_{n-\frac{1}{2}} + \mathbf{U}_{n+\frac{1}{2}} - \mathbf{B}_{n+\frac{1}{2}}\mathbf{U}_{n+\frac{3}{2}} = \mathbf{\Sigma}_{n+\frac{1}{2}}, \tag{6.8.8}$$

where

$$\mathbf{A}_{n+\frac{1}{2}} = \begin{pmatrix} \mathbf{T}(n+1, n) & 0 \\ \mathbf{R}(n+1, n) & 0 \end{pmatrix}, \quad \mathbf{B}_{n+\frac{1}{2}} = \begin{pmatrix} 0 & \mathbf{R}(n, n+1) \\ 0 & \mathbf{T}(n, n+1) \end{pmatrix},$$

and

$$\mathbf{\Sigma}_{n+\frac{1}{2}} = \begin{pmatrix} \mathbf{\Sigma}^+_{n+\frac{1}{2}} \\ \mathbf{\Sigma}^-_{n+\frac{1}{2}} \end{pmatrix}, \quad \mathbf{U}_{n+\frac{1}{2}} = \begin{pmatrix} \mathbf{U}^+_{n+\frac{1}{2}} \\ \mathbf{U}^-_{n+\frac{1}{2}} \end{pmatrix}. \tag{6.8.9}$$

We need to specify the boundary conditions for equation (6.8.8) for $n = 1$ and $n = N$ since $U_{\frac{1}{2}}$ and $U_{N+\frac{3}{2}}$ are undefined. We couple the output elements $U^+_{\frac{1}{2}}$ and $U^-_{N+\frac{3}{2}}$ or I^+_1 and I^-_{N+1}. These will be the only non-zero elements of $U^+_{\frac{1}{2}}$ and $U^-_{N+\frac{3}{2}}$; $U^-_{\frac{1}{2}}$ and $U^+_{N+\frac{3}{2}}$ are irrelevant.

Equation (6.8.8) can be solved by the method of Gaussian elimination or factorization of the block matrix equation (see Isaacson and Keller (1966)). For $n = N, N-1, \ldots, 1$ and $\mathbf{U}_{N+\frac{3}{2}} = \left[0 \ \mathbf{U}^-_{N+\frac{3}{2}}\right]^T$, we have

$$\mathbf{U}_{n+\frac{1}{2}} = \mathbf{C}_{n+\frac{3}{2}} + \mathbf{V}_{n+\frac{1}{2}}. \tag{6.8.10}$$

Substituting this in equation (6.8.8) by letting n be replaced by $n-1$ gives

$$-\mathbf{A}_{n-\frac{1}{2}}\left[C_{n-\frac{1}{2}}\mathbf{U}_{n+\frac{1}{2}} + \mathbf{V}_{n-\frac{1}{2}}\right] + \mathbf{U}_{n+\frac{1}{2}} - \mathbf{B}_{n+\frac{1}{2}}\mathbf{U}_{n+\frac{3}{2}} = \mathbf{\Sigma}_{n+\frac{1}{2}}. \tag{6.8.11}$$

Solving for $\mathbf{U}_{n+\frac{1}{2}}$, we get

$$\begin{aligned}\mathbf{U}_{n+\frac{1}{2}} = {} & \left[\mathbf{E} - \mathbf{A}_{n+\frac{1}{2}}\mathbf{C}_{n-\frac{1}{2}}\right]^{-1}\mathbf{B}_{n+\frac{1}{2}}\mathbf{U}_{n+\frac{3}{2}} + \left[\mathbf{E} - \mathbf{A}_{n+\frac{1}{2}}\mathbf{C}_{n-\frac{1}{2}}\right]^{-1} \\ & \times \left[\mathbf{\Sigma}_{n+\frac{1}{2}} + \mathbf{A}_{n+\frac{1}{2}}\mathbf{V}_{n-\frac{1}{2}}\right].\end{aligned} \tag{6.8.12}$$

Comparing the terms in equations (6.8.10) and (6.8.11), we can write, if we set $\mathbf{C}_{\frac{1}{2}} = 0$,

$$\mathbf{D}_{n-\frac{1}{2}} = \left[\mathbf{E} - \mathbf{A}_{n+\frac{1}{2}}\mathbf{C}_{n+\frac{1}{2}}\right]^{-1}, \quad n = 1, 2, \ldots, N, \tag{6.8.13}$$

then

$$\mathbf{C}_{n+\frac{1}{2}} = \mathbf{D}_{n+\frac{1}{2}}\mathbf{B}_{n+\frac{1}{2}}, \quad n = 1, 2, \ldots, N. \tag{6.8.14}$$

If we let

$$\mathbf{F}_{n+\frac{1}{2}} = \mathbf{D}_{n+\frac{1}{2}}\mathbf{A}_{n+\frac{1}{2}}, \quad \text{and } \mathbf{V}_{\frac{1}{2}} = \mathbf{U}_{\frac{1}{2}} = \begin{pmatrix} U^+_{\frac{1}{2}} \\ 0 \end{pmatrix}, \tag{6.8.15}$$

then

$$\mathbf{V}_{n+\frac{1}{2}} = \mathbf{F}_{n+\frac{1}{2}}\mathbf{V}_{n+\frac{1}{2}} + \mathbf{D}_{n+\frac{1}{2}}\mathbf{\Sigma}_{n+\frac{1}{2}}, \quad n = 1, 2, \ldots, N. \tag{6.8.16}$$

The block 2×2 matrices **C**, **D** and **F** are

$$\left.\begin{aligned} & \mathbf{C}_{n+\frac{1}{2}} = \begin{pmatrix} 0 & \mathbf{R}(1, n+1) \\ 0 & \hat{\mathbf{T}}(n, n+1) \end{pmatrix}, \quad \mathbf{F}_{n+\frac{1}{2}} = \begin{pmatrix} \hat{\mathbf{T}}(n+1, n) & 0 \\ \hat{\mathbf{R}}(n+1, n) & 0 \end{pmatrix} \\ & \mathbf{D}_{n+\frac{1}{2}} = \begin{pmatrix} \mathbf{E} & \mathbf{R}_{n+\frac{1}{2}} \\ 0 & \mathbf{T}_{n+\frac{1}{2}} \end{pmatrix}. \end{aligned}\right\} \tag{6.8.17}$$

The components of $\mathbf{C}_{n-\frac{1}{2}}$ are

$$\left.\begin{aligned} & \mathbf{R}(1, n+1) = \begin{pmatrix} 0 & r(1, n)\hat{t}(n, n+1) \\ 0 & r(1, n+1) \end{pmatrix}, \\ & \hat{\mathbf{T}}(n, n+1) = \begin{pmatrix} 0 & \mathbf{E} \\ 0 & \hat{t}(n, n+1) \end{pmatrix}, \end{aligned}\right\} \tag{6.8.18}$$

the component of $\mathbf{D}_{n+\frac{1}{2}}$ are

$$\mathbf{R}_{n+\frac{1}{2}} = \begin{pmatrix} 0 & r(1,n)T_{n+\frac{1}{2}} \\ 0 & R_{n+\frac{1}{2}} \end{pmatrix}, \quad \mathbf{T}_{n+\frac{1}{2}} = \begin{pmatrix} \mathbf{E} & 0 \\ 0 & T_{n+\frac{1}{2}} \end{pmatrix} \tag{6.8.19}$$

and the component of $\mathbf{F}_{n+\frac{1}{2}}$ are

$$\hat{\mathbf{R}}(n+1,n) = \begin{pmatrix} 0 & 0 \\ 0 & \hat{r}(n+1,n) \end{pmatrix}, \quad \hat{\mathbf{T}}(n+1,n) = \begin{pmatrix} 0 & T'_{n+\frac{1}{2}} \\ 0 & \hat{t}(n+1,n) \end{pmatrix}. \tag{6.8.20}$$

These quantities are given as follows. The elements of the matrices in equations (6.8.18)–(6.8.20) will satisfy the following recurrence relations. For $n = 1, 2, \ldots, n$,

$$r(1,1) = 0, \tag{6.8.21}$$

$$T_{n+\frac{1}{2}} = [E - r(n+1,n)r(1,n)]^{-1}, \tag{6.8.22}$$

$$R_{n+\frac{1}{2}} = t(n+1,n)r(1,n)T_{n+\frac{1}{2}}, \tag{6.8.23}$$

$$r(1,n+1) = r(n,n+1)R_{n+\frac{1}{2}}t(n,n+1), \tag{6.8.24}$$

or

$$r(1,n+1) = r(n,n+1) + t(n+1,n)r(1,n)\,[E - r(n+1,n)r(1,n)]^{-1} \times t(n,n+1), \tag{6.8.25}$$

$$\hat{t}(n,n+1) = T_{n+\frac{1}{2}}t(n,n+1), \tag{6.8.26}$$

$$T'_{n+\frac{1}{2}} = [E - r(1,n)r(n+1,n)]^{-1}, \tag{6.8.27}$$

$$\hat{t}(n+1,n) = t(n+1,n)T'_{n+\frac{1}{2}}, \tag{6.8.28}$$

$$\hat{r}(n+1,n) = r(n+1,n)T'_{n+\frac{1}{2}}. \tag{6.8.29}$$

We have

$$V_{n+\frac{1}{2}} = \begin{pmatrix} \mathbf{V}^{+}_{n+\frac{1}{2}} \\ \mathbf{V}^{-}_{n+\frac{1}{2}} \end{pmatrix}, \quad \mathbf{V}^{+}_{n+\frac{1}{2}} = \begin{pmatrix} \mathbf{V}^{+}_{n+\frac{1}{2}} \\ \mathbf{V}^{+}_{n+\frac{1}{2}} \end{pmatrix}, \quad \mathbf{V}^{-}_{n+\frac{1}{2}} = \begin{pmatrix} 0 \\ \mathbf{V}^{-}_{n+\frac{1}{2}} \end{pmatrix}. \tag{6.8.30}$$

With the initial condition $V_{\frac{1}{2}} = I_1^+$, we can write

$$V^{+}_{n+\frac{1}{2}} = \ddot{t}(n+1,n)V^{+}_{n-\frac{1}{2}} + \left[\Sigma^{+}_{n+\frac{1}{2}} + R_{n+\frac{1}{2}}\Sigma^{-}_{n+\frac{1}{2}}\right], \tag{6.8.31}$$

$$V^{-}_{n+\frac{1}{2}} = \hat{r}(n+1,n)V^{+}_{n-\frac{1}{2}} + T_{n+\frac{1}{2}}\Sigma^{-}_{n+\frac{1}{2}} \tag{6.8.32}$$

for $n = 1, 2, \ldots, N$ successively. Similarly from equation (6.8.10), we can have for the output intensities I^+_{n+1}, I^-_n of $\mathbf{U}_{n+\frac{1}{2}}$ with the initial conditions I^-_{N+1}

$$I^+_{n+1} = r(1, n+1)I^-_{n+1} + V^+_{n+\frac{1}{2}}, \tag{6.8.33}$$

$$I^-_n = \hat{t}(n, n+1)I^-_{n+1} + V^-_{n+\frac{1}{2}} \tag{6.8.34}$$

for $n = N, N-1, \ldots, 1$. Furthermore,

$$V^+_{n+\frac{1}{2}} = T_{n+\frac{1}{2}}V^+_{n-\frac{1}{2}} + r(1, n)T_{n+\frac{1}{2}}\Sigma^-_{n+\frac{1}{2}}, \tag{6.8.35}$$

$$I^+_n = r(1, n+1)\hat{t}(n, n+1)I^-_{n+1} + V^+_{n+\frac{1}{2}}. \tag{6.8.36}$$

The results presented here can be applied in plane parallel or spherically symmetric geometries. The quantity $r(1, n)$ is the diffuse reflection matrix for the compound layer 1 to layer n. From equations (6.8.33) and (6.8.34), if we set $I^+_1 = 0$, $\Sigma^+_{n+\frac{1}{2}} = \Sigma^-_{n+\frac{1}{2}} = 0$ for all n, then the Vs vanish and we get

$$I^+_{n+1} = r(1, n+1)I^-_{n+1} \tag{6.8.37}$$

and

$$I^-_n = \hat{t}(n, n+1)I^-_{n+1} = T_{n+\frac{1}{2}}t(n, n+1)I^-_{n+1}. \tag{6.8.38}$$

The quantity $T_{n+\frac{1}{2}}$ can be interpreted as the transmission from multiple reflections from layer 1 to layer n across the interface $[n, n+1]$. For example, in the case of the rod we have

$$T_{n+\frac{1}{2}} = [1 - r(n+1, n)r(1, n)]^{-1} = \sum_{k=0}^{\infty}[r(n+1, n)r(1, n)]^k$$
$$= 1 + r(n+1, n)r(1, n) + r(n+1, n)r(1, n)r(n+1, n)r(1, n) + \cdots$$

(see equations (6.6.14) and (6.6.15) and the following discussion on the several orders of scattering). As the operators act on vectors to the right, each term in the expansion of $\hat{t}(n, n+1)$ such as

$$r(n+1, n)r(1, n)\cdots r(n+1, n)r(1, n)t(n, n+1)$$

gives the transmission from the $(n+1)$th boundary to the nth boundary followed by a sequence of reflections as shown in figure 6.5. The other operators in equations (6.8.33) and (6.8.34) can be interpreted similarly. We can thus find the internal radiation field through equations (6.8.33)–(6.8.36).

6.9 Reflecting surface

If the surface at $N+1$ is a reflecting surface, such as a planetary atmosphere, the above procedure can be modified slightly. One can write

$$I_{N+1}^{-} = r_R I_{N+1}^{+}, \tag{6.9.1}$$

where r_R is the reflection operator which can be defined as required. Substituting this into equation (6.8.33) and setting $n = N$, we obtain

$$I_{N+1}^{+} = [E - r(1, N+1)r_R]^{-1} V_{N+\frac{1}{2}}^{+}, \tag{6.9.2}$$

$$I_{N}^{-} = \hat{t}(N, N+1)r_R \, [E - r(1, N+1)r_R]^{-1} V_{N+\frac{1}{2}}^{+}. \tag{6.9.3}$$

The remaining intensities can be computed from relations (6.8.33) and (6.8.34).

We need now to construct the cell matrices Ss or the sources Σs. For this we need to focus our attention on a particular problem of radiative transfer. In the following sections we shall concentrate on the procedure for deriving the r and t matrices or the S-matrices. In section 6.11 we will derive the condition of non-negativity, stability, existence, uniqueness problems and conservation of radiant flux in a purely scattering medium.

6.10 Monochromatic equation of transfer

We assume monochromatic radiation and an isotropic phase function (axisymmetric)

$$P(\tau; \mu, \mu') = P(\tau; -\mu, -\mu') \quad (1 \le \mu \le 1; -1 \le \mu' \le 1), \tag{6.10.1}$$

which can be normalized to

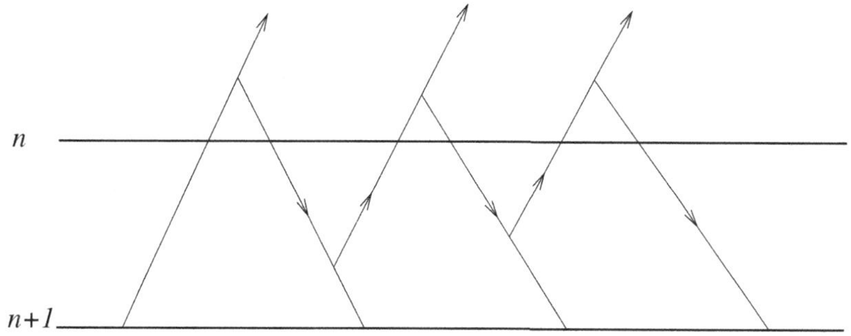

Figure 6.5 Diffuse reflection due to multiple reflections and transmissions. If we let $R = r(1, n)$, $r = r(n+1, n)$, $t = t(n+1, n)$, then $\hat{t}(n, n+1)$ will be equal to $t + (rR)t + (rR)^2 t + \cdots$. The net effect is multiple reflection and transmission from the layer bounded by n and $n+1$.

$$\frac{1}{2}\int_{-1}^{1} P(\tau;\mu,\mu')\,d\mu' = 1. \tag{6.10.2}$$

In this case the transfer equation can be written as (see chapter 2)

$$\mu\frac{dI(\tau,\mu)}{d\tau} + I(\tau,\mu) = [1-\varpi(\tau)]\,B^{+}(\tau) + \frac{1}{2}\varpi(\tau)\int_{-1}^{1} P(\tau;\mu,\mu')I(\tau,\mu')\,d\mu', \tag{6.10.3}$$

$$-\mu\frac{dI(\tau,-\mu)}{d\tau} + I(\tau,-\mu) = [1-\varpi(\tau)]\,B^{-}(\tau) + \frac{1}{2}\varpi(\tau)\int_{-1}^{1} P(\tau;-\mu,\mu')I(\tau,\mu')\,d\mu', \tag{6.10.4}$$

where $0 < \mu < 1$ and the Bs are source terms. As we saw earlier, the integrals can be replaced by quadratures, for example,

$$\int_{a}^{b} f(\mu')\,d\mu' = \sum_{j=1}^{m} c_j f(\mu_j) + e_m, \tag{6.10.5}$$

where e_m denotes the error which tends to zero for sufficiently large m. Gaussian quadratures are most useful in this regard and can be written as

$$0 \le |\mu_1| < |\mu_2| < |\mu_3| < \cdots < |\mu_m| < 1, \tag{6.10.6}$$

with the associated weights c_js positive ($c_j > 0$, $j = 1, \ldots, m$). The limits of the integral in equation (6.10.5) are taken to be -1 to 1, in which case the μ_js will be the zeros of the Legendre polynomial $P_m(\mu)$. However, as suggested by J. B. Sykes in 1951, it is convenient to split this interval into two halves and use the μs and cs corresponding to the interval [0,1]. In this case the μs are zeros of $P_m(2\mu - 1)$ in [0,1] and $\mu_j > 0$ and $\mu_{-j} = -\mu_j$. In the light of the above arguments, the phase function in equations (6.10.3) and (6.10.4) can be written in the discrete mode

$$\left.\begin{aligned} P(\tau;\mu_i,\mu_j) &= P_{ij}^{++}(\tau) = P(\tau;-\mu_i,-\mu_j) = P_{ij}^{--}, \\ P(\tau;\mu_i,-\mu_j) &= P_{ij}^{+-}(\tau) = P(\tau;-\mu_i,\mu_j) = P_{ij}^{-+}, \quad 0 < \mu_i < \mu_j < 1. \end{aligned}\right\} \tag{6.10.7}$$

The normalization condition (6.10.2) becomes in discrete form

$$\frac{1}{2}\sum_{i=1}^{m} c_i (P_{ij}^{++} + P_{ij}^{+-}) = \frac{1}{2}\sum_{j=1}^{m} (P_{ij}^{-+} + P_{ij}^{-})c_j = 1, \tag{6.10.8}$$

where

$$P_{ij}^{++} = P(+\mu_j, +\mu_j) \text{ etc.} \tag{6.10.9}$$

We shall write the discrete equation for the slab following Carlson (1963) and Lathrop and Carlson (1967). We define the 'cell' of discretization on the mesh as

$$\left[\tau_n, \tau_{n+1}\right]\left[\mu_{j-\frac{1}{2}}, \mu_{j+\frac{1}{2}}\right], \tag{6.10.10}$$

over τ and μ and

$$\mu_{j+\frac{1}{2}} = \sum_{k=1}^{j} c_k, \tag{6.10.11}$$

or

$$c_j = \mu_{j+\frac{1}{2}} - \mu_{j-\frac{1}{2}}. \tag{6.10.12}$$

Equation (6.10.3) can be integrated over this 'cell' (6.10.10) and can be written as

$$\frac{\mu_j(I_{n+1,j} - I_{n,j})}{\tau_{n+1} - \tau_n} + I_{n+\frac{1}{2}} = (1 - \varpi_{n+\frac{1}{2}})B^+_{n+\frac{1}{2}} + \frac{1}{2}\varpi_{n+\frac{1}{2}} \sum_k \left[P_{jk}^{++} c_k I_{n+\frac{1}{2},k} + P_{jk}^{+-} c_k I_{n+\frac{1}{2},-k}\right], \tag{6.10.13}$$

where c_k is defined as in equation (6.10.12), which is also the quadrature weight, and $I_{n+\frac{1}{2},j}$ and $I_{n+\frac{1}{2},-j}$ are the mean values over $[\tau_n, \tau_{n+1}]$. For example,

$$I_{n+\frac{1}{2},j} = \int_{\tau_n}^{\tau_{n+1}} d\tau \int_{\mu_{j-\frac{1}{2}}}^{\mu_{j+\frac{1}{2}}} d\mu\, I(\tau, \mu)/(\tau_{n+1} - \tau_n)c_j \tag{6.10.14}$$

and

$$I_{n,j} = \int_{\mu_{j-\frac{1}{2}}}^{\mu_{j+\frac{1}{2}}} \mu I(\tau_n, \mu)/c_j. \tag{6.10.15}$$

The quantities P_{jk}^{++} etc., are given by

$$\varpi_{n+\frac{1}{2}} P_{jk}^{++} c_k I_{n+\frac{1}{2},k} = \int_{\tau_n}^{\tau_{n+1}} d\tau \int_{\mu_{j-\frac{1}{2}}}^{\mu_{j+\frac{1}{2}}} d\mu \int_{\mu_{k-\frac{1}{2}}}^{\mu_{k+\frac{1}{2}}} d\mu' \frac{P(\tau\mu, \mu')I(\tau, \mu')\varpi(\tau)}{(\tau_{n+1} - \tau_n)c_j}. \tag{6.10.16}$$

Therefore the two equations (6.10.3) and (6.10.4) can be written for m angles in the discrete form:

$$\mathbf{M}\left(\mathbf{I}^+_{n+1} - \mathbf{I}^+_n\right) + \tau_{n+\frac{1}{2}}\mathbf{I}^+_{n+\frac{1}{2}} = \tau_{n+\frac{1}{2}}\left\{(1 - \varpi_{n+\frac{1}{2}})\mathbf{B}^+_{n+\frac{1}{2}} + \frac{1}{2}\varpi_{n+\frac{1}{2}}\left[\mathbf{P}^{++}_{n+\frac{1}{2}}\mathbf{C}\mathbf{I}^+_{n+\frac{1}{2}} + \mathbf{P}^{+-}_{n+\frac{1}{2}}\mathbf{C}\mathbf{I}^-_{n+\frac{1}{2}}\right]\right\}, \tag{6.10.17}$$

and

$$\mathbf{M}\left(\mathbf{I}_n^- - \mathbf{I}_{n+1}^-\right) + \tau_{n+\frac{1}{2}} I^-_{n+\frac{1}{2}} = \tau_{n+\frac{1}{2}} \Big\{ (1 - \varpi_{n+\frac{1}{2}})\mathbf{B}^-_{n+\frac{1}{2}} + \frac{1}{2}\varpi_{n+\frac{1}{2}} \left[\mathbf{P}^{-+}_{n+\frac{1}{2}} \mathbf{C}\mathbf{I}^+_{n+\frac{1}{2}} + \mathbf{P}^{--}_{n+\frac{1}{2}} \mathbf{C}\mathbf{I}^-_{n+\frac{1}{2}} \right] \Big\}. \tag{6.10.18}$$

Here the intensities are vectors defined as

$$\mathbf{I}_n^+ = \begin{pmatrix} I(\tau_n; \mu_1) \\ \vdots \\ I(\tau_n; \mu_m) \end{pmatrix}, \quad \mathbf{I}_n^- = \begin{pmatrix} I(\tau_n; -\mu_1) \\ \vdots \\ I(\tau_n; -\mu_m) \end{pmatrix}. \tag{6.10.19}$$

Furthermore,

$$\mathbf{M} = \left[\mu_j \delta_{jk}\right], \quad \mathbf{C} = \left[c_j \delta_{jk}\right], \; j = k = m, \quad \tau_{n+\frac{1}{2}} = \tau_{n+1} - \tau_n. \tag{6.10.20}$$

Equations (6.10.17) and (6.10.18) express the conservation of energy over a cell $(n, n+1)$ in terms of the intensities $\mathbf{I}_n^+$ and $\mathbf{I}_{n+1}^+$ and the mean values $\mathbf{I}^{\pm}_{n+\frac{1}{2}}$. We need to express these mean values in terms of $\mathbf{I}_n^+$ and $\mathbf{I}_{n+1}^+$ and similarly $\mathbf{I}^-_{n+\frac{1}{2}}$. We will make the simple linear assumption:

$$\left.\begin{aligned} \mathbf{I}^+_{n+\frac{1}{2}} &= (\mathbf{E} - \mathbf{X}_{n+\frac{1}{2}})\mathbf{I}_n^+ + \mathbf{X}_{n+\frac{1}{2}}\mathbf{I}^+_{n+1}, \\ \mathbf{I}^-_{n+\frac{1}{2}} &= \mathbf{X}_{n+\frac{1}{2}}\mathbf{I}_n^- + (\mathbf{E} - \mathbf{X}_{n+\frac{1}{2}})\mathbf{I}^-_{n+1}, \end{aligned}\right\} \tag{6.10.21}$$

where $\mathbf{X}_{n+\frac{1}{2}}$ are diagonal $m \times m$ matrices. For reasons of stability (Grant 1968b) we need to choose

$$\frac{1}{2}\mathbf{E} \le \mathbf{X}^{\pm}_{n+\frac{1}{2}} \le \mathbf{E}. \tag{6.10.22}$$

The matrix inequality stated above means that if the matrices **A** and **B** follow the relationship $\mathbf{A} \le \mathbf{B}$, then all the elements a_{ij} and b_{ij} of A and B will follow the relationship $a_{ij} \le b_{ij}$ for all i and j.

The best choice in this case (Grant 1968b) is

$$\mathbf{X}_{n+\frac{1}{2}} = \frac{1}{2}\mathbf{E}. \tag{6.10.23}$$

Therefore, the average values $\mathbf{I}^+_{n+\frac{1}{2}}$ and $\mathbf{I}^-_{n+1}$ can be written, using equations (6.10.23) and (6.10.21), as

$$\mathbf{I}^-_{n+1} = \frac{1}{2}(\mathbf{I}_n^+ + \mathbf{I}^+_{n+1}), \tag{6.10.24}$$

$$\mathbf{I}^-_{n+\frac{1}{2}} = \frac{1}{2}(\mathbf{I}_n^- + \mathbf{I}^-_{n+1}). \tag{6.10.25}$$

Substituting equations (6.10.24) and (6.10.25) in equations (6.10.17) and (6.10.18) we get

$$\begin{pmatrix} \mathbf{A}_1 & \mathbf{B}_1 \\ \mathbf{C}_1 & \mathbf{D}_1 \end{pmatrix} \begin{pmatrix} \mathbf{I}_{n+1}^+ \\ \mathbf{I}_n^- \end{pmatrix} = \begin{pmatrix} \mathbf{A}_2 & \mathbf{B}_2 \\ \mathbf{C}_2 & \mathbf{D}_2 \end{pmatrix} \begin{pmatrix} \mathbf{I}_n^+ \\ \mathbf{I}_{n+1}^- \end{pmatrix} + \tau_{n+\frac{1}{2}} \left(1 + \varpi_{n+\frac{1}{2}}\right) \begin{pmatrix} \mathbf{B}_{n+\frac{1}{2}}^+ \\ \mathbf{B}_{n+\frac{1}{2}}^- \end{pmatrix}, \tag{6.10.26}$$

where

$$\mathbf{A}_1 = \mathbf{M} + \frac{1}{2}\tau_{n+\frac{1}{2}}(\mathbf{E} - \mathbf{Q}_{n+\frac{1}{2}}^{++}), \tag{6.10.27}$$

$$\mathbf{B}_1 = -\frac{1}{2}\tau_{n+\frac{1}{2}}\mathbf{Q}_{n+\frac{1}{2}}^{+-}, \tag{6.10.28}$$

$$\mathbf{C}_1 = -\frac{1}{2}\tau_{n+\frac{1}{2}}\mathbf{Q}_{n+\frac{1}{2}}^{-+}, \tag{6.10.29}$$

$$\mathbf{D}_1 = \mathbf{M} + \frac{1}{2}\tau_{n+\frac{1}{2}}(\mathbf{E} - \mathbf{Q}_{n+\frac{1}{2}}^{--}), \tag{6.10.30}$$

$$\mathbf{A}_2 = \mathbf{M} - \frac{1}{2}\tau_{n+\frac{1}{2}}(\mathbf{E} - \mathbf{Q}_{n+\frac{1}{2}}^{++}), \tag{6.10.31}$$

$$\mathbf{B}_2 = \frac{1}{2}\tau_{n+\frac{1}{2}}\mathbf{Q}_{n+\frac{1}{2}}^{+-}, \tag{6.10.32}$$

$$\mathbf{C}_2 = \frac{1}{2}\tau_{n+\frac{1}{2}}\mathbf{Q}_{n+\frac{1}{2}}^{-+}, \tag{6.10.33}$$

$$\mathbf{D}_2 = \mathbf{M} - \frac{1}{2}\tau_{n+\frac{1}{2}}(\mathbf{E} - \mathbf{Q}_{n+\frac{1}{2}}^{--}), \tag{6.10.34}$$

$$\mathbf{Q}_{n+\frac{1}{2}}^{++} = \frac{1}{2}\varpi_{n+\frac{1}{2}}\mathbf{P}_{n+\frac{1}{2}}^{++}\mathbf{C}. \tag{6.10.35}$$

$\mathbf{Q}^{+-}$, $\mathbf{Q}^{-+}$, $\mathbf{Q}^{--}$ are defined similarly. The subscript $n + \frac{1}{2}$ represents the average of the subscripted quantities. Equation (6.10.26) can be written in the form of the interaction principle for the cell $(n, n+1)$ as

$$\begin{pmatrix} \mathbf{I}_{n+1}^+ \\ \mathbf{I}_n^- \end{pmatrix} = \mathbf{K}^{-1}\tau_{n+\frac{1}{2}}\left(1 - \varpi_{n+\frac{1}{2}}\right) \begin{pmatrix} \mathbf{B}_{n+\frac{1}{2}}^+ \\ \mathbf{B}_{n+\frac{1}{2}}^- \end{pmatrix} + \mathbf{K}^{-1} \begin{pmatrix} \mathbf{A}_2^+ & \mathbf{B}_2^+ \\ \mathbf{C}_2^- & \mathbf{D}_2^- \end{pmatrix} \begin{pmatrix} I_n^+ \\ \mathbf{I}_{n+1}^- \end{pmatrix}, \tag{6.10.36}$$

where

$$\mathbf{K} = \begin{pmatrix} \mathbf{A}_1 & \mathbf{B}_1 \\ \mathbf{C}_1 & \mathbf{D}_1 \end{pmatrix}. \tag{6.10.37}$$

The interaction principle for a cell with boundaries n and $n+1$ is

$$\begin{pmatrix} \mathbf{I}_{n+1}^{+} \\ \mathbf{I}_{n}^{-} \end{pmatrix} = \begin{pmatrix} \mathbf{t}(n+1,n) & \mathbf{r}(n,n+1) \\ \mathbf{r}(n+1,n) & \mathbf{t}(n,n+1) \end{pmatrix} \begin{pmatrix} \mathbf{I}_{n}^{+} \\ \mathbf{I}_{n+1}^{-} \end{pmatrix} + \begin{pmatrix} \mathbf{\Sigma}^{+}(n+1,n) \\ \mathbf{\Sigma}^{-}(n,n+1) \end{pmatrix}. \tag{6.10.38}$$

Comparison of equations (6.10.36) and (6.10.38) gives us the cell transmission and reflection operators $\mathbf{t}(n+1,n)$, $\mathbf{r}(n+1,n)$, $\mathbf{r}(n+1,n)$ and $\mathbf{r}(n,n+1)$ and the source vectors $\mathbf{\Sigma}^{+}(n+1,n)$ and $\mathbf{\Sigma}^{-}(n,n+1)$:

$$\mathbf{t}(n+1,n) = \mathbf{t}^{+}\left[\mathbf{\Delta}^{+}\mathbf{S}^{++} + \mathbf{r}^{+-}\mathbf{r}^{-+}\right], \tag{6.10.39}$$

$$\mathbf{t}(n,n+1) = \mathbf{t}^{-}\left[\mathbf{\Delta}^{-}\mathbf{S}^{--} + \mathbf{r}^{-+}\mathbf{r}^{+-}\right], \tag{6.10.40}$$

$$\mathbf{r}(n+1,n) = 2\mathbf{t}^{-}\mathbf{r}^{-+}\mathbf{\Delta}^{+}\mathbf{M}, \tag{6.10.41}$$

$$\mathbf{r}(n,n+1) = 2\mathbf{t}^{+}\mathbf{r}^{+-}\mathbf{\Delta}^{-}\mathbf{M}, \tag{6.10.42}$$

$$\mathbf{\Sigma}^{+}_{n+\frac{1}{2}} = \tau_{n+\frac{1}{2}}\left(1-\varpi_{n+\frac{1}{2}}\right)\mathbf{t}^{+}\left[\mathbf{\Delta}^{+}\mathbf{B}^{+} + \mathbf{r}^{+-}\mathbf{\Delta}^{-}\mathbf{B}^{-}\right], \tag{6.10.43}$$

$$\mathbf{\Sigma}^{-}_{n+\frac{1}{2}} = \tau_{n+\frac{1}{2}}\left(1-\varpi_{n+\frac{1}{2}}\right)\mathbf{t}^{-}\left[\mathbf{\Delta}^{-}\mathbf{B}^{-} + \mathbf{r}^{-+}\mathbf{\Delta}^{+}\mathbf{B}^{+}\right]. \tag{6.10.44}$$

The different terms in the above relations are explained below:

$$\mathbf{t}^{+} = \left[\mathbf{E} - \mathbf{r}^{+-}\mathbf{r}^{-+}\right]^{-1}, \quad \mathbf{t}^{-} = \left[\mathbf{E} - \mathbf{r}^{-+}\mathbf{r}^{+-}\right]^{-1}, \tag{6.10.45}$$

$$\mathbf{r}^{+-} = \mathbf{\Delta}^{+}\mathbf{S}^{+-}, \quad \mathbf{r}^{-+} = \mathbf{\Delta}^{-}\mathbf{S}^{-+}, \tag{6.10.46}$$

$$\mathbf{\Delta}^{+} = \left[\mathbf{M} + \frac{1}{2}\tau_{n+\frac{1}{2}}\left(\mathbf{E} - \mathbf{Q}^{++}_{n+\frac{1}{2}}\right)\right]^{-1}, \tag{6.10.47}$$

$$\mathbf{\Delta}^{-} = \left[\mathbf{M} + \frac{1}{2}\tau_{n+\frac{1}{2}}\left(\mathbf{E} - \mathbf{Q}^{--}_{n+\frac{1}{2}}\right)\right]^{-1}, \tag{6.10.48}$$

$$\mathbf{S}^{++} = \mathbf{M} - \frac{1}{2}\tau_{n+\frac{1}{2}}\left(\mathbf{E} - \mathbf{Q}^{++}_{n+\frac{1}{2}}\right), \tag{6.10.49}$$

$$\mathbf{S}^{--} = \mathbf{M} - \frac{1}{2}\tau_{n+\frac{1}{2}}\left(\mathbf{E} - \mathbf{Q}^{--}_{n+\frac{1}{2}}\right), \tag{6.10.50}$$

$$\mathbf{S}^{-+} = \frac{1}{2}\tau_{n+\frac{1}{2}}\mathbf{Q}^{-+}_{n+\frac{1}{2}}, \tag{6.10.51}$$

$$\mathbf{S}^{+-} = \frac{1}{2}\tau_{n+\frac{1}{2}}\mathbf{Q}^{+-}_{n+\frac{1}{2}}. \tag{6.10.52}$$

6.11 Non-negativity and flux conservation in cell matrices

In the previous section we have derived the transmission and reflection operators in a given cell with boundaries n and $n+1$. We have to show that these operators conserve radiant flux in a conservatively scattering medium. The analysis given here can be found in Grant and Hunt (1969b). Let us neglect all Σs, the source vectors. We should have all the intensities positive ($I^{\pm} \geq 0$, or all its elements are non-negative). When $I^{\pm}$ is incident on a detector it can be characterized by a set $\{d_j : 1 \leq j \leq m\}$ with non-negative weights such that

$$\sum_{j=1}^{m} d_j \left| I^{\pm}(\mu_j) \right|, \tag{6.11.1}$$

or, in the sense of modified vector norm (see Collatz (1966)),

$$S = \left\| I^{\pm} \right\|_D = \left\| D I^{\pm} \right\|_1, \tag{6.11.2}$$

where D is the diagonal matrix whose diagonal elements are $\{d_j\}$. If we set

$$D = 2\pi M c, \tag{6.11.3}$$

then

$$S = \left\| I^{\pm} \right\| = 2\pi \sum_{j=1}^{m} \mu_j c_j \left| I^{\pm}(\mu_j) \right| \tag{6.11.4}$$

is the total flux crossing the plane of stratification through the detectors in the positive sense. The matrix norm consistent and subordinate with equation (6.11.2) is

$$\begin{aligned} \|A\| &= \max_k \sum_{j=1}^{m} \left| \left(D A D^{-1} \right)_{jk} \right| \\ &= \max_k \sum_{j=1}^{m} d_j \left| A_{jk} \right| d_k^{-1}. \end{aligned} \tag{6.11.5}$$

We apply this to the interaction principle (6.5.5) without the source terms. From equation (6.11.2) we can write

$$\left\| \begin{pmatrix} I_n^+ \\ I_{n+1}^- \end{pmatrix} \right\| = \left\| I_n^+ \right\| + \left\| I_{n+1}^- \right\|, \tag{6.11.6}$$

(we have omitted the subscript D for simplicity), where the m-vector norms were defined earlier. Equation (6.11.6) gives the total flux entering the slab with boundaries at n and $n+1$ and $\left[\left\| I_{n+1}^- \right\| + \left\| I_n^- \right\| \right]$ is the total flux emerging from this set. Using equation (6.11.5), we can write,

$$\|S(n, n+1)\| = \max\{\|t(n+1, n) + r(n+1, n)\|,\\ \|t(n, n+1) + r(n, n+1)\|\}. \tag{6.11.7}$$

If $\|S(n, n+1)\| < 1$, then we have

$$\left\{\left\|I_{n+1}^{+}\right\| + \left\|I_{n}^{-}\right\|\right\} < \|S(n, n+1)\| \left\{\left\|I_{n}^{+}\right\| + \left\|I_{n+1}^{-}\right\|\right\}, \tag{6.11.8}$$

which means that the flux emerging from the slab will be less than that entering it, in which case we say that $S(n, n+1)$ is strictly dissipative, as the energy is dissipated in the slab. Furthermore, if $\|S(n, n+1)\| = 1$, we have

$$\left\{\left\|I_{n+1}^{+}\right\| + \left\|I_{n}^{-}\right\|\right\} \le \left\{\left\|I_{n}^{+}\right\| + \left\|I_{n+1}^{-}\right\|\right\}. \tag{6.11.9}$$

When the inequality holds, we say that $S(n, n+1)$ is dissipative and when equality holds it is said to be conservative. When it is conservative, a necessary and sufficient condition is that every column sum s_k, s'_k, such as

$$s_k = \sum_{j=1}^{m} d_j \{t(n+1, n) + r(n+1, n)\}_{jk} \, d_k^{-1} \quad (k = 1, \ldots, m) \tag{6.11.10}$$

and

$$s'_k = \sum_{j=1}^{m} d_j \{t(n, n+1) + r(n, n+1)\}_{jk} \, d_k^{-1} \quad (k = 1, \ldots, m), \tag{6.11.11}$$

should be unity.

Furthermore, if $0 \le \varpi < 1$, then the transmission and reflection operators of the cell will have non-negative elements if (a) the cell thickness τ is such that

$$\tau < \min\left(\frac{\mu_k}{1 - \varpi P_{kk}^{++} c_k}, \frac{\mu_k}{1 - \varpi P_{kk}^{++} c_k}\right). \tag{6.11.12}$$

and (b) $S(n, n+1)$ cannot be anything but dissipative. Property (a) can be seen from the quantities Δ^{+}, Δ^{-} in the equations (6.10.47) and (6.10.48). The matrices should have non-negative diagonally dominant elements and negative off-diagonal elements. This condition also leads to relation (6.11.12). Property (b) can be proved from the following arguments:

$$\|t(n+1, n) + r(n+1, n)\| = \max_k \sum_{j=1}^{m} 2\pi \mu_j c_j \left[\left(1 - \frac{\tau}{\mu_j}\right)\delta_{jk} \right.\\ \left. + \frac{1}{2}\varpi \frac{\tau}{\mu_j}\left(P_{jk}^{++} + P_{jk}^{-+}\right) c_k\right] (2\pi \mu_k c_k)^{-1} + O(\tau)$$

$$= \max_k \left\{1 - \frac{\tau}{\mu_k}\left[1 - \frac{1}{2}\sum c_j \left(P_{jk}^{++} + P_{jk}^{-+}\right)\right]\right\} + O(\tau). \quad (6.11.13)$$

Conservation of flux on scattering gives us relation (6.10.8):

$$\frac{1}{2}\sum_{j=1}^{m}\left(p_{jk}^{++} + p_{jk}^{-+}\right) c_j = 1, \quad (6.11.14)$$

which in turn gives us

$$\|t(n+1,n) + r(n+1,n)\| = \max_k \left[1 - (1-\varpi)\frac{\tau}{\mu_k}\right] + O(\tau) < 1, \quad (6.11.15)$$

which proves property (b). Similarly, we can prove that $\|t(n, n+1) + r(n, n+1)\|$ is also < 1, therefore

$$\|S(n, n+1)\| < 1. \quad (6.11.16)$$

When $\varpi = 1$, we have (the conservative case)

$$\|S(n, n+1)\| = 1 + O(\tau). \quad (6.11.17)$$

The inverses that occur in the star product operators (6.6.8)–(6.6.13) should follow the following property (see Varga (1963)). If A is an arbitrary complex $m \times m$ matrix with spectral radius (minimum of the absolute value of the eigenvalues) $\rho(A)$ less than unity, then $E - A$ is non-singular and

$$(E - A)^{-1} = E + A + A^2 + \cdots; \quad (6.11.18)$$

the series on the RHS converges. One should ensure that this result is satisfied.

6.12 Solution of the spherically symmetric equation

In the previous two sections we have obtained the solution of the radiative transfer equation in a plane parallel medium and studied some of the numerical aspects of the solution. In this section we study a spherically symmetric atmosphere (as shown in figure 6.6) and derive the solution of the radiative transfer equation following Peraiah and Grant (1973).

The transfer equation in spherical symmetry is written (in divergence form) as (see chapter 2)

$$\frac{\mu}{r^2}\frac{\partial}{\partial r}\left\{r^2 I(\tau,\mu)\right\} + \frac{1}{r}\frac{\partial}{\partial \mu}\left\{\left(1-\mu^2\right) I(\tau,\mu)\right\} + \sigma(r) I(r,\mu)$$

$$= \sigma(r)\left\{[1-\varpi(r)]\,b(r) + \frac{1}{2}\varpi(r)\int_{-1}^{1} p(r,\mu,\mu')\,d\mu'\right\}, \tag{6.12.1}$$

where $\sigma(r)$ is the absorption coefficient, $b(r)$ are the sources inside the medium and all other symbols have their usual meanings. The quantities $\sigma(r)$, $\varpi(r)$, $b(r)$ and $p(r, \mu, \mu')$ are generally piecewise continuous functions of their arguments, and

$$b(r) \geq 0,\ \sigma(r) \geq 0, \quad 0 \leq \varpi \leq 1. \tag{6.12.2}$$

The phase function $p(r, \mu, \mu')$ is normalized such that

$$\frac{1}{2}\int_{-1}^{1} p(r,\mu,\mu')\,d\mu' = 1; \quad p(r,\mu,\mu') \geq 0 \quad -1 \leq \mu,\ \mu' \leq 1. \tag{6.12.3}$$

If we write

$$\left.\begin{aligned} U(r,\mu) &= 4\pi r^2 I(r,\mu), \\ B(r) &= 4\pi r^2 b(r), \end{aligned}\right\} \tag{6.12.4}$$

the transfer equation (6.12.1) can be rewritten as

$$\mu\frac{\partial U(r,\mu)}{\partial r} + \frac{1}{r}\frac{\partial}{\partial\mu}\left[(1-\mu^2)U(r,\mu)\right] + \sigma(r)U(r,\mu) =$$
$$\sigma(r)\left\{[1-\varpi(r)]\,B(r) + \frac{1}{2}\varpi(r)\int_{-1}^{1} p(r,\mu,\mu')I(r,\mu')\,d\mu'\right\}, \tag{6.12.5}$$

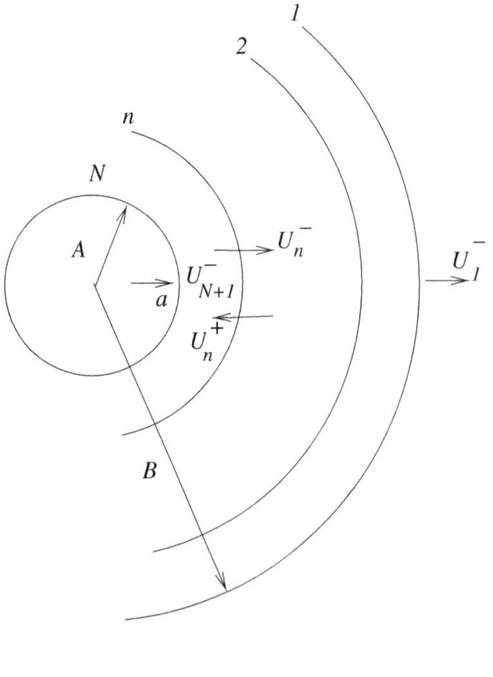

Figure 6.6 Schematic diagram of the diffuse radiation field in spherical symmetry. A and B are the inner and outer radii of the spherically symmetric atmosphere. $B/A = 1$ is the plane parallel case.

and for the oppositely directed beam, the transfer equation (6.12.1) can then be rewritten as

$$-\mu\frac{\partial U(r,-\mu)}{\partial r}-\frac{1}{r}\frac{\partial}{\partial\mu}\left[(1-\mu^2)U(r,-\mu)\right]+\sigma(r)U(r,-\mu)=$$
$$\sigma(r)\left\{[1-\varpi(r)]\,B(r)+\frac{1}{2}\varpi(r)\int_{-1}^{1}p(r,-\mu,\mu')I(r,\mu')\,d\mu'\right\}. \tag{6.12.6}$$

Here μ lies in the interval (0, 1). We shall employ the same 'cell' type integration that we used in the previous section. This integration is done on the 'cell' $\left[r_n, r_{n+1}\right]$ $\left[\mu_{j-\frac{1}{2}}, \mu_{j+\frac{1}{2}}\right]$ defined on a two-dimensional grid (which should not be confused with X–Y geometry) (Carlson 1963, Lathrop and Carlson 1967). We have

$$\mu_{j+\frac{1}{2}}=\sum_{k=1}^{j}c_k,\quad j=1,2,\ldots,J. \tag{6.12.7}$$

Here the μs and cs are the roots and weights of the Gauss–Legendre quadrature formula. We define the cell boundary by writing $\mu_{\frac{1}{2}}=0,\ \mu_0=-\mu_1$. It can been seen that $\mu_{j-\frac{1}{2}}\le\mu_j\le\mu_{j+\frac{1}{2}}$. By performing integration on the μ grid, we obtain

$$\begin{aligned}c_j\mu_j\frac{\partial U_j^+(r)}{\partial r}&+\frac{1}{r}\left\{\left(1-\mu^2_{j+\frac{1}{2}}\right)U^+_{j+\frac{1}{2}}(r)\right.\\&\left.-\left(1-\mu^2_{j-\frac{1}{2}}\right)U_{j-\frac{1}{2}}(r)\right\}+c_j\sigma(r)U_j^+(r)\\&=\sigma(r)c_j\left\{[1-\varpi(r)]\,B^+(r)\right.\\&\left.+\frac{1}{2}\varpi(r)\sum_{j'=1}^{J}\left[p^{++}(r)_{jj'}c_{j'}U^+_{j'}(r)+p^{+-}(r)_{jj'}c_{j'}U^-_{j'}(r)\right]\right\},\end{aligned} \tag{6.12.8}$$

where $U_j^+(r)=U(r,\mu_j)$, $U_j^-(r)=U(r,-\mu_j)$, $p^{++}(r)_{jj'}=p(r,\mu_j,\mu_{j'})$, $p^{-+}(r)_{jj'}=p(r,-\mu_j,\mu_{j'})$ etc. A similar expression for equation (6.12.6) can be obtained after integration on the μ grid.

The quantities $U^+_{j+\frac{1}{2}}$, $U^+_{j-\frac{1}{2}}$, $U^-_{j+\frac{1}{2}}$ and $U^-_{j-\frac{1}{2}}$ are defined by the interpolation formula with some loss of accuracy:

$$U^{\pm}_{j+\frac{1}{2}}=\frac{\left(\mu_{j+1}-\mu_{j+\frac{1}{2}}\right)U^{\pm}_j+\left(\mu_{j+\frac{1}{2}}-\mu_j\right)U^{\pm}_{j+1}}{m_{j+1}-\mu_j},\quad j=1,2,\ldots,J-1. \tag{6.12.9}$$

We shall also set $U^+_{\frac{1}{2}}=U^-_{\frac{1}{2}}$ by interpolation or

$$U^+_{\frac{1}{2}}=U^-_{\frac{1}{2}}=\frac{1}{2}\left(U^+_1+U^-_1\right). \tag{6.12.10}$$

Let us write

$$\mathbf{U}^{\pm}(r) = \left[U_1^{\pm}(r), \ldots, U_J^{\pm}(r)\right]^T . \tag{6.12.11}$$

Using the matrices (6.12.9), (6.12.10), (6.12.11) we can write equation (6.12.8) in matrix form as

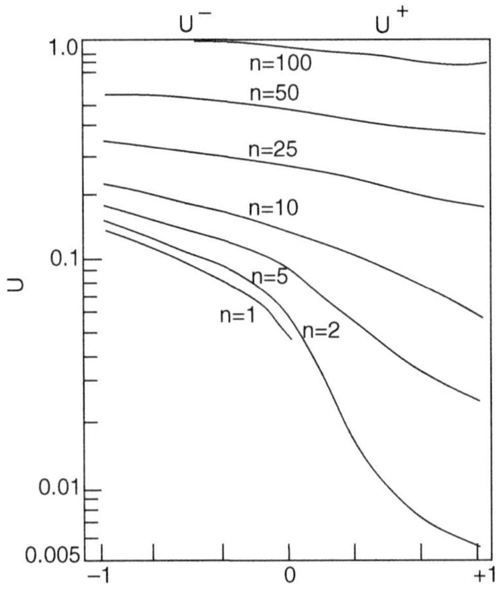

Figure 6.7 Angular distribution of the run of specific intensities from bottom ($n = 100$) to the top ($n = 1$)of the atmosphere in the plane parallel case. Here $\tau = 10$, $B/A = 1.0$, where B and A are the outer and inner radii of the atmosphere, $N = 100$, $\varpi = 1$ $p_{jk} = 1$ for j and k (isotropic scattering). The initial conditions are $U_1^+ = 0$ and $U_{N+1}^- = 1$ (from Peraiah (1971), with permission).

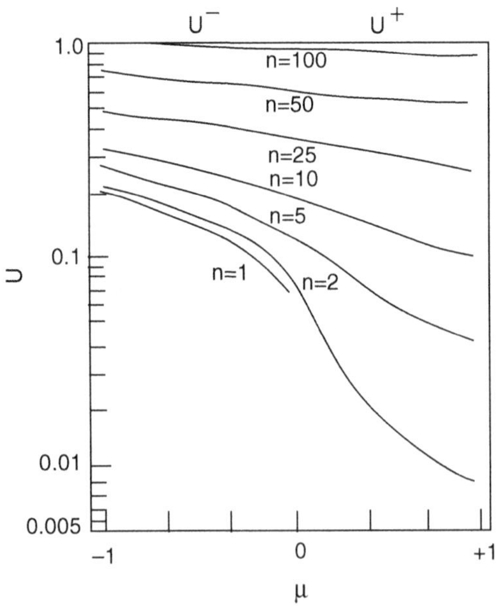

Figure 6.8 Angular distribution of U in a spherically symmetric atmosphere at different boundaries inside the atmosphere ($n = 100$ to $n = 1$), $\tau = 10$, $B/A = 1.5$, $N = 100$, $\varpi = 1$ (from Peraiah (1971), with permission). The initial conditions are $U_1^+ = 0$ and $U_{N+1}^- = 1$.

$$\mathbf{M}\frac{\partial \mathbf{U}^{+}(r))}{\partial r} + \frac{1}{r}\left[\boldsymbol{\Lambda}^{+}\mathbf{U}^{+}(r) + \boldsymbol{\Lambda}^{-}\mathbf{U}^{-}(r)\right] + \sigma(r)\mathbf{U}^{+}(r) =$$
$$\sigma(r)\left\{[1-\varpi(r)]\,\mathbf{B}^{+}(r) + \frac{1}{2}\varpi(r)\left[\mathbf{p}^{++}(r)\mathbf{C}\mathbf{U}^{+}(r) + \mathbf{p}^{+-}(r)\mathbf{C}\mathbf{U}^{-}(r)\right]\right\}, \tag{6.12.12}$$

and similarly

$$-\mathbf{M}\frac{\partial \mathbf{U}^{+}(r))}{\partial r} - \frac{1}{r}\left[\boldsymbol{\Lambda}^{+}\mathbf{U}^{-}(r) + \boldsymbol{\Lambda}^{-}\mathbf{U}^{+}(r)\right] + \sigma(r)\mathbf{U}^{-}(r) =$$
$$\sigma(r)\left\{[1-\varpi(r)]\,\mathbf{B}^{-}(r) + \frac{1}{2}\varpi(r)\left[\mathbf{p}^{-+}(r)\mathbf{C}\mathbf{U}^{+}(r) + \mathbf{p}^{--}(r)\mathbf{C}\mathbf{U}^{-}(r)\right]\right\}, \tag{6.12.13}$$

where C and M are the diagonal matrices given by

$$\mathbf{C} = \left[c_j\delta_{jj'}\right], \quad \mathbf{M} = \left[\mu_j\delta_{jj'}\right], \tag{6.12.14}$$

and B^{+}, B^{-} are the source vectors similarly defined as in equation (6.12.11). The matrices Λ^{+} and Λ^{-} are $J \times J$ matrices, which we call curvature scattering matrices. These are given by

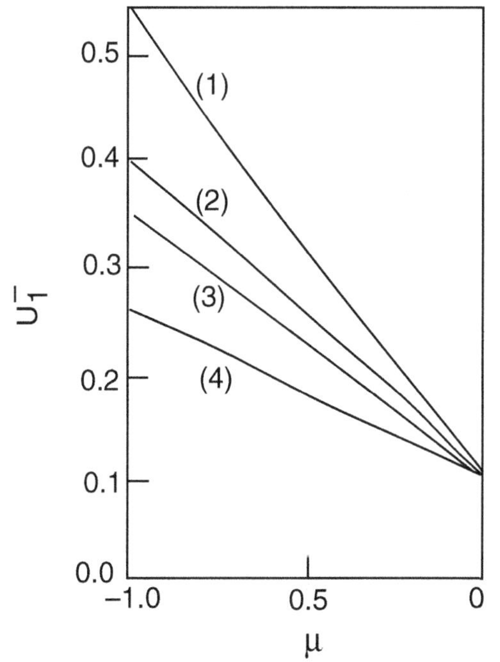

Figure 6.9 Angular distribution of the emergent intensities for $\tau = 5$, $\varpi = 1$ isotropic scattering: (1) $B/A = 2.0$; (2) $B/A = 1.5$; (3) $B/A = 1.3$; (4) $B/A = 1.0$. The last one is the plane parallel case. The initial conditions are $U_1^{+} = 0$ and $U_{N+1}^{-} = 1$ (from Peraiah (1971), with permission).

$$c_j \Lambda^+_{jk} = \begin{cases} \dfrac{\left(1-\mu^2_{j+\frac{1}{2}}\right)\left(\mu_{j+\frac{1}{2}}-\mu_j\right)}{(\mu_{j+1}-\mu_j)}, & k=j+1,\ j=1,2,\dots,J-1 \\ \dfrac{\left(1-\mu^2_{j+\frac{1}{2}}\right)\left(\mu_{j+1}-\mu_{j+\frac{1}{2}}\right)}{(\mu_{j+1}-\mu_j)} - \dfrac{\left(1-\mu^2_{j-\frac{1}{2}}\right)\left(\mu_{j-\frac{1}{2}}-\mu_{j-1}\right)}{\left(\mu_j-\mu_{j-1}\right)}, & k=j,\ j=1,2,\dots,J \\ -\dfrac{\left(1-\mu^2_{j-\frac{1}{2}}\right)\left(\mu_j-\mu_{j-\frac{1}{2}}\right)}{\left(\mu_j-\mu_{j-1}\right)}, & k=j-1,\ j=2,3,\dots,J \end{cases} \tag{6.12.15}$$

and

$$c_j \Lambda^-_{jk} = -\frac{1}{2}\delta_{j,1}\delta_{k,1}. \tag{6.12.16}$$

Equations (6.12.12) and (6.12.13) are then integrated on the radial grid giving

$$\mathbf{M}\left[\mathbf{U}^+_{n+1}-\mathbf{U}^+_n\right]+\tau_{n+\frac{1}{2}}\mathbf{U}^+_{n+\frac{1}{2}} = \tau_{n+\frac{1}{2}}\Bigg[\left(1-\varpi_{n+\frac{1}{2}}\right)B^+_{n+\frac{1}{2}} + \left(\frac{1}{2}\varpi_{n+\frac{1}{2}}p^{++}_{n+\frac{1}{2}}\mathbf{C}-\frac{\rho\Lambda^+}{\tau_{n+\frac{1}{2}}}\right)\mathbf{U}^+_{n+\frac{1}{2}} + \left(\frac{1}{2}\varpi_{n+\frac{1}{2}}p^{+-}_{n+\frac{1}{2}}-\frac{\rho\Lambda^-}{\tau_{n+f12}}\right)\mathbf{U}^-_{n+\frac{1}{2}}\Bigg], \tag{6.12.17}$$

$$\mathbf{M}\left[\mathbf{U}^-_{n}-\mathbf{U}^-_{n+1}\right]+\tau_{n+\frac{1}{2}}\mathbf{U}^-_{n+\frac{1}{2}} = \tau_{n+\frac{1}{2}}\Bigg[\left(1-\varpi_{n+\frac{1}{2}}\right)B^-_{n+\frac{1}{2}} + \left(\frac{1}{2}\varpi_{n+\frac{1}{2}}p^{-+}_{n+\frac{1}{2}}\mathbf{C}-\frac{\rho\Lambda^-}{\tau_{n+\frac{1}{2}}}\right)\mathbf{U}^+_{n+\frac{1}{2}} + \left(\frac{1}{2}\varpi_{n+\frac{1}{2}}p^{--}_{n+\frac{1}{2}}-\frac{\rho\Lambda^+}{\tau_{n+\frac{1}{2}}}\right)\mathbf{U}^-_{n+\frac{1}{2}}\Bigg], \tag{6.12.18}$$

where $U^+_n = U^+(rn)$ and the variables with subscript $n+\frac{1}{2}$ are averages over the cell. Furthermore,

$$\left.\begin{aligned} \Delta r_{n+\frac{1}{2}} &= r_{n+1}-r_n, \\ \tau_{n+\frac{1}{2}} &= \sigma_{n+\frac{1}{2}}\Delta r_{n+\frac{1}{2}}, \\ \rho &= \Delta r_{n+\frac{1}{2}}/r_{n+\frac{1}{2}}, \end{aligned}\right\} \tag{6.12.19}$$

where $r_{n+\frac{1}{2}}$ is a suitable average of r_n and r_{n+1} such as $\frac{1}{2}(r_n+r_{n+1})$. The quantities $U^\pm_{n+\frac{1}{2}}$ are expressed in terms of $U^\pm_n$ and $U^\pm_{n+1}$ using the diamond difference scheme (see equations (6.10.23)–(6.10.25)) and are given by

$$U^+_{n+\frac{1}{2}} = \frac{1}{2}\left(U^+_n+U^+_{n+1}\right), \quad U^-_{n+\frac{1}{2}} = \frac{1}{2}\left(U^-_n+U^-_{n+1}\right). \tag{6.12.20}$$

After substituting equations (6.12.20) into equations (6.12.17) and (6.12.18) and comparing the resulting equations with the principle of interaction

$$\begin{pmatrix} U^+_{n+1} \\ U^-_n \end{pmatrix} = \begin{pmatrix} t(n+1,n) & r(n,n+1) \\ r(n+1,n) & t(n,n+1) \end{pmatrix} \begin{pmatrix} U^+_n \\ U^-_{n+1} \end{pmatrix} + \begin{pmatrix} \Sigma^+_{n+\frac{1}{2}} \\ \Sigma^-_{n+\frac{1}{2}} \end{pmatrix}, \tag{6.12.21}$$

we obtain the transmission and reflection operators of the cell. These are given in equations (6.10.39)–(6.10.52) with changes in the Qs. The Qs in spherical symmetry are

$$\left.\begin{aligned} \mathbf{Q}^{++} &= \frac{1}{2}\varpi_{n+\frac{1}{2}}\mathbf{p}^{++}\mathbf{C} - \frac{\rho\mathbf{\Lambda}^+}{\tau_{n+\frac{1}{2}}}, \\ \mathbf{Q}^{--} &= \frac{1}{2}\varpi_{n+\frac{1}{2}}\mathbf{p}^{--}\mathbf{C} + \frac{\rho\mathbf{\Lambda}^+}{\tau_{n+\frac{1}{2}}}, \\ \mathbf{Q}^{-+} &= \frac{1}{2}\varpi_{n+\frac{1}{2}}\mathbf{p}^{-+}\mathbf{C} + \frac{\rho\mathbf{\Lambda}^-}{\tau_{n+\frac{1}{2}}}, \\ \mathbf{Q}^{++} &= \frac{1}{2}\varpi_{n+\frac{1}{2}}\mathbf{p}^{+-}\mathbf{C} - \frac{\rho\mathbf{\Lambda}^-}{\tau_{n+\frac{1}{2}}}. \end{aligned}\right\} \tag{6.12.22}$$

$p_{ij} = 1$ for isotropic scattering. The curvature scattering matrices Λ^+ and Λ^- are given below for $j = 2$ and 4. For $j = 2$, $\mu_1 = 0.211\,32$, $\mu_2 = 0.788\,68$, $c_1 = c_2 = 0.5$ (see Abramowitz and Stegun (1965), page 921).

$$\Lambda^+(J=2) = \begin{pmatrix} -0.25 & 0.75 \\ -0.75 & -0.75 \end{pmatrix}, \quad \Lambda^-(J=2) = \begin{pmatrix} -1 & 0 \\ 0 & 0 \end{pmatrix}: \tag{6.12.23}$$

for $J = 4$,

$$\mu_1 = 0.069\,43, \quad \mu_2 = 0.330\,01 \quad \mu_3 = 0.669\,99, \quad \mu_4 = 0.930\,57,$$

$$c_1 = 0.173\,93, \quad c_2 = 0.326\,07 \quad c_3 = 0.326\,07, \quad c_4 = 0.173\,93,$$

$$\Lambda^+(J=4) = \begin{pmatrix} 0.464\,94 & 2.235\,90 & 0 & 0 \\ -1.781\,39 & -0.042\,58 & 1.150\,05 & 0 \\ 0 & -1.150\,05 & -0.759\,45 & 0.583\,43 \\ 0 & 0 & -0.732\,228 & -1.093\,79 \end{pmatrix}$$

and

$$\Lambda^-_{jk} = -2.874\,76\delta_{j1}\delta_{k1}. \tag{6.12.24}$$

We now present the two pairs of transmission and reflection operators for a basic cell for a given curvature factor $\rho = \Delta r/\bar{r}$, $\Delta\tau <$ critical optical depth. For $\rho = 2.83 \times 10^{-3}$, $\Delta\tau = 0.05$, $B^+ = B^- = 0.0$, $U^-_1(\mu_j) = 1.0$, $\varpi = 1.0$,

$$t(n+1,n)=\begin{pmatrix} .2501E+00 & .7727E-01 & .1233E+00 & .6737E-01 \\ .1839E-01 & .7768E+00 & .3549E-01 & .2299E-01 \\ .6657E-02 & .2355E-01 & .8856E+00 & .1018E-01 \\ .4886E-02 & .1488E-01 & .1782E-01 & .9094E+00 \end{pmatrix},$$

$$t(n,n+1)=\begin{pmatrix} .2589E+00 & .1472E+00 & .1209E+00 & .6583E-01 \\ .6683E-02 & .7758E+00 & .4741E-01 & .2218E-01 \\ .6855E-02 & .1729E-01 & .8808E+00 & .1374E-01 \\ .5053E-02 & .1514E-01 & .1450E-01 & .9043E+00 \end{pmatrix},$$

$$r(n+1,n)=\begin{pmatrix} .8392E-02 & .1157E+00 & .1230E+00 & .6683E-01 \\ .1288E-01 & .3774E-01 & .4084E-01 & .2224E-01 \\ .6736E-02 & .2020E-01 & .2189E-01 & .1193E-01 \\ .4941E-02 & .1486E-01 & .1611E-01 & .8779E-02 \end{pmatrix},$$

$$r(n,n+1)=\begin{pmatrix} .7684E-01 & .1212E+00 & .1267E+00 & .6864E-01 \\ .1384E-01 & .3999E-01 & .4250E-01 & .2308E-01 \\ .7070E-02 & .2084E-01 & .2217E-01 & .1204E-01 \\ .5169E-02 & .1527E-01 & .1625E-01 & .8834E-02 \end{pmatrix}.$$

There are no sources as $\varpi = 1$ and therefore $\Sigma^+ = \Sigma^- = 0$.

From the condition of stability that we derived in the previous section, we must have $\Delta^+ \geq 0$, $\Delta^- \geq 0$, $S^{+-} \geq 0$, $S^{-+} \geq 0$. This is possible if

$$\tau \leq \tau_{crit} = \min_j \left| \frac{\left(\mu_j \pm \frac{1}{2}\rho\Lambda^+_{jj}\right)}{\frac{1}{2}\left(1 - \frac{1}{2}\varpi P^{++}_{jj} c_j\right)} \right| \tag{6.12.25}$$

for diagonal elements, and for off-diagonal elements

$$\frac{\rho}{\tau} < \min_j \left[\min_{k=j+1} \left| \frac{\frac{1}{2}\varpi p_{jk} c_k}{\Lambda^+_{jk}} \right| \right], \tag{6.12.26}$$

where p_{jk} is either p^{+-}_{jk} or p^{-+}_{jk}. The conservation of flux requires that

$$\frac{1}{2}\sum_{j=1}^{m} c_j \left(p^{++}_{jk} + P^{-+}_{jk}\right) = 1 \tag{6.12.27}$$

and

$$\sum_{j=1}^{J} c_j \left(\Lambda^+_{jk} - \Lambda^-_{jk}\right) = 0. \tag{6.12.28}$$

The solution of the transfer equation is obtained by dividing the medium into a number of shells or 'cells' in each of which the optical depth satisfies relation (6.12.25) and (6.12.26). The cell operators given in equations (6.10.39)–(6.10.44) are computed for all the cells or shells. In all cases we choose a shell whose thickness is larger than the τ_{crit} (see equation (6.12.25)), then divide this into smaller shells each of whose optical thickness satisfies relation (6.12.25) and compute the r and t operators for the composite shell by using the star algorithm given in equations (6.6.8)–(6.6.13). When the r and t operators for the cell are known, one can compute the intensities at the boundaries of any shell inside the atmosphere using the algorithm of the internal field given by equations (6.8.33) and (6.8.34). We present some results for the spherical case in figures 6.7, 6.8 and 6.9.

Table 6.1 clearly shows that the flux in a conservatively scattering atmosphere is conserved exactly.

6.13 Solution of line transfer in spherical symmetry

We shall now apply the discrete space theory to compute non-LTE spectral lines in plane parallel and spherically symmetric atmospheres (see Grant and Peraiah (1972)). The transfer equation for lines for a two-level atom is written as (see Jefferies (1970), Mihalas (1978))

$$\mu\frac{\partial I(r,x,\mu)}{\partial \mu} + \frac{1-\mu^2}{r}\frac{\partial I(r,x,\mu)}{\partial \mu}$$
$$= k_L(r)\,[\beta + \phi(x)]\,[S(r,x) - I(r,x,\mu)] \tag{6.13.1}$$

and

$$-\mu\frac{\partial I(r,x,-\mu)}{\partial \mu} - \frac{1-\mu^2}{r}\frac{\partial I(r,x,-\mu)}{\partial \mu}$$
$$= k_L(r)\,[\beta + \phi(x)]\,[S(r,x) - I(r,x,-\mu)], \tag{6.13.2}$$

Table 6.1 Global conservation for a conservative isotropically scattering shell illuminated at the inner most boundary $r = A$. The columns give the total flux from the shell at radii A and B and satisfy the equation $F^+(A) + F^-(B) = F^-(A) = \pi$. No incident radiation is given at B, that is, $F^+(B) = 0$. Calculations are based on eight-point Gauss–Legendre quadrature (Peraiah 1971).

B/A	$\tau = 2$		$\tau = 5$		$\tau = 10$	
	$\frac{F^-(B)}{2\pi}$	$\frac{F^+(A)}{2\pi}$	$\frac{F^-(B)}{2\pi}$	$\frac{F^+(A)}{2\pi}$	$\frac{F^-(B)}{2\pi}$	$\frac{F^+(A)}{2\pi}$
1.0	0.195 03	0.304 97	0.103 83	0.346 17	0.053 87	0.441 63
1.3	0.246 17	0.253 83	0.133 54	0.366 46	0.075 50	0.424 50
1.5	0.273 25	0.226 75	0.151 77	0.348 23	0.086 50	0.413 50
1.7	0.296 11	0.203 89	0.168 81	0.331 19	0.097 16	0.402 84
2.0	0.324 39	0.175 61	0.192 27	0.307 73	0.112 54	0.387 46

where $I(r, x, \mu)$ is the specific intensity at the radial point r for the normalized frequency x given by

$$x = \frac{\nu - \nu_0}{\Delta_s}, \tag{6.13.3}$$

where Δ_s is some standard frequency interval. The quantity $\phi(x)$ is the profile function of the line (this can be Doppler, Lorentz, Voigt or any other shape (see chapter 1)) normalized such that

$$\int_{-\infty}^{\infty} \phi(x)\, dx = 1. \tag{6.13.4}$$

The quantity β is the ratio k_c/k_L of opacity due to continuous absorption per unit frequency interval Δ_s to that in the line. The source function $S(x, r)$ is given by

$$S(x, r) = \frac{\phi(x)}{\beta + \phi(x)} S_L(r) + \frac{\beta}{\beta + \phi(x)} S_C(r), \tag{6.13.5}$$

where $S_C(r)$ is the continuum source function which in the stellar atmospheric situation can be written

$$S_C(r) = \rho(r) B(\nu_0, T(r)), \tag{6.13.6}$$

where $B(\nu_0, T(r))$ is the Planck function at frequency ν_0 and temperature $T(r)$ at radius r. In the present calculations, we assume that both ρ (which is arbitrary < 1) and B are known functions of r. The line source function $S_L(r)$ is

$$S_L(r) = \frac{A_{21} N_2(r)}{(B_{12} N_1(r) - B_{21} N_2(r))}, \tag{6.13.7}$$

where A_{21}, B_{12}, B_{21} are the Einstein coefficients and $N_1(r)$, $N_2(r)$ are the population densities of the lower and upper states respectively. The statistical equilibrium equation for the two-level atom is

$$N_1 \left[B_{12} \int_{-\infty}^{\infty} \phi(x) J(x)\, dx + C_{12} \right] = N_2 \left[A_{21} + C_{21} + B_{21} \int_{-\infty}^{\infty} \phi(x) J(x)\, dx \right], \tag{6.13.8}$$

where the Cs are collisional parameters and $J(x)$ is the mean intensity. If we combine equations (6.13.7) and (6.13.8) we get

$$S_L(r) = (1 - \epsilon) \int_{-\infty}^{\infty} \phi(x) J(x)\, dx + \epsilon B, \tag{6.13.9}$$

where

$$\epsilon = \frac{C_{21}}{\left\{ C_{21} + A_{21} \left[1 - \exp(-h\nu_0/kT) \right]^{-1} \right\}} \tag{6.13.10}$$

is the probability per scatter that a photon will be destroyed by collisional de-excitation. The optical depth can be written as

$$d\tau = k_L(r)\,dr = \frac{h\nu_0}{4\pi\,\Delta_s}(N_1 B_{12} - N_2 B_{21})\,dr. \tag{6.13.11}$$

We need to find the solution of equations (6.13.1) and (6.13.2) and we assume that the quantities $\beta,\ \rho,\ \epsilon,\ \phi$ are known. We write

$$\left.\begin{aligned} \mathbf{h} &= [1, 1, \ldots]^{-1}, \\ \mathbf{U}^{\pm}_{i,n} &= 4\pi r_n^2 \left[I(\pm\mu_1, x_i, \tau_n), \ldots, I(\pm\mu_m, x_i; \tau_n)\right]^T. \end{aligned}\right\} \tag{6.13.12}$$

Then, we integrate equation (6.13.1) as was done in the previous sections and get

$$\begin{aligned} &\mathbf{M}_m\left(\mathbf{U}^{+}_{i,n+1} - \mathbf{U}^{+}_{i,n}\right) + \rho_c\left[\mathbf{\Lambda}^{+}_m \mathbf{U}^{+}_{i,n+\frac{1}{2}}(r) + \mathbf{\Lambda}^{-}_m \mathbf{U}^{-}_{i,n+\frac{1}{2}}(r)\right] \\ &\quad + \tau_{n+\frac{1}{2}}(\beta+\phi)_{i,n+\frac{1}{2}}\mathbf{U}^{+}_{i,n+\frac{1}{2}} = \tau_{n+\frac{1}{2}}(\rho\beta + \epsilon\phi_i)_{n+\frac{1}{2}} B_{n+\frac{1}{2}}\mathbf{h} \\ &\quad + \frac{1}{2}\tau_{n+\frac{1}{2}}\sigma_{n+\frac{1}{2}}\phi_{i,n+\frac{1}{2}}\sum_{i'=-I}^{I} a_{i',n+\frac{1}{2}}(\mathbf{h}\mathbf{h}^T)\mathbf{b}\left[\mathbf{U}^{+} + \mathbf{U}^{-}\right]_{i',n+\frac{1}{2}}, \end{aligned} \tag{6.13.13}$$

where $\mathbf{b}$ and $\mathbf{M}_m$ are the $m \times m$ matrices of the quadrature weights and roots of the angle quadrature given by

$$\mathbf{b} = \left[b_j\delta_{ij}\right], \quad \mathbf{M}_m = \left[\mu_j\delta_{ij}\right] \tag{6.13.14}$$

and ρ_c is the curvature factor given by

$$\rho_c = \frac{\Delta r}{r}. \tag{6.13.15}$$

The quantities Λ^{+}_m, Λ^{-}_m are the curvature scattering matrices. Furthermore,

$$\sigma_{n+\frac{1}{2}} = 1 - \epsilon_{n+\frac{1}{2}}, \quad \tau_{n+\frac{1}{2}} = \tau_{n+1} - \tau_n. \tag{6.13.16}$$

A similar expression for equation (6.13.2) can be written after integration. We now define an index k corresponding to each (i, j) as

$$(i, j) \equiv k = j + (i-1)m, \quad 1 \le K = mI, \tag{6.13.17}$$

where i, j refer to the running indices of the frequency and the angle and m and I refer to the total numbers of angles and frequencies respectively. For each value of k, we define a coefficient $W_k = W_{k,n+\frac{1}{2}}$ by

$$\phi_k W_k = a_i b_j, \tag{6.13.18}$$

where the a_is are the weights of the frequency quadrature formula. Let

$$\mathbf{U}^{\pm}_n(L) = \left[\mathbf{U}^{\pm}_{1,n}, \ldots, \mathbf{U}^{\pm}_{-I,n}\right]^T, \quad \mathbf{U}^{\pm}_n(R) = \left[\mathbf{U}^{\pm}_{1,n}, \ldots, \mathbf{U}^{\pm}_{I,n}\right]^T. \tag{6.13.19}$$

The K-vectors correspond to frequencies to the left (L) and the right (R) of the line centre. Let $\phi_{n+\frac{1}{2}}$, $g_{n+\frac{1}{2}}$ be K-vectors, where the elements of $g_{n+\frac{1}{2}}$ are defined by

$$g_{n+\frac{1}{2}} = (\rho\beta + \epsilon\phi_k)_{n+\frac{1}{2}} B_{n+\frac{1}{2}}. \tag{6.13.20}$$

Let $\mathbf{\Phi}$, $\mathbf{W}$, $\mathbf{M}$ be $K \times K$ matrices defined by

$$\mathbf{\Phi}_{n+\frac{1}{2}} = [\Phi_{kk'}] = \left[(\beta + \phi_k)_{n+\frac{1}{2}} \delta_{kk'}\right], \tag{6.13.21}$$

$$\mathbf{W}_{n+\frac{1}{2}} = \left[W_{k,n+\frac{1}{2}} \delta_{kk'}\right] \tag{6.13.22}$$

and

$$\mathbf{M} = \begin{pmatrix} \mathbf{M}_m & & & \\ & \mathbf{M}_m & & \\ & & \ddots & \\ & & & \mathbf{M}_m \end{pmatrix} \quad \text{and} \quad \mathbf{\Lambda}^{\pm} = \begin{pmatrix} \mathbf{\Lambda}_m^{\pm} & & & \\ & \mathbf{\Lambda}_m^{\pm} & & \\ & & \ddots & \\ & & & \mathbf{\Lambda}_m^{\pm} \end{pmatrix}. \tag{6.13.23}$$

Now the line transfer equation (6.13.13) and another equation similar to it can be written after radial integration

$$\begin{aligned} &\mathbf{M}\left[\mathbf{U}_{n+1}^{+}(R) - \mathbf{U}_n^{+}(R)\right] + \rho_c\left[\mathbf{\Lambda}^{+}\mathbf{U}_{n+\frac{1}{2}}^{+}(R) + \mathbf{\Lambda}^{-}\mathbf{U}_{n+\frac{1}{2}}^{-}(R)\right] \\ &\quad + \tau_{n+\frac{1}{2}} \Phi_{n+\frac{1}{2}} \mathbf{U}_{n+\frac{1}{2}}^{+}(R) = \tau_{n+\frac{1}{2}} \mathbf{g}_{n+\frac{1}{2}}(R) \\ &\quad + \frac{1}{2}\left[\tau\sigma\boldsymbol{\phi}\boldsymbol{\phi}^T W\right]_{n+\frac{1}{2}} \left[\mathbf{U}^{+}(L) + \mathbf{U}^{+}(R) + \mathbf{U}^{-}(L) + \mathbf{U}^{-}(R)\right]_{n+\frac{1}{2}}, \end{aligned} \tag{6.13.24}$$

and

$$\begin{aligned} &\mathbf{M}\left[\mathbf{U}_n^{-}(R) - \mathbf{U}_{n+1}^{-}(R)\right] - \rho_c\left[\mathbf{\Lambda}^{+}\mathbf{U}_{n+\frac{1}{2}}^{-}(R) + \mathbf{\Lambda}^{-}\mathbf{U}_{n+\frac{1}{2}}^{+}(R)\right] \\ &\quad + \tau_{n+\frac{1}{2}} \Phi_{n+\frac{1}{2}} \mathbf{U}_{n+\frac{1}{2}}^{-}(R) = \tau_{n+\frac{1}{2}} \mathbf{g}_{n+\frac{1}{2}}(R) \\ &\quad + \frac{1}{2}\left[\tau\sigma\boldsymbol{\phi}\boldsymbol{\phi}^T W\right]_{n+\frac{1}{2}} \left[\mathbf{U}^{+}(L) + \mathbf{U}^{+}(R) + \mathbf{U}^{-}(L) + \mathbf{U}^{-}(R)\right]_{n+\frac{1}{2}}, \end{aligned} \tag{6.13.25}$$

with two similar equations with L and R interchanged. If the line profile is symmetric, we have $\mathbf{U}^{\pm}(R) = \mathbf{U}^{\pm}(L)$, $\mathbf{g}(L) = \mathbf{g}(R)$ and therefore we need to solve only half the number of equations. By using the diamond scheme of interpolation to replace $U_{n+\frac{1}{2}}^{+}$ etc.:

$$\mathbf{U}_{n+\frac{1}{2}}^{\pm} = \frac{1}{2}\left(\mathbf{U}_n^{\pm} + \mathbf{U}_{n+1}^{\pm}\right), \tag{6.13.26}$$

equations (6.13.24) and (6.13.25) can be written as

$$\begin{pmatrix} \mathbf{U}^+_{n+1} \\ \mathbf{U}^-_n \end{pmatrix} = \mathbf{H}^{-1} \begin{pmatrix} \mathbf{M} - \frac{\tau}{2} Z_+ & \frac{\tau}{2}\mathbf{Y}_- \\ \frac{\tau}{2}\mathbf{Y}_+ & \mathbf{M} - \frac{\tau}{2} Z_- \end{pmatrix} \begin{pmatrix} \mathbf{U}^+_n \\ \mathbf{U}^-_{n+1} \end{pmatrix} + \mathbf{H}^{-1} \begin{pmatrix} \mathbf{1} \\ \mathbf{1} \end{pmatrix} \mathbf{g}, \tag{6.13.27}$$

where

$$\mathbf{H} = \begin{pmatrix} \mathbf{M} + \frac{\tau}{2} Z_+ & -\frac{\tau}{2}\mathbf{Y}_- \\ -\frac{\tau}{2}\mathbf{Y}_+ & \mathbf{M} + \frac{\tau}{2} Z_- \end{pmatrix} \tag{6.13.28}$$

and

$$\left.\begin{aligned} Z_+ &= \left(\mathbf{Z} + \frac{\rho_c}{\mathbf{\Lambda}^+}\tau\right), \quad Z_- = \left(\mathbf{Z} - \frac{\rho_c}{\mathbf{\Lambda}^+}\tau\right), \\ Y_+ &= \left(\mathbf{Y} + \frac{\rho_c}{\mathbf{\Lambda}^-}\tau\right), \quad Y_- = \left(\mathbf{Y} - \frac{\rho_c}{\mathbf{\Lambda}^-}\tau\right), \\ \mathbf{Y} &= [\sigma \phi \phi^T]\mathbf{W} \quad \text{and} \quad \mathbf{Z} = \mathbf{\Phi} - \mathbf{Y}. \end{aligned}\right\} \tag{6.13.29}$$

Now, we write the sets of transmission and reflection matrices as

$$\left.\begin{aligned} \mathbf{t}(n+1, n) &= \mathbf{R}^{+-}\left[\mathbf{\Delta}^+\mathbf{A} + \mathbf{r}^{+-}\mathbf{\Delta}^-\mathbf{C}\right], \\ \mathbf{t}(n, n+1) &= \mathbf{R}^{-+}\left[\mathbf{\Delta}^-\mathbf{D} + \mathbf{r}^{-+}\mathbf{\Delta}^+\mathbf{B}\right], \\ \mathbf{r}(n+1, n) &= \mathbf{R}^{-+}\left[\mathbf{\Delta}^-\mathbf{C} + \mathbf{r}^{-+}\mathbf{\Delta}^+\mathbf{A}\right], \\ r(n, n+1) &= \mathbf{R}^{+-}\left[\mathbf{\Delta}^+\mathbf{B} + \mathbf{r}^{+-}\mathbf{\Delta}^-\mathbf{D}\right], \end{aligned}\right\} \tag{6.13.30}$$

where

$$\left.\begin{aligned} \mathbf{\Delta}^\pm &= \left[\mathbf{M} + \frac{\tau}{2} Z_\pm\right]^{-1}, \\ \mathbf{r}^{+-} &= \frac{1}{2}\tau\Delta^\pm Y_\mp, \\ \mathbf{A} &= \mathbf{M} - \frac{\tau}{2}\mathbf{Z}_+, \\ R^\pm &= \left[\mathbf{E} - r^{\pm\mp}\right]^{-1}, \end{aligned}\right\} \tag{6.13.31}$$

where $\mathbf{E}$ is the unit matrix.

The source vectors are

$$\left.\begin{aligned} \mathbf{\Sigma}^+_{n+\frac{1}{2}} &= \mathbf{\Sigma}^+(n+1, n) = \tau\mathbf{R}^{+-}\left[\mathbf{\Delta}^+ + \mathbf{r}^{+-}\Delta^-\right]\mathbf{g}, \\ \mathbf{\Sigma}^-_{n+\frac{1}{2}} &= \mathbf{\Sigma}^-(n, n+1) = \tau\mathbf{R}^{-+}\left[\mathbf{\Delta}^- + \mathbf{r}^{-+}\Delta^+\right]\mathbf{g}. \end{aligned}\right\} \tag{6.13.32}$$

These operators in equations (6.13.30) will give us the transmission and reflection matrices for the cell with $\tau_{n+\frac{1}{2}} < \tau_{crit}$. For the whole atmosphere we need to use the solution given by the internal field calculations in the equations (6.8.33) and

(6.8.34). We now apply this theory to compute the line profiles formed in a non-LTE, spherically symmetric atmosphere. We use the Doppler profile given by

$$\phi(x_i) = \frac{1}{\delta\sqrt{\pi}} \exp\left[-(x_i/\delta)^2\right]. \tag{6.13.33}$$

Here δ is the Doppler width and

$$\left.\begin{aligned} x_i &= \bar{\alpha} X_i, \\ a_i &= \frac{A_i \phi(X_i)}{\sum_{i'=-I}^{I} A_{i'} \phi(x_{i'})} \end{aligned}\right\} \tag{6.13.34}$$

where $\bar{\alpha}$ is the band width and X_i and A_i are the zeros and weights of a suitable quadrature formula ($|i| = 1, 2, \ldots, I$) on the interval $[-1, 1]$. a_i are the weights in the frequency integration (see Averett and Hummer (1965)):

$$a_i = \frac{A_i \phi(x_i)}{\sum\limits_{i=-I}^{I} A_{i'} \phi\,(X_{i'})}, \tag{6.13.35}$$

where the a_is are the weights in

$$\int_{-\infty}^{\infty} \phi(X) f(X)\, dX \simeq \sum_{i=-I}^{I} a_i f(X_i), \quad \sum_{i=-I}^{I} a_i = 1. \tag{6.13.36}$$

For each value of k, we define $W_k = W_{k,n+\frac{1}{2}}$, where

$$\phi_k W_k = a_i b_j. \tag{6.13.37}$$

Fluxes can be calculated from the formula:

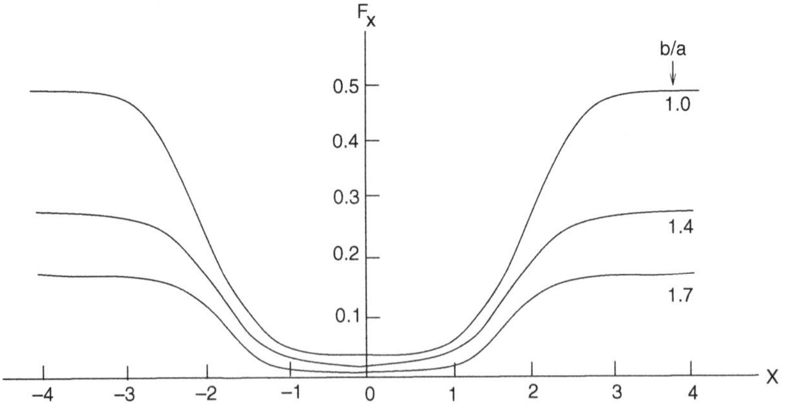

Figure 6.10 Line profiles for an effectively thin model atmosphere with total optical thickness $T = 100$. The boundary condition is $U^-(\tau = T, \mu_j, x_i) = 1$ and $U^+(\tau = 0, \mu_j, x_i) = 0$ $\epsilon = 10^{-6}$, $\beta = 10^{-5}$. The thermalization length $\Lambda = 100$, $B = (A/r)^2$, $A < r < B$ (from Grant and Peraiah (1972), with permission).

$$F = 2\pi \sum_{i=-I}^{I} A_i \sum_{j=1}^{m} b_j \mu_j \left| U_i(\mu_j) \right|. \tag{6.13.38}$$

The integral in the line source function S_L in equation (6.13.9) is

$$J_n' = \int_{-\infty}^{\infty} \phi(x) J(x)\, dx, \tag{6.13.39}$$

and is computed as follows:

$$J_n' = \left[\phi_{n-\frac{1}{2}}^T \mathbf{W}_{n-\frac{1}{2}} \mathbf{U}_n^+ + \phi_{n+\frac{1}{2}}^T \mathbf{W}_{n+\frac{1}{2}} \mathbf{U}_n^- \right]. \tag{6.13.40}$$

The equivalent width is computed from the formula:

$$EW = \int_{-\alpha}^{\alpha} \left(1 - \frac{F}{F_x} \right) dx, \tag{6.13.41}$$

where α is the band width large enough to contain the whole line profile.

We use the following data in the computation of an example: $k_L = 1$, $\rho = 1$, $\alpha = 4$, $\phi(x)$ is a Doppler profile with unit width in the frequency units. We use a seven-point Gauss–Legendre quadrature formula for frequency points ($i = 7$) and a two-point Gauss rule for angle quadrature ($m = 2$). In this case, the curvature matrices are

$$\mathbf{\Lambda}_m^+ = \begin{pmatrix} -0.25 & 0.75 \\ -0.75 & -0.75 \end{pmatrix}, \quad \mathbf{\Lambda}_m^- = \begin{pmatrix} -1 & 0 \\ 0 & 0 \end{pmatrix}. \tag{6.13.42}$$

The step size $\Delta\tau$ ranges between 0.5 and 0.67, and $N = 100$. Profiles are given in figure 6.10 and the corresponding equivalent widths in figure 6.11.

6.14 Integral operator method

The solution of the transfer equation discussed in the previous sections is stable and unique. In this section we present an alternative technique that follows Peraiah and Varghese (1985).

The monochromatic transfer equation in a spherically symmetric and isotropically scattering medium (see equations (6.12.5) and (6.12.6)) is written as

$$\mu \frac{\partial U(r,\mu)}{\partial r} + \frac{1}{r} \frac{\partial}{\partial \mu} \left[(1-\mu^2) U(r,\mu) \right] = K(r) \left[S(r,\mu) - U(r,\mu) \right] \tag{6.14.1}$$

and

$$-\mu \frac{\partial U(r,-\mu)}{\partial r} - \frac{1}{r} \frac{\partial}{\partial \mu} \left[(1-\mu^2) U(r,-\mu) \right] = K(r) \left[S(r,-\mu) - U(r,-\mu) \right], \tag{6.14.2}$$

where $K(r)$ is the absorption coefficient and all other symbols have their usual meanings defined earlier in this chapter. Equations (6.14.1) and (6.14.2) are integrated over the radius–angle mesh $[r_i, r_{i-1}] \times [\mu_j, \mu_{j-1}]$ (shown in figure 6.12). The quantity $U(r, \mu)$ is expressed by an interpolation formula given by

$$U(r, \mu) \approx (U_{00} + U_{01}\xi) + (U_{10} + U_{11}\xi)\eta, \tag{6.14.3}$$

where

$$\xi = \frac{(r - \bar{r})}{\Delta r/r}, \quad \eta = \frac{\mu - \bar{\mu}}{\Delta\mu/2}, \tag{6.14.4}$$

$$\bar{r} = \frac{1}{2}(r_i + r_{i-1}), \quad \bar{\mu} = \frac{1}{2}(\mu_j + \mu_{j-1}), \tag{6.14.5}$$

$$\Delta r = (r_i - r_{i-1}), \quad \Delta\mu = (\mu_j - \mu_{j-1}). \tag{6.14.6}$$

We estimate the nodal values of the Us in terms of the interpolation coefficients, U_{00}, U_{01}, U_{10}, and U_{11}. Thus,

$$\begin{pmatrix} U(r_i, \mu_j) & U_d \\ U(r_{i-1}, \mu_j) & U_b \\ U(r_i, \mu_{j-1}) & U_c \\ U(r_{i-1}, \mu_{j-1}) & U_a \end{pmatrix} = \begin{pmatrix} 1 & 1 & 1 & 1 \\ 1 & -1 & 1 & -1 \\ 1 & 1 & -1 & -1 \\ 1 & -1 & -1 & 1 \end{pmatrix} \begin{pmatrix} U_{00} \\ U_{01} \\ U_{10} \\ U_{11} \end{pmatrix}, \tag{6.14.7}$$

from which we obtain

$$\begin{pmatrix} U_{00} \\ U_{01} \\ U_{10} \\ U_{11} \end{pmatrix} = \frac{1}{4} \begin{pmatrix} 1 & 1 & 1 & 1 \\ -1 & -1 & 1 & 1 \\ -1 & 1 & -1 & 1 \\ 1 & -1 & -1 & 1 \end{pmatrix} \begin{pmatrix} U_a \\ U_b \\ U_c \\ U_d \end{pmatrix}. \tag{6.14.8}$$

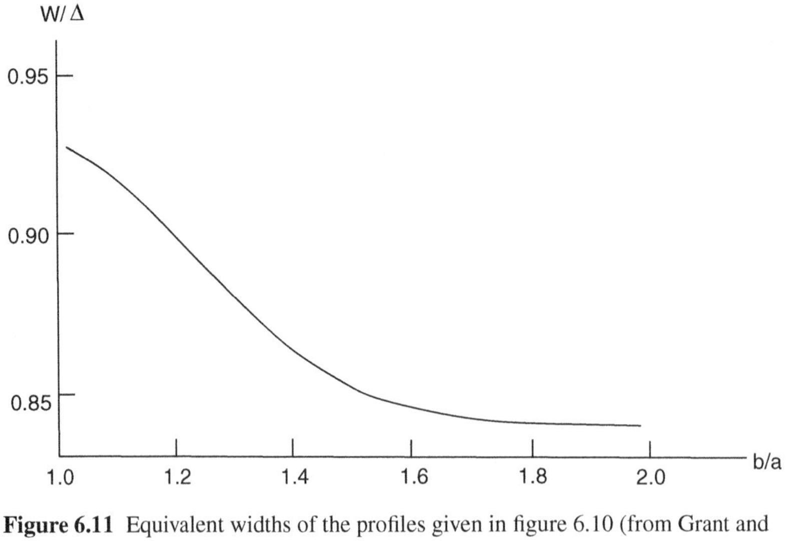

Figure 6.11 Equivalent widths of the profiles given in figure 6.10 (from Grant and Peraiah (1972), with permission).

We now define certain integral operators X and Y such that

$$X = \frac{1}{\Delta\mu}\int_{\Delta\mu} \cdots d\mu, \tag{6.14.9}$$

$$Y = \frac{1}{V}\int_{\Delta V} \cdots 4\pi r^2\, dr, \tag{6.14.10}$$

where

$$V = \frac{4\pi}{3}\left(r_i^3 - r_{i-1}^3\right). \tag{6.14.11}$$

We express the source function $S(r, \mu)$ by an interpolation formula similar to that in equation (6.14.3):

$$S(r, \mu) = S_{00} + S_{01}\xi + (S_{10} + S_{11}\xi)\eta. \tag{6.14.12}$$

The interpolation coefficients S_{00}, S_{01}, S_{10} and S_{11} are obtained analogously to equation (6.14.8).

Substituting equations (6.14.3) and (6.14.12) in equations (6.14.1) and (6.14.2) and applying the operators X and Y on equations (6.14.1) and (6.14.2) we obtain after some algebra

$$\begin{aligned}
&\left[\mathbf{M}_p^+ - \boldsymbol{\rho}^+ + \frac{1}{2}\tau^+\mathbf{Q}\left(\mathbf{E} - \boldsymbol{\gamma}^{++}\right)\right]\mathbf{U}_i^+ \\
&\quad - \left[\mathbf{M}_q^+ + \boldsymbol{\rho}^- - \frac{1}{2}\tau^+\mathbf{Q}\left(\mathbf{E} - \boldsymbol{\gamma}^{++}\right)\right]\mathbf{U}_{i-1}^+ \\
&\quad = (1-\varpi)\tau\mathbf{Q}\mathbf{B}^+ + \frac{1}{2}\tau^+\mathbf{Q}\boldsymbol{\gamma}^{+-}\mathbf{U}_{i-1}^- + \frac{1}{2}\tau^-\mathbf{Q}\boldsymbol{\gamma}^{+-}\mathbf{U}_{i-1}^-
\end{aligned} \tag{6.14.13}$$

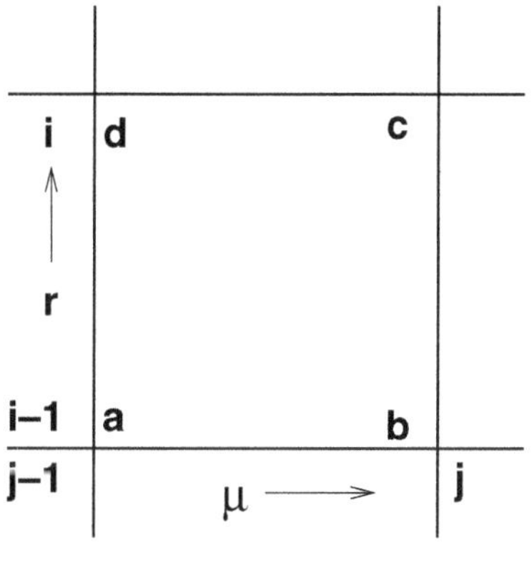

Figure 6.12 Schematic diagram of the radius–angle mesh.

and

$$-\left[\mathbf{M}_p^- - \boldsymbol{\rho}^+ - \frac{1}{2}\tau^+\mathbf{Q}\left(\mathbf{E} - \boldsymbol{\gamma}^{--}\right)\right]\mathbf{U}_i^-$$
$$-\left[\mathbf{M}_q^- + \boldsymbol{\rho}^- + \frac{1}{2}\tau^-\mathbf{Q}\left(\mathbf{E} - \boldsymbol{\gamma}^{--}\right)\right]\mathbf{U}_{i-1}^-$$
$$= (1-\varpi)\tau\mathbf{Q}\mathbf{B}^- + \frac{1}{2}\tau^+\mathbf{Q}\boldsymbol{\gamma}^{-+}\mathbf{U}_{i-1}^+ + \frac{1}{2}\tau^-\mathbf{Q}\boldsymbol{\gamma}^{-+}\mathbf{U}_{i-1}^+, \tag{6.14.14}$$

where

$$\mathbf{M}_{p,q}^{\pm} = (1 \pm p, q)\mathbf{M}, \tag{6.14.15}$$

$$\mathbf{M} = \begin{pmatrix} \mu_{\frac{1}{2}}^- & \mu_{\frac{1}{2}}^+ & & & \\ & \mu_{\frac{3}{2}}^- & \mu_{\frac{3}{2}}^+ & & \\ & & \ddots & \ddots & \\ & & & \mu_{j-\frac{1}{2}}^- & \mu_{j-\frac{1}{2}}^+ \\ & & & & \mu_j \end{pmatrix}, \tag{6.14.16}$$

$$\boldsymbol{\rho}^{\pm} = \begin{pmatrix} \rho_{\frac{1}{2}}^{\pm} & -\rho_{\frac{1}{2}}^{\pm} & & & \\ & \rho_{\frac{3}{2}}^{\pm} & -\rho_{\frac{3}{2}}^{\pm} & & \\ & & \ddots & \ddots & \\ & & & \rho_{j-\frac{1}{2}}^{\pm} & -\rho_{j-\frac{1}{2}}^{+} \\ & & & & 0 \end{pmatrix}, \tag{6.14.17}$$

$$\mathbf{Q} = \begin{pmatrix} 1 & 1 & & & \\ & 1 & 1 & & \\ & & \ddots & \ddots & 1 \\ & & & & 1 \end{pmatrix}, \tag{6.14.18}$$

$$\boldsymbol{\gamma}^{++} = \frac{1}{2}\varpi\,\mathbf{p}^{++}\mathbf{C}, \quad \boldsymbol{\gamma}^{+-} = \frac{1}{2}\varpi\,\mathbf{p}^{+-}\mathbf{C} \ \text{ etc.}, \tag{6.14.19}$$

$$\mu_{j-\frac{1}{21}}^{\pm} = \bar{\mu}\left(1 + \frac{1}{6}\frac{\Delta\mu}{\bar{\mu}}\right), \quad \tau = K \cdot \Delta r, \tag{6.14.20}$$

$$p = \frac{\Delta A}{\bar{A}}\left(\frac{1}{2} + \frac{\bar{r}}{\Delta r}\right) - 2, \tag{6.14.21}$$

$$q = 2 - \frac{\Delta A}{\bar{A}}\left(\frac{1}{2} + \frac{\bar{r}}{\Delta r}\right), \tag{6.14.22}$$

$$\frac{\Delta A}{\bar{A}} \approx 2t = 2\frac{\Delta r}{\bar{r}}, \quad \bar{A} = \frac{V}{\Delta r}, \quad \Delta A = 4\pi\left(r_i^2 - r_{i-1}^2\right), \tag{6.14.23}$$

$$\tau_{i-\frac{1}{2}}^{\pm} = \tau_{i-\frac{1}{2}}\left(1 \pm \frac{1}{6}\frac{\Delta A}{\bar{A}}\right), \tag{6.14.24}$$

$$\rho_{j-\frac{1}{2}}^{i-\frac{1}{2}} = \frac{1}{2}\frac{1-\bar{\mu^2}}{\Delta\mu}\frac{\Delta A}{\bar{A}}, \tag{6.14.25}$$

$$\bar{\mu^2} = \bar{\mu^2} = (\bar{\mu})^2 + \frac{(\Delta\mu)^2}{12}, \tag{6.14.26}$$

$$\rho_{j-\frac{1}{2}}^{i-\frac{1}{2},\pm} = \rho_{j-\frac{1}{2}}^{i-\frac{1}{2}}\left(1 + \pm\frac{1}{6}\frac{\Delta r}{\bar{r}}\right). \tag{6.14.27}$$

By setting equations (6.14.13) and (6.14.14) in the form of the interaction principle, we can get the transmission and reflection operators. The solution obtained satisfies the following basic tests:

1. invariance of the specific intensity in a medium in which radiation is neither absorbed nor emitted;
2. continuity of the solution in both angle and radial distibution;
3. uniqueness of the solution; and
4. the condition of zero net flux in a scattering medium with one boundary having a specular reflector and global conservation of energy.

Hunt and Grant (1969) applied discrete space theory to problems of planetary atmospheres. Hunt (1972a,b) developed a comprehensive theory of spectral lines in the Venus atmosphere in which the pressure broadening of the lines and the physical processes of scattering and absorption by the cloud particles are accurately taken into account. Wehrse (1981) applied discrete space theory in computing the effects of abundance changes in the atmospheres of M supergiants. Peraiah and Varghese (1990) studied Compton scattering in a spherical medium by taking the first three terms in the Taylor's expansion of the intensity (see Chandrasekhar (1960)). Mohan Rao *et al.* (1990) studied the time spent by a photon during an act of scattering using discrete space theory. Mohan Rao *et al.* (1995) have compared discrete space theory with Auers Hermitian methods (AHM). They found that discrete space theory is stable with respect to logarithmic spacing of optical depth and gives less error for the specific intensity at the surface than that given by AHM. Analytical solutions of the difference equations of discrete space theory have been studied and it was found that the solution gives the correct surface value and the diffusion limit in a semi-infinite atmosphere. Plass *et al.* (1973) applied the matrix operator theory to Rayleigh scattering media and Waterman (1981) used the matrix-exponential operator technique to obtain the r and t operators. Wehrse and Kalkofen (1985) studied the formation of resonance lines in dusty gaseous nebulae using a slight variant of discrete space theory. Magnan (1976) applied the line transfer and an approximate solution in spatially correlated random velocity fields. Cassinelli (1971) used the S_n discrete scheme to compute the extended model atmospheres for the central stars of planetary nebulae.

Exercises

6.1 Derive equations (6.4.7)–(6.4.13).

6.2 Prove the principle of invariance of Ambarzumian from relations (6.4.9)–(6.4.10) for the rod. Furthermore, show that the reflection factor r_∞ in this case is given by

$$r_\infty = \frac{q - 1 + \varpi}{q + 1 - \varpi},$$

where $q = (1 - \varpi)\,[1 + \varpi(1 - 2p)]$.

6.3 Find the explicit solution for the internal field by expanding equations (6.8.33) and (6.8.34).

6.4 If the surface at $N + 1$ is a perfect reflecting surface, show that

$$I_N^- = T_{N+\frac{1}{2}} t(N, N+1)\,[E - r(1, N+1)]^{-1}\, V_{N+\frac{1}{2}}^+.$$

6.5 Show that

$$c_j = \mu_{j+\frac{1}{2}} - \mu_{j-\frac{1}{2}}.$$

6.6 Prove the relation

$$\frac{1}{2}\sum_{j=1}^{m} c_j \left(P_{jk}^{++} + P_{jk}^{-+}\right) = 1,$$

where the Ps are the phase function elements for conservative isotropic or Rayleigh scattering atmospheres.

6.7 In the case of spherically symmetric atmospheres prove that the relation

$$\sum_{j=1}^{J} c_j \left(\Lambda_{jk}^+ - \Lambda_{jk}^-\right) = 0,$$

must be satisfied for the radiant flux to be conserved.

6.8 Compute the critical optical depth in each layer of a plane parallel atmosphere whose total optical depth is 10 and whose atmosphere is divided into 50 layers assuming isotropic conservative scattering with $J = 2, 3, 4, 5$.

6.9 Repeat exercise 6.8 with $B/A = 2, 5, 10$. Change the number of shells if necessary to get a meaningful τ_{crit}.

6.10 If the lower boundary at $n = N$ is reflecting (specular reflection) with r_R as the reflection matrix, show that

$$\begin{aligned} I_1^- &= r(N+1, 1) + t(1, N+1) r_R\,[E - r(1, N+1) r_R]^{-1}\, t(N+1, 1) I_1^+ \\ &\quad + \Sigma^-(1, N+1) + t(1, N+1) r_R\,[E - r_R(E - r(1, N+1) r_R]^{-1} \\ &\quad \times \Sigma^+(1, N+1). \end{aligned}$$

6.11 Approximate the specific intensity by

$$I(\mu) \simeq \sum_{m=0}^{M} a_m P_m(\mu),$$

and let the $(m + 1)$ coefficients be chosen so that

$$I(\mu_j) = \sum_{m=0}^{M} a_m P_m(\mu_j), \quad j = 0, 1, \ldots, M,$$

then assuming the $P_m(\mu)$s are orthonormal derive the r and t operators of the transfer equation in spherical symmetry.

6.12 Derive equations (6.14.13) and (6.14.14) and the r and t operators from these equations.

REFERENCES

Abramowitz, M., Stegun, I.A., (eds.), 1965, *Handbook of Mathematical Functions*, Dover, New York.

Ambarzumian, V.A., 1943, *Dokl. Akad. Nawuk. SSSR*, **38**, 229.

Auer, L.H., 1976, *JQSRT*, **16**, 931.

Averett, E.H., Hummer, D.G., 1965, *MNRAS*, **135**, 295.

Carlson, B.G., 1963, *Meth. Computational Phys.*, **1**, 1.

Cassinelli, J.P., 1971, *ApJ*, **165**, 265.

Chandrasekhar, S., 1960, *Radiative Transfer*, Dover, New York.

Collatz, L., 1966, *Functional Analysis and Numerical Mathematics*, Academic Press, New York.

Grant, I.P., 1968a, Lecture Notes on New Methods in Radiative Transfer, Pembroke College, Oxford (unpublished).

Grant, I.P., 1968b, *J. Comp. Phys.*, **2**, 381.

Grant, I.P., Hunt, G.E., 1968, *MNRAS*, **141**, 27.

Grant, I.P., Hunt, G.E., 1969a, *Proc. Roy. Soc. London A*, **313**, 183.

Grant, I.P., Hunt, G.E., 1969b, *Proc. Roy. Soc. London A*, **313**, 199.

Grant, I.P., Peraiah, A., 1972, *MNRAS*, **160**, 237.

Hulst, van de, 1965, *A New Look at Multiple Scattering*, NASA Institute of Space Studies, New York.

Hunt, G.E., 1972a, *JQSRT*, **12**, 387.

Hunt, G.E., 1972b, *JQSRT*, **12**, 405.

Hunt, G.E., Grant, I.P., 1969, *J. Atmospheric Sciences*, **26**, 963.

Isaacson, E., Keller, H.B., 1966, *Analysis of Numerical Methods*, Wiley, New York.

Jefferies, J.T., 1970, *Proc, Astr. Soc. Australia*, **1**, No. 8, 356.

Lathrop, K., Carlson, B.G., 1967, *J. Comp. Phys.*, **2**, 173.

Magnan, C., 1976, *JQSRT*, **16**, 281.

Mihalas, D., 1978, *Stellar Atmospheres*, 2nd Edition, Freeman, San Francisco.

Mohan Rao, D., Rangarajan, K.E., Peraiah, A., 1990, *ApJ*, **358**, 622.

Mohan Rao, D., Varghese, B.A., Srinivasa Rao, M., 1995, *JQSRT*, **53**, 639.

Peraiah, A., 1971, D.Phil. thesis, Oxford University.

Peraiah, A., Grant, I.P., 1973, *J. Inst. Math. Applic.*, **12**, 75.

Peraiah, A., Varghese, B.A., 1985, *ApJ*, **290**, 411.

Peraiah, A., Varghese, B.A., 1990, *Pub. Astron. Soc. Japan*, **42**, 675.

Plass, G.N., Kattawar, G.W., Catchings, F.E., 1973, *Appl. Opt.*, **12**, 314.

Preisendorfer, R.W., 1965, *Radiative Transfer on Discrete Spaces*, Pergamon, Oxford.

Redheffer, R.M., 1962, *J. Math. Phys.*, **41**, 1.

Rybicki, G.B., Usher, P.D., 1966, *Astrophys. J.*, **146**, 871.

Sobolev, V.V., 1963, *A Treatise on Radiative Transfer* (translated by S.I. Gaposchkin), Van Nostrand Company Inc., New York.

Sykes, J.B., 1951, *MNRAS*, **11**, 377.

Ueno, S., 1965, *J. Math. Anal. Applics.*, **11**, 11.

Varga, R.S., 1963, *Matrix Iterative Analysis*, Prentice Hall, Englewood Cliffs.

Waterman, P.C., 1981, *J. Opt. Soc. America*, **71**, 410.

Wehrse, R., 1981, *MNRAS*, **195**, 553.

Wehrse, R., Kalkofen, W., 1985, *A&A*, **147**, 71.

Wing, G.M., 1962, *An Introduction to Transport Theory*, John Wiley, New York.

Chapter 7

Transfer equation in moving media: the observer frame

7.1 Introduction

Rapid expansion in nova, stellar atmospheres of supernovae and similar objects is established through several spectroscopic observations. In spectra of these objects, the absorption lines shift towards the violet side from the rest position indicating matter outflow in their atmospheres. These lines are accompanied by red shifted emission characteristics of P Cygni type as seen in figure 7.1 (see Beals (1950), Kuan and Kuhi (1975)). Beals (1929, 1931) interpreted the large widths in the lines of WR spectra to be due to the velocities of expansion of the order of 3000 km s^{-1} indicative of a rapid outflow of the matter in the outer layers of these stars. He suggested that this outflow of matter is influenced by the radiation pressure in the medium.

It is difficult to obtain the solution of the transfer equation in such spherical media. Beals (1929, 1930, 1931, 1934), Chandrasekhar (1934), Gerasimovič (1934) and Wilson (1934) investigated this problem assuming the medium to be optically thin, neglecting the transfer effects. Struve and Elvey (1934) found that the Doppler widths derived from the flat part of the growth curve were much larger than the thermal value, which they attributed to the 'turbulent' motion in the atmosphere which is non-thermal. Struve's observations (1946) showed large scale velocities through the fact that the line profile widths in certain stars were larger than the Doppler widths obtained from curve of growth analysis of their spectra. The velocities appear to be directed systematically along the line of sight. Pulsating stars were found to exhibit periodic changes in Doppler shifts.

We need to study the radiation field in the atmospheres of stars with extended and expanding atmospheres such as WR stars, P Cygni stars. We shall study the solution of the radiative transfer equation without going into the details of the fluid flow equations. We shall study the transfer equation in two frames:

1. The observer's frame (inertial or laboratory frame) in which the observer is fixed with respect to the centre of the star and the radiation field is estimated. It is generally restricted to velocities of the order of a few Doppler widths. However, Sobolev's method gives an approximate solution for the medium moving with large *velocity gradients*.
2. The comoving frame, that is the frame moving with the fluid. The advantage of this frame is that the opacities and emissions in it are isotropic. We shall study the comoving frame solution in chapter 8.

7.2 Observer's frame in plane parallel geometry

The photon frequencies are changed, due to Doppler shifts when the photons are received in the observer's frame, from the frame of the atoms contained in the moving matter. If the matter moves with a velocity of $\mathbf{v}(r)$ relative to the external observer (or the centre of the star) and if the frequency of the photon in the observer's frame is ν, then the frequency (in the atom's frame) with which the photon moving in the direction $\mathbf{n}$ is emitted or absorbed is

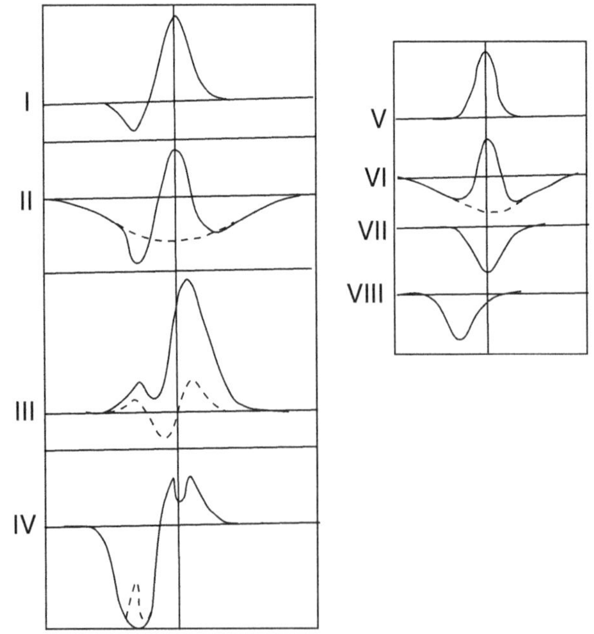

Figure 7.1 Beals classification of P Cygni type profiles (Beals, 1950).

$$\nu' = \nu - \nu_0 \left(\frac{\mathbf{n} \cdot \mathbf{v}}{c} \right), \tag{7.2.1}$$

where, ν_0 is the line centre frequency and c is the velocity of light. Therefore, the opacity and emmissivity of the material as measured in the observer's frame become angle dependent. In plane parallel geometry, the time independent transfer equation is given by (Mihalas 1978)

$$\mu \frac{\partial I(z, \mu, \nu)}{\partial z} = j(z, \mu, \nu) - \kappa_\nu(z, \mu, \nu) I(z, \mu, \nu). \tag{7.2.2}$$

It is ususal to measure the frequency changes in terms of Doppler widths and we define x as

$$x' = x - \mu V, \tag{7.2.3}$$

where $x = (\nu - \nu_0)/\Delta\nu_D$, with $\Delta\nu_D$ the Doppler width given by $\Delta\nu_D = \nu_0 v_{th}/c$ and v_{th} the thermal velocity parameter. The velocities are measured in units of thermal velocity $v = V/v_{th}$. The changes in the photon frequencies due to the Doppler effect are negligible in continuum radiation, while they are substantial in line radiation. Therefore the opacity and emission are affected through the line profile and can be written as

$$\kappa(x, \mu, z) = \kappa_c(z) + \kappa_l(z)\phi(x, \mu, z) \tag{7.2.4}$$

and

$$j(x, \mu, z) = j_c(z) + j_l(z)\phi(x, \mu, z), \tag{7.2.5}$$

where κ_c and κ_l are the continuum and line centre opacities, j_c and j_l are the continuum and line centre emissions and ϕ is the normalized profile function given by

$$\phi(x, \mu, z) \rightarrow \phi(x - \mu v, \mu, z). \tag{7.2.6}$$

The Doppler profile is given by

$$\phi(x, \mu, z) = \pi^{-\frac{1}{2}} \delta^{-1}(z) \exp\left\{ \frac{-[x - \mu v(z)]^2}{\delta^2(z)} \right\}, \tag{7.2.7}$$

where

$$\delta(z) = \frac{\Delta\nu_D(z)}{\Delta\nu_d}. \tag{7.2.8}$$

The total source function is (see chapter 1)

$$S(x, \mu, z) = \frac{\phi(x, \mu, z)}{\phi(x, \mu, z) + \beta(z)} S_l(z) + \frac{\beta(z)}{\phi(x, \mu, z) + \beta(z)} S_c(z), \tag{7.2.9}$$

where

$$\left.\begin{aligned} S_L &= \frac{j_l(z)}{\kappa_l(z)}, \\ S_C(z) &= \frac{j_c(z)}{\kappa_c(z)}, \\ \beta(z) &= \kappa_c(z)/\kappa_l(z). \end{aligned}\right\} \tag{7.2.10}$$

The transfer equation (7.2.2) can be written as

$$\frac{\partial I(x,\mu,z)}{\partial \tau(x,\mu,z)} = I(x,\mu,z) - S(x,\mu,z), \tag{7.2.11}$$

where the optical depth $\tau(x,\mu,z)$ is given by

$$\tau(x,\mu,z) = \frac{1}{\mu}\int_z^{z_{max}} \kappa(x,\mu,z)\,dz, \tag{7.2.12}$$

with z_{max} the upper surface of the atmosphere. The formal solution of equation (7.2.11) is written as

$$\begin{aligned} I(x,\mu,z_{max}) &= I(x,\mu,0)\exp[-(x,\mu,0)] \\ &\quad + \int^{\tau(x,\mu,0)} S(x,\mu,z)\exp[-\tau(x,\mu,z)]\,dt(x,\mu,z) \\ &= I(x,\mu,0)\exp[-\tau(x,\mu,0)] + \int_0^{z_{max}} \mu^{-1}\,[\phi(x,\mu,z)S_L(z) \\ &\quad +\beta(z)S_c(z)]\exp[-\tau(x,\mu,z)]\kappa_l(z)\,dz. \end{aligned} \tag{7.2.13}$$

Through equation (7.2.13) the emergent intensity can be computed for a known source function. If LTE prevails we can use $S_c = S_l = B$. The line source function generally consists of scattering terms which depend on the radiation field which in turn depends on the moving matter with which it interacts. Moving matter can shift the line substantially into the neighbouring continuum, thus changing the radiation field in the line. The mean intensity J which occurs in the line source function will not be the same as it is in the static case and the line source function has to be written as

$$S_l(z) = \frac{1-\epsilon}{2}\int_{-\infty}^{+\infty} dx \int_{-1}^{+1} d\mu\, I(x,\mu,z)\phi(x,\mu,z) + \epsilon B(z). \tag{7.2.14}$$

The profile function ϕ will have to be estimated for each angle–velocity dependent normalized frequency point, as the profile is no longer symmetric. The approximation of complete redistribution of photons in the line is inconsistent in a moving medium. Rybicki (1970) explained several aspects of transfer in moving media (this paper is highly recommended to anyone who is interested in the transfer of radiation in moving media). Partial frequency redistribution of photons in the line is appropriate and this requires the full angle–frequency dependent redistribution. This problem can be handled only in the comoving frame of the gas as the opacities and emissions can be used in their isotropic form.

The scattering integral is computed by quadrature sums in the observer's frame. However, the line profile $\phi(x - \mu v)$ $(-1 \leq \mu \leq 1)$ shifts between $+v$ and $-v$ so that the total length of the angle–frequency mesh is $2|x + v|$, which is quite large. If the material motion is of the order of 200 or 300 v_{th}, we need a mesh twice as large in the angle–frequency discretization. Furthermore, there is a complex coupling between the angle and the frequency and therefore we need to use a large number of angles. This makes the mesh size very large and we are restricted to smaller material velocities. If we want to deal with higher velocities, it is necessary to work with the comoving frame.

Equation (7.2.11) can be written as the second order differential equation form by writing

$$u(x, \mu, z) = \frac{1}{2}[I(x, \mu, z) + I(-x, -\mu, z)] \tag{7.2.15}$$

and

$$v(x, \mu, z) = \frac{1}{2}[I(x, \mu, z) - I(-x, -\mu, z)] \tag{7.2.16}$$

for a symmetric line profile, that is $\phi(x - \mu v) = \phi(-x + \mu V)$, $d\tau(x, \mu, z) = d\tau(-x, -\mu, z)$ and $S(x, \mu, z) = S(-x, -\mu, z)$. We then obtain (see chapter 4)

$$\mu^2 \frac{\partial^2 t(x, \mu, z)}{\partial \tau(x, \mu, z)^2} = u(x, \mu, z) - S(x, \mu, z). \tag{7.2.17}$$

This can be solved with the boundary conditions described in chapter 4. At the lower boundary when the diffusion approximation is assumed one must ensure that the velocity gradient satisfies the relation $dv/d\tau \ll 1$ which is small over a photon mean free path. The discrete depth mesh (d) and the angle–frequency mesh (l) are chosen as described in chapter 4. The mesh on x should include the whole profile which can vary over the range $\pm 2v_{max}$. Then the source function is written as (equation (7.2.14))

$$S_{dl} = S(x_l, \mu_l, z_d) = \alpha_{dl}\bar{J}_d + \beta'_{dl}, \tag{7.2.18}$$

where $l = m + (n - 1)M$, n is the number of frequency points, m is the number of angle points $(m = 1, 2, \ldots, M)$, the αs and βs contain the various quantities in equation (7.2.14) and $\bar{J}_d$ is given by

$$\bar{J}_d = \sum_{l=1}^{L} W_l \phi_{dl} u_{dl}, \tag{7.2.19}$$

where $\phi_{dl} = \phi(x_l - \mu_l v_d, z_l)$. One can use the standard Rybicki type scheme (see chapter 4):

$$\mathbf{T}\mathbf{u}_i + \mathbf{u}_i\mathbf{J} = \mathbf{K}_i \quad (i = 1, \ldots, I). \tag{7.2.20}$$

Many authors have studied the numerical computation of the line in a moving medium using the plane parallel approximation. These studies were exploratory in nature rather than aimed at computing accurate solutions to fit the observations. Kulander (1968) studied an atmosphere with constant physical properties through a semi-analytical approach that divided the atmosphere into several layers, and the solution for the discrete ordinate intensities in each layer was then a linear combination of elementary exponential functions. Continuity of the intensity components was assumed at the boundaries and at the interface between two layers. However, in general, a discontinuous source function was found across the boundary between the layers with enhancements near the surface. Kulander's (1967) differential equation method dealt with media with a continuous variation of properties. However, Hummer and Rybicki (1967) reasoned that this method is unstable for large optical depths, although the method of slabs can be used for arbitrarily thick atmospheres. There is a limitation imposed by the number of frequency and angle points. The single point angle quadrature used by Kulander is equivalent to the Eddington approximation. Hummer and Rybicki (1968) extended the Riccati method (Rybicki and Hummer 1967) to an effectively thin isothermal, differentially expanding, plane parallel atmosphere to compute the non-LTE source function and the emergent line profile (see figure 7.2). They used a total optical depth $T = 20$ and $\epsilon = 10^{-3}$, $b = 1$ and $\delta = 1$ (see equation (7.2.8)) with a linear velocity law with a velocity gradient v_1 of the form

$$v = v_0 + \tau v_1. \tag{7.2.21}$$

The line profile shows the central reversal which develops from the material nearest to the observer with red shifted emission. The shift is due to the optical depth effect, the actual changes in the source function being irrelevant. The emission and absorption from the material nearest to the observer is violet shifted and an optical depth of unity on the violet side of the line occurs much closer to the surface where the excitation is smaller, which reduces the violet emission and the line appears red shifted. The Riccati method has a limitation (see chapter 4) on the angle–frequency mesh size – not more than 60 components in a hemisphere can be included in the computation. Kalkofen (1970) applied the integral equation method to the velocity gradient problem. This method has an advantage over the differential equation and difference equation methods in that a large number of frequencies and angles can be taken as the computation time increases only linearly, rather than quadratically or cubically. Kalkofen computed several profiles (see figure 7.3) generated by an atmosphere which has a Planck function similar to that of a chromosphere (see figure 7.3(a)). He used the velocity law

$$v(\tau) = \frac{10}{1 + \tau/T}, \tag{7.2.22}$$

with $\epsilon = 10^{-2}$ and $\beta = 10^{-4}$. The emergent intensity profiles are shown in figure 7.3(b).

The important result shown in figure 7.3(a) is that the velocity field has little effect on the the line source function in spite of the fact that the line profiles show substantial changes for different values of T. More photons escape through the red wing while exactly the opposite happens through the blue wing thus the source function remains fairly unchanged. The photon escape probability increases in an expanding atmosphere and this explains the low values of the source function compared to those for the static atmosphere. A physical discussion of these profiles is given in Athay (1972) and Mihalas (1978).

7.3 Wave motion in the observer's frame

Shine (1975) studied the effect of mesoscale velocity fields on line formation with a velocity law of the form

$$v(\tau, t) = \beta \sin\left[2\pi\left(\frac{\log\tau}{\lambda} + t\right)\right]$$

for the sawtooth waves using the Rybicki type solution to solve the transfer equation. Cram (1972) and Heasly (1975) studied the effects of acoustic waves on the formation of solar Ca II lines taking account of the pulse produced changes in temperature and density. Rangarajan (1997) studied the polarization in the resonance lines in the presence of wave motion in a plane parallel medium. The transfer equation is given by (see Chandrasekhar (1960))

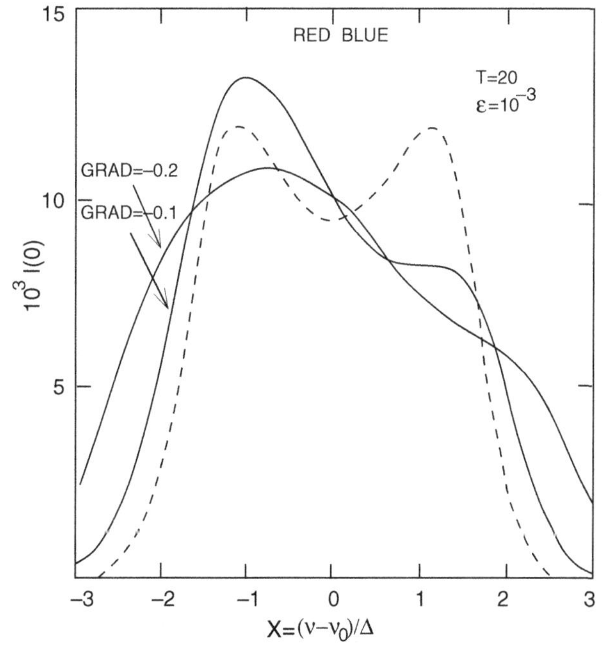

Figure 7.2 Normally emergent intensity at $\tau = 0$ for $T = 20$, $\epsilon = 10^{-3}$, $b = 1$, $\delta = 1$. The dashed line is the symmetric profile produced by a static atmosphere (see equation (7.2.21)). The line becomes broader with increasing velocity gradient and the peak intensity is shifted towards the red (from Hummer and Rybicki (1968), with permission).

$$\mu \frac{dI(x, \mu, z)}{dz} = -\kappa(x, \mu, z)\mathbf{I}(x, \mu, z) + j(x, \mu, z), \tag{7.3.1}$$

where κ and j are absorption and emission coefficients and $\mathbf{I} = (I_l, I_r)^T$ with I_l and I_r the intensities in the two polarization states. The polarization p is given by

$$p = \frac{Q}{I}, \quad I = I_l + I_r, \quad Q = I_l - I_r, \tag{7.3.2}$$

with the usual meanings of the symbols. The absorption coefficient $\kappa(x, \mu, z)$ consists of line and continuum parts given by

$$\kappa(x, \mu, z) = \kappa_l \phi(x, \mu, z) + \kappa_c(x, \mu, z), \tag{7.3.3}$$

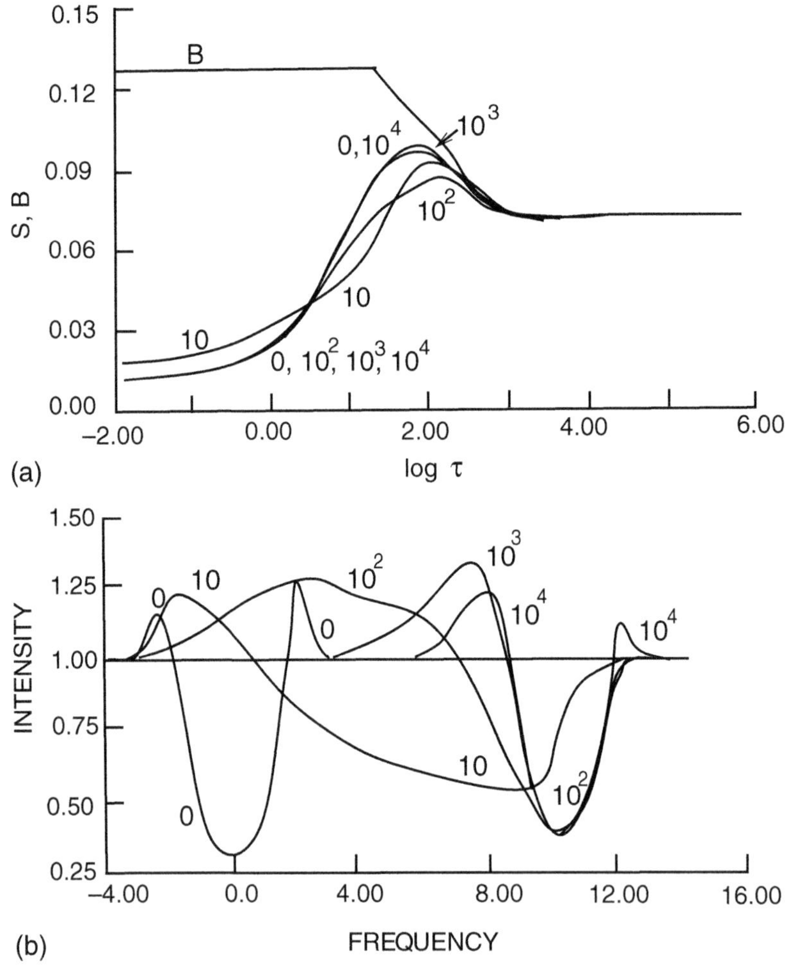

Figure 7.3 (a) Planck function and source function in an expanding atmosphere with $\epsilon = 10^{-2}$, $\beta = 10^{-4}$, versus static optical depth. (b) Normally emergent intensity versus $x (= \Delta\nu/\Delta\nu_D)$ (from Kalkofen (1970), with permission).

where $\phi(x, \mu, z)$ is the profile function. At the observer, the Doppler shifted frequency is $x' = x - \mu v$, where x is the frequency of the atom and v being the velocity of the gas. In this case, the profile function becomes

$$\phi(x, \mu, z) = \phi(x - \mu v, z). \tag{7.3.4}$$

Equation (7.3.1) is solved assuming complete redistribution and using the discrete space theory described in chapter 6. The continuum source function that has been used is

$$B(\tau_c) = 1 + 10\tau_c^{0.9} + 100 \exp\left(-70.7\tau_c^{\frac{1}{2}}\right), \tag{7.3.5}$$

where τ_c is the continuum optical depth. Three different types of waves are used:

1. Damped sine wave with $v = v' \sin\phi'$, where the amplitude v' is given by

$$v' = v_0 / \left[1 + (\tau_c/\tau^\star)^{\frac{1}{2}}\right], \tag{7.3.6}$$

   where $\tau^\star$ is approximately the depth at which v' approaches v_0. The phase ϕ' is given by

$$\phi' = 2\pi(z - ct)/cP, \tag{7.3.7}$$

   where P is the period of the wave and z is the height. P is chosen to be 100 s, $\tau^\star = 10^{-3}$, $c = 7$ km s^{-1}, and $v_0 = 3$ km s^{-1} and the amplitude of the wave = 3 Doppler units.

2. Undamped sine wave. This is obtained by setting $\nu' = \nu_0$ and assuming it is constant throughout the atmosphere.

3. Sawtooth wave. This is of the form (Shine 1975)

$$\nu = \nu_0 \left[2(z - ct)/cP_{modl} - 1\right]. \tag{7.3.8}$$

7.4 Observer's frame and spherical symmetry

There have been many attempts to solve the problems of line transfer in the observer's frame in a spherically symmetric atmosphere. We describe some of the techniques and their results below. These methods apply in low velocity regimes.

7.4.1 Ray-by-ray method

Kunasz and Hummer (1974) applied this method when studying line formation in deeper layers of the expanding atmosphere (see also Tam and Schwartz (1976)). We have already described this method for static atmosphere in the (p, z) coordinate system in chapter 4. Kunasz and Hummer solved the line transfer problem with complete redistribution, assuming the radial velocity to be a function of radius, using a velocity of expansion that was a few times the mean thermal velocity. The transfer equation in the observer's frame is differenced along the rays and the resulting coupled linear equations are solved by Rybicki's scheme. The transfer equation along the ray is given by (Mihalas 1978)

$$\pm\frac{\partial I^{\pm}(x, p, z)}{\partial z} = j(x, p, z) - \kappa(x, p, z)I^{\pm}(x, p, z), \tag{7.4.1}$$

where $j(x, p, z)$ is the emission coefficient and the absorption coefficient (consisting of line and continuum coefficients) is given by

$$\kappa(x, p, z) = \kappa_l(r)\phi(x, p, z) + \kappa_c(r) \tag{7.4.2}$$

and

$$\left.\begin{aligned} r(p, z) &= \left(p^2 + z^2\right)^{\frac{1}{2}}, \\ \mu(p, z) &= \frac{z}{(p^2 + z^2)^{\frac{1}{2}}}. \end{aligned}\right\} \tag{7.4.3}$$

The profile ϕ in equation (7.4.2) is given in the observer's frame as

$$\phi(x, p, z) = \phi\,[x - \mu(p, z)V(r), r(p, z)], \tag{7.4.4}$$

where the velocity $V(r)$ measured in mean thermal units is positive for increasing r. The optical depth along the ray is given by

$$\tau(x, p, z) = \int_z^{z_{max}} \kappa(x, p, z')\,dz'. \tag{7.4.5}$$

By defining the mean-intensity- and flux-like quantities u and v as (from chapter 4)

$$u(x, p, z) = \frac{1}{2}\left[I^+(x, p, z) + I^-(x, p, z)\right] \tag{7.4.6}$$

and

$$v(x, p, z) = \frac{1}{2}\left[I^+(x, p, z) - I^-(x, p, z)\right], \tag{7.4.7}$$

we obtain the second order differential equation (see chapter 4)

$$\mu^2\frac{\partial^2(x, p, z)}{\partial\tau(x, p, z)^2} = u(x, p, z) - S(x, p, z), \tag{7.4.8}$$

where $S(x, p, z)$ is the source function and is equal to $j(x, p, z)/\kappa(x, p, z)$ (see equation (7.2.9)). The line source function is given by

$$S_L(r) = [1 - \epsilon(r)]\,\bar{J}(r) + \epsilon(r)B(r), \tag{7.4.9}$$

where $\epsilon(r)$ is the probability that a photon is absorbed in the act of scattering, $B(r)$ is the Planck function at the line centre for the local electron temperature and $\bar{J}(r)$ is the intensity averaged over the line profile given by

$$\bar{J}(r) = \frac{1}{2}\int_{-\infty}^{\infty} dx \int_{-1}^{1} d\mu\, \phi(x', r) I(x, \mu, r). \tag{7.4.10}$$

In terms of the (p, z) coordinate system, equation (7.4.10) becomes

$$\bar{J}(r) = \frac{1}{2r}\int_{-\infty}^{+\infty} dx \int_{0}^{r} dp\, p(r^2 - p^2)^{-\frac{1}{2}} \left[\phi(x - \mu V(r) I^{+}(x, z, p) \right.$$
$$\left. + \phi(x + \mu V(r) I^{-}(x, z, p)\right] \tag{7.4.11}$$

(see Kunasz and Hummer (1974)). In terms of equations (7.4.6) and (7.4.7), $\bar{J}(r)$ can be written for the positive half of the profile (restricting the integration of x to from 0 to ∞) as

$$\bar{J}(r) = \frac{1}{2}\int_{0}^{\infty} dx \int_{0}^{r} dp \quad p(r^2 - p^2)^{-\frac{1}{2}} [\phi(x - \mu V(r)) u(x, z, p)$$
$$+ \phi(x + \mu V(r)) u(x, -z, p)]. \tag{7.4.12}$$

We now need the boundary conditions:

1. The incident intensities at the upper and lower boundaries ($\pm z_{max}$) for the rays that do not intersect the core are to be specified. This means

$$\left[\frac{\partial u(x, p, z)}{\partial \tau(x, p, z)}\right]_{z=\pm z_{max}} = \pm\, u(x, p, z)|_{z=\pm z_{max}}\,. \tag{7.4.13}$$

2. For rays that intersect the central core ($p \le r_c$) we have

(a) $v(x, p, z_{min}) = 0$ for an opaque core, (7.4.14)

(b) $$\left.\frac{\partial u(x, p, z)}{\partial \tau(x, p, z)}\right|_{z=\pm z_{min}} = \pm\, u(x, p, z)|_{z=\pm z_{min}} \quad \text{for a hollow core.} \tag{7.4.15}$$

The Feautrier technique can be used to solve the above problem (see chapter 4). The same discrete meshes $\{r_{d_i}\}$ and $\{p_{d_i}\}$ are used. We need to include the whole set of frequency points from $-x_N$ to $+x_N - \{x_n\}$ with $n = \pm 1, \ldots, \pm N$ and $x_{-N} = -x_N$. We obtain equations of the form given in chapter 4, and can apply Rybicki's method for getting $\bar{J}$ which needs to be defined only for $\{r_d, 1 \le D\}$, whereas $u_{din} = u(z_d, p_i, x_n)$ corresponds to the mesh $\{z_{di}; d_i = 1, \ldots, D\}$ for the whole length of the ray. Now **T** is a square matrix and **U** is a rectangular Chevron matrix. For every choice of (i, n) we have the relation,

$$\mathbf{u}_{in} = \mathbf{A}_{in}\bar{\mathbf{J}} + \mathbf{B}_{in}, \tag{7.4.16}$$

where $\bar{\mathbf{J}}$ is given in the discrete form as

$$\bar{J}(r_d) = \sum_{n=-N}^{N} \omega_n \sum_{i=1}^{I_d} a_{d_i} \phi\,[r_d; x_n - \mu(r_d, p_i) V(r_d)]\, u_{din}. \tag{7.4.17}$$

We can use the spherical symmetry property and write $I^{\pm}(x, p, z) = I^{\pm}(x, p - z)$ and this gives us

$$\left.\begin{aligned} u(-x, p, z) &= u(x, p - z), \\ v(-x, p, z) &= -v(x, p, -z). \end{aligned}\right\} \quad (7.4.18)$$

These allow the elimination of the values of u at $-x$ and $+z$ in terms of $+x$ and $-z$. With this in mind, we can rewrite equation (7.4.17) as

$$\bar{J} = \sum_{n=1}^{N} \omega_n \sum_{i=1}^{i_d} a_{di} \{\phi\,[r_{d'}, x_n - \mu_{di} V_d]\, u_{din} + \phi\,[r_{d'}, x_n + \mu_{di} V_d]\, u_{d'in}\}, \quad (7.4.19)$$

where $d' = D_i + 1 - d$. When Rybicki's method is applied, equation (7.4.19) gives V-matrices which are rectangular and Chevron matrices. Equations (7.4.16) and (7.4.19) are jointly solved for all values of i and n. This gives the system for $\bar{J}$ which is used to obtain solution. This theory has been applied to the lines formed in expanding atmospheres in the observer's frame (see figures 7.4 and 7.5).

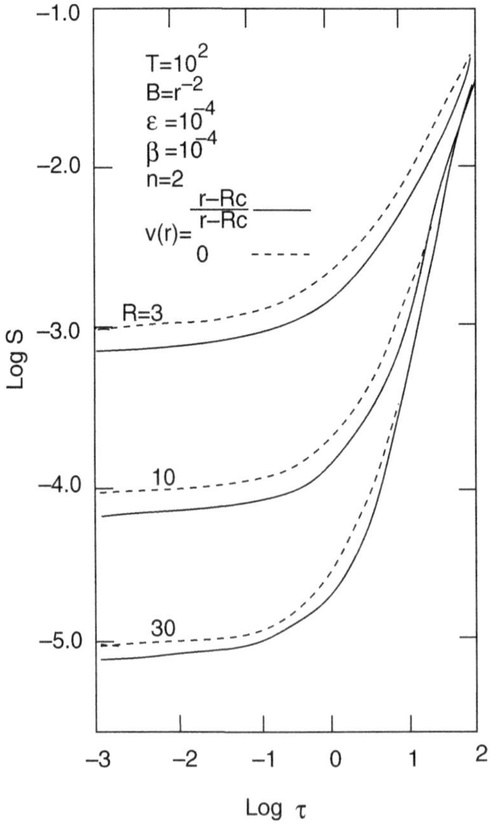

Figure 7.4 Line source functions for $T = 10^3$, $R = r^{-2}$, $\epsilon = \beta = 10^{-2}$ with the linear velocity law. The curves are labelled by $V(r)$. $V(R) = 2$. The broken lines show the corresponding plane parallel static results (from Kunasz and Hummer (1974), with permission).

7.4.2 Observer's frame and discrete space theory

The line transfer equation is solved in the (x, μ, r) system using the discrete theory formalism given in chapter 6. The frequency changes in the observer's frame from $-x - \mu V$ to $+x + \mu V$, where V is the velocity of the moving gas in units of the mean thermal velocity, $x = (\nu - \nu_0)/\Delta\nu_D$, $\Delta\nu_d$ being the Doppler width. We shall present a general formalism to include complete and partial redistribution functions in spherical symmetry.

The transfer equation for a non-LTE, two-level atom in a spherically symmetric atmosphere (Peraiah 1978)

$$\pm\mu\frac{\partial I(x, \pm\mu, r)}{\partial r} \pm \frac{1-\mu^2}{r}\frac{\partial I(x, \pm\mu, r)}{\partial \mu} = \\ k_L\left[\beta + \phi(x, \pm\mu, r)\right]\left[S(x, \pm\mu, r) - I(x, \pm\mu, r)\right] \quad (\mu \in (0, 1)), \quad (7.4.20)$$

where all the symbols have their usual meanings (see chapter 1). The profile function contains x and this changes as mentioned above in the interval of $2\,|x + \mu V|$. The source function $S(x, \pm\mu, r)$ is given by

$$S(x, \pm\mu, r) = (\phi(x, \pm\mu, r) + \beta)^{-1}\left[\phi(x, \pm\mu, r)S_L(x, \pm\mu, r) + \beta S_C(r)\right], \quad (7.4.21)$$

where $S_L(x, \pm\mu, r)$ and $S_C(r)$ are the line and continuum source functions respectively. We have retained the functional dependence of $\pm\mu$ in S_L as this is needed for angle dependent redistribution functions and is

$$S_L(x, \pm\mu, r) = \frac{1-\epsilon}{\phi(x, \pm\mu, r)}\int_{-\infty}^{+\infty} dx' \int_{-1}^{+1} R(x, \pm\mu; x', \mu)I(x', \mu, r)\,d\mu' \\ + \epsilon B(r). \quad (7.4.22)$$

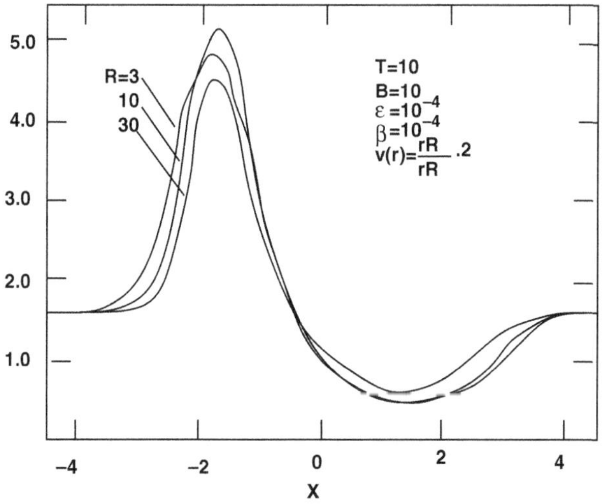

Figure 7.5 Luminosity profile for the source functions shown in figure 7.4 (from Kunasz and Hummer (1974), with permission).

Here $R(x, \pm\mu; x', \mu)$ represents the angle dependent partial frequency redistribution function (see chapter 1). The profile function $\phi(x, \mu)$ is given by

$$\phi(x, \mu) = \int_{-\infty}^{\infty} dx' \int_{-1}^{+1} R(x, \mu'; x', \mu')\, d\mu'. \tag{7.4.23}$$

Integration of equation (7.4.20) according to discrete space theory (see chapter 6) in the r–μ–x mesh gives the equations as (Peraiah and Wehrse 1978)

$$\mathbf{M}\left[\mathbf{U}^+_{n+1} - U^+_n\right] + \rho_c\left[\mathbf{\Lambda}^+\mathbf{U}^+_{n+\frac{1}{2}} - \mathbf{\Lambda}^-\mathbf{U}^-_{n+\frac{1}{2}}\right] + \tau_{n+\frac{1}{2}}\mathbf{\Phi}^+_{n+\frac{1}{2}} =$$

$$\tau_{n+\frac{1}{2}}\mathbf{S}^+_{n+\frac{1}{2}} + \frac{1}{2}(1-\epsilon)\tau_{n+\frac{1}{2}}\left[\mathbf{R}^{++}\mathbf{W}^{++}\mathbf{U}^+ + \mathbf{R}^{+-}\mathbf{W}^{+-}\mathbf{U}^-\right]_{n+\frac{1}{2}} \tag{7.4.24}$$

and

$$\mathbf{M}\left[\mathbf{U}^-_n - U^-_{n+1}\right] - \rho_c\left[\mathbf{\Lambda}^+\mathbf{U}^-_{n+\frac{1}{2}} + \mathbf{\Lambda}^-\mathbf{U}^+_{n+\frac{1}{2}}\right] + \tau_{n+\frac{1}{2}}\mathbf{\Phi}^-_{n+\frac{1}{2}}\mathbf{U}^-_{n+\frac{1}{2}} =$$

$$\tau_{n+\frac{1}{2}}\mathbf{S}^-_{n+\frac{1}{2}} + \frac{1}{2}(1-\epsilon)\tau_{n+\frac{1}{2}}\left[\mathbf{R}^{-+}\mathbf{W}^{-+}\mathbf{U}^+ + \mathbf{R}^{--}\mathbf{W}^{--}\mathbf{U}^-\right]_{n+\frac{1}{2}}. \tag{7.4.25}$$

The various quantities in equations (7.4.24) and (7.4.25) are explained thus:

$\rho_c = (\Delta r/\bar{r})_{n+\frac{1}{2}}$ is the curvature factor, where $r_{n+\frac{1}{2}}$ is the average radius of the shell bounded by the radii r_n and r_{n+1} and $\Delta r = r_{n+1} - r_n$;

$$S^+_{n+\frac{1}{2}} = \left[\rho\beta + \epsilon\phi^+_k\right]_{n+\frac{1}{2}} B'_{n+\frac{1}{2}}\delta_{kk'}; \tag{7.4.26}$$

$$\mathbf{M} = [\mathbf{M}_m\delta_{mm'}], \quad \mathbf{M}_m = \left[\mu_j\delta_{jj}\right]; \tag{7.4.27}$$

$$\mathbf{\Lambda}^\pm = \left[\mathbf{\Lambda}^\pm_m\delta_{mm'}\right], \tag{7.4.28}$$

where $\Lambda^\pm$ are the curvature matrices given in chapter 6;

$$\mathbf{U}^+_n = 4\pi r_n^2\left[I^+_{1,n}, I^+_{2,n}, \ldots I^+_{I,n}\right]^T; \tag{7.4.29}$$

$$\Phi^+_{n+\frac{1}{2}} = \left[\Phi^+_{kk'}\right]_{n+\frac{1}{2}} = \left[\beta + \phi^+_k\right]_{n+\frac{1}{2}}\delta_{kk'}; \tag{7.4.30}$$

$(i, j) = k = j + (i-1)J, \quad 1 \le k \le K, K = IJ, I =$ total number of frequency points and J is the total number of angle points;

$$(\phi_i W_k)_{n+\frac{1}{2}} = a_{i,n+\frac{1}{2}} c_j, \tag{7.4.31}$$

where the c_js are the angle quadrature weights and

$$a_i = A_i\phi(x_i) / \sum_{i'=-I}^{I} A_i\phi(x_{i'});$$

the quantities $R^{++}, R^{-+}, R^{-+}, R^{--}$ are defined as follows:

$$R^{-+}_{i,i'}(r) = \left[R(x_i - \mu_1 V(r), x_{i'} + \mu_1 V(r), r), \ldots,\right.$$

$$\left.R(x_i - \mu_m V(r), x'_{i'} + \mu_m V(r); r\right]^T. \tag{7.4.32}$$

The procedure described in chapter 6 is applied and one gets the cell transmission and reflection operators, which are given by

$$\left.\begin{aligned}
\mathbf{t}(n+1,n) &= \mathbf{G}^{+-}\left[\Delta^{+}\mathbf{A}+\mathbf{g}^{+-}\mathbf{g}^{-+}\right],\\
\mathbf{t}(n,n+1) &= \mathbf{G}^{-+}\left[\Delta^{-}\mathbf{D}+\mathbf{g}^{-+}\mathbf{g}^{+-}\right],\\
\mathbf{r}(n+1,n) &= \mathbf{G}^{-+}\mathbf{g}^{-+}\left[\mathbf{E}+\Delta^{+}\mathbf{A}\right],\\
\mathbf{r}(n,n+1) &= \mathbf{G}^{+-}\mathbf{g}^{+-}\left[\mathbf{E}+\Delta^{-}\mathbf{D}\right],
\end{aligned}\right\} \tag{7.4.33}$$

with the source terms

$$\left.\begin{aligned}
\mathbf{\Sigma}^{+}_{n+\frac{1}{2}} &= \mathbf{G}^{+-}\tau\left[\Delta^{+}S^{+}+\mathbf{g}^{+-}\Delta^{-}S^{-}\right],\\
\mathbf{\Sigma}^{-}_{n+\frac{1}{2}} &= \mathbf{G}^{-+}\tau\left[\Delta^{-}S^{-}+\mathbf{g}^{-+}\Delta^{+}S^{+}\right],
\end{aligned}\right\} \tag{7.4.34}$$

where $\mathbf{E}$ is the unit matrix and

$$\left.\begin{aligned}
\mathbf{G}^{+-} &= \left[\mathbf{E}-\mathbf{g}^{+-}\mathbf{g}^{-+}\right]^{-1},\\
\mathbf{g}^{+-} &= \frac{\tau}{2}\Delta^{+}\mathbf{Y}_{-}.
\end{aligned}\right\} \tag{7.4.35}$$

$\mathbf{g}^{-+}$ and $\mathbf{G}^{-+}$ are obtained by interchanging the $+$ and $-$ signs in $\mathbf{G}^{+-}$ and $\mathbf{g}^{+-}$. Other matrices are

$$\left.\begin{aligned}
\mathbf{A} &= \mathbf{M}-\frac{\tau\mathbf{Z}_{+}}{2}, \quad \mathbf{D} = \mathbf{M}-\frac{\tau\mathbf{Z}_{-}}{2},\\
\Delta^{+} &= \left[\mathbf{M}+\frac{\tau\mathbf{Z}_{+}}{2}\right]^{-1}, \quad \Delta^{-} = \left[\mathbf{M}+\frac{\tau\mathbf{Z}_{-}}{2}\right]^{-1},\\
\mathbf{Y}_{+} &= \frac{\rho_c\mathbf{\Lambda}^{-}}{\tau}+\frac{\mathbf{R}^{-+}\mathbf{W}^{-+}}{2},\\
\mathbf{Y}_{-} &= -\frac{\rho_c\mathbf{\Lambda}^{-}}{\tau}+\frac{\mathbf{R}^{+-}\mathbf{W}^{+-}}{2},\\
\mathbf{Z}_{+} &= \phi^{+}-\frac{\mathbf{R}^{++}\mathbf{W}^{++}}{2}+\frac{\rho_c\mathbf{\Lambda}^{+}}{\tau},\\
\mathbf{Z}_{-} &= \phi^{-}-\frac{\mathbf{R}^{--}\mathbf{W}^{--}}{2}-\frac{\rho_c\mathbf{\Lambda}^{+}}{\tau}.
\end{aligned}\right\} \tag{7.4.36}$$

Using the above transmission and reflection operators and the scheme of the internal and emergent radiation field given in chapter 6, one gets the line transfer solution in the observer's frame of the expanding atmospheres.

This method has been applied to a planetary nebula using the redistribution function R_{II-A} (see chapter 1). The angle independent redistribution function has been used: this is given by

$$R_{II-A}(x, x') = \pi^{-\frac{1}{2}} \int_{|x-x'|/2}^{\infty} \exp(-u^2) \times \left[\tan^{-1}\left(\frac{\underline{x} + u}{a}\right) - \tan^{-1}\left(\frac{\bar{x} - u}{a}\right) \right] du, \tag{7.4.37}$$

where $\underline{x}$ and $\bar{x}$ are the minimum and maximum of x and x' and a is the damping constant ($= 4.3 \times 10^{-4}$) corresponding to pure radiation damping. In a moving medium, we need to compute the four redistribution functions, namely, $R(x + \mu V, x'+\mu' V)$, $R(x+\mu V, x'-\mu' V)$, $R(x-\mu V, x'+\mu' V)$ and $R(x-\mu V, x'-\mu' V)$. For the ϕ profile we could integrate R over x' but we have used the Voigt profile. As the medium becomes inhomogeneous in a moving medium and the line centre optical depth is greater than the critical step size (see chapter 6), we need to divide the medium into shells and subshells to satisfy the conditions of stability and non-negativity of the solution. If the shell is halved p times, the optical depth and the curvature factors in each shell are given by

$$\tau_{SS} = \exp(-p)\tau_s, \tag{7.4.38}$$

and

$$\rho_{SS} = \frac{\rho_s 2^{-p}}{\left[1 - \rho_s \left(2^{-1} - 2^{-p}\right)\right]}, \tag{7.4.39}$$

where τ_s and ρ_s are the optical depth and curvature factor in each shell respectively.

The inner and outer radii of the nebula are taken to be 3.6×10^{16} cm and 1.2×10^{17} cm (see Wehrse and Peraiah (1979)). The ionization structure adopted is from Baschek and Wehrse (1975). The central star's temperature is $T_e = 10^5$ K and its radius $R = 0.4 R_\odot$. The resulting optical depth (for 1000 hydrogen atoms per cm^3) at the centre of the hydrogen Lyman alpha line is approximately 1000. This is not very high compared with many of those observed. However, this is good enough to show how the method works.

A velocity law that gives the velocity in the nth shell is used and is given by

$$v_n = v_N + \Delta v \left(N - n + \frac{1}{2}\right), \tag{7.4.40}$$

where $\Delta v = (v_1 - v_N)/N$, N being the number of shells into which the nebula has been divided, 1 and N correspond to the outer and inner shells respectively. The fraction 1/2 on the RHS of equation (7.4.40) signifies the fact that the physical properties of a shell are represented by those of the shell centre. We have considered expansion velocities v up to 26 km s^{-1} ($\approx$ twice the mean thermal units). The profiles are shown in figures 7.6, 7.7 and 7.8.

Vardavas (1974, 1976) applied the redistribution function in the laboratory frame and obtained line profiles in a semi-infinite plane parallel medium. He found that there are large differences in the cores of the line compared to those formed by the partial frequency redistribution. However, Prabhjot Singh (1994) used the angle-averaged R_{III} function in spherically symmetric expanding media and found that the differences between complete redistribution and partial frequency redistribution are smaller than those found in the plane parallel case.

7.4.3 Integral form due to Averett and Loeser

Averett and Loeser (1984) devised a method which uses the integral form of the transfer equation in moving media in the observer's frame. They considered two types of rays: shell rays which do not intercept the central region of the star and disc rays which intercept the central region of the star. Consequently, the boundary conditions of these two rays differ. We shall describe this method by following the above reference and Sen and Wilson (1998).

We consider a spherically symmetric, radially expanding atmosphere (see figure 7.9) with core radius r_0 and outer radius R. The atmosphere is divided into several shells r_i $(i = 0, 1, 2, \ldots, N)$. The optical depth τ_i increases from $r_i = R$ to $r_i = r_0$.

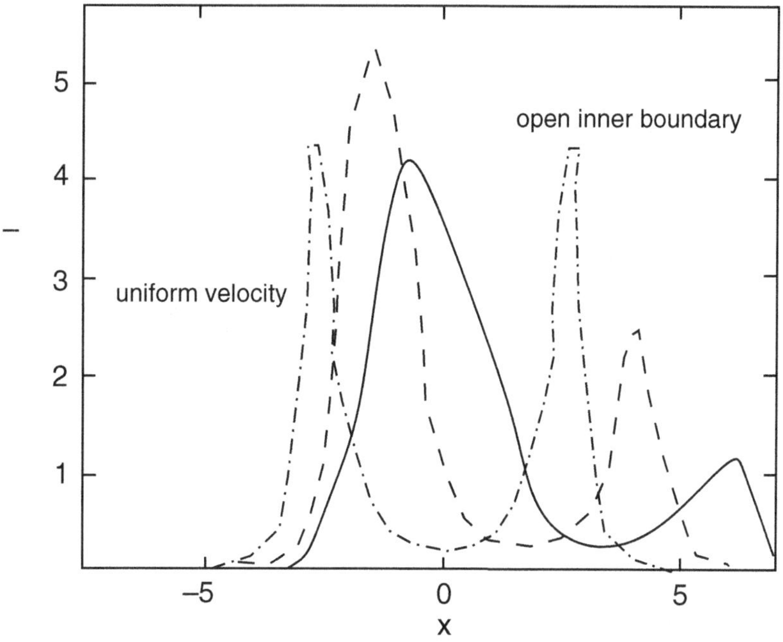

Figure 7.6 Profiles of the emergent H Lyα lines for nebulae expanding with uniform velocities and with the assumption of open inner boundaries. The full line refers to an expansion of $v = 2v_D$ ($\approx$ 26 km s^{-1}) and the dashed line to $v = v_D$ ($\approx$ 13 km s^{-1}). The static case is shown for comparison by the dash-dot curve. These are the profiles obtained by solving the line transfer equation in spherical symmetry in the observer's frame. The scale is arbitrary (from Wehrse and Peraiah (1979), with permission).

There are two types of rays: disc rays and shell rays (see figure 7.9). The shell ray intersects the circle with radius r_i at the point P and becomes tangent to the circle with radius r_m. Then if μ_{im} is the cosine of the angle between the outward ray with

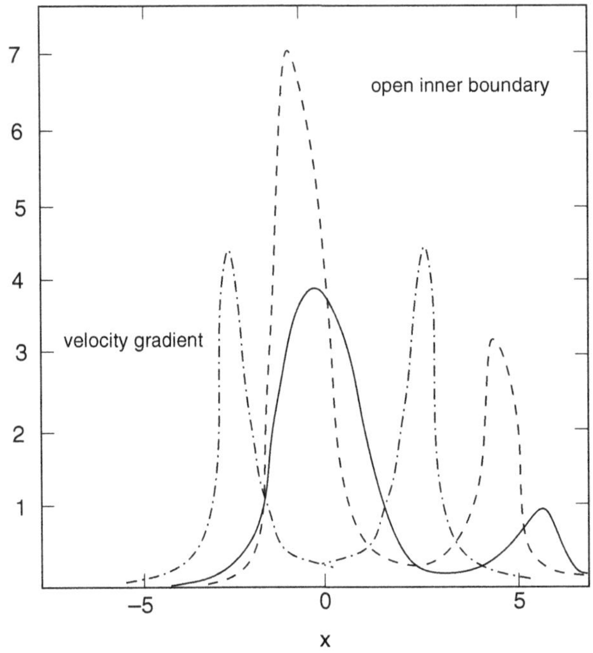

Figure 7.7 Emergent line profiles for nebulae expanding with velocity gradients $1v_D$ (dashed line) and $2v_D$ (full line). The inner boundary is assumed to be at rest. Photons are allowed to leave the system through the inner surface. The static case is indicated by the dash-dot curve (from Wehrse and Peraiah (1979), with permission).

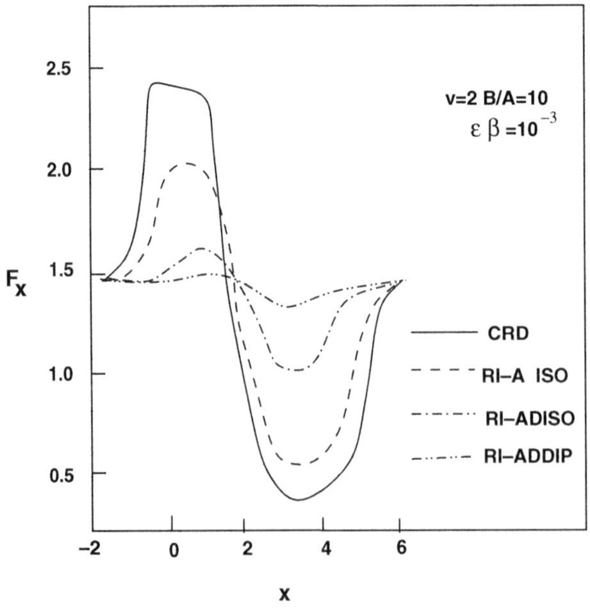

Figure 7.8 Profiles due to complete redistribution, and other redistribution functions (from Peraiah (1978), with permission).

intensity I^+ (I^- is the intensity of the inward ray) specified by r_m at r_i and the outward normal $\mathbf{n}$ at r_i (at P exactly),

$$\mu_{im} = \text{sign}\left[1 - \left(\frac{r_m}{r_i}\right)^2\right]^{\frac{1}{2}}. \tag{7.4.41}$$

Here $m < N$ for the shell ray, $m = cr_n,\ 0 \le c \le 1$ and

$$\text{sign} = \begin{cases} +1 & \text{for the disc ray,} \\ +1 & \text{for the near half of the shell ray,} \\ -1 & \text{for the far half of the shell ray.} \end{cases}$$

The line profile function $\phi(x)$ (where $x = \Delta\nu/\Delta\nu_D,\ \Delta\nu = \nu - \nu_0$) is generally taken to be Doppler, Lorentz or Voigt. In a static atmosphere the Doppler profile is $\phi(x) = \left(1/\pi^{1/2}\Delta\nu_D\right)\exp(-x^2)$. In terms of wavelength units this becomes $x = (\lambda - \lambda_0)/\Delta\lambda_D$ and $\Delta\lambda_D$, that is, the Doppler half width is given by

$$\Delta\lambda_D = \frac{\lambda_0}{c}\left(\frac{2kT}{M} + V_b^2\right)^{\frac{1}{2}}, \tag{7.4.42}$$

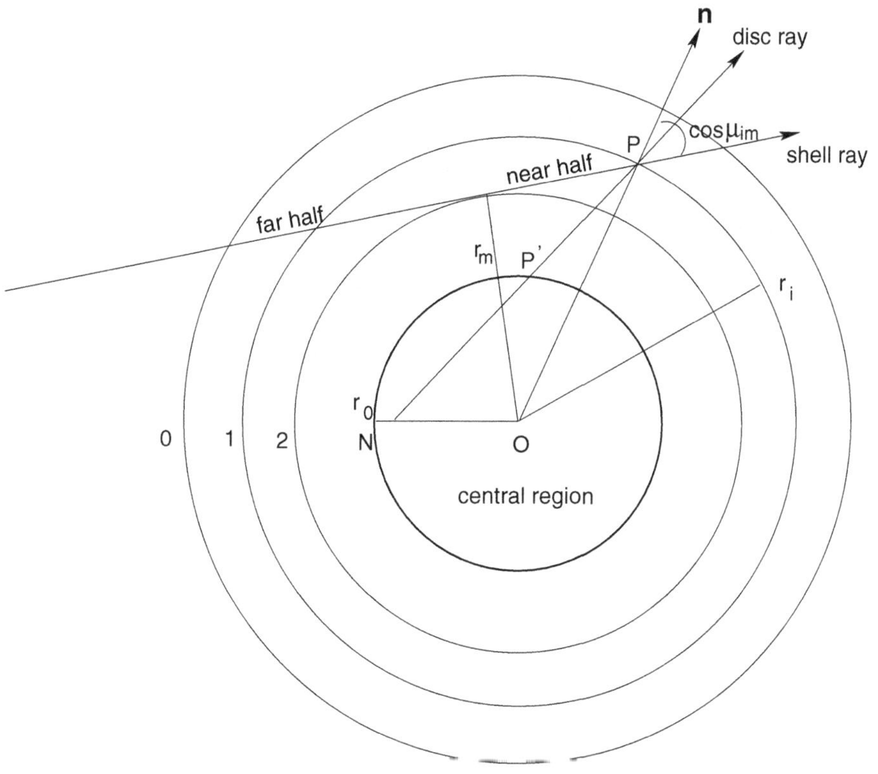

Figure 7.9 The shell rays through the point P does not intersect the innermost boundary of the spherical medium (or the the boundary of the central region). While the disc ray through the point P intersects the inner boundary at the point P'.

where V_b is the turbulent velocity, M is the atomic mass and T is the temperature.

Consider a point r_i moving with a radial velocity $V(r)$ then if μ_{im} is the cosine of the angle made by the ray and the normal at r_i

$$x_{ikm} = \frac{(\lambda_k - \lambda_0) + \frac{\lambda_0}{c}\mu_{im} V_i}{\Delta\lambda_D}. \tag{7.4.43}$$

The quantity $\mu_{im} V_i$ is the velocity component along the given ray. The subscript k represents the frequency grid point. If we consider the profile at r_l as seen in the comoving frame at r_l moving radially with velocity V_l, then x_{ikm} becomes

$$x^l_{ikm} = \frac{(\lambda_k - \lambda_0) + \frac{\lambda_0}{c}(\mu_{im} V_i - \mu_{lm} V_l)}{(\Delta\lambda_D)_i}. \tag{7.4.44}$$

This gives the optical depth along a ray from point r_l to any other point r_i as seen in the comoving frame at r_l. We need not specify the index l and simply write x_{ikm} in equation (7.4.44). Normally V_l is set equal to 0. Let us define $\bar{V}_i$ in terms of V_i at r_i as

$$\bar{V}_i = \frac{V_i}{c}\left(\frac{\lambda_0}{\Delta\lambda_D}\right) \quad \text{and set} \quad x_k = \frac{\lambda_k - \lambda_0}{\Delta\lambda_D}. \tag{7.4.45}$$

We can write, using the relation that $x = (\lambda - \lambda_0)/\Delta\lambda_D$,

$$x_{ikm} = x_k + \left(\mu_{im}\bar{V}_i - \mu_{lm}\bar{V}_l\right). \tag{7.4.46}$$

We can compute the optical depth τ_{ikm} of the shell of radius r_i as seen from r_l for the wavelength λ_k in the direction μ_{im}:

$$\tau_{ikm} = -\int \kappa(r)\phi(x_{ikm})\,(dr/\mu_{im}), \tag{7.4.47}$$

where $\kappa(r)$ is the absorption coefficient at r.

The transfer equation along the ray is

$$\frac{dI(r, \mathbf{s}, \nu)}{\kappa(r, \mathbf{s})\,ds} = -I(r, \mathbf{s}, \nu) + S(r, \nu). \tag{7.4.48}$$

If we write

$$d\tau_\nu = -\kappa(r)\phi(\nu)\,ds, \tag{7.4.49}$$

where $\phi(\nu)$ is the line profile function, equation (7.4.48) becomes

$$\frac{dI_\nu}{\tau_\nu} = I_\nu - S_\nu. \tag{7.4.50}$$

The formal solution of equation (7.4.50) is

$$I_\nu = \int S(\tau_{\nu'}) \exp(-\tau_{\nu'})\,d\tau_{\nu'} + c. \tag{7.4.51}$$

The constant c is zero for shell rays and depends on the physics at the inner boundary for the disc rays. The source function S contains the mean intensity and this is given at r_i by

$$\bar{J}_i = \frac{1}{2}\int_{-1}^{+1} d\mu \int_{-\infty}^{+\infty} \phi(x_{ikm}) I(r_i, \mu, \lambda)\, d\lambda. \tag{7.4.52}$$

The wavelength integral is replaced by the quadrature:

$$\int_{-\infty}^{+\infty} \phi(x_{ikm}) I(r_i, \mu_m, \lambda)\, d\lambda = \sum_k A_k \phi_k(u_{ikm}) I_{ikm} = \bar{I}_{im}, \quad \text{say}, \tag{7.4.53}$$

where the A_ks are appropriate quadrature weights such that the normalization over the profile is satisfied. Thus

$$\sum_k A_k \phi_k(x_{ikm}) = 1. \tag{7.4.54}$$

The mean intensity is now given by

$$\bar{J}_i = \sum_{m=1}^{M} C_{im} \bar{I}_{im}. \tag{7.4.55}$$

In the above relation, it is assumed that $\bar{J}$ varies linearly with μ between I_{ikm} at μ_{im} $(0 < \mu_{im} < 1)$ with $\mu_{1m} = 0$, $\mu_{iM} = 1$, where $m = 1, 2, \ldots, M$. The coefficients C_{im} are given by

$$C_{im} = \begin{cases} (\mu_{i2} - \mu_{i1}), & m = 1, \\ (\mu_{i(m+1)} - \mu_{i(m-1)}, & m \neq 1, \\ (\mu_{iM} - \mu_{i(M-1)}, & m = M, \end{cases} \tag{7.4.56}$$

while the angle quadrature weights satisfy the relation

$$\sum_m^M C_m = 1. \tag{7.4.57}$$

To obtain the auxiliary equation for S and $\bar{J}$, assume a standard two-level atom with complete redistribution.

From the formal solution (7.4.51) we get

$$I_{ikm} = \sum_{j=1}^{N} \bar{W}_{ijkm} S_j, \tag{7.4.58}$$

assuming S is isotropic. Or,

$$I_{ikm} - S_i = \sum_{j=1}^{N} W_{ijkm} S_j, \tag{7.4.59}$$

where

$$W_{ijkm} = \bar{W}_{ijkm} - \delta_{ij}. \tag{7.4.60}$$

From equations (7.4.46), (7.4.53) and (7.4.59) we obtain

$$\bar{J}_i = \sum_m^M C_{im}\left[\sum_k A_k\phi_k\left(x_{ikm}\right)\left(S_j + \sum_j^N W_{ijkm}S_j\right)\right], \tag{7.4.61}$$

$$\bar{J}_i = \sum_k A_k\left(\sum_j \bar{W}_{ijk}S_j\right) + S_i, \tag{7.4.62}$$

where

$$\bar{W}_{ijk} = \sum_m C_{im}W_{ijkm}\phi_k(x_{ikm}). \tag{7.4.63}$$

If we let

$$W'_{ij} = \sum_m C_m W_{ijkm}\phi_k(x_{ikm}), \tag{7.4.64}$$

then

$$\bar{J}_i = \sum_i W'_{ij}S_j + S_i. \tag{7.4.65}$$

The source function S and the mean intensity $\bar{J}$ are connected by

$$S_i = \frac{\bar{J}_i + \epsilon_i B_i}{1 + \epsilon_i} \tag{7.4.66}$$

(see chapter 1). Then from the preceding analysis, we obtain

$$S_i - \frac{1}{\epsilon_i}\sum W'_{ij}S_i = B_i, \tag{7.4.67}$$

where the B_is are the Planck functions. If the quadrature weighting coefficients W_{ijkm} are known in equation (7.4.59), equation (7.4.67) can be computed easily. From the formal solution, we know that $C = 0$ for the shell ray. Then,

$$\sum_{j=1}^N W_{ijkm}S_j = \frac{1}{2}\int_0^{\tau_{N'km}} S(t)\exp\left(-|t - \tau_{ikm}|\right)dt - S_i, \tag{7.4.68}$$

where τ_{ikm} is the optical depth and $N' = 2N$ for the present shell ray. If $S(t)$ is assumed to vary linearly between each pair of successive grid points τ_{jkm} and $\tau_{(j+1)km}$ and we define

$$\Delta_{jkm} = \tau_{(j+1),km} - \tau_{j,km}, \quad j = 1, 2, \ldots, N \tag{7.4.69}$$

and $\Delta_{0km} = \Delta_{Nkm} = 0$, W_{ijkm} becomes

$$W_{ijkm} = \frac{1}{2} \begin{cases} \exp[-(\tau_i - \tau_{j+1})]q(\Delta_j) + \exp[-(\tau_i - \tau_j)]m(\Delta_{j-1}), & j < 1 \\ m(\Delta_{j-1}) + m(\Delta_{i-2}), & j = 1 \\ \exp[-(\tau_{j-1} - \tau_i)]q(\Delta_{j-1}) + \exp[-(\tau_j - \tau_i)]m(\Delta_j), & j > 1 \end{cases} \tag{7.4.70}$$

where

$$m(\Delta) = 1 - \frac{1 - \exp(-\Delta)}{\Delta} \tag{7.4.71}$$

and

$$q(\Delta) = 1 - \frac{1 - \exp(-\Delta)}{\Delta} - \exp(-\Delta). \tag{7.4.72}$$

Here the indices km are suppressed on the RHS. In a semi-infinite medium, the disc ray will have $N' = N$ and the contributions $W'_{iN-1} = -K_i$ and $W'_{iN} = (1 + \Delta_{N-1})K_i$ are added to the last two columns of j obtained above with

$$K_i = (2\Delta_{N-1})^{-1} \exp[-(\tau_N - \tau_i)]. \tag{7.4.73}$$

Further improvements of the weights have been given in Averett and Loeser (1984).

Exercises

7.1 An atmosphere is expanding with a radial velocity of 2 Doppler units. Examine how the Doppler profile changes with $x = \pm 5$ Doppler units for different angles (take $\mu = 0.1, -0.5, 1.0$).

7.2 In exercise 7.1, instead of the Doppler profile, consider the angle independent R_I, R_{II} and R_{III}. Trace these functions for $\mu = 0.1, 0.5$ and 1.0.

7.3 Write a computer code for obtaining the r and t operators given in equation (7.4.33).

7.4 Obtain an expression for the radial velocity in an expanding atmosphere when the absorption depth of a weak line is known and show that the line depth as fraction of continuum is independent of μ.

REFERENCES

Athay, R.G., 1972, *Radiation Transport in Spectral Lines*, D. Reidel, Dordrecht, p 53.

Averett, E.H., Loeser, R., 1984, in *Methods in Radiative Transfer*, ed. W. Kalkofen, Cambridge University Press, Cambridge, p 341.

Baschek, B., Wehrse, R., 1975, *A&A*, **43**, 29.

Beals, C., 1929, *MNRAS*, **90**, 202.

Beals, C., 1930, *Publ. Dominion Astrophys. Obs., Victoria*, **4**, 271.

Beals, C., 1931, *MNRAS*, **9**, 966.

Beals, C., 1934, *Publ. Dominion Astrophys. Obs. Victoria*, **6**, 95.

Beals, C., 1950, *Publ. Dominion Astrophys. Obs. Victoria*, **9**, 1.

Chandrasekhar, S., 1934, *MNRAS*, **94**, 522.

Chandrasekhar, S., 1960, *Radiative Transfer*, Dover, New York.

Cram, L., 1972, *Solar Phys.*, **22**, 375.

Gerasimovič, B., 1934, *Z. f. Astrophys.*, **7**, 335.

Heasley, J., 1975, *Solar Phys.*, **44**, 275.

Hummer, D.G., Rybicki, G.B., 1967, *Methods of Computational Physics*, eds. B. Adler, S. Fernbach and M. Rotenberg, Academic Press, New York, p 53.

Hummer, D.G., Rybicki, G.B., 1968, *ApJ*, **153**, L107.

Kalkofen, W., 1970, in *Spectrum Formation in Stars with Steady State Extended Atmospheres*, eds. H. Groth and P. Wellmann, US Department of Commerce, Washington DC.

Kuan, P., Kuhi, L., 1975, *ApJ*, **199**, 148.

Kulander, J.L., 1967, *ApJ*, **147**, 1063.

Kulander, J.L., 1968, *JQSRT*, **8**, 273.

Kunasz, P.B., Hummer, D.G., 1974, *MNRAS*, **166**, 57.

Mihalas, D., 1978, *Stellar Atmospheres*, 2nd Edition, W.H. Freeman and Company, San Francisco.

Peraiah, A., 1978, *Kodaikanal Obs. Bull. Ser. A*, **2**, 115.

Peraiah, A., Wehrse, R., 1979, *A&A*, **70**, 213.

Prabhjot Singh, 1994, *MNRAS*, **269**, 442.

Rangarajan, K.E., 1997, *A&A*, **320**, 265.

Rybicki, G.B., 1970, in *Spectrum Formation in Stars with Steady State Extended Atmospheres*, US Department of Commerce, Washington DC.

Rybicki, G.B., Hummer, D.G., 1967, *ApJ*, **150**, 607.

Sen, K.K., Wilson, S.J., 1998, *Radiative Transfer in Moving Media*, Springer, Singapore.

Shine, R., 1975, *ApJ*, **202**, 543.

Struve, O., Elvey, C., 1934, *ApJ*, **79**, 409.

Struve, O., 1946, *ApJ*, **104**, 138.

Tam, R.E., Schwartz, E.D., 1976, *ApJ*, **204**, 842.

Vardavas, I.M., 1974, *JQSRT*, **14**, 909.

Vardavas, I.M., 1976, *JQSRT*, **16**, 901.

Wehrse, R., Peraiah, A., 1979, *A&A*, **71**, 289.

Wilson, O., 1934, *ApJ*, **80**, 259.

Chapter 8

Radiative transfer equation in the comoving frame

8.1 Introduction

In chapter 7, we studied the solution of the transfer equation in the rest frame of the observer. There are two difficulties in this way of treating the transfer equation: (1) the absorption and emission coefficients become angle dependent due to Doppler shifts in the frequency of the photon and hence become anisotropic and aberration of light is generated; and (2) the coupling between angle and frequency creates the practical problem of dealing with an unmanageably large mesh size for computation in scattering problems. This restricts the expansion velocities to a few times the mean thermal velocities. When the expansion velocities are very high one needs to use the comoving frame or the moving frame of the material. As the observer is with the moving frame, no Doppler shifts in the frequency of the photon occur and the opacity and emissivity are isotropic. One can use the redistribution functions for a static atmosphere. For problems involving scattering integrals, one can use a line profile with a band width which contains the full profile and which does not contain the velocity components. This makes the angle–frequency mesh small enough to contain the full profile of a static medium.

Lorentz transformations are used to describe the change (in the relevant physical variables) between the rest and comoving frames. Lorentz transforms are applicable when the relative velocity of the two frames is uniform and constant. In reality we find that the velocities are functions of not only radii but also time. This means that the fluid frame is not an inertial frame. We shall assume instantaneous transformation between uniformly moving frames and the moving fluid. In two uniformly moving frames the transfer equation is covariant if proper account is taken of Doppler shifts and the aberration of photons when atomic properties are calculated. In unsteady or steady differential flows, new terms appear to represent changes in the Lorentz transformation from one point to another in the medium. These terms can be derived by using the differential operator $(1/c)\partial/\partial t + \mathbf{n} \cdot \nabla$ in the transfer equation to the transformation coefficient of the specific intensity.

Velocity fields generate Doppler shifts and aberrations of photons giving rise to advection which describes the 'sweeping up' of radiation by the moving fluid. These effects contain terms of order v/c. However, in the case of line profiles, the effect of the frequency shift $\Delta\nu$ is significant only when $\Delta\nu/\Delta\nu_D = v/v_{th}$ is important and not when $\Delta\nu/\nu = v/c$. The Doppler effects due to velocity fields are amplified by a factor of c/v_{th} by the rapid variation of the line profile with frequency. Therefore one can ignore the aberration and advection terms and include only the terms due to Doppler shifts in the transfer equation.

Several authors have computed lines in expanding plane parallel and spherical media. Hewitt and Noerdlinger (1974) studied the transfer of resonance line radiation in differentially expanding atmospheres. Surdej (1979) computed line profiles in an expanding envelope using a Sobolev type approximation. Puetter and Hubbard (1983) considered the effects of source function variation and the total thickness on the formation of quasar emission lines. Bertout (1984) developed an efficient method for computing line profiles in stellar envelopes. Abbott (1978) derived the terminal velocities of stellar winds from early type stars.

8.2 Transfer equation in the comoving frame

If ν and ν_0 are the frequencies of a photon seen in the observer's and the comoving frame of the fluid, then

$$\nu = \nu_0 \left(1 + \frac{\mu_0 V}{c}\right). \tag{8.2.1}$$

If we consider the time independent radiative transfer equation in a spherically symmetric moving medium, the differential operator $\partial/\partial s$ in the observer's frame is given by

$$\frac{\partial I}{\partial s} = \mu \frac{\partial I(r, \mu, \nu)}{\partial r} + \frac{1 - \mu^2}{r} \frac{\partial I(r, \mu, \nu)}{\partial \mu}, \tag{8.2.2}$$

where the radial derivative $\mu \partial I(r, \mu, \nu)/\partial r$ is evaluated at constant ν. However, if we change the position from r to $r + \Delta r$ keeping ν constant, $\nu_0 = \nu_0(\nu, r)$ will change because the medium is moving with velocity $V(r)$ at r. The variables that are measured in the comoving frame are labelled by a superscript 0 such as $I^0(r, \mu_0, \nu_0)$, $j^0(r, \nu_0)$, $\kappa^0(r, \nu_0)$ etc. and the quantities μ_0, ν_0 by a subscript 0. The specific intensity in the comoving frame is

$$I^0 = I^0(r, \mu_0, \nu_0). \tag{8.2.3}$$

The differential operator dI^0/ds is given by

$$\frac{dI^0}{ds} = \frac{\partial I(r, \mu_0, \nu_0)}{\partial r}\frac{\partial r}{\partial s} + \frac{\partial I^0(r, \mu_0, \nu_0)}{\partial \mu_0}\frac{d\mu_0}{ds} + \frac{\partial I^0(r, \mu_0, \nu_0)}{\partial \nu_0}\frac{d\nu_0}{ds}. \tag{8.2.4}$$

Now, $\mu_0 = \cos\theta_0$, $dr = \cos\theta_0\, ds$, $rd\theta_0 = -\sin\theta_0\, ds$ (see chapter 2). Therefore

$$\frac{dr}{ds} = \mu_0, \qquad \frac{d\mu_0}{ds} = \frac{d\mu_0}{d\theta_0}\cdot\frac{d\theta_0}{ds} = \frac{\sin^2\theta_0}{r} = \frac{1-\mu_0^2}{r}. \tag{8.2.5}$$

From the Doppler effect (see equation (8.2.2)) we evaluate $d\nu_0/ds$ from

$$\frac{d\nu_0}{ds} = \frac{\partial \nu_0}{\partial r}\cdot\frac{dr}{ds} + \frac{\partial \nu_0}{\partial \mu_0}\cdot\frac{d\mu_0}{ds}. \tag{8.2.6}$$

From equation (8.2.1), we have

$$\frac{\partial \nu_0}{\partial r} = -\nu_0\frac{\mu_0}{c}\frac{dV}{dr} \tag{8.2.7}$$

and

$$\frac{\partial \nu_0}{\partial \mu_0} = \frac{\nu_0 V}{c}. \tag{8.2.8}$$

Substituting equations (8.2.7) and (8.2.8) into equation (8.2.6) we obtain

$$\frac{d\nu_0}{ds} = -\frac{\mu_0^2}{c}\frac{\partial V}{\partial r} - \nu_0\left(\frac{V}{c}\right)\frac{1-\mu_0^2}{r}. \tag{8.2.9}$$

Using equations (8.2.4), (8.2.5) and (8.2.9), the time independent transfer equation in the comoving frame in a spherically symmetric medium is obtained as

$$\mu_0\frac{\partial I^0(r, \mu_0, \nu_0)}{\partial r} + \frac{1-\mu_0^2}{r}\cdot\frac{\partial I^0(r, \mu_0, \nu_0)}{\partial \mu_0} =$$
$$j^0(r, \nu_0) - \kappa^0(r, \mu_0, \nu_0) + \frac{\nu_0 V}{rc}\left[\left(1-\mu_0^2\right) + \mu_0^2\frac{d\ln V}{d\ln r}\right]\frac{\partial I^0(r, \mu_0, \nu_0)}{\partial \nu_0}. \tag{8.2.10}$$

In planar geometry this is

$$\mu_0\frac{\partial I^0(z, \mu_0, \nu_0)}{\partial z} - \left[\left(\frac{\mu_0^2\nu_0}{c}\right)\frac{\partial V}{\partial z}\right]\left[\frac{\partial I^0(z, \mu_0, \nu_0)}{\partial \nu_0}\right] =$$
$$j^0(z, \nu_0) - \kappa^0(z, \nu_0)I^0(z, \mu_0, \nu_0). \tag{8.2.11}$$

We should note that:

1. The emission and absorption coefficients j^0 and κ^0 are isotropic in the comoving frame (unlike in the rest or the observer's frame).

2. Scattering terms (which have coupling between angle and frequency) can be evaluated in a small region around ν_0.
3. Equations of the type (8.2.10) lead to hyperbolic equations. The solution requires two boundary conditions in the space variable and one boundary condition in the frequency variable.
4. The frequency derivative gives the change in the photon frequency ν_0 at a given spatial point in the comoving frame as seen by an external observer or for the frequency shift of photons as seen in the comoving frame.
5. In the plane parallel case only velocity gradients are important but in spherical symmetry, a net effect occurs for constant velocities which mean transverse velocity gradients also occur.

Equation (8.2.10) for spherical symmetry in the fluid frame is more complicated. As we noted above, equation (8.2.10) is a mixed initial and boundary value problem for the coupled partial integro-differential equation which was first derived by McCrea and Mitra (1936). In spite of these complexities the physical advantages outweigh the mathematical difficulties. Chandrasekhar (1945a,b) obtained solutions in plane parallel geometry assuming complete redistribution over a rectangular profile in the comoving frame, treating all radiation flowing at the same angle to the radius vector and using a linear velocity law. Abhyankar (1964a,b, 1965) generalized this approach using coherent scattering and a two-stream approach. Lucy (1971) solved this problem in plane parallel geometry by neglecting the spatial derivative and including only the frequency derivative. Simonneau (1973) applied the integral method restricted to linear velocity laws which appears to be stable at large velocities. Sobolev (1947, 1957) exploited the concept of escape probability in the comoving frame in the presence of large velocity gradients. This was generalized to a small extent by Rublev (1961, 1964) and by Lyong (1967). Castor (1970) developed a more useful escape probability method.

Nordlinger and Rybicki (1974) developed a method similar to that of Feautrier (see chapter 4) in the comoving frame in planar geometry. Use of a complete redistribution function is less accurate when large velocity gradients exist. The angle-averaged partial redistribution function give inaccurate results (Hummer 1968, Magnan 1974) when used in the rest frame. It may not give inaccurate results when used in comoving frame calculations. In the rest frame, one needs to use the angle dependent redistribution functions.

8.3 Impact parameter method

If complete redistribution is used one can solve the transfer equation in the comoving frame using a Rybicki type solution (Mihalas *et al.*, 1975a) and if partial redistribution is used, then this means using moment equations which eliminate the angle variable and one can use the Feautrier type solution. We describe below a method following Mihalas *et al.* (1975a,b), Mihalas (1978)) (see also Sen and Wilson (1998)).

In spherical geometry the (p, z) coordinate system is adopted in place of (r, μ) (see chapters 4 and 7). By using the relations $r = r(p, z) = (p^2 + z^2)^{1/2}$ and $\mu = z/(p^2 + z^2)^{1/2}$ we can write equation (8.2.10) (suppressing the 0 super and subscripts to avoid confusion) as

$$\pm\left[\frac{\partial I^{\pm}(z, p, \nu)}{\partial z}\right] - \bar{\gamma}(z, p)\left[\frac{\partial I^{\pm}(z, p, \nu)}{\partial \nu}\right] = j(r, \nu) - \kappa(r, \nu)I^{\pm}(z, p, \nu), \tag{8.3.1}$$

where

$$\bar{\gamma}(z, p) = \frac{\nu V(r)}{cr}\left[1 - \mu^2 + \mu^2 \frac{d \ln V}{d \ln r}\right]. \tag{8.3.2}$$

It is understood that all the quantities are measured in the comoving frame.

We now introduce the Feautrier quantities, that is the mean-intensity- and flux-like variables defined as

$$u(z, p, \nu) = \frac{1}{2}\left[I^{+}(z, p, \nu) + I^{-}(z, p, \nu)\right] \tag{8.3.3}$$

and

$$v(z, p, \nu) = \frac{1}{2}\left[I^{+}(z, p, \nu) - I^{-}(z, p, \nu)\right]. \tag{8.3.4}$$

Furthermore, the optical depth is given by

$$d\tau(z, p, \nu) = -\kappa(z, p, \nu)\, dz. \tag{8.3.5}$$

Setting $\gamma(z, p, \nu) = \bar{\gamma}(z, p, \nu)/\kappa(z, p, \nu)$ and using equations (8.3.1)–(8.3.5), we obtain

$$\frac{\partial u(z, p, \nu)}{\partial \tau(z, p, \nu)} + \gamma(z, p, \nu)\frac{\partial v(z, p, \nu)}{\partial \nu} = v(z, p, \nu) \tag{8.3.6}$$

and

$$\frac{\partial v(z, p, \nu)}{\partial \tau(z, p, \nu)} + \gamma(z, p, \nu)\frac{\partial u(z, p, \nu)}{\partial \nu} = u(z, p, \nu) - S(z, p, \nu), \tag{8.3.7}$$

where the source function S corresponds to that of an equivalent two-level atom with complete redistribution or is of the form (see chapter 1)

$$S(z, p, \nu) = S\left(r(z, p), \nu\right) = (1 - \epsilon)\bar{J}(r) + \epsilon B(r), \tag{8.3.8}$$

where

$$\bar{J} - \int_{\nu_{min}}^{\nu_{max}} \phi(\nu)\, d\nu \int_0^1 d\mu\, (u(z(r, \mu), p(r, \mu), \nu)), \tag{8.3.9}$$

where ν_{min} and ν_{max} contain the whole line profile seen in the comoving frame. We can see from equation (8.3.9) that it is sufficient if we compute the line profile function ϕ only once, unlike in the observer's frame where we need to compute ϕ when ν changes to $\nu \to \nu_0(1 \pm [v_0(r)/c]\mu)$.

Boundary conditions. At the outer radius $r = R$, $I^- = 0$, hence $u = v$ and we therefore have

$$\left.\frac{\partial u(z, p, \nu)}{\partial \tau(z, p, \nu)}\right|_{z_{max}} + \gamma(z_{max}, p, \nu)\left[\frac{\partial u(z_{max}, p, \nu)}{\partial \nu}\right] = u(z_{max}, p, \nu). \tag{8.3.10}$$

At the plane of symmetry $z = 0$

$$v(0, p, \nu) = 0. \tag{8.3.11}$$

Therefore, for the rays that do not intersect the central portion (core) of the star:

$$\left.\frac{\partial u(z, p, \nu)}{\partial \tau(z, p, \nu)}\right|_{\tau=0} = 0. \tag{8.3.12}$$

For the rays that do intersect the core with $p \leq r_c$ there are two options: (1) if it is a stellar core then we apply the diffusion approximation giving v and (2) if the core is hollow (in the planetary nebula case) then $v = 0$ by symmetry. We need an initial condition for the frequency where $V(r) = 0$ and $dV(r)/dr > 0$. The high frequency edge at ν_{max} of the line profile in the comoving frame cannot intercept line photons from any other point in the atmosphere, as they will be red shifted. Any photon incident at the high frequency edge must be a continuum photon. We can write the initial condition as

$$u(z, p, \nu_{max}) = u_{continuum}. \tag{8.3.13}$$

This is obtained from equations (8.3.6) and (8.3.7) with the frequency derivative $\partial/\partial\nu$ set equal to zero, which gives $u_{continuum}$ radiation or we must have

$$\left(\frac{\partial u}{\partial \nu}\right)_{\nu_{max}} = 0, \tag{8.3.14}$$

which means that the slope of the continuum at ν_{max} is zero.

We now discretize the system of equations with the grid $\{r_d, p_i, z_{d,i}\}$ (see chapters 4 and 7) where

$$\begin{aligned} &\{r_d\}, \quad d = 1, 2, \ldots, D, \\ &\{p_i\}, \quad i = 1, 2, \ldots, I, \\ &\{z_{d,i}\}, \quad d, i = 1, 2, \ldots, DI. \end{aligned}$$

The frequency grid $\{\nu_n\}$ is chosen in order of decreasing values ($\nu_1 > \nu_2 > \cdots > \nu_N$) as n increases from 1 to N because the initial condition is set at the highest frequency. The quantity $\bar{J}$ in equation (8.3.9) can be written in terms of quadratures as

$$\bar{J}(r_i) = \sum_{n=1}^{N} \omega_n \sum a_{d,i} \phi(r_d, \nu_n) u[z(r_d, p_i), p_i, \nu_n]. \tag{8.3.15}$$

Equations (8.3.6) and (8.3.7) are replaced with their corresponding finite-difference equations, but before doing this we write $U,\ \nu,\ \kappa,\ \Delta\tau,\ \delta$ as follows:

(1) $$u_{d,in} = u(z_d, p_i, \nu_n); \tag{8.3.16}$$

(2) $$v_{(d\pm\frac{1}{2}),i,n} = v(z_{d\pm\frac{1}{2}}, p_i, \nu_n), \tag{8.3.17}$$

where

$$z_{d\pm\frac{1}{2}} = \frac{1}{2}(z_d \pm z_{d+1}); \tag{8.3.18}$$

(3) $$\kappa_{d\pm\frac{1}{2},i,n} = \frac{1}{2}\left[\kappa_{d,i,n} + \kappa_{(d\pm 1),i,n}\right]; \tag{8.3.19}$$

(4) $$\Delta\tau_{d,i,n} = \frac{1}{2}\left[\Delta\tau_{d+\frac{1}{2},i,n} + \Delta\tau_{d-\frac{1}{2},i,n}\right], \tag{8.3.20}$$

where

$$\Delta\tau_{d+\frac{1}{2},i,n} = \kappa_{d+\frac{1}{2},i,n}\left[z_d - z_{d+1}\right], \tag{8.3.21}$$

$$\Delta\tau_{d-\frac{1}{2},i,n} = \kappa_{d-\frac{1}{2},i,n}\left[z_{d-1} - z_d\right]; \tag{8.3.22}$$

and

(5) $$\delta_{d,i,n-\frac{1}{2}} = \gamma_{d,i,n}/(\nu_{n+1} - \nu_n). \tag{8.3.23}$$

Using equations (8.3.15)–(8.3.23), we can write equations (8.3.6) and (8.3.7) as follows:

$$\frac{\left(u_{d+1,i,n} - u_{d,i,n}\right)}{\Delta\tau_{d+\frac{1}{2},i,n}} = v_{d+\frac{1}{2},i,n} + \delta_{d+\frac{1}{2},i,n-\frac{1}{2}}\left(v_{d+\frac{1}{2},i,n} - v_{d+\frac{1}{2},i,n-1}\right) \tag{8.3.24}$$

and

$$\frac{\left(v_{d+\frac{1}{2},i,n} - v_{d-\frac{1}{2},i,n}\right)}{\Delta\tau_{d,i,n}} = u_{d,i,n} - S_{d,i,n} + \delta_{d,i,n-\frac{1}{2}}\left(u_{d,i,n} - u_{d,i,n-1}\right). \tag{8.3.25}$$

From equation (8.3.24), we obtain $v_{d+\frac{1}{2},i,n}$:

$$v_{d+\frac{1}{2},i,n} = \frac{\left\{\left[\frac{\left(u_{d+1,i,n} - u_{d,i,n}\right)}{\Delta\tau_{d+\frac{1}{2},i,n}}\right] + \delta_{d+\frac{1}{2},i,n-\frac{1}{2}} v_{d+\frac{1}{2},i,n-1}\right\}}{1 + \delta_{d+\frac{1}{2},i,n-\frac{1}{2}}}. \tag{8.3.26}$$

Using equations (8.3.25) and (8.3.26) $v_{d\pm\frac{1}{2},i,n}$ can be eliminated and a set of second order equation for $u_{d,i,n}$ obtained. This is given by

$$\begin{aligned}
&\frac{\left\{\frac{\left(u_{d+1,i,n} - u_{d,i,n}\right)}{\left[\Delta\tau_{d-\frac{1}{2},i,n}\left(1+\delta_{i-\frac{1}{2},i,n-\frac{1}{2}}\right)\right]} - \frac{\left(u_{d,i,n} - u_{d-1,i,n}\right)}{\left[\Delta_{d-\frac{1}{2},i,n}\left(1+\delta_{i-\frac{1}{2},i,n-\frac{1}{2}}\right)\right]}\right\}}{\Delta\tau_{d,i,n}} \\
&= \left(1+\delta_{d,i,n-\frac{1}{2}}\right) u_{d,i,n} - S_{d,i,n} - \delta_{d,i,n-\frac{1}{2}} u_{d,i,n-1} \\
&\quad + \left[\delta_{d-\frac{1}{2},i,n-\frac{1}{2}}\left(1+\delta_{d-\frac{1}{2},i,n-\frac{1}{2}}\right)^{-1} v_{d-\frac{1}{2},i,n-1}\right. \\
&\quad \left.- \delta_{d+\frac{1}{2},i,n-\frac{1}{2}}\left(1+\delta_{d+\frac{1}{2},i,n-\frac{1}{2}}\right)^{-1} v_{d+\frac{1}{2},i,n-1}\right] / \Delta\tau_{d,i,n}.
\end{aligned} \tag{8.3.27}$$

With the boundary conditions, we obtain the following system:

$$\mathbf{T}_{i,n}\mathbf{U}_{i,n} + \mathbf{U}'_{i,n}\mathbf{U}_{i,n-1} + \mathbf{V}'_{i,n}\mathbf{V}_{i,n-1} + \mathbf{W}_{i,n}\mathbf{J} = \mathbf{X}_{i,n}, \tag{8.3.28}$$

where $\mathbf{T}_{i,n}$ is tri-diagonal, $\mathbf{U}'_{i,n}$ and $\mathbf{W}_{i,n}$ are diagonal, $\mathbf{V}'_{i,n}$ is tri-diagonal and $\mathbf{X}_{i,n}$ is a vector. Also

$$\mathbf{U}_{i,n} = \left[u_{1,i,n}, u_{2,i,n}, \ldots, u_{D_{i-\frac{1}{2}},i,n}\right]^T \tag{8.3.29}$$

and $\mathbf{V}_{i,n}$ is given through equation (8.3.26) as

$$\mathbf{V}_{i,n} = \mathbf{G}_{i,n}\mathbf{U}_{i,n} + \mathbf{H}_{i,n}\mathbf{V}_{i,n-1}. \tag{8.3.30}$$

Here $\mathbf{G}_{i,n}$ is bi-diagonal and $\mathbf{H}$ is diagonal. The system is solved by choosing a particular ray for a given p_i and performing frequency-by-frequency integration from $n = 1$ to $n = N$ with the initial condition that $\mathbf{U}_{i,1}$, $\mathbf{V}'_{i,1}$ and $\mathbf{H}_{i,1}$ are all zeros. Then we get

$$\mathbf{U}_{i,1} = \mathbf{A}_{i,1} - \mathbf{B}_{i,1}\bar{\mathbf{J}} \tag{8.3.31}$$

and

$$\mathbf{V}_{i,1} = \mathbf{C}_{i,1} - \mathbf{D}_{i,1}\bar{\mathbf{J}}, \tag{8.3.32}$$

where

$$\mathbf{A}_{i,1} = \mathbf{T}_{i,1}^{-1}\mathbf{X}_{i,1}, \tag{8.3.33}$$

$$\mathbf{B}_{i,1} = \mathbf{T}_{i,1}^{-1}\mathbf{W}_{i,1}, \tag{8.3.34}$$

$$\mathbf{C}_{i,1} = \mathbf{G}_{i,1}\mathbf{A}_{i,1}, \tag{8.3.35}$$

$$\mathbf{D}_{i,1} = \mathbf{G}_{i,1}\mathbf{B}_{i,1}. \tag{8.3.36}$$

Successive values of n are given by the scheme

$$\mathbf{U}_{i,n} = \mathbf{A}_{i,n} - \mathbf{B}_{i,n}\bar{\mathbf{J}} \tag{8.3.37}$$

and

$$\mathbf{V}_{i,n} = \mathbf{C}_{i,n} - \mathbf{D}_{i,n}\bar{\mathbf{J}}, \tag{8.3.38}$$

where

$$\mathbf{A}_{i,n} = \mathbf{T}_{i,n}^{-1}\left(\mathbf{X}_{i,n} - \mathbf{U}'_{i,n}\mathbf{A}_{i,n-1} - \mathbf{V}'_{i,n}\mathbf{C}_{i,n-1}\right), \tag{8.3.39}$$

$$\mathbf{B}_{i,n} = \mathbf{T}_{i,n}^{-1}\left(\mathbf{W}_{i,n} - \mathbf{U}'_{i,n}\mathbf{B}_{i,n-1} - \mathbf{V}'_{i,n}\mathbf{D}_{i,n-1}\right), \tag{8.3.40}$$

$$\mathbf{C}_{i,n} = \mathbf{G}_{i,n}\mathbf{A}_{i,n} + \mathbf{H}_{i,n}\mathbf{C}_{i,n-1}, \tag{8.3.41}$$

$$\mathbf{D}_{i,n} = \mathbf{G}_{i,n}\mathbf{B}_{i,n} + \mathbf{H}_{i,n}\mathbf{D}_{i,n-1}. \tag{8.3.42}$$

For a given ray (p_i) with frequency ν_n equation (8.3.38) is substituted in equation (8.3.15) to obtain a final system of the form

$$\left(\mathbf{I} + \sum_{i,n}\mathbf{F}_{i,n}\mathbf{B}_{i,n}\right)\bar{\mathbf{J}} = \sum_{i,n}\mathbf{F}_{i,n}\mathbf{A}_{i,n}, \tag{8.3.43}$$

where the **F**s contain the quadrature weights. The solution of equation (8.3.43) gives **J** and therefore $S(r, \nu)$. The quantities $U(z, p, \nu)$ and $v(z, p, \nu)$ are obtained from equations (8.3.37) and (8.3.38). It is now possible to calculate $I^0(r, \nu)$ and $K^0(r, \nu)$ in the comoving frame. We can calculate the flux $H^0(r, \nu)$ from $v(r, \mu, \nu)$. Thus we obtain a complete radiation field in the comoving frame. These profiles should be computed in the frame of the observer at infinity (see figures 8.1 and 8.2 for the source functions and their corresponding line profiles).

8.4 Application of discrete space theory to the comoving frame

We shall now describe the solution of line transfer in the comoving frame using the discrete space theory described in chapter 6 (Peraiah 1980a,b). Unlike in the previous sections, the solutions are obtained by directly solving the transfer equations in the (r, μ) coordinate system in the spherically symmetric atmosphere. We obtain the source function and this source function is utilized in obtaining the line profile at infinity.

The transfer equation in the comoving frame of a spherically symmetric, expanding atmosphere (see equation (8.2.10)) for a non-LTE two-level atom with complete redistribution is

$$\begin{aligned}\pm\mu\frac{\partial I(x,\mu,r)}{\partial r} \pm \frac{1-\mu^2}{r}\frac{\partial I(x,\mu,r)}{\partial\mu} &= K(x,r)S_L(r)+K_c(r)S_c(r)\\ &-[K(x,r)+K_c(r)]I(x,\pm\mu,r)\\ &+\left[(1-\mu^2)\frac{V(r)}{r}+\mu^2\frac{dV(r)}{dr}\right]\frac{\partial I(x,\pm\mu,r)}{\partial x}.\end{aligned} \tag{8.4.1}$$

Here all variables are measured in the comoving frame. K and K_c are the absorption

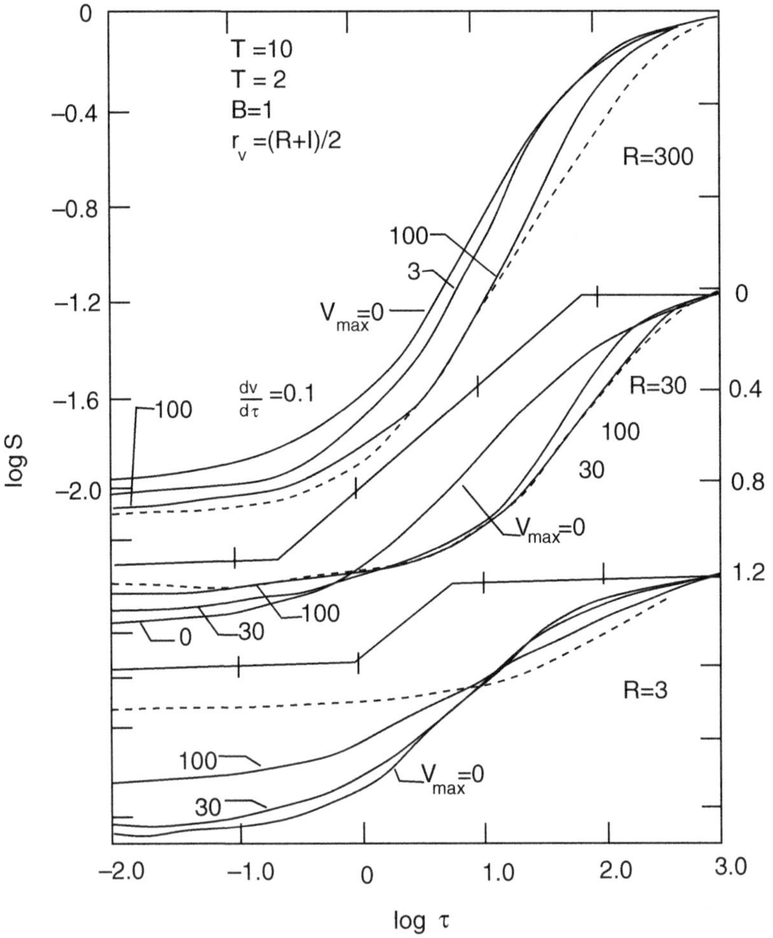

Figure 8.1 Line source functions versus line optical depths for isothermal models with various values of R and V_{max}. For each value of R, the continuum mean intensity J_c is drawn as a dashed curve. For $R = 300$, an additional source function is included for $V_{max} = 100$ and $dV/d\tau = -0.1$ as indicated (from Mihalas *et al.* (1975a), with permission).

coefficients per unit frequency interval in the line and continuum respectively. S_L and S_C are the line and continuum source functions respectively and are given by

$$S_L(r) = \frac{1-\epsilon}{2}\int_{-1}^{+1} d\mu \int_{-\infty}^{+\infty} dx\, \phi(x) I(x,\mu,r) + \epsilon B(r), \tag{8.4.2}$$

and

$$S_C(r) = \rho(r) B(\nu, T_e(r)), \tag{8.4.3}$$

$$K(x,r) = \phi(x) K_L(r), \tag{8.4.4}$$

where x is the frequency width in terms of a standard frequency interval (say Doppler) and is given by

$$x = \frac{\nu - \nu_0}{\Delta_d}, \tag{8.4.5}$$

and K_L is the line centre absorption coefficient. $\phi(x)$ is the profile function normalized such that

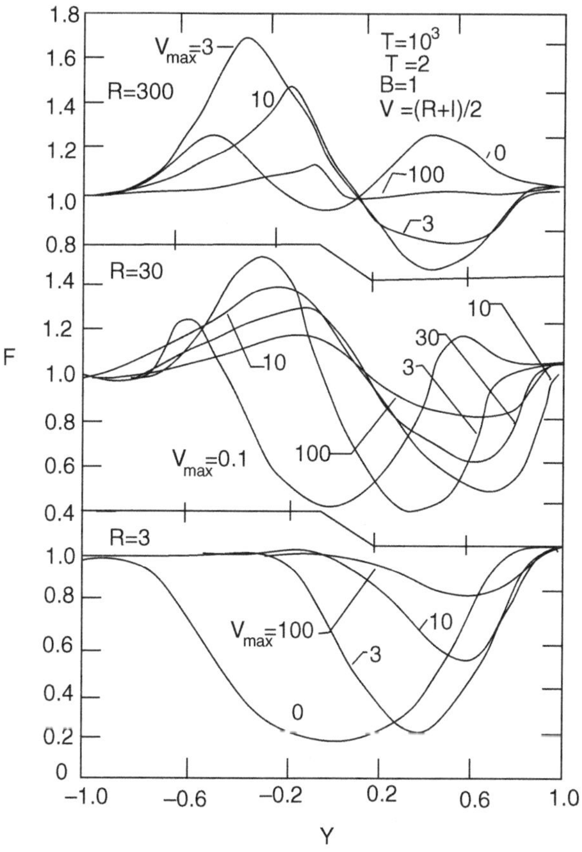

Figure 8.2 Emergent fluxes for the models of figure 8.1, normalized to unity in the continuum versus displacement $Y = X/X_{max}$, where X_{max} is chosen so that F is nearly unity at $Y = \pm 1$ (from Mihalas *et al.* (1975a), with permission).

$$\int_{-\infty}^{+\infty} \phi(x)\, dx = 1. \tag{8.4.6}$$

$\rho(r)$ is an arbitrary factor which is normally less than unity and $B(x, T(r))$ is the Planck function. $V(r)$ is the velocity of the gas in the atmosphere in units of thermal velocity. The quantity ϵ is the probability per scatter that a photon is lost by collisional de-excitation. We shall integrate equations (8.4.1) on the lines described in chapter 6, together with the comoving frame terms given by

$$\left[(1-\mu^2)\frac{V(r)}{r} + \mu^2 \frac{dV(r)}{dr}\right] \frac{\partial I(x, \pm\mu, r)}{\partial x}. \tag{8.4.7}$$

In discrete space theory, we approximate integrals by the appropriate quadrature sums and differentials by weighted differences. For example integration over the angle variable is effected by the formula

$$\int_0^1 f(\mu)\, d\mu \approx \sum_{j=1}^{J} c_j f(\mu_j), \; \sum_{j=1}^{J} c_j = 1, \tag{8.4.8}$$

where μ_j and c_j are the zeros and weights of the angle quadrature formula over [0,1] and J is the total number of angles. We define the matrices

$$\mathbf{c} = \left[c_j \delta_{jk}\right], \quad \mathbf{M}_m = \left[\mu_j \delta_{jk}\right], \tag{8.4.9}$$

and the corresponding intensities as

$$\mathbf{U}^{\pm}_{i,n} = 4\pi r^2 \begin{bmatrix} I(r_n, \pm\mu_1, x_i) \\ I(r_n, \pm\mu_2, x_i) \\ \vdots \\ I(r_n, \pm\mu_J, x_i) \end{bmatrix} \tag{8.4.10}$$

and

$$U^{\pm}_n = \left[\mathbf{U}^{\pm}_{1,n}, \mathbf{U}^{\pm}_{2,n}, \ldots, \mathbf{U}^{\pm}_{I,n}\right]^T. \tag{8.4.11}$$

where T indicates the transpose. We shall further write

$$\mathbf{\Phi}_{n+\frac{1}{2}} = [\Phi_{kk'}]_{n+\frac{1}{2}} = (\beta + \phi_k)_{n+\frac{1}{2}} \delta_{kk'}, \tag{8.4.12}$$

where

$$\beta = K_c / K_L \tag{8.4.13}$$

and

$$k = j + (i-1)J, \quad 1 \leq k \leq K = IJ, \tag{8.4.14}$$

with i and j the running indices of the frequency and angle quadrature points respectively and I the total number of frequency points. The subscript $n + \frac{1}{2}$ on a variable indicates that it represents the average of that variable over the shell bounded by radii r_n and r_{n+1}. We shall write

$$\phi_k = \phi(x_i, \mu_j), \tag{8.4.15}$$

$$S_{n+\frac{1}{2}} = (\rho\beta + \epsilon\phi_k)\, B'_{n+\frac{1}{2}}\delta_{kk'}, \tag{8.4.16}$$

$$B'_{n+\frac{1}{2}} = 4\pi r^2_{n+\frac{1}{2}} B(r_{n+\frac{1}{2}}), \tag{8.4.17}$$

$$\phi_i w_k = a_i c_j, \tag{8.4.18}$$

$$a_i = A_i\phi_i / \sum_{i'=-I}^{I} A_{i'}\phi(x_{i'}), \tag{8.4.19}$$

where the As are the weights of the frequency quadrature. Using equations (8.4.10)–(8.4.19), equations (8.4.1) are integrated over the angle–frequency–radius mesh and we obtain (see chapter 6)

$$\mathbf{M}\left[\mathbf{U}^+_{n+1} - \mathbf{U}^+_n\right] + \rho_c\left[\Lambda^+\mathbf{U}^+_{n+\frac{1}{2}} + \Lambda^-\mathbf{U}^-_{n+\frac{1}{2}}\right] + \tau_{n+\frac{1}{2}}\Phi_{n+\frac{1}{2}}\mathbf{U}^+_{n+\frac{1}{2}} =$$
$$\tau_{n+\frac{1}{2}}\mathbf{S}_{n+\frac{1}{2}} + \frac{1}{2}(1-\epsilon)\tau_{n+\frac{1}{2}}\left[\Phi\Phi^T\mathbf{W}\right]\left[\mathbf{U}^+ + \mathbf{U}^-\right]_{n+\frac{1}{2}} + \mathbf{M}_1\mathbf{dU}^+_{n+\frac{1}{2}} \tag{8.4.20}$$

and

$$\mathbf{M}\left[\mathbf{U}^-_{n} - \mathbf{U}^-_{n+1}\right] - \rho_c\left[\Lambda^+\mathbf{U}^-_{n+\frac{1}{2}} + \Lambda^-\mathbf{U}^+_{n+\frac{1}{2}}\right] + \tau_{n+\frac{1}{2}}\Phi_{n+\frac{1}{2}}\mathbf{U}^-_{n+\frac{1}{2}} =$$
$$\tau_{n+\frac{1}{2}}\mathbf{S}_{n+\frac{1}{2}} + \frac{1}{2}(1-\epsilon)\tau_{n+\frac{1}{2}}\left[\Phi\Phi^T\mathbf{W}\right]\left[\mathbf{U}^+ + \mathbf{U}^-\right]_{n+\frac{1}{2}} + \mathbf{M}_1\mathbf{dU}^-_{n+\frac{1}{2}}, \tag{8.4.21}$$

with

$$\mathbf{M} = \begin{pmatrix} \mathbf{M}_m & & 0 \\ & \mathbf{M}_m & \\ & & \ddots & \\ 0 & & & \mathbf{M}_m \end{pmatrix}, \quad \mathbf{\Lambda}^\pm = \begin{pmatrix} \mathbf{\Lambda}^\pm_m & & 0 \\ & \mathbf{\Lambda}^\pm_m & \\ & & \ddots & \\ 0 & & & \mathbf{\Lambda}^\pm_m \end{pmatrix}, \tag{8.4.22}$$

where $\mathbf{M}_m = [\mu_j\delta_{jj'}]$ and $\mathbf{\Lambda}^\pm_m$ are the curvature matrices. The comoving terms given in equation (8.4.7) are represented by the terms $M_1\mathbf{dU}^\pm_{n+\frac{1}{2}}$ in equations (8.4.20) and (8.4.21). $\mathbf{M}_1$ is given by

$$\mathbf{M}_1 = \left[\mathbf{M}^1 \Delta V_{n+\frac{1}{2}} + \mathbf{M}^2 \rho_c V_{n+\frac{1}{2}}\right], \tag{8.4.23}$$

where

$$\mathbf{M}^1 = \begin{pmatrix} M_m^1 & & & 0 \\ & M_m^1 & & \\ & & \ddots & \\ 0 & & & M_m^1 \end{pmatrix}, \quad \mathbf{M}_m^1 = \left[\mu_j^2 \delta_{jj'}\right] \tag{8.4.24}$$

and

$$\mathbf{M}^2 = \begin{pmatrix} M_m^2 & & & 0 \\ & M_m^2 & & \\ & & \ddots & \\ 0 & & & M_m^2 \end{pmatrix}, \quad \mathbf{M}_m^2 = \left[\mu_j^2 \delta_{jj'}\right], \tag{8.4.25}$$

where $j, j' = 1, 2, \ldots, J$, ρ_c is the curvature factor $\Delta r / r_{n+\frac{1}{2}}$, $\Delta V_{n+\frac{1}{2}} = V_{n+1} - V_n$ and $V_{n+\frac{1}{2}}$ is the average velocity over the shell bounded by the radii r_n and r_{n+1}. The matrix $\mathbf{d}$ is determined from the condition of flux conservation and is given by

$$\mathbf{d} = \begin{pmatrix} -d_1 & d_1 & 0 & & & & \\ -d_2 & 0 & d_2 & 0 & & & \\ 0 & -d_3 & 0 & d_3 & 0 & & \\ & & & & \ddots & \ddots & \\ 0 & & & & & -d_I & d_I \end{pmatrix}, \tag{8.4.26}$$

where

$$d_i = (x_{i-1} - x_i)^{-1} \quad \text{for } i = 2, 3, \ldots, I - 1 \tag{8.4.27}$$

and we shall set $d_i = d_I = 0$.

The average intensities are approximated by the diamond scheme (Grant 1968):

$$\left.\begin{aligned} \left(\mathbf{E} - \mathbf{X}_{n+\frac{1}{2}}\right)\mathbf{U}_n^+ + \mathbf{X}_{n+\frac{1}{2}} U_{n+1}^+ &= \mathbf{U}_{n+\frac{1}{2}}^+, \\ \left(\mathbf{E} - \mathbf{X}_{n+\frac{1}{2}}\right)\mathbf{U}_{n+1}^- + \mathbf{X}_{n+\frac{1}{2}} U_n^+ &= \mathbf{U}_{n+\frac{1}{2}}^-, \end{aligned}\right\} \tag{8.4.28}$$

where $\mathbf{X}_{n+\frac{1}{2}} = \frac{1}{2}\mathbf{E}$, $\mathbf{E}$ being the unit matrix. Introducing equation (8.4.28) into equations (8.4.20) and (8.4.21), we obtain

$$\begin{pmatrix} \mathbf{A} & \mathbf{B} \\ \mathbf{C} & \mathbf{D} \end{pmatrix} \begin{pmatrix} \mathbf{U}_{n+1}^+ \\ \mathbf{U}_{n+1}^- \end{pmatrix} = \begin{pmatrix} \mathbf{A}' & \mathbf{B}' \\ \mathbf{C}' & \mathbf{D}' \end{pmatrix} \begin{pmatrix} \mathbf{U}_n^+ \\ \mathbf{U}_{n+1}^- \end{pmatrix} + \tau \begin{pmatrix} \mathbf{S} \\ \mathbf{S} \end{pmatrix}, \tag{8.4.29}$$

where

$$\mathbf{A} = \mathbf{M} + \frac{1}{2}\rho_c \mathbf{\Lambda}^+ + \frac{1}{2}\tau \mathbf{\Phi} - \frac{1}{4}\sigma\tau \left[\boldsymbol{\phi}\boldsymbol{\phi}^T W\right] - \frac{1}{2}\mathbf{M}_1 \mathbf{d}, \tag{8.4.30}$$

$$\mathbf{B} = \frac{1}{2}\rho_c \mathbf{\Lambda}^- - \frac{1}{4}\tau\sigma \left[\boldsymbol{\phi}\boldsymbol{\phi}^T \mathbf{W}\right], \tag{8.4.31}$$

$$\mathbf{C} = -\frac{1}{2}\rho_c \mathbf{\Lambda}^- - \frac{1}{4}\tau\sigma \left[\boldsymbol{\phi}\boldsymbol{\phi}^T \mathbf{W}\right], \tag{8.4.32}$$

$$\mathbf{D} = \mathbf{M} - \frac{1}{2}\rho_c \mathbf{\Lambda}^+ + \frac{1}{2}\tau\mathbf{\Phi} - \frac{1}{4}\sigma\tau \left[\boldsymbol{\phi}\boldsymbol{\phi}^T \mathbf{W}\right] - \frac{1}{2}\mathbf{M}_1\mathbf{d}, \tag{8.4.33}$$

$$\mathbf{A}' = \mathbf{M} - \frac{1}{2}\rho_c \mathbf{\Lambda}^+ - \frac{1}{2}\tau\mathbf{\Phi} + \frac{1}{4}\sigma\tau \left[\boldsymbol{\phi}\boldsymbol{\phi}^T \mathbf{W}\right] + \frac{1}{2}\mathbf{M}_1\mathbf{d}, \tag{8.4.34}$$

$$\mathbf{B}' = -\frac{1}{2}\rho_c \mathbf{\Lambda}^- + \frac{1}{4}\tau\sigma \left[\boldsymbol{\phi}\boldsymbol{\phi}^T \mathbf{W}\right], \tag{8.4.35}$$

$$\mathbf{C}' = \frac{1}{2}\rho_c \mathbf{\Lambda}^- + \frac{1}{4}\tau\sigma \left[\boldsymbol{\phi}\boldsymbol{\phi}^T \mathbf{W}\right], \tag{8.4.36}$$

$$\mathbf{D}' = \mathbf{M} + \frac{1}{2}\rho_c \mathbf{\Lambda}^+ - \frac{1}{2}\tau\mathbf{\Phi} + \frac{1}{4}\sigma\tau \left[\boldsymbol{\phi}\boldsymbol{\phi}^T \mathbf{W}\right] + \frac{1}{2}\mathbf{M}_1\mathbf{d}, \tag{8.4.37}$$

where $\sigma = 1 - \epsilon$ and $\tau = \tau_{n+\frac{1}{2}}$.

A comparison of equation (8.4.29) with the interaction principle gives us the reflection and transmission operators and the source vectors for the basic cell (see chapter 6 for the procedure). The reflection and transmission operators are

$$\mathbf{T}(n+1, n) = \boldsymbol{\alpha}^{+-} \left[\mathbf{\Delta}^- \mathbf{\Gamma}^+ + \boldsymbol{\beta}^{+-}\boldsymbol{\beta}^{-+}\right], \tag{8.4.38}$$

$$\mathbf{R}(n+1, n) = \boldsymbol{\alpha}^{-+}\boldsymbol{\beta}^{-+} \left[\mathbf{E} + \mathbf{\Delta}^+ \mathbf{\Gamma}^+\right]. \tag{8.4.39}$$

$\mathbf{T}(n, n+1)$ and $\mathbf{R}(n, n+1)$ can be obtained by interchanging $+$ and $-$ signs in the corresponding $\mathbf{T}$ and $\mathbf{R}$ operators. The various quantities in equations (8.4.38) and (8.4.39) are given below:

$$\mathbf{Y} = \frac{1}{2}\sigma \left[\boldsymbol{\phi}\boldsymbol{\phi}^T \mathbf{W}\right], \quad \mathbf{Z} = \mathbf{\Phi} - \mathbf{Y}, \tag{8.4.40}$$

$$\mathbf{Z}_+ = \mathbf{Z} + \frac{\rho_c \mathbf{\Lambda}^+}{\tau} - \frac{\mathbf{M}_1\mathbf{d}}{\tau}, \quad \mathbf{Y}_+ = \mathbf{Y} + \frac{\rho_c \mathbf{\Lambda}^-}{\tau}, \tag{8.4.41}$$

$$\mathbf{Z}_- = \mathbf{Z} - \frac{\rho_c \mathbf{\Lambda}^+}{\tau} - \frac{\mathbf{M}_1\mathbf{d}}{\tau}, \quad \mathbf{Y}_+ = \mathbf{Y} - \frac{\rho_c \mathbf{\Lambda}^-}{\tau}, \tag{8.4.42}$$

$$\mathbf{\Delta}^\pm = \left[\mathbf{M} + \frac{1}{2}\tau\mathbf{Z}_\pm\right]^{-1}, \quad \mathbf{\Gamma}^\pm = \left[\mathbf{M} - \frac{1}{2}\tau\mathbf{Z}_\pm\right]^{-1}, \tag{8.4.43}$$

$$\boldsymbol{\beta}^{+-} = \frac{1}{2}\tau\mathbf{\Delta}^+\mathbf{Y}_-, \quad \boldsymbol{\alpha}^{+-} = \left[\mathbf{E} - \boldsymbol{\beta}^{+-}\boldsymbol{\beta}^{-+}\right]^{-1}. \tag{8.4.44}$$

$\boldsymbol{\beta}^{-+}$, $\boldsymbol{\alpha}^{-+}$ are written by interchanging the signs. The source vectors are

$$\mathbf{\Sigma}^{+}_{n+\frac{1}{2}} = \tau\boldsymbol{\alpha}^{+-}\left[\mathbf{\Delta}^{+} + \boldsymbol{\beta}^{+-}\mathbf{\Delta}^{-}\right]\mathbf{S}, \tag{8.4.45}$$

$$\mathbf{\Sigma}^{-}_{n+\frac{1}{2}} = \tau\boldsymbol{\alpha}^{-+}\left[\mathbf{\Delta}^{-} + \boldsymbol{\beta}^{-+}\mathbf{\Delta}^{+}\right]\mathbf{S}. \tag{8.4.46}$$

We need to obtain the non-negative **R** and **T** operators given in equations (8.4.38) and (8.4.39). To achieve this we need to see that the matrices $\mathbf{\Delta}^{\pm}$ in equation (8.4.43) at positive in which case the following conditions on the elements of $(\mathbf{\Delta})^{-1}$ matrices are required. The diagonal elements of $(\mathbf{\Delta})^{-1}$ should be positive and dominant. This condition is obtained if

$$\tau_{k,k} < \left| \frac{2\mu_k \pm \rho_c \Lambda^{+}_{k,k} - d_{k,k}\left[\mu^2_{kk}\Delta V_{n+\frac{1}{2}} + (1-\mu^2_{kk})\rho_c V_{n+\frac{1}{2}}\right]}{(\beta+\phi_k) - \frac{1}{2}\sigma(\phi\phi^T W)_{kk}} \right|. \tag{8.4.47}$$

The off-diagonal elements of $(\mathbf{\Delta}^{\pm})^{-1}$ should be negative. This happens if

$$\tau_{k,k+1} < \left| \frac{2\rho_c \Lambda^{+}_{k,k+1} - 2d_{k,k+1}\left[\Delta V_{n+\frac{1}{2}}\mu^2_{k,k+1} + \rho_c V_{n+\frac{1}{2}}\left(1-\mu^2_{k,k+1}\right)\right]}{\sigma\left(\phi\phi^T W\right)_{k,k+1}} \right| \tag{8.4.48}$$

for the upper diagonal elements and

$$\tau_{k+1,k} < \left| \frac{2\rho_c \Lambda^{+}_{k+1,k} + 2d_{k+1,k}\left[\Delta V_{n+\frac{1}{2}}\mu^2_{k+1,k} + \rho_c V_{n+\frac{1}{2}}\left(1-\mu^2_{k+1,k}\right)\right]}{\sigma\left(\phi\phi^T W\right)_{k+1,k}} \right| \tag{8.4.49}$$

for the lower diagonal elements. We have to select $\tau_{crit}(=\tau_{cell})$ such that

$$\tau_{crit} = \min\left\{\tau_{k+1,k}, \tau_{k,k}, \tau_{k,k+1}\right\}. \tag{8.4.50}$$

We use τ_{crit} to obtain the **R** and **T** operators and the source vectors. τ_{crit} depends on the number of angles, the number of frequencies, the velocity gradients, the local velocities, the curvature factors, the profile functions, and the thermalization parameter ϵ. The atmosphere is divided into N shells so that each shell has an optical depth $\leq \tau_{crit}$. To solve the problem, we need two spatial boundary conditions and one initial condition on frequency. The radiation field is incident on either side of the atmosphere as given below:

(a) $$U^{-}_{N+1}\left(x_i, \tau = T, \mu_j\right) = f \text{ (say)}. \tag{8.4.51}$$

In this case one can use the diffusion approximation on the surface of the star or as in the Schuster problem, we can specify the intensity emerging from the star at $\tau = T$:

(b) $$U^{+}_{1}(x_i, \tau = 0, \mu_j) = 0, \tag{8.4.52}$$

which means that no radiation is incident from outside the atmosphere.

In the case of the frequency boundary condition we set $\partial I/\partial x|_{continuum} = 0$ (see section 8.3). This leads to

$$\left[\mathbf{d}U^{\pm}_{n+\frac{1}{2}}\right]_{i=1 \text{ and } I} = 0. \tag{8.4.53}$$

In addition to the above conditions, we need to specify the boundary conditions on velocity. Several velocity laws are available and here we use a linear velocity law, given by

$$V(r) = V_A + \frac{V_B - V_A}{B - A}(r - A), \tag{8.4.54}$$

where A and B are the inner and outer radii of the atmosphere and V_A, V_B and $V(r)$ are the velocities at radii A, B and r respectively. It should be noted that other velocity laws can be used as easily as the above law.

The **R** and **T** operators are computed for all the shells, each of which has an optical depth $\leq \tau_{crit}$. Then by the internal radiation field scheme (see chapter 6), we can obtain the intensities $U^{\pm}$ at all the boundaries of the shells. Then the source functions are calculated at different radial points or the shell boundaries.

The line source function is computed from the following:

$$S_L(r_n) = \frac{1-\epsilon}{2}\sum a_i\phi_i \sum c_j\left(U_n^+ + U_n^-\right), \tag{8.4.55}$$

from which one can get the total source function:

$$S(r_i, x_i) = \frac{\phi(x_i)}{\beta + \phi(x_i)}S_L(r) + \frac{\beta}{\beta + \phi(x_i)}S_c(r). \tag{8.4.56}$$

We can use the profile function corresponding to a static medium. Consequently, we need to use only a small number of frequency points which will reduce the size of angle-frequency mesh. Equally spaced trapezoidal points are used for the frequency grid. To test the accuracy, 9, 11, 13, 15, and 19 points on the frequency grid are chosen, odd numbers being selected in order to include the centre of the line at $x = 0$. The source functions S_9, S_{11}, S_{13}, S_{15}, S_{19} have been computed corresponding to 9, 11, 13, 15 and 19 frequency points. S_9 differs from the others in the fourth or fifth place, whereas S_{11}, S_{13}, S_{15} and S_{19} agree to about ninth place. In order to save computer time and storage space, one can use 11 frequency points and 2–4 angle points.

In the rest frame, the application of complete redistribution would lead to inaccurate results (Magnan 1974) but the results may not be as inaccurate in the comoving frame. However, one needs to employ angle dependent partial redistribution functions in both the frames of moving media. However, according Mihalas *et.al.* (1975a) the use of the angle-averaged redistribution function in the comoving frame may be 'sufficiently accurate'.

One can use the angle-averaged redistribution function in the discrete space theory method (see Peraiah (1980a)). The line source function in this case becomes

$$S_L(x,r) = \frac{(1-\epsilon)}{\phi(x)} \int_{-\infty}^{\infty} dx' \int_{-1}^{+1} R(x,x') I(x',\mu')\, d\mu' + \epsilon B(x, T_e(r)), \tag{8.4.57}$$

where $R(x, x')$ is the angle-averaged partial frequency redistribution and $\phi(x)$ is obtained from the relation

$$\phi(x) = \int_{-\infty}^{+\infty} R(x,x')\, dx'. \tag{8.4.58}$$

The integration of the transfer equation in the comoving frame equation (8.4.1) with the line source function with angle-averaged partial frequency redistribution given in equation (8.4.57) can be over the discrete radius–angle–frequency mesh similar to that of complete redistribution. The only change that is introduced into the discrete equations (8.4.29)–(8.4.37) is that the matrix $[\boldsymbol{\phi\phi}^T\mathbf{W}]$ is replaced by $[\mathbf{RW}]_{n+\frac{1}{2}}$, where $\mathbf{W}$ is defined in equation (8.4.18). Once we obtain the source function $S(r, x)$ in the comoving frame, we need to compute line fluxes in the frame of the observer at infinity.

In figure 8.3(a), we show a spherically symmetric, radially expanding atmosphere surrounding a star. One can explain qualitatively the absorption and emission features of a spectral line formed in these objects. The formation of lines in these atmospheres depends on several factors: density distribution, geometrical extension, expansion velocities etc. Let us consider a transparent envelope so that all photons reach the observer. For simplicity, we consider a line centre frequency for either emission or absorption for which the external observer will receive the line centre frequency from any part of the medium Doppler shifted by an amount corresponding to the velocity of expansion along the line of sight.

Region (4) in figure 8.3(a) is the region occulted from the observer by the stellar disc and therefore cannot be seen by the observer. Region (5), the region projected onto the stellar disc, can emit radiation either without any major reabsorption (similar to the formation of a forbidden line in a nebula) or when the electron temperature is much greater than the colour temperature of the radiation from the background photosphere. In this case, one would obtain a violet shifted emission feature. Another possibility that can occur in this region is that the incident photospheric radiation is absorbed or scattered out of the line of sight resulting in the formation of a violet shifted absorption feature. Region (3) (the emission lobes of the sides of the disc) generates photons which are emitted thermally or scattered by stellar and diffuse radiations from the envelope. In these emission lobes, the line of sight velocities lie between positive and negative values producing a symmetric emission feature covering from the violet to the red side of the central frequency of the line. As region (4) is occulted from the observer by the stellar disc, we cannot observe the maximum red shift, therefore we can obtain information about the velocities only from the blue shift of the absorption or emission feature.

We shall now describe a method to obtain the flux profiles once the source function is known as a function of frequency and radius. The envelope is divided into N shells (strictly speaking sectors) with radii r_n, $r_1 = A$ (the inner radius) and $r_N = B$ (the outer radius). We know in advance the values of the absorption coefficient $\kappa(r)$, $S(r)$, $V(r)$. We select a set of N rays parallel to the direction of the observer at infinity. First we consider the side lobes (region (3)). Each ray is defined by an impact parameter p_n corresponding to the ray n tangential to the sector with

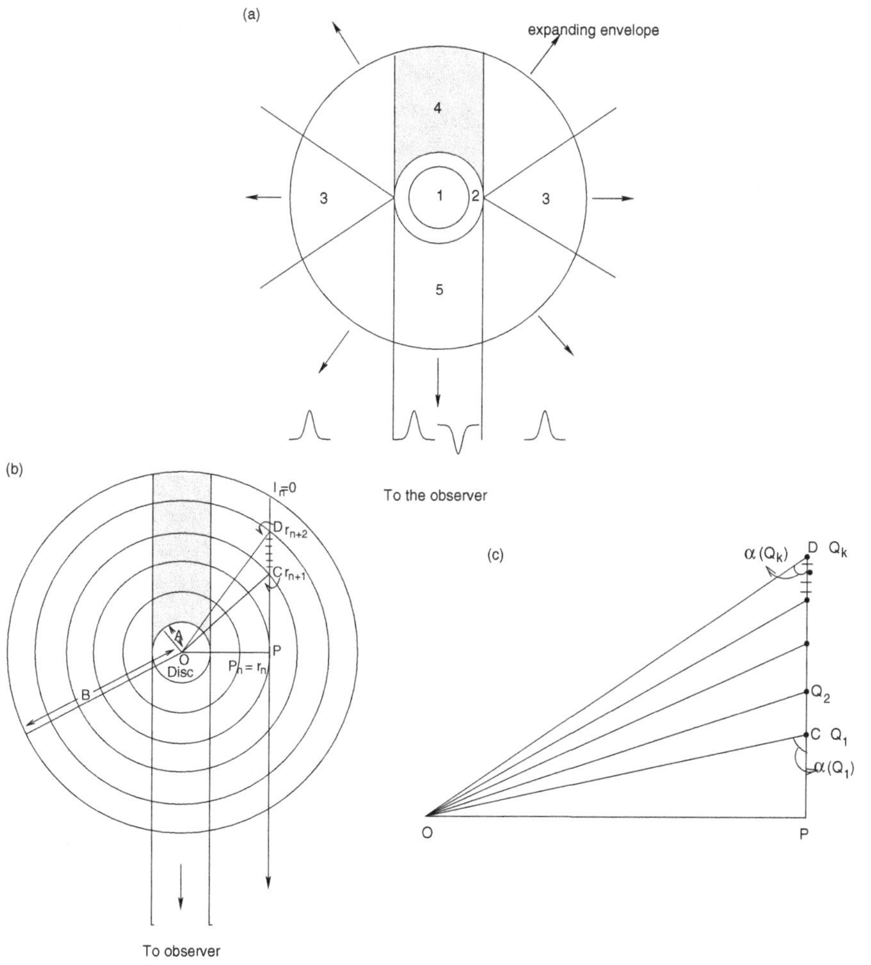

Figure 8.3 (a) Schematic diagram of an expanding stellar envelope: (1) central star, (2) photospheric surface, (3) emission lobe, (4) region occulted from the observer, (5) absorption feature region (sometimes even emission). (b) Schematic diagram showing how the line of sight fluxes are computed. (c) Expansion of part of the ray segment PD in (b) to compute optical depths and intensities.

radius r_n. This ray will intersect $2(N-n)+1$ sectors, the number 1 represents the sector with which the ray is a tangent. We need to evaluate the transfer in each segment such as CD (see figure 8.3(b)). We apply the formal solution

$$I^n(C) = I^{n+1}(B)\exp(-\tau) + \int_0^\tau S_{n+\frac{1}{2}}(t)\exp\{-[-(\tau-t)]\}\,dt, \qquad (8.4.59)$$

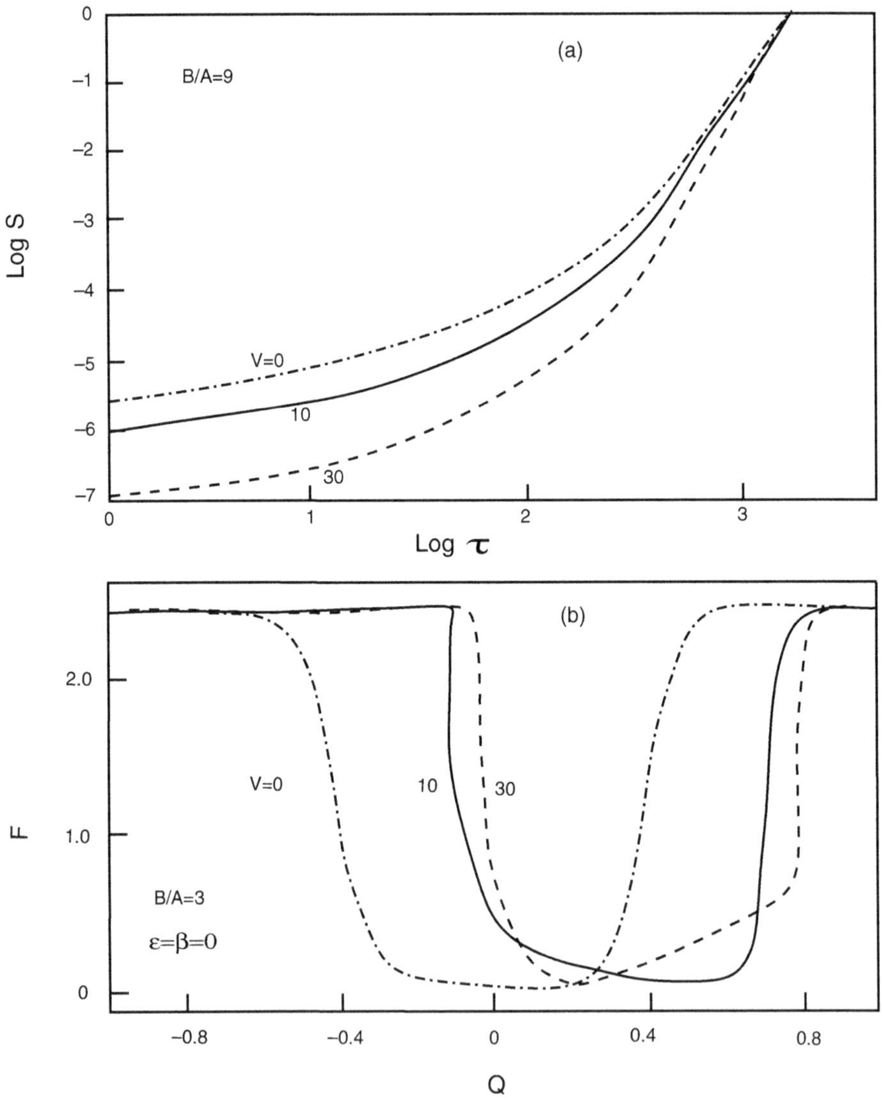

Figure 8.4 (a) Frequency and angle independent source function for $B/A = 9$; (b) flux profiles of the lines received at the observer $F = F(x)/Fx_{max}$ versus $Q = x/x_{max}$ corresponding to the source function given in figure 8.4(a) (from Peraiah (1980b), with permission).

where τ is the optical depth of each segment such as CD. In the observer's frame, the line frequencies will have Doppler shifts from $(-x$ to $+x)$ to $(x \pm \mu V)$ due to velocities. Therefore the source functions calculated for the lines with frequencies $-x$ to x in the comoving frame cannot cover the red and blue shifts corresponding to velocities $\pm V$ in the line frequencies in the rest frame. We integrate the source function over the line with respect to frequency so that it can be used over the whole range of line frequencies in the observer's frame. The source function is then

$$S_n = \sum_{i=-I}^{+I} A_i S_n(x_i, \tau_n), \tag{8.4.60}$$

$$S_{n+\frac{1}{2}} = \frac{1}{2}(S_n + S_{n+1}). \tag{8.4.61}$$

We need to estimate the optical depth τ_{CD} accurately to compute $I^n(C)$ in region (3). We divided the segment CD into K smaller segments. Let $OD = r_{n+2}$ and $OC = r_{n+1}$. Then we divide the segment CD into k smaller segments each of equal length at point Q_1 (this is at C), $Q_2, \ldots, Q_k$ (see figure 8.3(c)). Let each segment be equal to $\Delta Q = CD/k$. We need to find the cosines of the angles $\alpha(Q_k)$made by the lines such as $OQ_1, OQ_2, \ldots, OQ_k$ with the ray at $Q_1, Q_2, \ldots, Q_k$. We then obtain $\mu(Q_k)$ as

$$\mu(Q_k) = \cos\alpha(Q_k) = \frac{PQ_k}{OQ_k}, \tag{8.4.62}$$

where

$$\left.\begin{aligned} OQ_k^2 &= OP^2 + PQ_k^2, \quad OP = r_n, \\ PQ_k &= PC + k \cdot \Delta Q, \quad k = 0, 1, \ldots, k, \\ PC &= (r_{n+1}^2 - r_n^2)^{\frac{1}{2}}, \\ \Delta Q &= CD/k, \\ CD &= (r_{n+2}^2 - r_n^2)^{\frac{1}{2}} - (r_{n+1}^2 - r_n^2)^{\frac{1}{2}}. \end{aligned}\right\} \tag{8.4.63}$$

Now we can compute the frequencies in the observer's frame using the relation

$$x_0 = x \pm \mu(Q_k) \cdot V(Q_k), \tag{8.4.64}$$

where $V(Q_k)$ is the radial velocity along the radius vector OQ_k at the point Q_k. The source functions at the points $Q_1, Q_2, \ldots$ are obtained by a simple linear interpolation of the values at r_{n+1} and r_{n+2}. For this purpose, one should take the maximum number of shells which is the same as the number of rays . As the calculation does not need much storage space or computational time, one can use a large number of rays. The boundary condition where the ray crosses the Nth shell is given as $I^N = 0$. Then $I^{(N-1)}$ is estimated using equation (8.4.59) repeatedly along a segment such as CD. The profile function is computed using equation (8.4.64). For redistribution functions one can use equation (8.4.58). We then compute the intensities for each ray corresponding to each frequency in the range $x_0 \rightarrow x \pm \mu(Q_k)V(Q_k)$.

The optical depth in each segment such as $(Q_k Q_{k-1})$ along the line of sight is computed using the relation

$$\tau(x, Q_k, Q_{k-1}) = \kappa(Q_k, Q_{k-1})\phi(X_0), \tag{8.4.65}$$

where $\phi(x_0)$ is the profile function and x_0 is given in equation (8.4.64). $\kappa(Q_k, Q_{k-1})$ is the absorption coefficient between the points Q_k and Q_{k-1}. Thus, the monochromatic flux is computed by using the formula

$$F_x = 2\pi \int_{p=r_1}^{p=r_N} I(p)p\,dp. \tag{8.4.66}$$

A similar calculation is performed of the intensities in region (5) (of figure 8.3(a)).

Mallik (1986) applied this theory to solve problem of H alpha line profiles. Peraiah and Ingalgi (1990) studied the effects of dust on the equivalent widths of spectral lines formed in expanding spherically symmetric shells.

8.5 Lorentz transformation and aberration and advection

Lorentz transformation is used for clear understanding of the physical variables between the rest and the comoving frames. In this section we closely follow Mihalas (1978) and Sen and Wilson (1998). Lorentz transformation for four vectors corresponds to proper rotation in four-dimensional spacetime. If velocity gradients occur in the medium, local Lorentz transformation has to be used for every $v(r)$ in spherical symmetry. A point in the laboratory frame is represented by the four coordinates

$$(x^1, x^2, x^3, x^4) = (x, y, z, ict) \tag{8.5.1}$$

and in the comoving frame by

$$(x_0^1, x_0^2, x_0^3, x_0^4) = (x_0, y_0, z_0, ict_0). \tag{8.5.2}$$

The comoving frame is moving relative to the rest system with a velocity v in the z-direction. The transformation is effected through the equation

$$x_0^\alpha = L_\beta^\alpha x^\beta, \quad (\alpha = 1, \ldots, 4), \tag{8.5.3}$$

where the Einstein convention of summing over repeated indices is used. **L** is given by

$$\mathbf{L} = \begin{pmatrix} 1 & 0 & 0 & 0 \\ 0 & 1 & 0 & 0 \\ 0 & 0 & \gamma & i\beta\gamma \\ 0 & 0 & -i\beta\gamma & \gamma \end{pmatrix}, \tag{8.5.4}$$

where $\gamma = (1 - v^2/c^2)^{-\frac{1}{2}}$, $\beta = (v/c)$, $i^2 = -1$.

It is to be understood that four vectors and four tensors are covariant under Lorentz transformations, and so are the physical laws which are written in terms of four vectors or four tensors under this transformation. Here $\mathbf{L}$ is Hermitian, which means that $\mathbf{L} = L^+$, where '+' denotes the adjoint (or conjugate) transpose and $\mathbf{L}^{-1} = \mathbf{L}^T$, where T denotes the transpose. If we write (see equation (8.5.3))

$$\mathbf{X}_0 = \mathbf{LX}, \tag{8.5.5}$$

where $\mathbf{X}_0$ and $\mathbf{X}$ are continuum vectors, then

$$\mathbf{X} = \mathbf{L}^{-1}\mathbf{X}_0 = \mathbf{L}^T X_0, \tag{8.5.6}$$

or

$$X^\alpha = \left(L^{-1}\right)^\alpha_\beta X_0^\beta. \tag{8.5.7}$$

Applying the transformation rules to the coordinates, we can show that the measurement of intervals particular to the z-axis is unaffected by the relative motion of the frames, that is,

$$\Delta X_0 = \Delta X, \quad \Delta Y_0 = \Delta Y, \tag{8.5.8}$$

and Δz at rest in the fixed frame is measured by an observer in the moving frame to the length,

$$\Delta z_0 = \Delta z/\gamma, \tag{8.5.9}$$

which is the famous Lorentz–Fitzgerald contraction effect. Similarly the time interval Δt in the fixed frame measured by an observer in the moving frame is (by the time-dilation effect)

$$\Delta t_0 = \gamma \Delta t. \tag{8.5.10}$$

From equations (8.5.8)–(8.5.10), we conclude that the spacetime volume is invariant, or

$$dV\,dt = dV_0\,dt_0. \tag{8.5.11}$$

A covariant four-gradient transformation is given by

$$\left(\frac{\partial}{\partial x}, \frac{\partial}{\partial y}, \frac{\partial}{\partial z}, \frac{1}{ic}\frac{\partial}{\partial t}\right) = \left[\frac{\partial}{\partial x_0}, \frac{\partial}{\partial y_0}, r\left(\frac{\partial}{\partial z_0} - \frac{\beta}{c}\frac{\partial}{\partial t_0}\right), \frac{\gamma}{ic}\left(\frac{\partial}{\partial t_0} - \beta c\frac{\partial}{\partial z_0}\right)\right]. \tag{8.5.12}$$

We shall now consider the transformation of the radiation field and transfer related variables. If P^α is the four-momentum of any particle, then

$$P^\alpha = \left(p_x, p_y, p_z, \frac{iE}{c}\right), \tag{8.5.13}$$

where p_x, p_y, p_z are the components of the ordinary momentum and E is the total energy of the particle. Now

$$p^2 = p_x^2 + p_y^2 + p_z^2, \tag{8.5.14}$$

and

$$E^2 = p^2c^2 + (m_0c^2)^2, \tag{8.5.15}$$

where m_0 is the rest mass of the particle. For photons

$$m_0 = 0, \quad E = h\nu, \quad p = h\nu/c, \tag{8.5.16}$$

where h is Planck's constant and

$$P^\alpha = \frac{h\nu}{c}\left(n_x^0, n_y^0, n_z^0, i\right)^T = \frac{h\nu}{c}(\mathbf{n}, i)^T, \tag{8.5.17}$$

where $\mathbf{n}$ is the direction of propagation and is given by

$$\mathbf{n} = (n_x, n_y, n_z) = (\sin\theta\cos\varphi, \sin\theta\sin\varphi, \cos\theta). \tag{8.5.18}$$

If Lorentz transformation is applied to equation (8.5.17), we get

$$(\nu_0 n_x^0, \nu_0 n_y^0, \nu_0 n_z^0, i\nu_0) = \left[\nu n_x + \nu n_y, \nu\gamma(n_z - \beta), i\nu\gamma(1 - n_z\beta)\right], \tag{8.5.19}$$

or equivalently

$$\left[\phi_0; (1-\mu_0^2)^{\frac{1}{2}}; \mu_0; \nu_0\right] = \left[\phi; \frac{(1-\mu^2)^{\frac{1}{2}}}{\gamma(1-\mu\beta)}; \frac{\mu-\beta}{1-\mu\beta}; \nu\gamma(1-\mu\beta)\right]. \tag{8.5.20}$$

The inverse transformation gives (replace β by $-\beta$)

$$\left[(1-\mu^2)^{\frac{1}{2}}; \mu; \nu\right] = \left[\frac{(1-\mu_0^2)^{\frac{1}{2}}}{\gamma(1-\mu_0\beta)}; \frac{\mu_0+\beta}{1+\mu_0\beta}; \nu_0\gamma(1+\mu_0\beta)\right]. \tag{8.5.21}$$

We can obtain the Doppler shift and aberration from equations (8.5.19)–(8.5.21); we can easily see from equation (8.5.20) that

$$\nu_0 = \nu\gamma(1-\mu\beta), \quad \text{Doppler shift}, \tag{8.5.22}$$

and

$$\mu_0 = \frac{\mu-\beta}{1-\mu\beta}, \quad \text{aberration effect}. \tag{8.5.23}$$

By retaining terms only up to $O(v/c)$ and setting $\gamma = 1$, we obtain the expressions for the Doppler shift and aberration in the classical Galilean approximation. These are

$$\nu_0 = \nu(1 - \mu\beta), \tag{8.5.24}$$

$$\mu_0 = \frac{\mu - \beta}{1 - \mu\beta}, \tag{8.5.25}$$

$$d\nu_0 = (1 - \mu\beta)\, d\nu, \tag{8.5.26}$$

$$\begin{aligned} d\mu_0 &= d\mu\,(1 - \mu\beta)^{-1} + (\mu - \beta)\beta(1 - \mu\beta)^{-2}\, d\mu \\ &\approx (1 - \mu\beta)^{-1}\, d\mu. \end{aligned} \tag{8.5.27}$$

From equations (8.5.22) and (8.5.23), we get

$$d\nu_0 = \gamma(1 - \mu\beta)\, d\nu, \tag{8.5.28}$$

$$d\mu_0 = (1 - \mu\beta)^{-2}\, d\mu. \tag{8.5.29}$$

From equations (8.5.24), (8.5.26), (8.5.27), (8.5.22), (8.5.28) and (8.5.29), we get

$$d\nu_0 = \left(\frac{\nu_0}{\nu}\right) d\nu \tag{8.5.30}$$

and

$$d\mu_0 = \left(\frac{\nu}{\nu_0}\right)^2 d\mu. \tag{8.5.31}$$

Here terms up to $O(v/c)$ are retained. The elementary solid angle $d\omega$ is given by

$$d\omega = \sin\theta\, d\theta\, d\varphi = d\mu\, d\varphi, \tag{8.5.32}$$

then from equations (8.5.22) and (8.5.30)–(8.5.32) we obtain

$$\nu\, d\nu\, d\omega = \nu_0\, d\nu_0\, d\omega_0, \tag{8.5.33}$$

which means that $\nu\, d\nu\, d\omega$ is Lorentz invariant.

Now we shall see how the specific intensity is transformed. By the definition of the specific intensity (see chapter 1), the number of photons passing through an elementary area $d\sigma$ oriented perpendicular to the z-axis, in the frequency interval $d\nu$, into the solid angle $d\omega$, moving at an angle $\theta = \cos^{-1}\mu$ to the z-axis

$$N = \frac{I(r, t, \mu, \nu)}{h\nu}\, d\omega\, d\nu\, d\sigma \cos\theta\, dt. \tag{8.5.34}$$

If N_0 is number of photons counted by an observer in the comoving frame,

$$N_0 = \frac{I^0(r, t, \mu_0, \nu_0)}{h\nu_0}\, d\omega_0\, d\nu_0 \left(d\sigma \cos\theta_0\, dt_0 + \frac{d\sigma\, v\, dt_0}{c}\right), \tag{8.5.35}$$

where the first term gives the number that would have been counted if $d\sigma$ had been stationary in the comoving frame and the second term gives the density of photons

$(I^0/h\nu_0 c)$ times the volume $(d\sigma\ v\ dt_0)$ swept out by $d\sigma$ in time dt_0. Therefore from equations (8.5.8) to (8.5.11), (8.5.22), (8.5.23) (8.5.34) and (8.5.35), we obtain

$$\frac{I(r,t,\mu,\nu)}{\nu}\, d\omega\, d\nu\, d\sigma\ \mu\, dt = \frac{I^0(r,t,\mu_0,\nu_0)}{\nu_0}\, d\omega_0\, d\nu_0\, d\sigma\, dt_0(\mu_0+\beta). \tag{8.5.36}$$

Multiplying both sides by $\nu^2\nu_0^2$ and using equations (8.5.10), (8.5.23) and (8.5.33), we obtain

$$I(r,t,\mu,\nu) - \left(\frac{\nu}{\nu_0}\right)^3 I^0(r,t,\mu_0,\nu_0). \tag{8.5.37}$$

The emissivity can be transformed as follows. The number of photons emitted from a volume in a specified time interval into a specified solid angle and frequency interval is the same in both frames. If $j(r,t,\mu,\nu)$ is the emissivity then from equations (8.5.11) and (8.5.33) we get

$$\frac{j(r,t,\mu,\nu)\, d\omega\, d\nu\, dV\, dt}{h\nu} = \frac{j_0(r,t,\mu,\nu_0)\, d\omega_0\, d\nu_0\, dV_0\, dt_0}{h\nu_0}. \tag{8.5.38}$$

From equations (8.5.11), (8.5.33) and (8.5.38) we obtain

$$j(r,t,\mu,\nu) = \left(\frac{\nu}{\nu_0}\right)^2 j_0(r,t,\nu_0). \tag{8.5.39}$$

Notice that j becomes isotropic in the comoving frame.

Emission losses must be matched by the absorption in any frame to maintain the energy balance. Then in terms of the number of photons, we have

$$\frac{\kappa(r,t,\mu,\nu) I(r,t,\mu,\nu)\, d\omega\, d\nu\, dV\, dt}{h\nu} = \frac{\kappa^0(r,t,\mu_0,\nu_0) I(r,t,\mu_0,\nu_0)\, d\omega_0\, d\nu_0\, dV_0\, dt_0}{h\nu_0}. \tag{8.5.40}$$

Using relations (8.5.38) and (8.5.37), we obtain

$$\kappa(r,t,\mu,\nu) = \left(\frac{\nu_0}{\nu}\right)\kappa^0(r,t,\nu_0), \tag{8.5.41}$$

where κ^0 is isotropic in the fluid frame. The differential operator in the transfer equation transforms (between the two frames moving uniformly with respect to each other) after using equations (8.5.12) and (8.5.19) to

$$\frac{1}{c}\frac{\partial}{\partial t} + \mathbf{n}\cdot\boldsymbol{\nabla} = \left(\frac{\nu_0}{\nu}\right)\left[\frac{1}{c}\frac{\partial}{\partial t_0} + \mathbf{n}^0\cdot\boldsymbol{\nabla}^0\right]. \tag{8.5.42}$$

The transfer equation is covariant. For, from the equations (8.5.37), (8.5.38), (8.5.41) and (8.5.42), we see that

$$\frac{1}{c}\frac{\partial I_\nu}{\partial t} + (\mathbf{n}\cdot\boldsymbol{\nabla})I_\nu = j_\nu - \kappa_\nu I_\nu, \tag{8.5.43}$$

transforms to

$$\left(\frac{\nu_0}{\nu}\right)\left[\frac{1}{c}\frac{\partial}{\partial t_0} + \left(\mathbf{n}^0\cdot\boldsymbol{\nabla}^0\right)\right]\left[\left(\frac{\nu}{\nu_0}\right)^3 I^0(\mu_0,\nu_0)\right]$$
$$= \left(\frac{\nu}{\nu_0}\right)^2\left[j^0(\nu_0) - \kappa^0(\nu_0)I^0(\mu_0,\nu_0)\right]. \tag{8.5.44}$$

If the two frames are in uniform motion with respect to each other then (ν/ν_0) is constant and only then can equations (8.5.44) be written as

$$\left[\frac{1}{c}\left(\frac{\partial}{\partial t_0}\right) + \left(\mathbf{n}^0\cdot\boldsymbol{\nabla}^0\right)\right]I^0(\mu_0,\nu_0) = j^0(\nu_0) - \kappa^0(\nu_0)I^0(\mu_0,\nu_0), \tag{8.5.45}$$

which is of the same form as equation (8.5.43).

We need to recognize two important points: (1) although equations (8.5.43) and (8.5.45) look similar, the latter equation which is at rest relative to the fluid is much simpler because of the isotropy of the $j^0(\nu_0)$ and $\kappa(\nu_0)$; (2) transformation of the former to the latter is possible if the two frames do not move uniformly with respect to each other, that is if they are covariant. This implies that the latter equation does not apply in an expanding or pulsating atmosphere.

We shall now turn to the transformation of the moments of the radiation field, J, H and K, between the rest and the comoving frames. Using equations (8.5.20), (8.5.30) and (8.5.37), setting $\gamma = 1$ and expanding to the first order in v/c, we get

$$I_\nu^0\,d\nu_0\,d\omega_0 = \left(\frac{\nu_0}{\nu}\right)^2 I_\nu\,d\nu\,d\omega \approx (1-2\mu\beta)I_\nu\,d\nu\,d\omega. \tag{8.5.46}$$

This gives us

$$J^0 = J - 2\beta H. \tag{8.5.47}$$

In a similar way

$$I_\nu^0\mu_0\,d\nu_0\,d\omega_0 = (\mu-\beta)(1-\mu\beta)I_\nu\,d\nu\,d\omega$$
$$\approx \left[\mu - \beta(1-\mu^2)\right]I_\nu\,d\nu\,d\omega \tag{8.5.48}$$

and

$$I_\nu^0\mu_0^2\,d\nu_0\,d\omega_0 = (\mu-\beta)^2 I_\nu\,d\nu\,d\omega$$
$$\approx (\mu^2 - 2\mu\beta)I_\nu\,d\nu\,d\omega. \tag{8.5.49}$$

From the above, it follows that

$$H^0 = H - \beta(J+K),$$

$$K^0 = K - 2\beta H, \tag{8.5.50}$$

where J^0, H^0, K^0 and J, H, K are the frequency integrated angular moments in the comoving and rest frames respectively. The inverse of this transformation is given by (changing the sign of β)

$$[J, H, K] = [J^0 + 2\beta H^0, H^0 + \beta(J^0 + K^0), K^0 + 2\beta H^0]. \tag{8.5.51}$$

8.6 The equation of transfer in the comoving frame

Covariant differentiation can be applied to the full transformation of the transfer equation, together with the differential operator for a non-uniform velocity field, as the comoving frame of a fluid consists of sets of inertial frames attached to the elements of the fluid, each of which is instantaneously associated with the velocity of the element of the fluid. Furthermore, Lorentz transformation is applied for the transformation of the variables between the Eulerian rest frame and Lagrangian comoving frame which applies strictly to frames in uniform relative motion to each other. In stellar atmospheres, $V = V(r, t)$. This makes the fluid frames non-inertial. One can set up an infinite set of local inertial frames and use the Lorentz transformation of the physical variables of the radiation field between the rest and local comoving frames. The transfer equations that we consider are

$$\left(\frac{\nu}{\nu_0}^3\right)\left[\frac{1}{c}\frac{\partial}{\partial t} + \mu\frac{\partial}{\partial r}\right] I^0(r_0, \mu_0, \nu_0, t_0) = \left(\frac{\nu}{\nu_0}\right)^2 \left[j^0(\nu_0) - \kappa^0(\nu_0) I^0(r_0, \mu_0, \nu_0, t_0)\right] \tag{8.6.1}$$

for plane parallel geometry, and

$$\left(\frac{\nu}{\nu_0}\right)^3\left[\frac{1}{c}\frac{\partial}{\partial t} + \mu\frac{\partial}{\partial r} + \frac{1-\mu^2}{r}\frac{\partial}{\partial \mu}\right] I^0(r_0, \mu_0, \nu_0, t_0) = \left(\frac{\nu}{\nu_0}\right)^2 \left[j^0(\nu_0) - \kappa^0(\nu_0) I^0(r_0, \mu_0, \nu_0, t_0)\right] \tag{8.6.2}$$

for spherical geometry.

Assuming one-dimensional flows, we apply Lorentz transformation to a frame that instantaneously coincides with the moving fluid. If we set $\gamma = 1$ and ignore terms of $O(v/c)^2$, then

$$r_0 = r, \tag{8.6.3}$$

$$ct_0(r, t) = ct - \frac{1}{c}\int_0^r v(r, t)\, dr'. \tag{8.6.4}$$

Equation (8.6.3) means that the observers in both the frames see the same space increments, that is there is no Lorentz contraction. Equation (8.6.4) represents that retardation effect.

We apply the chain rule to evaluate the derivatives in equations (8.6.1) and (8.6.2)

$$\left(\frac{\partial}{\partial\mu}\right) = \left(\frac{\partial}{\partial\mu}\right)_{r\nu t} = \left(\frac{\partial r_0}{\partial\mu}\right)_{r\nu t}\frac{\partial}{\partial r_0} + \left(\frac{\partial\mu_0}{\partial\mu}\right)_{r\nu t}\frac{\partial}{\partial\mu_0} + \left(\frac{\partial\nu_0}{\partial\mu}\right)_{r\nu t}\frac{\partial}{\partial\nu_0} + \left(\frac{\partial t_0}{\partial\mu}\right)_{r\nu t}\frac{\partial}{\partial t_0}, \tag{8.6.5}$$

$$\left(\frac{\partial}{\partial t}\right) = \left(\frac{\partial}{\partial t}\right)_{r\mu\nu} = \left(\frac{\partial r_0}{\partial t}\right)_{r\mu\nu}\frac{\partial}{\partial r_0} + \left(\frac{\partial\mu_0}{\partial t}\right)_{r\mu\nu}\frac{\partial}{\partial\mu_0} + \left(\frac{\partial\nu_0}{\partial t}\right)_{r\mu\nu}\frac{\partial}{\partial\nu_0} + \left(\frac{\partial t_0}{\partial t}\right)_{r\mu\nu}\frac{\partial}{\partial t_0}, \tag{8.6.6}$$

$$\left(\frac{\partial}{\partial r}\right) = \left(\frac{\partial}{\partial r}\right)_{\mu\nu t} = \left(\frac{\partial r_0}{\partial r}\right)_{\mu\nu t}\frac{\partial}{\partial r_0} + \left(\frac{\partial\mu_0}{\partial r}\right)_{\mu\nu t}\frac{\partial}{\partial\mu_0} + \left(\frac{\partial\nu_0}{\partial r}\right)_{\mu\nu t}\frac{\partial}{\partial\nu_0} + \left(\frac{\partial t_0}{\partial r}\right)_{\mu\nu t}\frac{\partial}{\partial t_0}. \tag{8.6.7}$$

We can use expressions of the first order in β:

$$\left.\begin{aligned} &\nu = \nu_0(1+\beta\mu_0), \quad \nu_0 = \nu(1-\beta\mu), \\ &\mu_0 = (\mu-\beta)/(1-\beta\mu), \quad \mu = (\mu_0+\beta)/(1+\beta\mu_0). \end{aligned}\right\} \tag{8.6.8}$$

In addition we will make the assumption that the acceleration of the fluid, which is zero for the steady flow, within the flight of photon mean free path is negligible compared to the velocity, which means that we neglect $\partial v/\partial t$, $\partial r_0/\partial t$, $\partial\mu_0/\partial t$, $\partial\nu_0/\partial t$ and $\partial t_0/\partial t = 1$. We get the following up to $O(v/c)$:

$$\left(\frac{\partial}{\partial r}\right)_{\mu\nu t}(r_0, \mu_0, \nu_0, t_0) = \left[1, \frac{\mu_0^2-1}{c}\frac{\partial v}{\partial r_0}, -\frac{\mu_0\nu_0}{c}\frac{\partial v}{\partial r_0}, -\frac{\beta}{c}\right] \tag{8.6.9}$$

and

$$\left(\frac{\partial}{\partial\mu}\right)_{r\nu t}(r_0, \mu_0, \nu_0, t_0) = \left[0, (1+2\mu_0\beta), -\nu_0\beta, -0\right]. \tag{8.6.10}$$

Equation (8.6.1) can be written with the above approximation as

$$\left\{\left(\frac{\nu}{\nu_0}\right)\left[\frac{1}{c}\frac{\partial}{\partial t} + \mu\frac{\partial}{\partial r}\right] + 3\mu\left[\partial\left(\frac{\nu}{\nu_0}\right)/\partial r\right]\right\} I^0(r_0, \mu_0, \nu_0, t_0) = j^0(\nu_0) - \kappa(\nu_0)I^0(r_0, \mu_0, \nu_0, t_0), \tag{8.6.11}$$

using equations (8.6.5)–(8.6.10) and the first order expressions for (ν/ν_0) etc. Then retaining the terms up to the first order in (v/c), we get

$$\left[\frac{1}{c}\frac{\partial}{\partial t_0} + \left(\mu_0 + \frac{v}{c}\right)\frac{\partial}{\partial r_0} + \frac{\mu_0(\mu_0^2 - 1)}{c}\frac{\partial v}{\partial r_0}\frac{\partial}{\partial \mu_0} - \frac{\nu_0 \mu_0^2}{c}\frac{\partial v}{\partial r_0}\frac{\partial}{\partial \nu_0} + \frac{3\mu_0^2}{c}\left(\frac{\partial v}{\partial r_0}\right)\right] I^0(r_0, \mu_0, \nu_0, t_0)$$
$$= j^0(\nu_0) - \kappa^0(\nu_0) I^0(r_0, \mu_0, \nu_0, t_0). \tag{8.6.12}$$

In spherical symmetry an extra term is required. This is given by

$$\left(\frac{\nu_0}{\nu}\right)^2 \frac{1-\mu^2}{r}\frac{\partial}{\partial \mu}\left(\frac{\nu}{\nu_0}\right)^3 I^0 =$$
$$\left(\frac{\nu}{\nu_0}\right)\frac{(1-\mu^2)}{r}\frac{\partial I_0}{\partial \mu} + \frac{3(1-\mu^2)}{r}\left[\frac{\partial(\nu/\nu_0)}{\partial \mu}\right] I^0. \tag{8.6.13}$$

Using equation (8.5.20) and the facts that $\nu^2(1-\mu^2) = \nu_0^2(1-\mu_0^2)$ and $\partial(\nu/\nu_0)/\partial\mu = v/c$, these extra terms in equation (8.6.13) give

$$\frac{1-\mu_0^2}{r_0}\{[(1+\beta\mu_0)(\partial/\partial\mu_0) - \beta\nu_0(\partial/\partial\nu_0)] + 3\beta\} I^0. \tag{8.6.14}$$

Therefore, the comoving frame transfer equation up to $O(v/c)$ in spherical geometry is

$$\left\{\frac{1}{c}\frac{\partial}{\partial t_0} + \left(\mu_0 + \frac{v}{c}\right)\frac{\partial}{\partial r_0} + \frac{1-\mu_0^2}{r_0}\left[1 + \frac{\mu_0 v}{c}\left(1 - \frac{d\ln v}{d\ln r_0}\right)\right]\frac{\partial}{\partial \mu_0} - \left(\frac{\nu_0 v}{c r_0}\right)\left[1 - \mu_0^2\left(1 - \frac{d\ln v}{d\ln r_0}\right)\right]\frac{\partial}{\partial \nu_0} + \left(\frac{3v}{c r_0}\right)\left[1 - \mu_0^2\left(1 - \frac{d\ln v}{d\ln r_0}\right)\right]\right\}$$
$$\times I^0(r_0, \mu_0, \nu_0, t_0) = j^0(\nu_0) - \kappa^0(\nu_0) I^0(r_0, \mu_0, \nu_0, t_0). \tag{8.6.15}$$

Equations (8.6.12) and (8.6.15) are the comoving frame equations up to terms of the order $O(v/c)$. They were first derived by Castor (1972). The time derivative in the above equation is still in the fixed frame allowing for the retardation. The Lagrangian time derivative consists of two terms: $D/DT = \partial/\partial t + (v/c)\partial/\partial r$, the second term representing the advection term.

8.7 Aberration and advection with monochromatic radiation

We shall now consider how aberration and advection change the radiation field. We study these effects in a plane parallel medium with coherent and isotropic scattering (Peraiah 1987) with no creation or destruction of photons within the medium. The monochromatic, plane parallel, steady state radiative transfer equation in a fluid frame (see equation (8.6.12)) is

$$(\mu+\beta)\frac{\partial I(z,\mu)}{\partial z}+\frac{\mu(\mu^2-1)}{c}\frac{\partial v}{\partial z}\frac{\partial I(z,\mu)}{\partial \mu}+\frac{3\mu^2}{c}\frac{\partial v}{\partial z}I(z,\mu)$$
$$=K\,[S-I(z,\mu)], \tag{8.7.1}$$

where we have replaced r_0 by z and, as we are considering only coherent scattering, we set $\partial/\partial v=0$, $\mu=(\mu'-\beta)/(1-\mu'\beta)$ and $0\leq\mu'\leq 1$ is the cosine of the angle made by the ray with z-axis; $\beta=v/c$. K is the absorption coefficient and S is the source function which in this case is given by

$$S=\frac{1}{2}\int_{-1}^{+1}P(z,\mu_1',\mu_2')I(z,\mu_2')\,d\mu_2', \tag{8.7.2}$$

where $P(z,\mu_1',\mu_2')$ represents the isotropic phase function. Equation (8.7.1) is integrated by expanding the specific intensity $I(z,\mu)$ as

$$I=I_0+\xi I_z+\eta I_\mu+\xi\eta I_{z\mu}, \tag{8.7.3}$$

where I_0, I_z, I_μ and $I_{z\mu}$ are the interpolation coefficients and

$$\xi=\frac{z-\bar{z}}{\Delta z/2},\quad \bar{z}=\frac{1}{2}(z_i+z_{i+1}),\quad \Delta z=(z_i-z_{i-1}) \tag{8.7.4}$$

and

$$\eta=\frac{\mu-\bar{\mu}}{\Delta\mu/2},\quad \bar{\mu}=\frac{1}{2}\left(\mu_j+\mu_{j-1}\right),\quad \Delta\mu=\mu_j-\mu_{j-1}, \tag{8.7.5}$$

with z_i, z_{i-1} and μ_j, μ_{j-1} the discrete points along the z–μ discrete grid. Substituting equation (8.7.3) into equation (8.7.1) and integrating the resulting equation over the z–μ grid, we obtain

$$\frac{2}{\Delta z}(\bar{\mu}+\beta)\,I_z+(g+\Delta\mu\bar{\mu})\frac{d\beta}{dz}I_\mu+\frac{1}{3}\Delta\mu+I_{z\mu}+\left(K+3\bar{\mu^2}\frac{d\beta}{dz}\right)I_0$$
$$=KS,\quad 0\leq\mu\leq 1. \tag{8.7.6}$$

A similar equation is obtained for the range $-1\leq\mu\leq 0$. Here we assume that $d\beta/dz$ is constant over the interval (z_i, z_{i-1}) and

$$\left.\begin{aligned}
\bar{\mu^2} &= (\bar{\mu}^2) + \frac{1}{12}(\Delta\mu)^2,\\
g &= \frac{2\bar{\mu}}{\Delta\mu}\left(\langle\mu^2\rangle - 1\right),\\
\langle\mu^2\rangle &= \frac{1}{2}\left(\mu_j^2 + \mu_{j-1}^2\right).
\end{aligned}\right\} \quad (8.7.7)$$

The interpolation coefficients $I_0, I_z, I_\mu, I_{z\mu}$ are replaced by their corresponding nodal values (see Peraiah and Varghese (1985)). Thus equation (8.7.6) becomes

$$A_a I_{j-1}^{i-1,+} + A_b I_j^{i-1,+} + A_c I_{j-1}^{i,+} + A_d I_j^{i,+} = \tau\left(S_{j-1}^{i-1,+} + S_j^{i-1,+} + S_{j-1}^{i,+} + S_j^{i,+}\right), \quad (8.7.8)$$

with a similar equation for $-1 \le \mu \le 0$:

$$A'_a I_{j-1}^{i-1,-} + A'_b I_j^{i-1,-} + A'_c I_{j-1}^{i,-} + A'_d I_j^{i,-} = \tau\left(S_{j-1}^{i-1,-} + S_j^{i-1,-} + S_{j-1}^{i,-} + S_j^{i,-}\right), \quad (8.7.9)$$

where

$$I_{j-1}^{i-1,+} = I(z_{i-1}, \mu_{j-1}) \text{ etc.} \quad (8.7.10)$$

The coefficients $A_a, A_b, \ldots, A'_a, A'_b \ldots$ are functions of $\Delta\mu$, $\bar{\mu}$, $\Delta\beta$, β, g, $\bar{\mu^2}$; $\Delta\beta = \beta_{n+1} - \beta_n$. Furthermore,

$$S_{j-1}^{i-1,+} = \sum \frac{1}{2}\left(P^{++} C I^{i-1,+} + P^{+-} C I^{i-1,-}\right)_{j-1}, \quad (8.7.11)$$

where the Ps are the phase matrices and the Cs are the angle quadrature weights, and τ is the optical depth given by

$$\Delta\tau = K\Delta z. \quad (8.7.12)$$

We can write other Ss similarly. Equations (8.7.8) and (8.7.9) can be written for j angles. Thus we have

$$\left(\mathbf{A}^{cd} - \tau\mathbf{Q}\boldsymbol{\gamma}^{++}\right)\mathbf{I}_i^+ + \left(\mathbf{A}^{ab} - \tau\mathbf{Q}\boldsymbol{\gamma}^{++}\right)\mathbf{I}_{i-1}^+ = \tau\mathbf{Q}\boldsymbol{\gamma}^{+-}\left(\mathbf{I}_i^- + \mathbf{I}^- i - 1\right) \quad (8.7.13)$$

and

$$\left(\mathbf{A}'^{cd} - \tau\mathbf{Q}\boldsymbol{\gamma}^{--}\right)\mathbf{I}_i^- + \left(\mathbf{A}'^{ab} - \tau\mathbf{Q}\boldsymbol{\gamma}^{--}\right)\mathbf{I}_{i-1}^- = \tau\mathbf{Q}\boldsymbol{\gamma}^{-+}\left(\mathbf{I}_i^+ + \mathbf{I}_{i-1}^+\right), \quad (8.7.14)$$

where

$$\left\{Q_{j,j},\ Q_{j,j+1}\right\} = 1 \quad (8.7.15)$$

and

$$\mathbf{A}^{ab} = \begin{pmatrix} A_a^{j-1} & A_b^j & & & \\ & A_a^j & A_b^{j+1} & & \\ & & \ddots & \ddots & \\ & & & A_a^{J-1} & A_b^J \\ & & & & A_a^J \end{pmatrix}. \tag{8.7.16}$$

The matrices $\mathbf{A}^{cd}$, $\mathbf{A}'^{ab}$, $\mathbf{A}'^{cd}$ are defined similarly. Furthermore,

$$\left.\begin{aligned} \gamma^{++} &= \mathbf{P}^{++}\mathbf{C}, \quad \text{etc.,} \\ \mathbf{P}^{++} &= \mathbf{P}(+\mu, +\mu), \text{ etc.} \end{aligned}\right\} \tag{8.7.17}$$

Equations (8.7.13) and (8.7.14) give us the transmission and reflection operators (see chapter 6 for the procedure and for computations of the internal radiation field). We set the boundary conditions

$$\left.\begin{aligned} I^-(\tau = \tau_{max}, \mu_j) &= 1, \\ I^+(\tau = 0, \mu_j) &= 0. \end{aligned}\right\} \tag{8.7.18}$$

The velocity gradient $dv/d\tau$ is assumed to be constant with $dv/d\tau < 0$, and the boundary conditions on the velocity are

$$\left.\begin{aligned} v(\tau = \tau_{max}) &= 0, \\ v(\tau = 0) &= v \end{aligned}\right\} \tag{8.7.19}$$

where $v = 0$–5000 km s^{-1} (in steps of 1000 km s^{-1}) ($\beta = 0$–0.0167).

The mean intensities are computed using the formula

$$J = \frac{1}{2}\int_{-1}^{+1} I(\mu)\, d\mu. \tag{8.7.20}$$

The changes $\bar{J}$ in Js are computed as follows:

$$\bar{J} = \frac{\Delta J}{J(v=0)} \times 100, \tag{8.7.21}$$

where

$$\Delta J = J(v = 0) - J(v > 0). \tag{8.7.22}$$

We need to examine the individual effects of aberration and advection on the radiation field in the moving medium. It would be interesting to know the effects of aberration and advection separately, but it is difficult to separate the terms corresponding to these phenomena in equation (8.7.1). In this equation, the first term on the LHS represents aberration, while the second and third terms represent both aberration and advection. Here, we shall set the following conditions for aberration and advection even though they are artificial:

$$\left.\begin{aligned} &\frac{d\beta}{dz} = 0, \quad \beta \geq 0, \text{ for aberration and no advection,} \\ &\frac{d\beta}{dz} \gtrless 0, \ \beta = 0, \text{ for advection and no aberration.} \end{aligned}\right\} \tag{8.7.23}$$

We show $\bar{J}$ for various effects in the medium with $v = 5000$ km s^{-1} and $\tau = 50$ in figure 8.5(a). We plot the amplification factors defined as $\bar{J}/100\beta$ as a function of the total thickness of the slab for positive velocities in figure 8.5(b). The amplification factors increase with increasing maximum optical thickness (τ_{max} of the plane parallel slab when positive velocities are used. Similar characteristics are seen for negative velocities (see Peraiah (1987)).

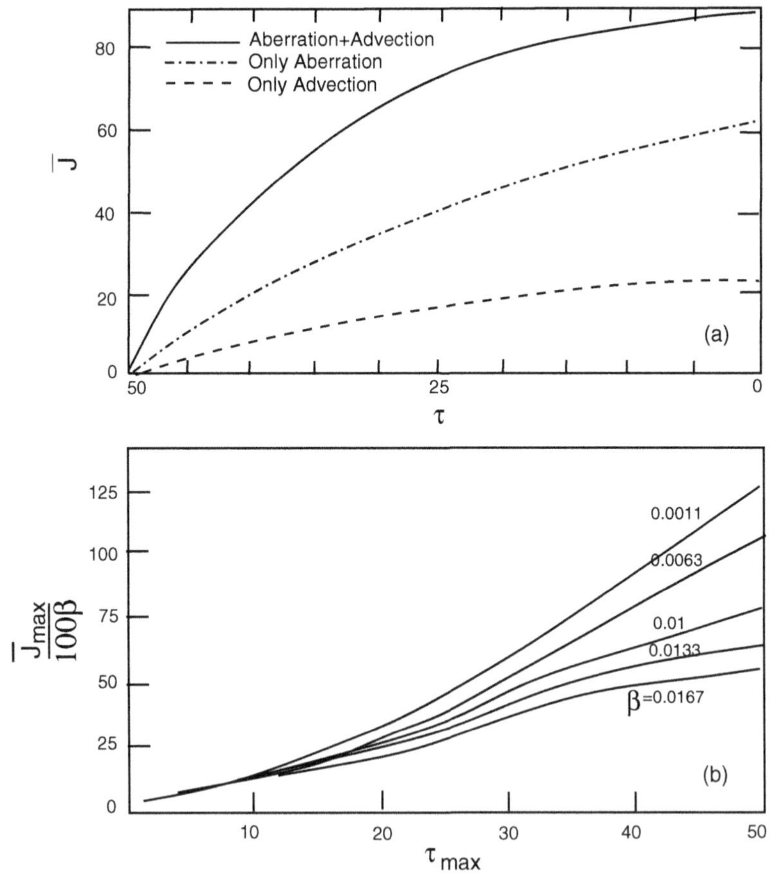

Figure 8.5 (a) $\bar{J}$ versus the optical depth (Peraiah 1987, Figure 3). Here we have shown the individual effects of aberration and advection and their combined effect on the radiation field. (b) Amplification factor $\bar{J}_{max}/100\beta$ vs τ_{max} for positive velocities (from Peraiah (1987), with permission).

We shall now turn to a spherically symmetric atmosphere. Assuming the same physical situation that is isotropic and coherent scattering, as in the plane parallel geometry above, we add the sphericity and assume neither creation nor destruction of photons. The transfer equation in spherical geometry in the fluid frame with terms up to $O(v/c)$ is (see equation (8.6.15))

$$(m+\beta)\frac{\partial U(r,m)}{\partial r} + \frac{1-m^2}{r}\left[1 + m\beta\left(1 - \frac{r}{\beta}\frac{d\beta}{dr}\right)\right]\frac{\partial U(r,m)}{\partial m} =$$
$$K(r)\left[S(r) - U(r,m)\right] + \frac{2(m+\beta)}{r}U(r,m)$$
$$-3\left[\frac{\beta(1-m^2)}{r} + m^2\frac{d\beta}{dr}\right]U(r,m), \tag{8.7.24}$$

where $\beta = v/c$, $m = (\mu - \beta)/(1 - \mu\beta)$, $U(r,m) = 4\pi r^2 I(r,m)$, $0 \le \mu \le 1$. A similar equation can be written for $-1 \le \mu \le 0$. $K(r)$ is the absorption coefficient and $S(r)$ is the source function. Equation (8.7.24) and another equation similar to it (for $-1 \le \mu \le 0$) can be integrated on the angle–radius grid following the procedure of Peraiah and Varghese (1985). We shall not describe this as it is too long and instead we refer the reader to Peraiah (1991a). However, we shall quote some results in figures 8.6(a) and (b).

It is quite obvious from figure 8.5 and 8.6 that the phenomena of aberration and advection change the radiation field in both the plane parallel and spherical geometries.

8.8 Line formation with aberration and advection

We shall consider the changes that can occur in the lines due to aberration and advection in a spherically symmetric expanding medium. We solve equation (8.6.15) without the term $c^{-1}\partial I_0/\partial t_0$. This equation is solved for a purely scattering medium in which no thermal emission occurs (Peraiah 1991b). Complete redistribution will be employed as a starting point of investigation of the effects of aberration and advection. Although angle-averaged and angle dependent redistribution functions are more appropriate, the use of complete redistribution gives the direction of changes in the line formation in expanding media with aberration and advection taken into account. We consider an isothermal scattering atmosphere. We use as the line source function

$$S_L = \frac{1}{2}\int_{-1}^{+1}\int_{-\infty}^{\infty}\phi(x', \mu_0, r)U(x', \mu_0, r)\,dx'\,d\mu', \tag{8.8.1}$$

where we have put $\epsilon = 0$ and no photon is either created or absorbed, that is there are no internal sources and sinks. The boundary conditions are

$$U^-(\mu_j, \tau = T, x_i) = 1, \quad U^+(\mu_j, \tau = 0, x_i) = 0 \tag{8.8.2}$$

and the frequency boundary condition is

$$dU^{\pm}_{n+\frac{1}{2}}\Big|_{i=1,I} = 0. \tag{8.8.3}$$

Here T is the total optical depth, $T = \tau_{max}$. It is assumed that the density changes as r^2. The velocity boundary conditions are the same as those given in the case of coherent, isotropic scattering, that is

$$\left.\begin{aligned} v(r = A, \tau = T) &= 0, \\ v(r = B, \tau = 0) &= v. \end{aligned}\right\} \tag{8.8.4}$$

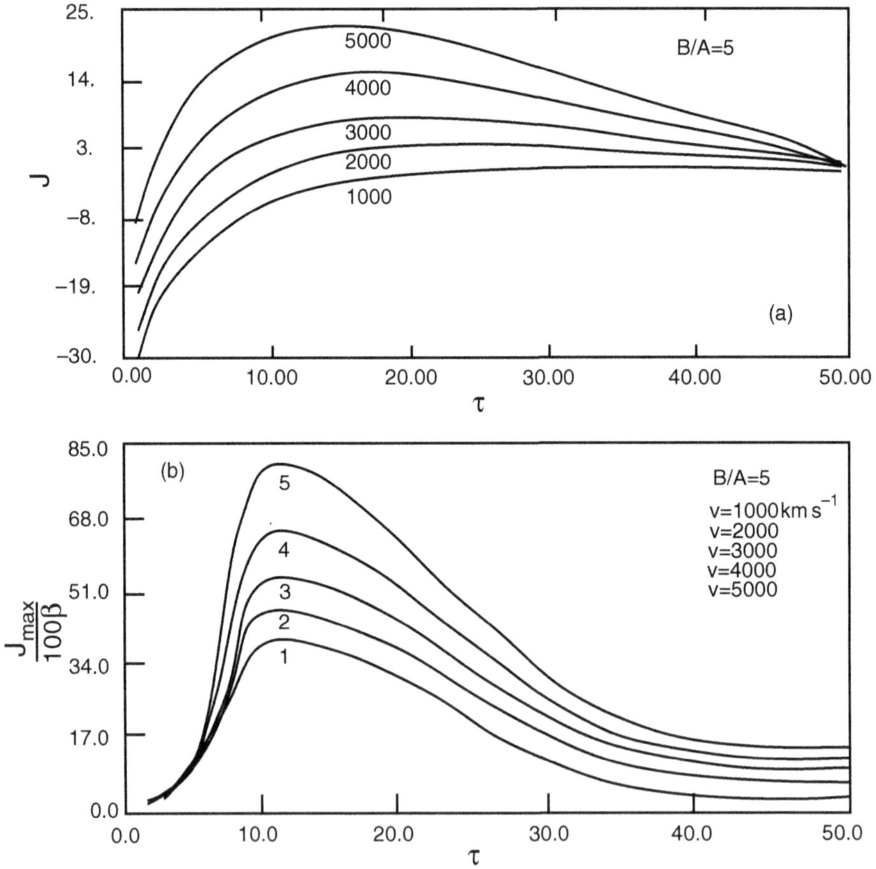

Figure 8.6 (a) The quantity $\bar{J}$ defined in equation (8.7.21) is plotted against the total optical depth $T = 50$, for an atmosphere whose ratio of outer to inner radius $B/A = 5$ for velocities $v = 0$–5000 km s^{-1} (Peraiah 1991a, Figure 10). (b) The amplification factor $\bar{J}_{max}/100\beta$, with $B/A = 5$ (from Peraiah (1991a), with permission).

where v is taken to be 0–5000 km s^{-1} (in steps of 1000 km s^{-1}) and A and B are the inner and outer radii of the spherical medium.

In plane parallel geometry, Doppler radial velocity gradients are operative, while in spherical geometry, two additional phenomena, transverse velocities and sphericity, become operative in changing the radiation field. We need to consider all these effects together in the expanding atmospheres. The source function in equation (8.8.1) has been computed and is given in figure 8.7(a) for planar geometry and in figure 8.7(b) for spherical geometry with $B/A = 10$. The source function is calculated for an isothermal atmosphere using a temperature of 30 000 K to calculate $v_{\rm th}$ through the formula $v_{\rm th} = (2kT/m_H)^{\frac{1}{2}}$, where m_H is the mass of the hydrogen

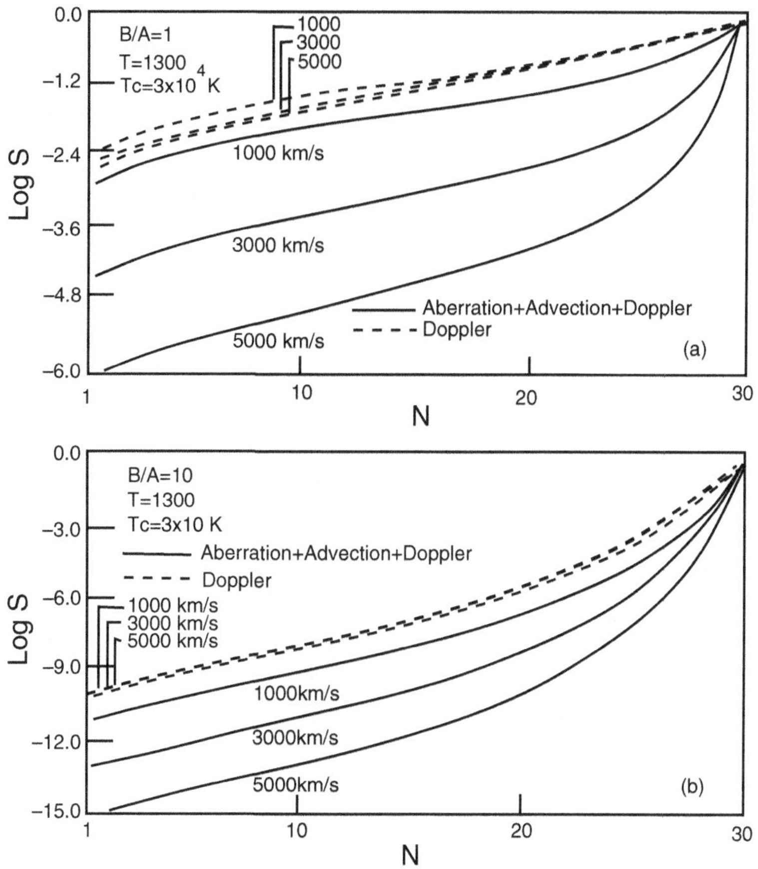

Figure 8.7 (a) A plot of log S (see equation (8.81) versus N, the shell number, where $N = 1$ corresponds to $\tau = 0$ and $N = 30$ to $\tau = T$ in planar geometry. The broken curves represent the variation of the source functions for Doppler velocity gradients and the continuous curves represent those with additional aberration and advection effects. (b) Same as (a) but for the spherical case with $B/A = 10$ (from Peraiah (1991b), with permission).

atom, k is the Boltzmann constant and T is the temperature. The velocities v are given above.

In planar geometry, transverse velocity gradients do not exist (as $r \to \infty$, $v/r \to 0$). But in spherical geometry, we have curvature effects and transverse velocity gradients. These are represented by the terms $[(1-\mu_0^2/r](\partial U/\partial\mu_0)$ and $(1-\mu_0^2)(v/r)(\partial U/\partial x)$, where $U = 4\pi r^2 I(x, r, \mu_0)$. In addition to these, the terms due to aberration and advection given by

$$(\mu_0 - \beta)\frac{\partial U}{\partial r},$$

$$\frac{(1-\mu_0^2)}{r}\mu_0\beta\left[1-\left(\frac{r}{\beta}\right)\left(\frac{d\beta}{dr}\right)\right]\left(\frac{\partial U}{\partial\mu_0}\right),$$

$$\left\{3\left[\frac{\beta}{r}(1-\mu_0^2)+\mu_0^2\frac{d\beta}{dr}\right]-2\left(\frac{\mu_0+\beta}{r}\right)\right\}U$$

are introduced into the transfer equation. Figures 8.7(a) and (b) show these different effects in the planar and spherical geometries. The combined effect is to reduce the source function considerably at $\tau = 0$.

In addition to the linear velocity law, the velocity laws due to Lucy (1971) and Castor and Lamers (1979) have been used and these are given by

$$v(r) = v_t\left[1-(1-\alpha)\left(\frac{P}{r}\right)-\alpha\left(\frac{P}{r}\right)^2\right]^{\frac{1}{2}} \tag{8.8.5}$$

and

$$\omega = 0.01 + 0.99\left(1-\frac{1}{x'}\right)^{\beta}, \tag{8.8.6}$$

where $v(r)$ is the velocity at r, v_t is the terminal velocity, $\omega = v/v_\infty$ and $x' = r/A$, with A the photospheric radius or the inner radius of the spherical shell. In figure 8.8 we have put $v(r = A) = v_A = 0$ and $v(r = B) = v_B = 50 = v_{\text{th}}$. The spherical shell is divided into $N(= 100)$ shells of unequal optical thickness. Discontinuities in the velocities and other aspects have been discussed in Peraiah (1991a,b).

8.9 Method of adaptive mesh

This method is due to Winkler and Norman (1983, 1985) and Mihalas *et al.* (1984a,b). The mesh is attached neither to the rest frame nor to the comoving frame. It is allowed to evolve freely so that one can follow the main physical characteristics of the flow. This gives accurate and efficient computational schemes. We shall summarize this method briefly following Sen and Wilson (1998).

In a spherically symmetric medium, the evolution of the mesh is controlled by: (1) the radial function f^r which maintains the global consistency of the mesh; and

(2) the structure function f^s which gives the physical structure of the flow problem. The main difficulty lies in the formulation of the problem in a conservative form for the adaptive coordinate system. We shall discuss f^r and f^s.

The equation that gives the grid motion is

$$\left(\Delta f^r\right)_i + \left(\Delta f^s\right)_i - \sigma_g \left(\frac{\delta r_i}{\delta t}\right) = 0, \tag{8.9.1}$$

where f^r is the radial function, f^s is the structure function and σ_g is a positive constant. The spatial difference Δ is zone centred, or $(\Delta f)_i = f_{i+1} - f_i$. The radial function is

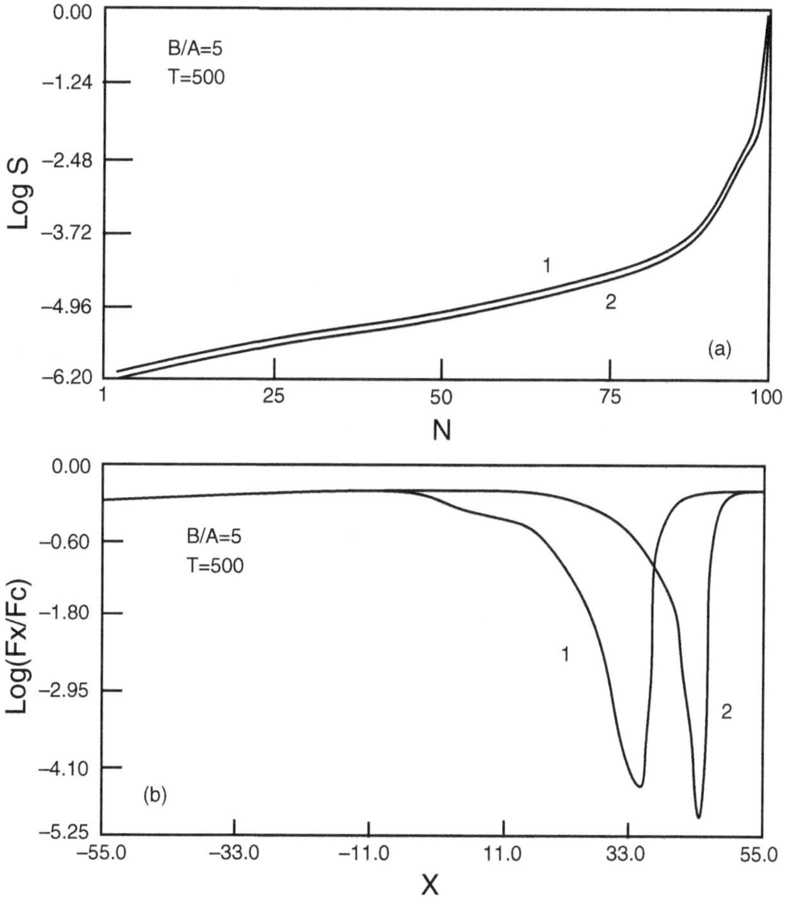

Figure 8.8 (a) Source functions according to the velocity laws given in equations (8.8.5) and (8.8.6): (1)$\alpha = -1$, (2)$\alpha = 0.9$ (Peraiah 1991a, Figure 7a). (b) Line profiles corresponding to source functions given in (a) (from Peraiah (1991b), with permission).

$$f_i^r = W_r \Delta r_i + W_{rz} \frac{(\Delta r_i)^2}{(\Delta r_{i-1} \Delta r_{i+1})} + W_{r_{max}} \left(\frac{\Delta r_i}{\Delta r_{max}} \right)^{\bar{n}} - W_{r_{min}} \left(\frac{\Delta r_{min}}{\Delta r_i} \right)^{\bar{n}}, \tag{8.9.2}$$

where $\Delta r_i = (r_i - r_{i-1})/R_{scale}$. The Ws are the weights assigned to each term and the parameters R_{scale}, Δr_{max}, Δr_{min} respectively are the scale of the given problem and the maximum and the minimum zone sizes; $\bar{n}$ is typically equal to 4. In the above equation, the first term represents the spatial grid and the second represents the stability of the mesh and maintains the monotonicity of the grid spacing (see Winkler and Norman (1985)). The constant $\sigma_g = 0$, under which condition the change is instantaneous. The time-filtering constant is defined as

$$\sigma_g = CVMGP(0., 1., \delta f), \tag{8.9.3}$$

where

$$\delta f = f^s - f^n \tag{8.9.4}$$

and

$$CVMGP(x, y, z) = \begin{cases} x & \text{if} \quad z > 0 \\ y & \text{if} \quad z < 0, \end{cases} \tag{8.9.5}$$

f^n being the reference value. This can be changed with each time step n to

$$f^{n+1} = f^s + (1 - \varepsilon)(f^n - f^s) CVMGP(0., 1., \delta f). \tag{8.9.6}$$

The quantity ε retains the memory of the structure of the flow after several time steps. f^s is given by,

$$\begin{aligned} f_i^s = {} & W_m \Delta m_i + W_{ml} \Delta ml_i + W_{\rho l} \Delta \rho l_i + W_{Pl} \Delta Pl_i + Wk\rho l \Delta k\rho l_i \\ & + W_{El} \Delta El_i + W_v \Delta \bar{V}_i + W_{vl} \Delta \bar{V} l_i, \end{aligned} \tag{8.9.7}$$

where

$$\left. \begin{aligned} \Delta m_i &= (m_i - m_{i+1})/M_{scale}, \\ \Delta ml_i &= (m_i - m_{i-1})/(m_i + m_{i+1}), \\ \Delta \rho l_i &= \left[(\rho_{i-1} - \rho_{i+1})/\rho_i \right]^2, \\ \Delta Pl_i &= \left[(P_{i-1} - P_{i+1})/P_i \right]^2, \\ \Delta k\rho l_i &= \left[(k_{i-1}\rho_{i-1} - k_{i+1}\rho_{i+1})/k_i \rho_i \right]^2, \\ \Delta El_i &= \left[(E_{i-1} - E_{i+1})/E_i \right]^2, \\ \Delta \bar{V}_i &= (\bar{V}_i - \bar{V}_{i+1})^2 CVMGP(1, 0, \bar{V}_i - \bar{V}_{i+1}), \\ \Delta \bar{V} l_i &= \frac{(\bar{V}_i - \bar{V}_{i+1})^2}{[\bar{V}_i^2 + \bar{V}_{i+1}^2 + C\bar{V}l]} CVMGP(1, 0, \bar{V}_i - \bar{V}_{i+1}). \end{aligned} \right\} \tag{8.9.8}$$

In the above set of equations the natural logarithm of a variable X is denoted by Xl and $\bar{V} = V_{rel}$. The adaptive radiative transfer equation can be obtained as follows (see Mihalas *et al.* (1984a)). If $\mathbf{V}$ is the fluid velocity and f is any differentiable function, these can be connected to the Eulerian derivative and the Lagrangian derivative (D/Dt) by the relation

$$\frac{Df}{Dt} = \frac{\partial f}{\partial t} + (\mathbf{V} \cdot \boldsymbol{\nabla}) f. \tag{8.9.9}$$

We define the grid velocity as

$$\mathbf{V}_g = \frac{dr}{dt}, \tag{8.9.10}$$

where d/dt is the adaptive mesh derivative taken with respect to the fixed values of the moving adaptive mesh coordinates. Then from equation (8.9.9), we get

$$\frac{df}{dt} = \frac{\partial f}{\partial t} + \left(\mathbf{V}_g \cdot \boldsymbol{\nabla}\right) f. \tag{8.9.11}$$

We define $\mathbf{V}_{rel}$ as the relative velocity of the fluid with respect to the adaptive mesh grid, which is given by

$$\mathbf{V}_{rel} = \mathbf{V} - \mathbf{V}_g. \tag{8.9.12}$$

From equation (8.9.9) one can get the Reynolds transport theorem as

$$\frac{D}{Dt}\left(\int_{V_{fluid}} f \, dV_{fluid}\right) = \int_{V_{fluid}} \left[\frac{\partial f}{\partial t} + \boldsymbol{\nabla} \cdot (\mathbf{V} f)\right] dV_{fluid}. \tag{8.9.13}$$

Using equations (8.9.9) and (8.9.10), we get

$$\int_V \rho \frac{D}{Dt}\left(\frac{f}{\rho}\right) dV = \frac{d}{dt}\int_V f dV + \int_{\partial V} f \mathbf{V}_{rel} \cdot d\mathbf{S}, \tag{8.9.14}$$

where V is the volume of the adaptive mesh enclosed by the surface ∂V with an outward pointing surface element $d\mathbf{S}$. Using the above results, we can write the transfer equation in the adaptive mesh coordinates. From equation (8.6.15), we have for an expanding spherically symmetric shell,

$$\begin{aligned} &\frac{1}{c}\left(\frac{\partial I}{\partial t} + v\frac{\partial I}{\partial r}\right) + \frac{\mu}{r^2}\frac{\partial}{\partial r}(r^2 I) \\ &+ \frac{\partial}{\partial \mu}\left\{(1-\mu^2)\left[\frac{1}{r} + \frac{\mu}{c}\left(\frac{v}{r} - \frac{\partial v}{\partial r}\right) - \frac{a}{c^2}\right] I\right\} \\ &- \frac{\partial}{\partial \nu}\left\{\nu\left[(1-\mu^2)\frac{v}{cr} + \frac{\mu^2}{c}\frac{\partial v}{\partial r} + \frac{\mu a}{c^2}\right] I\right\} \\ &+ \left[(3-\mu^2)\frac{v}{cr} + \left(\frac{1+\mu^2}{c}\right)\frac{\partial v}{\partial r} + \frac{2\mu a}{c^2}\right] I = j - \kappa I, \end{aligned} \tag{8.9.15}$$

where $a = \partial v/\partial t$. Introducing the density ρ and $dvol = \frac{1}{3}d(r^3)$, equation (8.9.15) can be written in terms of the Lagrangian derivative D/Dt as

$$\left(\frac{\rho}{c}\right)\frac{D}{Dt}\left(\frac{I}{\rho}\right)+\frac{\partial}{\partial vol}(\mu r^2 I)$$
$$+\frac{\partial}{\partial\mu}\left\{(1-\mu^2)\left[\frac{1}{r}+\frac{\mu}{c}\left(\frac{3v}{r}-\frac{\partial(r^2v)}{\partial vol}\right)-\frac{a}{c^2}\right]I\right\}$$
$$-\frac{\partial}{\partial\nu}\left\{\nu\left[(1-3\mu^2)\frac{v}{cr}+\frac{\mu^2}{c}\frac{\partial(r^2v)}{\partial vol}+\frac{\mu a}{c^2}\right]I\right\}-j$$
$$+\left\{\kappa+\left[(1-3\mu^2)\frac{v}{cr}+\frac{\mu^2}{c}\frac{\partial(r^2v)}{\partial vol}+\frac{2\mu a}{c^2}\right]\right\}I=0. \tag{8.9.16}$$

We write that

$$\tilde{\mathbf{V}}=(v,0,0),\quad v=\frac{Dr}{Dt},$$
$$\tilde{\mathbf{V}}_g=\left(V_g,\frac{d\mu}{dt},\frac{d\nu}{dt}\right)\quad\text{with } V_g=\frac{dr}{dt},$$
$$\tilde{\mathbf{V}}_{rel}=\tilde{\mathbf{V}}-\tilde{\mathbf{V}}_g=\left(v-\frac{dr}{dt},-\frac{d\mu}{dt},-\frac{d\nu}{dt}\right)$$

and

$$\tilde{\nabla}=\left(\frac{\partial}{\partial r},\frac{\partial}{\partial\mu},\frac{\partial}{\partial\nu}\right).$$

Using equations (8.9.9) and (8.9.10), equation (8.9.16) can be written in the adaptive mesh coordinates as

$$\frac{1}{c}\frac{dI}{dt}+\left(\tilde{\mathbf{V}}_{rel}\cdot\tilde{\nabla}\right)I+I(\nabla\cdot\mathbf{V})+\frac{\partial}{\partial vol}(\mu r^2 I)$$
$$+\frac{\partial}{\partial\mu}\left\{1-\mu^2)\left[\frac{1}{r}+\frac{\mu}{c}\left(\frac{3v}{r}-\frac{\partial(r^2v)}{\partial vol}\right)-\frac{a}{c^2}\right]I\right\}$$
$$-\frac{\partial}{\partial\nu}\left\{\nu\left[(1-3\mu^2)\frac{v}{cr}+\frac{\mu^2}{c}\frac{\partial(r^2v)}{\partial vol}+\frac{2\mu a}{c^2}\right]I\right\}$$
$$-j+\left\{\kappa+\left[(1-3\mu^2)\frac{v}{cr}+\frac{\mu^2}{c}\frac{\partial(r^2v)}{\partial vol}+\frac{2\mu a}{c^2}\right]\right\}I=0, \tag{8.9.17}$$

which reduces to that of the comoving frame equation if $\tilde{\mathbf{V}}_{rel}=0$.

Introducing the adaptive mesh volume $d\tilde{V}=d_{vol}\,d\mu\,d\nu$ with the corresponding surface element $\mathbf{S}$ and using relation (8.9.14), equation (8.9.17) can be integrated over the mesh volume dV to obtain the conservative form of adaptive mesh transfer equation, which is

$$\frac{d}{dt}\left[\int_{\tilde{V}}\left(\frac{I}{c}\right)d\tilde{V}\right]+\int_{\partial\tilde{V}}\left(\frac{I}{c}\right)\left(\tilde{\mathbf{V}}_{rel}\cdot d\tilde{\mathbf{S}}\right)+\int_{\tilde{V}}\frac{\partial}{\partial vol}(\mu r^2 I)\,d\tilde{V}$$

$$+\int_V \frac{\partial}{\partial\mu}\left\{(1-\mu^2)\left[\frac{1}{r}+\frac{\mu}{c}\left(\frac{3v}{r}-\frac{\partial(r^2v)}{\partial vol}\right)-\frac{a}{c^2}\right]I\right\}d\tilde{V}$$
$$-\int_V \frac{\partial}{\partial\nu}\left\{\nu\left[(1-3\mu^2)\frac{v}{cr}+\frac{\mu^2}{c}\frac{\partial(r^2v)}{\partial vol}+\frac{\mu a}{c^2}\right]I\right\}d\tilde{V}$$
$$+\int_V\left\{-j+\left[\kappa+(1-3\mu^2)\frac{v}{cr}+\frac{\mu^2}{c}\frac{\partial(r^2v)}{\partial vol}+\frac{2\mu a}{c^2}\right]I\right\}d\tilde{V}=0. \tag{8.9.18}$$

The discrete form of the above equation can be written for a hypersurface $dV = \Delta V\Delta\mu\Delta\nu$, using Δr, $\Delta\mu$ and $\Delta\nu$ as the space, angle and frequency differences. It is

$$\frac{\delta}{\delta t}\left[\left(\frac{I}{c}\right)\cdot\Delta V\Delta\mu\Delta\nu\right]+\left\{\Delta_r\left[r^2v_{rel}\left(\frac{I}{c}\right)+r^2\mu I\right]\right\}\Delta\mu\Delta\nu$$
$$+\left\{\Delta_\mu\left[-\mu\left(\frac{I}{c}\right)+(1-\mu^2)\frac{1}{r}+\frac{\mu}{c}\left(\frac{3v}{r}-\frac{\partial(r^2v)}{\partial vol}-\frac{a}{c^2}I\right)\right]\right\}\Delta V\Delta\nu$$
$$-\left\{\Delta_\nu\left[\nu\left(\frac{I}{c}\right)+\nu(1-3\mu^2)\frac{v}{cr}+\frac{\mu^2}{c}\frac{\partial(r^2v)}{\partial vol}+\frac{\mu a}{c^2}I\right]\right\}\Delta V\Delta\mu$$
$$+\left\{-j+\left[\kappa+(1-3\mu^2)\frac{v}{cr}+\frac{\mu^2}{r}\frac{\partial(r^2v)}{\partial vol}+\frac{2\mu a}{c^2}\right]I\right\}\Delta V\Delta\mu\Delta\nu=0. \tag{8.9.19}$$

The above equation is solved by using the zeroth and first moment equations together with the variable Eddington factors, so that three equations with three unknowns are solved. The Eddington factors are computed using equation (8.9.19).

The radiation energy $E(\nu)$, flux $\bar{F}(\nu)$ and pressure $P(\nu)$ (defined in chapter 1) are related to the moments J, F and K as follows:

$$E(\nu)=\frac{4\pi}{c}J(\nu),\quad \bar{F}(\nu)=\pi F(\nu),\quad P(\nu)=\frac{4\pi}{c}K(\nu). \tag{8.9.20}$$

The zeroth and first moments of equation (8.9.18) are integrated over frequency ν. We obtain the frequency integrated radiation energy E, flux $\bar{F}$ and pressure P from

$$\frac{d}{dt}\left[\int_V Edvol\right]+\int_{\partial v}E(\mathbf{V}_{rel}\cdot dS)+\int_V\frac{\partial}{\partial vol}(r^2\bar{F})\,dvol$$
$$+\int_V\left[(E-3P)\frac{v}{cr}+P\frac{\partial(r^2v)}{\partial vol}+\frac{2a\bar{F}}{c^2}\right]dvol$$
$$+\int_V\left\{\int_0^\infty[-4\pi j(\nu)+c\kappa(\nu)E(\nu)]\,d\nu\right\}dvol=0, \tag{8.9.21}$$

and

$$\frac{d}{dt}\left(\int_V\frac{\bar{F}}{c^2}dvol\right)+\int_{\partial V}\frac{\bar{F}}{c^2}(\mathbf{V}_{rel}\cdot dS)$$

$$+\int_V \left[\frac{\partial P}{\partial r} + \frac{(3P-E)}{r} + \frac{\bar{F}}{c^2}\frac{\partial v}{\partial r} + \frac{a}{c^2}(E+P)\right] dvol$$
$$+\left\{\int_V \frac{1}{c}\int_0^\infty \left[\kappa(\nu)\bar{F}(\nu)\right] d\nu\right\} dvol = 0. \tag{8.9.22}$$

The above two equations are solved together with the equations of mass and continuity given respectively by

$$dm - \rho\, dvol = 0, \tag{8.9.23}$$

and

$$\frac{d}{dt}\left[\int_V \rho\, dvol\right] + \int_{\partial V} \rho \mathbf{V}_{rel}\cdot d\mathbf{S} = 0. \tag{8.9.24}$$

Equations (8.9.1), (8.9.21), (8.9.22), (8.9.23) and (8.9.24) are written in their discrete forms as follows:

$$-\sigma_g \frac{\delta r}{\delta t} + \Delta[f^r + f^s] = 0, \tag{8.9.25}$$

$$\frac{\delta}{\delta t}\left(\frac{E}{\rho}\Delta\xi\right) - \Delta\left[\frac{\delta m}{\delta t}\left(\frac{E}{\rho}\right) - r\bar{F}\right] + P\Delta(rv)$$
$$+\left[(E-3P)\frac{v}{r} + \frac{2a\bar{F}}{c^2}\right]\Delta vol = \left(4\pi k_p B - ck_E E\right)\Delta\xi, \tag{8.9.26}$$

$$\frac{\delta}{\delta t}\left(\frac{\bar{F}}{\rho c^2}\Delta\xi\right) - \Delta\left[\frac{\delta m}{\delta t}\left(\frac{\bar{F}}{\rho c^2}\right)\right]$$
$$+r\left(\Delta P + \frac{\bar{F}}{c^2}\Delta v\right) + \left[\frac{(3P-E)}{r} + \frac{a}{c^2}(E+P)\right]\Delta vol = -\frac{k_F}{c}\bar{F}\Delta\xi, \tag{8.9.27}$$

$$\Delta m - \rho\Delta vol = 0, \tag{8.9.28}$$

$$\frac{\delta}{\delta t}(\rho\Delta vol) + \Delta(r v_{rel}\rho) = 0, \tag{8.9.29}$$

where $\xi = \rho\Delta vol$, $B =$ the Planck function and k_p, k_E and k_F are the Planck mean, the absorption mean and the flux mean given by

$$k_p = \int_0^\infty k^a(\nu)\frac{B(\nu)}{B}\, d\nu, \tag{8.9.30}$$

$$k_E = \int_0^\infty k^a(\nu)\frac{E(\nu)}{E}\, d\nu, \tag{8.9.31}$$

$$k_F = \int_0^\infty [k^a(\nu) + k^s(\nu)]\frac{\bar{F}(\nu)}{\bar{F}}\, d\nu, \tag{8.9.32}$$

where $k^a(\nu)$ and $k^s(\nu)$ are the true absorption and scattering coefficients respectively. Thus we have five equations, (8.9.25)–(8.9.29), for six variables r, m, ρ, E,

$\bar{F}$ and P. The sixth equation is obtained from the Eddington factors. The adaptive mesh scheme is found to be computationally efficient.

Exercises

8.1 Derive equation (8.2.10) and its equivalent for a time independent, plane parallel medium.

8.2 If we set $x = (\nu - \nu_0)/\Delta\nu_D$, where $\Delta\nu_D$ is the Doppler width, show that equation (8.2.10) becomes

$$\mu\frac{\partial I}{\partial r} + \frac{1-\mu^2}{r}\frac{\partial I}{\partial \mu} = j - \kappa I + \left[(1-\mu^2)\frac{v}{r} + \mu^2\frac{dv}{dr}\right]\frac{dI}{dx}.$$

8.3 Derive equations (8.3.26) and (8.3.28). Write down the forms of the matrices $\mathbf{G}$, $\mathbf{U}'$ and $\mathbf{V}'$.

8.4 Show that the matrix $\mathbf{d}$ in equation (8.4.26) satisfies the relation

$$\sum d_i a_i = 0,$$

using the condition of flux conservation and the a_is defined in equation (8.4.19). Explain the physical meaning of the boundary condition that $d_1 = d_I = 0$.

8.5 Prepare a table of τ_{crit} for different values of μ, x, ρ_c and v using relations (8.4.47)–(8.4.50).

8.6 If $dv/dr \geq 0$ or ≤ 0, show that in plane parallel atmospheres the choice of the frequency boundary condition is unique and depends on the sign of dv/dr.

8.7 In a spherically symmetric atmosphere, show that unique conditions can be found only if $v > 0$, $dv/dr \geq 0$ or $v < 0$, $dv/dr \leq 0$ and that due to projection effects along a ray at the plane of symmetry ($z = 0$), velocity distribution of the form ($v > 0$, $dv/dr < 0$) or ($v < 0$, $dv/dr > 0$), though monotonic in the radial direction, produces non-monotonic fields along tangent rays.

8.8 Show that in a spherical shell, the optical depth along the line of sight at an impact angle parameter p is given by

$$\tau(p) = \frac{2k_0}{\bar{v}p}\tan^{-1}\left(\frac{L}{p}\right),$$

where $L^2 = R^2 - p^2$, R being the outer radius of the shell and the absorption coefficient varying as $\kappa(r) = \kappa_0/\bar{v}r^2$ assuming a constant $\bar{v}$.

8.9 Show that the matrix $\mathbf{L}$ is Hermitian and that $\mathbf{L}^\star = \mathbf{L}$ where $\star$ denotes the adjoint or conjugate transpose. Show also that $\mathbf{L}^{-1} = \mathbf{L}^T$, where T denotes the transpose.

8.10 Derive equations (8.5.47), (8.5.50) and (8.5.51).

8.11 Derive equations (8.6.12) and (8.6.15), supplying the intermediate steps.

8.12 Show that the frequency dependent moment equations for equation (8.6.15) are

$$\frac{1}{c}\frac{\partial J_\nu^0}{\partial t} + \frac{v}{c}\frac{\partial J_\nu^0}{\partial r} + \frac{1}{r^2}\frac{\partial (r^2 H^0)}{\partial r} + \left(\frac{v}{cr}\right)(3J_\nu^0 - K^0) + \frac{1}{c}(J_\nu^0 + K_\nu^0)\frac{\partial v}{\partial r}$$
$$+ \frac{v}{cr}\frac{\partial}{\partial \nu_0}\left[\nu_0\left(3K_\nu^0 - J_\nu^0\right)\right] - \frac{1}{c}\left(\frac{2v}{r} + \frac{\partial v}{\partial r}\right)\frac{\partial(\nu_0 K_\nu^0)}{\partial \nu_0}$$
$$= j^0(\nu_0) - \kappa^0(\nu_0)J_\nu^0,$$

for the zero order moment and

$$\frac{1}{c}\frac{\partial H_\nu^0}{\partial t} + \frac{v}{c}\frac{\partial H_\nu^0}{\partial r} + \frac{\partial K_\nu^0}{\partial r} + \frac{1}{r}\left(3K_\nu^0 - J_\nu^0\right) + \frac{2}{c}\left(\frac{v}{r} + \frac{\partial v}{\partial r}\right)H_\nu^0$$
$$+ \left(\frac{v}{cr}\right)\frac{\partial}{\partial \nu_0}\left[\nu_0\left(3G_\nu^0 - H_\nu^0\right)\right] - \frac{1}{c}\left(\frac{2v}{r} + \frac{\partial v}{\partial r}\right)\frac{\partial(\nu_0 G_\nu^0)}{\partial \nu_0}$$
$$= -\kappa^0(\nu_0)H_\nu^0,$$

where $G_\nu^0 = \frac{1}{2}\int_{-1}^{+1} I^0(r, \mu_0, \nu_0, t)\mu_0^3\, d\mu_0$ and we have written $J_\nu^0 \equiv J^0(r, \nu_0, t)$ etc., suppressing the suffix 0 on r and t.

8.13 Derive the frequency integrated moment equations given in exercise 8.12.

8.14 Write out the coefficients $A_a, \ldots, A_d$; $A'_a, \ldots, A'_d$ in equations (8.7.8) and (8.7.9) in terms of $\Delta\mu$, $\bar{\mu}$, $\overline{\mu^2}$, β, $\Delta\beta$ and g.

8.15 Derive the two pairs of transmission and reflection operators using equations (8.7.13) and (8.7.14).

8.16 Solve equations (8.7.24) on the angle–radius grid $[m_{j-1}, m_j][r_{i-1}, r_i]$.

8.17 Using the procedure to compute the line profiles described in section 8.4, develop computer code to obtain lines in the different regions in figure 8.3(a).

REFERENCES

Abbott, D.C., 1978, *ApJ*, **225**, 893.

Abhyankar, K.D., 1964a, *ApJ*, **140**, 1353.

Abhyankar, K.D., 1964b, *ApJ*, **140**, 1368.

Abhyankar, K.D., 1965, *ApJ*, **141**,1056.

Bertout, C., 1984, *ApJ*, **285**, 269.

Castor, J.I., 1970, *MNRAS*, **149**, 111.

Castor, J.I., 1972, *ApJ*, **178**, 779.

Castor, J.I., Lamers, H.J.G.L.M., 1979, *ApJ Suppl.*, **39**, 481.

Chandrasekhar, S., 1945a, *ApJ*, **102**, 402.

Chandrasekhar, S., 1945b, *Rev. Mod. Phys.*, **17**, 138.

Grant, I.P., 1968, *J. Comp. Phys.*, **2**, 381.

Hewitt, T.G., Noerdlinger, D.D., 1974, *ApJ*, **188**, 315.

Hummer, D.G., 1968, *MNRAS*, **141**, 479.

Lucy, L., 1971, *ApJ*, **163**, 95.

Lyong, L.V., 1967, *Soviet Astr.*, **11**, 224.

Magnan, C., 1974, *A&A*, **35**, 233.

Mallik, S.V., 1986, *MNRAS*, **222**, 307.

McCrea, W., Mitra, K., 1936, *Z. für Astrophys.*, **11**, 359.

Mihalas, D., 1978. *Stellar Atmospheres*, 2nd Edition, Freeman, San Francisco.

Mihalas, D., Kunasz, P., Hummer, D., 1975a, *ApJ*, **202**, 465.

Mihalas, D., Kunasz, P., Hummer, D., 1975b, *ApJ*, **203**, 647.

Mihalas, D., Kunasz, P., Hummer, D., 1976a, *ApJ*, **206**, 515.

Mihalas, D., Kunasz, P., Hummer, D., 1976b, *ApJ*, **210**, 419.

Mihalas, D., Shine, R., Kunasz, P., 1976, *ApJ*, **205**, 492.

Mihalas, D., Winkler, K.H.A., Norman, M.L., 1984a, *JQSRT*, **31**, 473.

Mihalas, D., Winkler, K.H.A., Norman, M.L., 1984b, *JQSRT*, **31**, 479.

Nordlinger, P., Rybicki, G., *ApJ*, 1974, **193**, 651.

Peraiah, A., 1980a, *Acta Astronomica*, **30**, 525.

Peraiah, A., 1980b, *J. Astrophys. Astr.,* **1**, 3.

Peraiah, A., 1987, *ApJ*, **317**, 271.

Peraiah, A., Ingalgi, M.F., 1990, *Bull. Astron. Soc. India*, **18**, 17.

Peraiah, A., 1991a, *ApJ*, **371**, 673.

Peraiah, A., 1991b, *ApJ*, **380**, 212.

Peraiah, A., Varghese, B.A., 1985, *ApJ*, **290**, 411.

Puetter, R., Hubbard, E.N., 1983, *ApJ*, **273**, 36.

Rublev, S.V., 1961, *Soviet Astr.*, **4**, 780.

Rublev, S.V., 1964, *Soviet Astr.*, **7**, 492.

Sen, K.K., Wilson. S.J., 1998, *Radiative Transfer in Moving Media*, Springer, Singapore.

Simonneau, E., 1973, *A&A*, **29**, 357.

Sobolev, V.V., 1947, *Moving Atmospheres of Stars*, (English translation S. Gaposchkin, 1960, Harvard University Press, Cambridge, MA).

Sobolev, V.V., 1957, *Soviet Astr.*, **1**, 678.

Surdej, J., 1979, *A&A*, **73**, 1.

Winkler, K.H.A., Norman, M.L., 1983, *Astrophysical Radiation Hydrodynamics*, D. Riedel, Dordrecht.

Winkler, K.H.A., Norman, M.L., 1985, *Comp. Phys. Comm.*, **36**, 121.

Chapter 9

Escape probability methods

Exact numerical methods become costly in terms of computer time when the radiation field is coupled with hydrodynamics. In such situations one needs methods which are fast and give insight into the physics of the problem in an easy and quick manner. Escape probability methods satisfy these requirements to a large extent and therefore became popular. There are ‘first order methods’ due to Biberman, Holstein, Sobolev and Zanstra, which are reviewed by Irons (1979a,b). The methods due to Athay (1972a,b), Rybicki (1972), Frisch and Frisch (1975), Canfield *et al.* (1981, 1984), Scharmer (1981, 1983, 1984) and others are the so called ‘second order methods’. We shall describe these and others methods in this chapter. These methods have been reviewed by Rybicki (1984).

Nordlund (1984) developed a method for obtaining an iterative solution of radiative transfer in a spherically symmetric atmosphere using a single ray approximation. The convergence is achieved in 2–3 iterations to give an accuracy better than 1% in the source function.

The Monte Carlo technique has been used by several authors (see, for example, Magnan (1970), Panagia and Ranieri (1973); Pozdnyakov *et al.* (1976)).

9.1 Surfaces of constant radial velocity

The geometrical region from which most of the observed emission at a given frequency x comes is likely to be a thin zone centred on a surface of constant radial velocity in such a way that $v_z = \mu v_r = x$, where the term radial velocity means the velocity along the line of sight, which is different from v_r the velocity along the radius vector (see Mihalas (1978)) measured from the centre of the star.

In the limit as the width of the line becomes negligible as the thermal velocity becomes much smaller than the flow of the gases, these zones degenerate to the radial velocity surfaces. One can study these surfaces for simple velocity laws of the form $v = r^n$ (in units of $r_0 = v_0 = 1$). Some of these are: (1) $v(r) =$ constant; (2) $v = r$; (3) $v = v_\infty(1 - r_c/r)^{\frac{1}{2}}$, where v_∞ is the terminal velocity in stellar wind; (4) $v = r^{-\frac{1}{2}}$. In the last situation the material may be ejected with the escape velocity and is decelerated by gravity. All the above velocity laws have different sets of constant radial velocity surfaces. For the velocity laws (3) and (4) these are shown in figures 9.1(a) and 9.1(b). The fact that the velocity field of the medium can be represented by a succession of surfaces of constant radial velocities is used in the escape probability method of Sobolev, which takes care of the existence of velocity gradients in the medium. When $v_{th} \gg v_{gas}$, an observer at infinity (or in the laboratory frame) observes the radiation coming from a specific velocity surface.

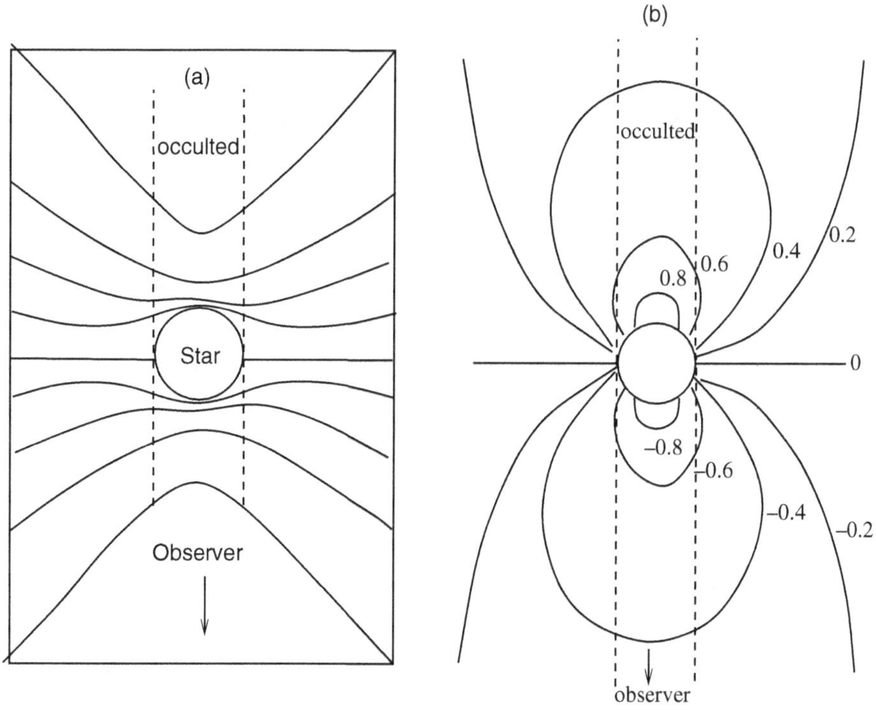

Figure 9.1 Surfaces of constant radial velocity $v_z =$ constant: (a) equal velocity surface of an accelerating atmosphere with the velocity law $v(r) = v_\infty(1 - r_c/r)^{1/2}$; (b) equal velocity surfaces of a decelerating atmosphere with the velocity law $v(r) = v_0(R/r)^{1/2}$. The numbers in the figure represent the ratio v_z/v_0 (from Kuan and Kuhi (1975), with permission).

9.2 Sobolev method of escape probability

The spectra of novae, supernovae, WR stars and of stars indicate large scale expansion of the outer layers with the velocity of expansion as large as 3000 km s^{-1}. The absorption lines show strong violet shifts from their rest positions which indicates that the material is flowing rapidly towards the observer. Some of these lines are accompanied by large red shifted emission features producing typical P Cygni profiles (see figure 9.2). Beals (1950) suggested that radiation pressure is the main cause of the outflow of the matter, which is supported by current thinking.

The escape of photons is greatly enhanced by the presence of macroscopic velocity fields – large velocity gradients. A simple theory was developed by Sobolev (1947, 1957), and is also known as the *large velocity gradient theory*. The definition of the escape probability of photons is the probability that a photon emitted at a given point will escape the medium in a single flight without suffering absorption or scattering on its way. The escape probability β_ν of a photon of frequency ν emitted at a point and with a given direction is given by

$$\beta_\nu = \exp(-|\tau_{\nu f}|), \tag{9.2.1}$$

where $|\tau_{\nu f}|$ is the optical depth along the ray in the forward direction for the frequency ν (Mihalas 1978, Rybicki 1984, Sen and Wilson 1998). If the average escape probability over angles and frequencies with a line profile $\phi(\nu)$ of the frequencies along the ray is β, then this is given by

$$\beta = \frac{1}{4\pi} \oint d\Omega \int_0^\infty \phi(\nu)\beta_\nu \, d\nu, \tag{9.2.2}$$

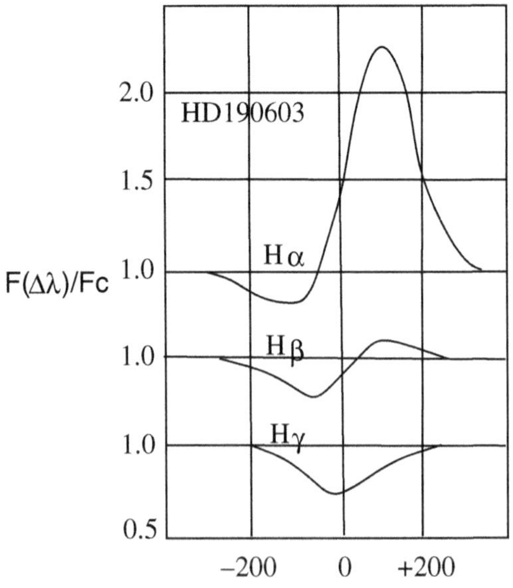

Figure 9.2 P Cygni profiles of the hydrogen lines in the spectrum of HD190603 (Beals, 1950). Ordinate: observed flux in units of the continuum flux. Abscissa: displacement from the line centre in velocity units $v = c\,(\Delta\lambda/\lambda)$.

with a normalized profile function and positive optical depths β between 0 and 1. As mentioned earlier, the velocity gradients must be large for the proper application of Sobolev's theory. The size of these velocity gradients can be measured in terms of what is called the *Sobolev length* L. This is defined as the length L over which the profile function of a line is shifted through a distance equal to its own width by the macroscopic velocity gradients that exist in the moving medium. If we have a line whose profile $\phi(\nu)$ is normalized, with a central frequency ν_0, then ν_0 changes by $\Delta\nu$ when the velocity of the matter changes by Δv (the velocity change along the line of sight). Therefore

$$\Delta\nu = \frac{\nu_0}{c}\Delta v = \frac{\nu_0}{c}\left(\frac{dv}{ds}\right)\Delta s, \tag{9.2.3}$$

where Δs is the distance over which ν_0, the central frequency of the line changes by $\Delta\nu$ to ν_0'. If this shift of frequency is equal to $\Delta\nu_D$, one Doppler width, then

$$\Delta\nu = \Delta\nu_D = \nu_0\frac{v_{th}}{c}, \tag{9.2.4}$$

where v_{th} is the mean thermal velocity ($= (2kT/m)^{\frac{1}{2}}$) and c is the velocity of light. Then from equations (9.2.3) and (9.2.4), we get

$$\frac{\nu_0}{c}\left(\frac{dv}{ds}\right)\Delta s = \frac{\nu_0}{c}v_{th}. \tag{9.2.5}$$

If we set $\Delta s = L$, the Sobolev length is

$$L = v_{th}\Big/\left|\frac{dv}{ds}\right|. \tag{9.2.6}$$

It is interesting to note that L is independent of ν and depends on the velocity gradient, temperature and composition of the medium and that L is same for all lines. If R is a typical scale of variation of the macroscopic quantities, then Sobolev's theory is applicable only when

$$L \ll R. \tag{9.2.7}$$

If dv/ds is estimated to be V/R, where V is the typical macroscopic velocity, then from equation (9.2.6) and (9.2.7), we obtain the condition that

$$v_{th} \ll V. \tag{9.2.8}$$

Normally v_{th} is of the order of the speed of sound. The Sobolev theory is also called the *supersonic approximation* and can be applied to a medium whose properties are approximately constant over the Sobolev length. It is the velocity gradient and not the velocity that is constant over the Sobolev length. When an emitted photon has travelled one Sobolev length to the point where the profile is Doppler shifted by one characteristic width, it can travel unimpeded and will easily escape the local neighbourhood of the initial point. It is obvious that the velocity field introduces a natural mechanism for the escape of photons. The region in which photons are emitted or scattered and affect the intensity in the line is limited to a small region around the test point. Thus the interaction region is small and may therefore be assumed to be homogeneous in its physical properties, such as temperature, density, ionization etc. This means that the theory can be formulated in terms of the local quantities and a parameter β that gives the escape probability of photons summed over all directions and line frequencies. If we define

$$\bar{J}(r) = \int J_\nu \phi(\nu)\, d\nu, \tag{9.2.9}$$

then, neglecting the transfer effects, we can write

$$\bar{J}(r) = (1-\beta)S(r) + \beta_c I_c, \tag{9.2.10}$$

where the first term is obtained from the value that $\bar{J}$ would have in the limit of no escapes, meaning that $\bar{J} = S$, corrected for velocity induced escapes. The quantity β_c is the probability of penetration of photons, integrated over angle and frequency, of the specific intensity I_c emitted by the core to the test point. If $\beta = \beta_c = 0$, then we have

$$\bar{J}(r) = S(r). \tag{9.2.11}$$

Calculation of β

The velocities are measured in units of mean thermal velocity, that is

$$V(r) = \frac{v(r)}{v_{th}}. \tag{9.2.12}$$

The frequency shifts are measured in units of Doppler widths, or

$$x = \frac{\nu - \nu_0}{\Delta\nu_D}, \qquad \Delta\nu_D = \frac{\nu_0 v_{th}}{c}, \tag{9.2.13}$$

where $\Delta\nu_D$ is the Doppler width.

The optical depth along the ray (in the z, p coordinate system) to the observer at infinity is

$$\tau(x, p, z) = \int_z^\infty \kappa(z', p, x)\, dz' = \int_z^\infty \kappa_l(r')\phi(x')\, dz', \tag{9.2.14}$$

where

$$\left.\begin{aligned} r' &= (z'^2 + p^2)^{\frac{1}{2}}, \quad \mu' = z'/r', \\ x' &= x'(x, z, p) = x - V_z(z') = x - \mu' V(r') \end{aligned}\right\} \tag{9.2.15}$$

Most of the contribution to the optical depth $\tau(x, p, z)$ in equation (9.2.14) arises from the region where $x' = 0$, from $z' = z_0(x, p)$, where

$$\left.\begin{aligned} z_0 V(r_0) &= x r_0, \\ r &= (r_0^2 + p^2)^{\frac{1}{2}}. \end{aligned}\right\} \tag{9.2.16}$$

The surface $z_0(x, p)$ is the surface of constant velocity. $\kappa(r')$ can be replaced by $\kappa_l(r_0)$ and we change the variable of integration from z' to x'. From equations (9.2.15) , we then have

$$\begin{aligned} -\left(\frac{\partial x'}{\partial z}\right)_p = \left(\frac{\partial V_z}{dz}\right)_p &= \left\{\frac{\partial(\mu(z, p)V[r(z, p)])}{\partial z}\right\}_p \\ &= \mu^2\frac{\partial V}{\partial r} + (1 - \mu^2)\frac{V}{r} = Q(r, \mu), \end{aligned} \tag{9.2.17}$$

where

$$\mu = \mu(z, p), \quad r = r(z, p). \tag{9.2.18}$$

As the region of interaction is small, the above transformation coefficients may be assumed constant and hence can be estimated at the resonance point $z = z_0(x, p)$. Defining

$$\Phi(x) = \int_{-\infty}^{x} \phi(y)\, dy, \tag{9.2.19}$$

the optical depth $\tau(x, p, z)$ in equation (9.2.14) can be written as

$$\tau(x, p, z) = \tau(x, p, -\infty)\Phi[x'(x, p, z)], \tag{9.2.20}$$

where $x'(x, p, z)$ is given in equation (9.2.15). Notice that from equation (9.2.19), we get

$$\left.\begin{aligned} \Phi(-\infty) &= 0, \\ \Phi(\infty) &= 1, \end{aligned}\right\} \tag{9.2.21}$$

$$\tau(x, p, -\infty) = \frac{\kappa_l(r_0)}{Q(r_0, \mu_0)} = \frac{\tau_0(r_0)}{\left[1 + \mu^2\left(\dfrac{d \ln V}{d \ln r}\right) - 1\right]}, \tag{9.2.22}$$

where $\kappa_l(r_0) = (\pi e^2/mc)\ f_{ij}[n_i(r_0) - (g_i/g_j)n_j(r_0)/\Delta\nu_D$, and

$$\tau_0(r_0) = \frac{\kappa_l(r_0)}{(V/r)_0}. \tag{9.2.23}$$

In equations (9.2.22) and (9.2.23), we should remember that

$$\left.\begin{aligned} r_0 &= r_0\,[z_0(x, p), p], \\ \mu_0 &= \mu_0\,[z_0(x, p), p]. \end{aligned}\right\} \tag{9.2.24}$$

We now choose a fixed r and calculate $\beta(r)$. The escape probability along the ray is just $\exp(-\Delta\tau_\infty)$, where $\Delta\tau_\infty$ is the optical path length from the point in question to infinity. Integrating over angles and frequencies, we get

$$\beta(r) = \frac{1}{2}\int_{-1}^{+1} d\mu \int_{-\infty}^{+\infty} dx\phi\left[x'(x, p, z)\right]\exp\{-\tau\left[x, p(r, \mu), z(r, \mu)\right]\}. \tag{9.2.25}$$

The photons that hit the core of the star are assumed to be absorbed and therefore lost. Equation (9.2.25) is evaluated using equations (9.2.19)–(9.2.24). The matter in the interacting region is assumed to be sufficiently homogeneous that the distinction between r and r_0 is negligible, then

$$\begin{aligned}\beta(r) &= \frac{1}{2}\int_{-1}^{+1} d\mu \int_{0}^{1} d\Phi \exp\left[-\kappa_l(r)\Phi/Q(r, \mu)\right] \\ &= \kappa_l^{-1}\int_0^1 \{1 - \exp\left[-\kappa_l(r)/Q(r, \mu)\right]\}\, Q(r, \mu)\, d\mu.\end{aligned} \tag{9.2.26}$$

If we set $V = rQ(r, \mu)$, equation (9.2.26) becomes

$$\beta(r) = 1 - \exp\left[-\tau_0(r)\right]/\tau_0(r), \tag{9.2.27}$$

where

$$\tau_0(r) = \kappa_l(r)/Q(r, \mu). \tag{9.2.28}$$

This result can also be obtained if the angle dependent terms are neglected in equation (9.2.17).

Calculation of β_c

This is defined as the probability of the penetration of photons of specific intensity I_c to the test point. The photons are emitted from the core and the probability is integrated over angles and frequencies. The test point is taken at a large distance from the core ($-\infty$). Then from the physical meaning, β_c can be written as

$$\begin{aligned}\beta_c(r) &= \frac{1}{2}\int_{-1}^{\mu_c} d\mu \int_0^1 d\Phi \exp\left[\kappa_l(r)\Phi/Q(r, \mu)\right] \\ &= \frac{1}{2\kappa_l(r)}\int_{\mu_c}^{1} \{1 - \exp\left[-\kappa_l(r)/Q(r, \mu)\right]\}\, Q(r, \mu)\, d\mu,\end{aligned} \tag{9.2.29}$$

where $\mu_c = \left[1 - (r_c/r)^2\right]^{\frac{1}{2}}$, r_c being the radius of the core. If the velocity varies as r, that is $V(r) = Ar$, where A is a constant, then

$$\begin{aligned}\beta_c &= \kappa_l^{-1}(r)\frac{1}{2}\int_{\mu_c}^{1} \{1 - \exp\left[-\kappa_l(r)/A\right]\}\, d\mu \\ &= \frac{A}{2\kappa_l(r)}\left(\{1 - \exp[-\kappa_l(r)/A]\} - \mu_c\{1 - \exp[-\kappa_l(r)/A]\}\right)\end{aligned}$$

$$= \frac{1}{2}\beta(r)(1-\mu_c) = \frac{1}{2}\beta\left[1-\left(1-\frac{r_c^2}{r^2}\right)^{\frac{1}{2}}\right]^{\frac{1}{2}} = W\beta(r), \tag{9.2.30}$$

where W is the dilution factor given by (see exercise 1.2c)

$$W = \frac{1}{2}\left[1-\left(1-r_c^2/r^2\right)^{\frac{1}{2}}\right]^{\frac{1}{2}}. \tag{9.2.31}$$

The above result can be understood from the fact that W is the fraction of the full sphere contained in the solid angle subtended by the disc and β measures the probability of penetration from the disc to the test point. It is interesting to note that both β and β_c are expressed in terms of local variables such as opacity and velocity gradients. If β and β_c are known, then using relation (9.2.10) $\bar{J}$ can be calculated without actually solving the transfer equation. In the case of a two-level atom with the assumption of complete redistribution, the source function can be written as

$$S = (1-\epsilon)\bar{J} + \epsilon B, \tag{9.2.32}$$

and, using equation (9.2.10), we get

$$S = [(1-\epsilon)\beta_c I_c + \epsilon B]\,[\epsilon + (1-\epsilon)\beta]^{-1}. \tag{9.2.33}$$

One can understand the effects of thermalization on the source function in a simple way in a uniformly expanding plane parallel medium. If τ is the integrated line optical depth in a medium at rest and the velocity gradient $g = \partial V/\partial\tau$ is constant everywhere, the specific intensity at the test point τ along the direction μ is

$$I(\tau,\mu,x) = \int_\tau^\infty S(\tau')\exp\left[\frac{1}{\mu}\int_0^{\tau'-\tau}\phi(x+g\mu t)\,dt\right] \times \phi[x+g\mu(\tau'-\tau)\mu^{-1}]\,dt'. \tag{9.2.34}$$

The source function for a two-level atom is given by the integral equation

$$S(\tau) = (1-\epsilon)\bar{J}(\tau) + \epsilon B(\tau) = (1-\epsilon)\int_{-\infty}^{+\infty} K_\beta|\tau'-\tau|S(\tau')\,dt' + \epsilon B(\tau), \tag{9.2.35}$$

where the Kernel function $K_\beta(S)$ is given by

$$K_\beta(S) = \frac{1}{2}\int_{-\infty}^{+\infty} dx \int_0^1 \frac{d\mu}{\mu}\phi(x)\phi(x+g\mu s)\exp\left[\frac{1}{\mu}\int_0^s \phi(x+g\mu t)\,dt\right]. \tag{9.2.36}$$

In the static case, the kernel is normalized to unity, but in the present situation of motion, the escape of photons leads to

$$\int_{-\infty}^{+\infty} K|\tau|\,d\tau = 1 - \beta, \tag{9.2.37}$$

where β is the escape probability in plane parallel geometry (see equations (9.2.17) and (9.2.26) and set the limit $r \to \infty$) given by

$$\beta = |g| \int_0^1 \left[1 - \exp\left(-1/\,|g|\,\mu^2\right)\right] \mu^2\,d\mu. \tag{9.2.38}$$

Furthermore, equation (9.2.35) for a two-level atom can be written as

$$S(\tau) = (1 - \epsilon') \int_{-\infty}^{+\infty} K' \left|\tau' - \tau\right| S(\tau')\,d\tau' + \epsilon' B'(\tau), \tag{9.2.39}$$

where $K'(\tau)$ is the renormalized kernel

$$K'(\tau) = \frac{K_\beta(\tau)}{(1-\beta)}, \quad (1 - \epsilon') = (1 - \beta)(1 - \epsilon) \text{ and } B'(\tau) = \epsilon B(\tau)/\epsilon'.$$

If thermalization is attained, S varies so slowly that it can be removed from within the integral and we obtain

$$S(\tau) = B'(\tau) = \epsilon B(\tau)/(\epsilon + \beta - \epsilon\beta), \tag{9.2.40}$$

where $\epsilon \gg \beta$; then $S(\tau) \to B(\tau)$ and if $\beta \gg \epsilon$ then $S(\tau) \to \epsilon B(\tau)$ which is the local creation rate of photons and can be understood on the physical grounds.

Flux profiles of the lines

The flux in the line at frequency x seen by an external observer consists of three parts: (1) the emission from the part of the envelope seen outside the disc ($p > r_c$); (2) the emission from the part of the envelope superposed on the core: and (3) the continuum contribution from the core. Thus the emergent flux F_x at frequency x is

$$\begin{aligned} F_x &= 2\pi \int_0^\infty I(x, p, -\infty) p\,dp \\ &= 2\pi \int_{r_c}^\infty S(r_0)\,\{1 - \exp[-\tau(x, p, -\infty)]\}\,p\,dp \\ &\quad + 2\pi \int_0^{r_c} S(r_0)\,\{1 - \exp[-\tau(x, p, -\infty)\Phi(x_c)]\}\,p\,dp \\ &\quad + 2\pi I_c \int_0^{r_c} \exp[-\tau(x, p, -\infty)\Phi(x_c)]\,p\,dp, \end{aligned} \tag{9.2.41}$$

where $r_0 = r$ at the surface of the constant radial velocity corresponding to x, x_c is the value of x' from equation (9.2.15) at $r = r_c$, that is $x' = x - \mu' V(r')$ and

$\mu' = \left[1 - (p/r_c)^2\right]^{\frac{1}{2}}$. The factor $\Phi(x_c)$ corrects for occultation of material by the core. For an expanding atmosphere

$$\left.\begin{aligned} \Phi(x_c) &= 0 \quad \text{for } x < 0, \\ \Phi(x_c) &\approx 1 \quad \text{for } x > 0. \end{aligned}\right\} \tag{9.2.42}$$

The continuum is unattenuated in the red wing and there is substantial absorption in the blue wing of the line. The flux F_c in the continuum just outside the line is

$$F_c = 2\pi I_c \int_0^{r_c} p\,dp = \pi r_c^2 I_c. \tag{9.2.43}$$

Transforming the variable of integration p to r on the surface $zV(r) = r_x$ and combining the above two equations for F_x and F_c, we can write the line profile R_x as

$$R_x = (F_x/F_c) - 1 \tag{9.2.44}$$

or

$$\begin{aligned} R_x = {} & \frac{2}{r_c^2 I_c} \int_{r_{min(x)}}^{\infty} S(r)\,[\tau_0(r)/\tau(x, p, -\infty)]\,\{1 - \exp[-\tau(x, p, -\infty)]\}\,r\,dr \\ & - \frac{2}{r_c^2 I_c} \int_0^{r_c} S(r_c) \exp\big\{-\tau(x, p, -\infty)\Phi(x_c) \\ & - \exp[-\tau(x, p, -\infty)]\big\}\,p\,dp \\ & - \frac{2}{r_c^2} \int_0^{r_c} \{1 - \exp[-\tau(x, p, -\infty)\Phi(x_c)]\}\,p\,dp, \end{aligned} \tag{9.2.45}$$

where $r_{min}(x) = r$ at which $V(r) = x$, $p = p(r, x)$. The quantity R_x (the residual flux) in equation (9.2.44) gives positive numbers for emission lines and negative numbers for absorption.

Castor (1970) employed a two-level atom source function (see equation (9.2.33)) with the given velocity law $V(r) = V_\infty(1 - r_c/r)^{\frac{1}{2}}$ and a given $\tau_0(r)$, a constant ϵ and B/I_c in the range $1.1 \le r/r_c \le 4$. Asymmetric profiles are obtained if monotonically decreasing functions are used. Characteristic profiles that are similar to those observed in WR stars are obtained (see figure 9.3). This theory has been applied to multi-level atoms (see Mihalas (1978), Castor and Nussbaumer (1972), Castor and van Blerkom (1970)) to explain the spectra formed in Wolf–Rayet stars .

Elitzur (1984) derived the escape probability expressions for the statistical rate equations in a homogeneous slab of photoionized hydrogen including the effects of diffuse and external ionizing radiation without using the probabilistic arguments. Williams *et al.* (1984) obtained the probabilities of photons escaping from a cold electron plasma after undergoing the process of multiple scattering. Finn (1971, 1972) studied the probability of photon exit in radiative transfer problems.

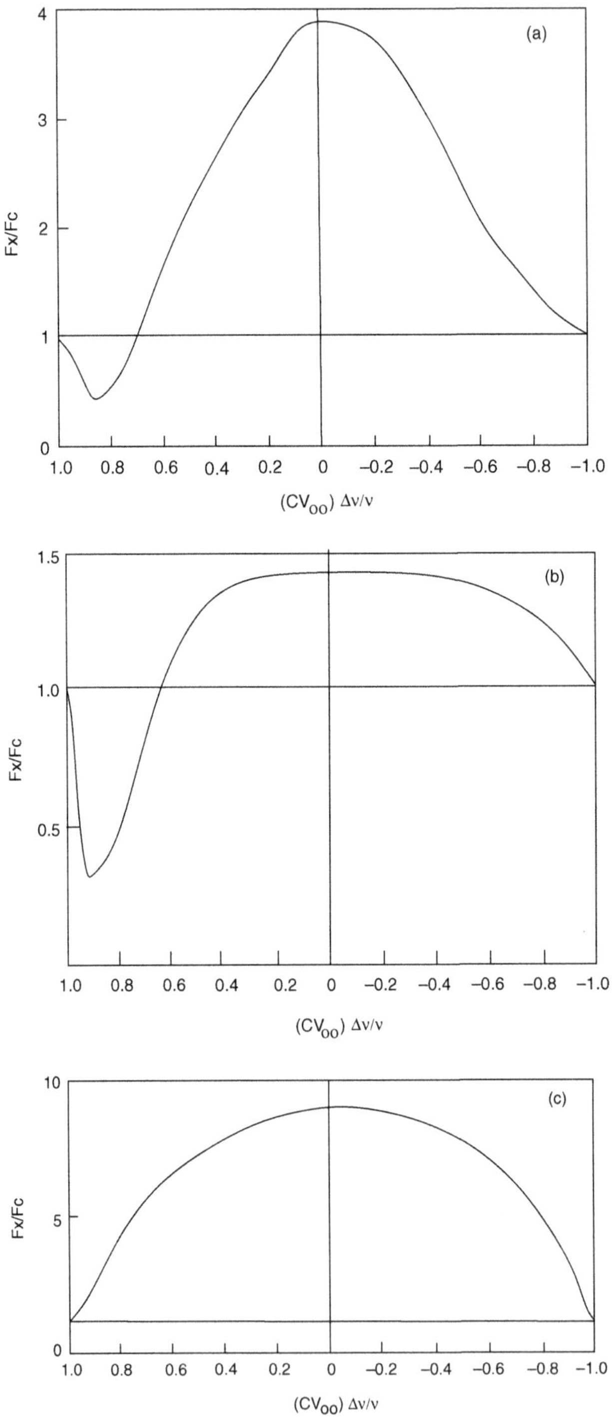

Figure 9.3 Calculated line profiles in expanding spherical atmospheres: (a) $\epsilon = 9.2 \times 10^{-3}$ $\tau_0(\max) = 15$; (b) $\epsilon = 2 \times 10^{-3}$, $\tau_0(\max) = 0.5$; (c) $\epsilon = 2.1 \times 10^{-2}$, $\tau_0(\max) \approx 2$ (from Castor (1970), with permission).

9.3 Generalized Sobolev method

Some of the velocity laws have surfaces of constant velocities where the line of sight may intersect these surfaces more than once (see figure 9.1(b)) such as in the case decelerating flows. The matter at points of intersection may have different physical properties. The radiation received by the observer will have passed through these points. Therefore this radiation must have interacted with the matter at these points and must contain information on the physical properties of the matter at these points. Sobolev's method requires modification to study this problem of the coupling of the radiation field at several points on surfaces of constant radial velocity.

The problem was studied by Rybicki and Hummer (1978). We shall describe this method briefly following Sen and Wilson (1998); for a detailed study one can refer to the original paper of Rybicki and Hummer. This method deals with the problem in spherically symmetric media and can be applied to three-dimensional velocity fields. The time independent radiative transfer equation is written as

$$\mathbf{n} \cdot \nabla I(\mathbf{r}, \mathbf{n}, \nu) = -\kappa(\mathbf{r})\phi\left[\nu - \frac{\nu_0}{c} \cdot \mathbf{n} \cdot \mathbf{V}(\mathbf{r})\right][I(\mathbf{r}, \mathbf{n}, \nu) - S(\mathbf{r}, \nu)], \quad (9.3.1)$$

where $I(\mathbf{r}, \mathbf{n}, \nu)$ is the usual specific intensity at $\mathbf{r}$ in the direction $\mathbf{n}$, with frequency ν and $\mathbf{V}(r)$ is the velocity of the gas. For a two-level atom, the opacity $\kappa(\mathbf{r})$ is given by

$$\kappa(\mathbf{r}) = \frac{h\nu_0}{4\pi} B_{12} n_1(r), \quad (9.3.2)$$

where B_{12} is the Einstein transition probability for the transition $1 \rightarrow 2$, $n_1(r)$ is the population density of level 1 and h and c are the Planck constant and the velocity of light respectively. The source function for a line in complete redistribution is

$$S(\mathbf{r}) = [1 - \epsilon(\mathbf{r})]\, \bar{J}(\mathbf{r}) + \epsilon(\mathbf{r}) B(\mathbf{r}), \quad (9.3.3)$$

where $\epsilon(\mathbf{r})$ is the probability per scatter that a photon will be destroyed by collisional de-excitation and $B(\mathbf{r})$ is the Planck function at $\mathbf{r}$ with the local temperature at line centre frequency ν_0. The mean intensity $\bar{J}$ is

$$\bar{J}(\mathbf{r}) = \frac{1}{4\pi} \int d\Omega(\mathbf{n}) \int_0^\infty d\nu\, \phi\left[\nu - \mathbf{n} \cdot \mathbf{V}(r)\left(\frac{\nu_0}{c}\right)\right] I(\mathbf{r}, \mathbf{n}, \nu), \quad (9.3.4)$$

where $\phi(\mathbf{r}, \nu)$ is the profile function and is normalized such that

$$\int_0^\infty \phi(\mathbf{r}, \nu)\, d\nu = 1. \quad (9.3.5)$$

The radiation force per unit volume $\bar{F}$ due to line photons is

$$\bar{F}(r) = \frac{4\pi\kappa(\mathbf{r})}{c} \bar{H}(r), \quad (9.3.6)$$

where $\bar{H}(r)$ is

$$\bar{H}(r) = \frac{1}{4\pi}\int d\Omega(\mathbf{n})\cdot\mathbf{n}\int_0^\infty d\nu\phi\left[\nu - \frac{\nu_0}{c}\mathbf{n}\cdot\mathbf{v}(\mathbf{r})\right] I(\mathbf{r},\mathbf{n},\nu). \tag{9.3.7}$$

The continuous absorption outside the line is assumed to be very small. The formal solution of equation (9.3.1) for the intensity is

$$I(\mathbf{r},\mathbf{n},\nu) = \int_0^R dl\,\kappa(\mathbf{r}-\mathbf{n}l)\phi\left[\nu - \frac{\nu_0}{c}\mathbf{n}\cdot\mathbf{v}(\mathbf{r}-\mathbf{n}l)\right]S(\mathbf{r}-\mathbf{n}l) + I_\nu^{inc}\exp\left\{-\int_0^R dl'\kappa(\mathbf{r}-\mathbf{n}l')\phi\left[\nu - \frac{\nu_0}{c}\mathbf{n}\cdot\mathbf{v}\left(\mathbf{r}-\mathbf{n}l'\right)\right]\right\}. \tag{9.3.8}$$

For the definition of the quantities in equation (9.3.8) see figure 9.4. The incident intensity I_ν^{inc} is specified at infinity (normally $I = 0$) on the stellar surface which is called the core. Given the opacity, the source function and the velocity along the line of sight, equation (9.3.8) can be solved. When the velocity gradients are large equation (9.3.8) simplifies considerably as Sobolev (1947, 1957) recognized.

The intensity at frequency ν of a given ray changes only at certain discrete points called the *resonance points*, where the material has just the right Doppler shift to allow it to absorb or emit at frequency ν. These points occur when the line of sight velocity $v_l = \mathbf{v}\cdot\mathbf{n}$ satisfies the resonance condition

$$\frac{(\nu-\nu_0)}{\nu_0} = \frac{v_l}{c}. \tag{9.3.9}$$

In Sobolev's theory, one determines the radiation field around a single resonance point. The generalized Sobolev method due to Rybicki and Hummer determines the radiation field where the line of sight meets the surfaces of constant radial velocity more than once. As mentioned earlier, the radiation field becomes a function of the local physical conditions at these points; this is non-local interlocking of the radiation field. The assumptions made by Rybicki and Hummer are: (1) the slowly

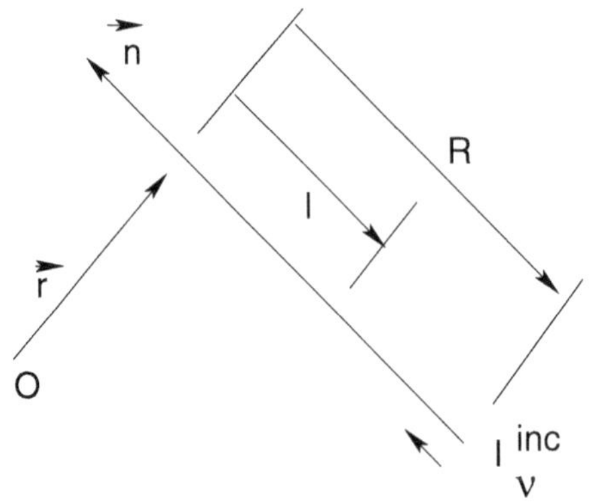

Figure 9.4 Variables appearing in the formal solution given in equation (9.3.8). l is the distance backwards along the ray passing through $\mathbf{r}$ with direction $\mathbf{n}$. The limit R is the distance at which an incident intensity I_ν^{inc} is specified (from Rybicki and Hummer (1978), with permission).

varying κ and S in the integral in equation (9.3.8) are kept outside the integral; and (2) the velocities are expanded to the first order in l, or in tensor notation

$$V_i(\mathbf{r} - \mathbf{n}l) = V_i(r) - \sum_j n_j \frac{\partial v_i}{\partial r_j} l. \tag{9.3.10}$$

Then

$$\mathbf{n} \cdot \mathbf{v}(\mathbf{r} - \mathbf{n}l) = \mathbf{n}\mathbf{v}(\mathbf{r}) - Q(\mathbf{r}, \mathbf{n})l, \tag{9.3.11}$$

where $Q(\mathbf{r}, \mathbf{n})$ is given by

$$Q(\mathbf{r}, \mathbf{n}) = \sum_i \sum_j n_i n_j \frac{\partial v_i}{\partial v_j} = \sum_i \sum_j n_i n_j e_{ij} = \frac{d v_l}{dl}. \tag{9.3.12}$$

The anti-symmetric part of dv_i/dr_j does not contribute to Q, and therefore e_{ij} is taken to be the symmetric rate of strain tensor and is given by

$$e_{ij} = \frac{1}{2}\left(\frac{\partial v_i}{\partial r_j} + \frac{\partial v_j}{\partial r_i}\right). \tag{9.3.13}$$

Now the intensity equation (9.3.8) can be written with the above assumptions (1) and (2):

$$\begin{aligned} I(\mathbf{r}, \mathbf{n}, \nu) = \kappa S \int_0^R dl\, \phi\left[\nu - \left(\frac{\nu_0}{c}\right)(\mathbf{n} \cdot \mathbf{v} + Ql)\right] \\ \times \exp\left\{-\kappa \int_0^l dl'\, \phi\left[\nu - \left(\frac{\nu_0}{c}\right)(\mathbf{n} \cdot \mathbf{v} + Ql')\right]\right\} \\ + I_\nu^{inc} \exp\left\{-\kappa \int_0^R dl'\, \phi\left[\nu - \left(\frac{\nu_0}{c}\right)(\mathbf{n} \cdot \mathbf{v} + Ql')\right]\right\}, \end{aligned} \tag{9.3.14}$$

where the quantities κ, v and Q are to be evaluated at $\mathbf{r}$. The above equation will now be subjected to a few more transformations. We introduce the dimensionless distance and frequency,

$$\lambda = \frac{l}{\Delta l} \quad \text{and} \quad \xi = \frac{\nu - \frac{\nu_0}{c}\mathbf{n} \cdot \mathbf{v}}{\Delta \nu}, \tag{9.3.15}$$

where Δl is obtained by differentiating equation (9.3.9), that is

$$\Delta l = \frac{c}{\nu_0} \Delta \nu \Big/ \left|\frac{d v_l}{dl}\right|, \tag{9.3.16}$$

where $\Delta \nu$ is the width of the line profile and can be taken as the Doppler width. We define the profile function

$$\psi(\xi) = \Delta \nu \phi(\xi \Delta \nu), \tag{9.3.17}$$

with the normalization condition that

$$\int_{-\infty}^{+\infty} \psi(\xi)\, d\xi = 1. \tag{9.3.18}$$

Then with these new variables, equation (9.3.14) can be written as

$$I = S\tau \int_0^{\infty} d\lambda\, \psi(\xi - \lambda) \exp\left[-\tau \int_0^{\lambda} d\lambda'\, \phi(\xi - \lambda')\right] + I_\nu^{inc} \exp\left[-\tau \int_0^{\infty} d\lambda'\, \psi(\xi - \lambda')\right], \tag{9.3.19}$$

where

$$\tau \equiv \kappa \frac{\Delta l}{\Delta \nu} = \frac{\kappa c}{\nu_0 |Q|}, \tag{9.3.20}$$

is the total optical depth of the velocity surface at point **r** measured in the direction **n**. The upper limits of the integrals in equation (9.3.19) are actually $R/\Delta l$ but are replaced by ∞ because $R \gg \Delta L$. A further change of variables is then introduced:

$$t = \xi - \lambda, \quad t' = \xi - \lambda'. \tag{9.3.21}$$

This results in

$$I = S\tau \int_{-\infty}^{\xi} dt\, \psi(t) \exp\left[-\tau \int_t^{\xi} dt'\, \phi(t')\right] + I_\nu^{inc} \exp\left[-\tau \int_{-\infty}^{\xi} dt' \psi(t')\right]. \tag{9.3.22}$$

Introducing another change of the variable of integration

$$W(t) = \int_{\infty}^{t} dt'\, \psi(t'), \tag{9.3.23}$$

gives us

$$I(\mathbf{r}, \mathbf{n}, \nu) = S\{1 - \exp[-\tau W(\xi)]\} + I_\nu^{inc} \exp[-\tau W(\xi)], \tag{9.3.24}$$

where $\tau W(\xi)$ is the optical depth to a particular point within the resonance region such that $W(\xi)$ gives a normalized optical depth scale changing between 0 and 1. We let the side on which $\xi \to \infty$ represent the side of the region on which the incident intensity I_ν^{inc} falls and define I_ν^{emg} as the limit of I as $\xi \to +\infty$ which is the intensity that emerges out of the resonance region. Equation (9.3.24) then gives us

$$I_\nu^{emg} = I_\nu^{inc} e^{-\tau} + S[1 - \exp(-\tau)]. \tag{9.3.25}$$

This gives the variation of intensities on the slow scale and it remains constant along a ray until it enters a resonance point. The photons escaping the resonance region meet the second resonant region, I_ν^{emg} acts as the I_ν^{inc} for the second resonant region, or

$$I_{\bar{\nu}}^{inc}\big)_2 = I_\nu^{emg}\big)_1, \tag{9.3.26}$$

where the subscripts 1 and 2 denote first and second resonant points. The frequency integrated intensity weighted by the profile function is given by

$$\bar{I}(\mathbf{r},\mathbf{n}) = \int_0^\infty d\nu\, \phi\left[\nu - \frac{\nu_0}{c}\mathbf{n}\cdot\mathbf{v}(\mathbf{r})\right] I(\mathbf{r},\mathbf{n},\nu). \tag{9.3.27}$$

$\bar{I}(\mathbf{r},\mathbf{n})$ can be used to calculate the local excitation of the material and the radiation force. The slow scale behaviour of the intensity is not sufficient to determine $\bar{I}$. We need to evaluate it by changing to fast variables. Using equation (9.3.25), we get

$$I(\mathbf{r},\mathbf{n}) = \int_{-\infty}^{+\infty} d\xi\, \psi(\xi)\left(S\{1 - \exp[-\tau W(\xi)]\} + I_{\bar{\nu}}^{inc}\exp[-\tau W(\xi)]\right). \tag{9.3.28}$$

Using $W(\xi)$ as the variable of integration, we obtain

$$I(\mathbf{r},\mathbf{n}) = S\left[1 - \frac{1-\exp(-\tau)}{\tau}\right] + I_\nu^{inc}\frac{1-\exp(-\tau)}{\tau}, \tag{9.3.29}$$

The δ-function nature of the profile requires that I_ν^{inc} should be found at $\nu = \bar{\nu}$, where

$$\bar{\nu} = \nu_0 + \left(\frac{\nu_0}{c}\right)\mathbf{n}\cdot\mathbf{v}(\mathbf{r}), \tag{9.3.30}$$

which is the line centre frequency, Doppler shifted by the velocity field along the line of sight. Equations (9.3.25) and (9.3.29) are the basic equations of the Sobolev approximation with complete redistribution. Equation (9.3.30) can be applied when $\mathbf{v}(\mathbf{r})$ is replaced by $-\mathbf{v}(\mathbf{r})$ as τ depends on the absolute value of Q and on the source function being the same.

The above theory is applied to multiple surfaces of equal radial velocity. For example, in a double surface problem we can apply equation (9.3.24) to each surface and use that the emergent intensity at the surface 1 becomes the incident intensity at the surface 2. Thus, the emergent intensity after passing through the two surfaces is

$$\begin{aligned} I_{\bar{\nu}}^{emg} &= \exp(-\tau_1)\exp(-\tau_2)I^{inc} + \exp(-\tau_1)\left[1 - \exp(-\tau_1)\right] S_2 \\ &\quad + \left[1 - \exp(-\tau_1)\right] S_1, \end{aligned} \tag{9.3.31}$$

where the quantities subscripted with 1 and 2 refer to surfaces 1 and 2. The frequency integrated intensity $\bar{I}(\mathbf{r},\mathbf{n})$ is given by

$$\bar{I}(\mathbf{r},\mathbf{n}) = \exp(-\tau_2)\left[\frac{1-\exp(-\tau_1)}{\tau_1}\right] I^{inc}$$
$$+ \left[1-\exp(-\tau_2)\right]\left[\frac{1-\exp(-\tau_1)}{\tau_1}\right] S_2(\mathbf{r})$$
$$+ \left[1 - \frac{1-\exp(-\tau)}{\tau_1}\right] S_1(\mathbf{r}). \qquad (9.3.32)$$

The source functions for different parameters are shown in figure 9.5. Flux profiles are given in figure 9.6 and the radiation forces H corresponding to the source functions in figure 9.5 are shown in figure 9.7. The reader is referred to the papers of Rybicki and Hummer (1978), Marti and Noerdlinger (1977) and Grachev and Grinin (1975) for more details.

The method due to Sobolev is sometimes called the first order probability method, while the methods due to Athay (1972a,b), Frisch and Frisch (1975), Canfield *et al.* (1981), Scharmer (1981, 1984) are called the 'second order probability methods'. These latter methods work well with partial frequency redistribution, while the former works only with complete redistribution. The second order methods are more economical in computer time. Rybicki's (1972, 1984) core-saturation method and the operator perturbation methods of Cannon (1973a,b, 1984), Kalkofen (1987) and

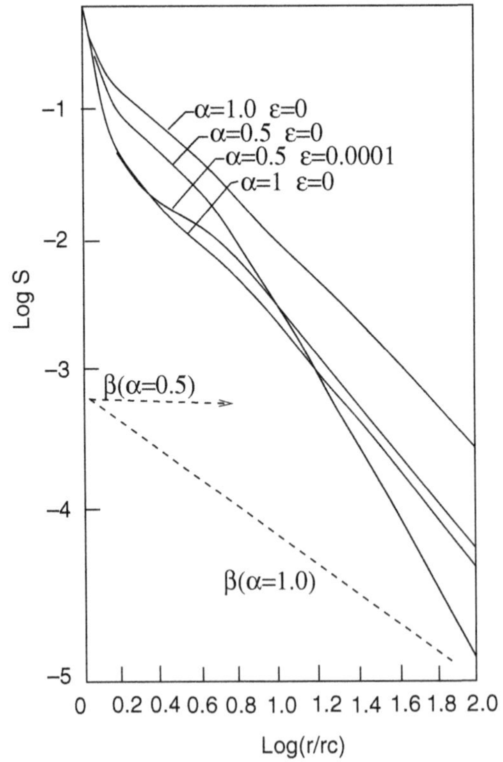

Figure 9.5 Source functions for various models with purely radiative excitation for different αs in the law of velocity $v(r) = v_0(r_0/r)^\alpha$, $\alpha > 0$ and for different τ_0s, where $\tau_0 = hcB_{12}r_0n_0/4\pi v_0$ is the tangential ($\mu = 0$) optical thickness of the velocity surface at the inner radius r_0. Broken lines show the variation of β (from Rybicki and Hummer (1978), with permission).

Scharmer (1981, 1984) are based on the escape probability of photons from the media.

Miyamoto (1949, 1952) and Kogure (1959, 1961, 1967) developed a diagnostic

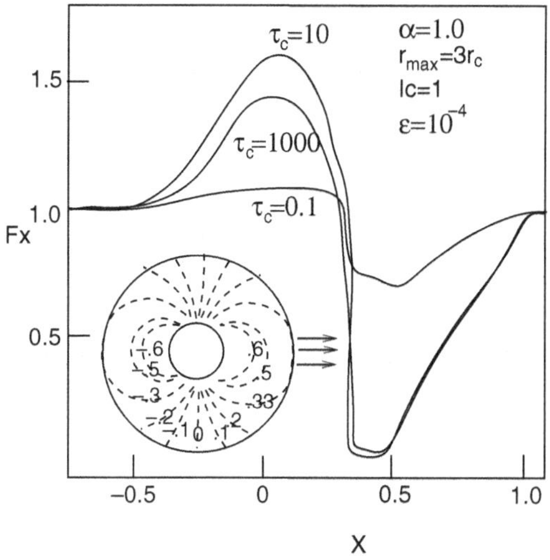

Figure 9.6 Flux profiles for various values of the parameter τ_0 for $r_{max}/r_c = 3$. The inset shows the location of the common direction velocity surfaces for the indicated value of frequency displacement x (from Rybicki and Hummer (1978), with permission).

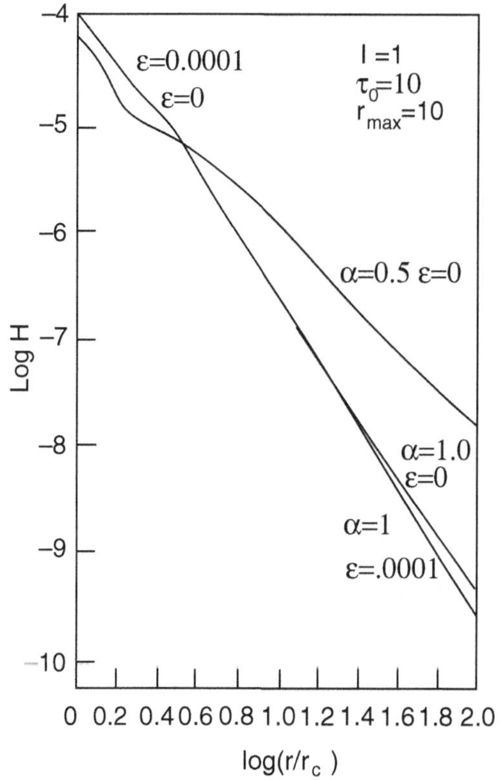

Figure 9.7 Radiation forces H for the source functions shown in figure 9.5 (from Rybicki and Hummer (1978), with permission).

method based on the Rosseland cycle to explain the abnormal characteristics of Be stars. They did not treat the problem of the effect of differential Doppler effect on the transfer of radiation in the stellar envelope. Ueno (1965) studied a new probabilistic approach to derive the integral equations of the scattering functions for slab media which contain emitting sources.

9.4 Core-saturation method of Rybicki (1972)

Osterbrock (1962) discussed the escape of resonance line radiation with complete redistribution using the fact that most of the photon transfer takes place in the wings of the line while the core plays mainly a passive role. According to him, the mean number of scatterings can be connected to a partial integral over the line profile and this integral gives the probability of emission sufficiently in the wings for escape to occur in a single photon flight. Rybicki and Hummer (1969) showed that the thermalization length is mainly determined by the single longest flight in a series of random flights of the photons, and that this longest flight is connected with the rare event in which the photon is emitted far in the wings. An improvement in the study of this problem arises if we eliminate the passive components of intensity in the core leaving only the active wing components.

The above ideas can be derived as follows. The transfer equation in plane parallel geometry is

$$\frac{\mu}{\kappa_\nu}\frac{\partial I_\nu}{\partial \tau} = I_\nu - S_\nu, \tag{9.4.1}$$

where I_ν is the specific intensity, S_ν is the total source function, κ_ν is the total opacity divided by the integrated line opacity and τ is the integrated line optical depth. When the opacity κ_ν is very large in the core of a line and $\mu = 1$, the term on the LHS of equation (9.4.1) can be neglected and we then have

$$I_\nu = S_\nu. \tag{9.4.2}$$

In the case of an isotropic source function this gives us the mean intensity J_ν as

$$J_\nu = S_\nu. \tag{9.4.3}$$

Equation (9.4.2) or equation (9.4.3) is regarded as the core-saturation approximation due to Rybicki (1972).

If the equation (9.4.3) holds true at all line frequencies, the situation is called complete line saturation; this was discussed by Kalkofen (1966). Core-saturation is a generalization of the complete line saturation.

We have to find the critical frequency which separates the saturated core and the unsaturated core frequencies. The unsaturated components are called transfer or wing components or simply wing. The critical frequency may be decided in such a way that the monochromatic optical depth τ_ν from the given point to the nearest boundary is of the order of unity or $\tau_\nu \sim 1$. This determines the critical frequency.

If γ is the parameter which represents the core–wing separation, then

$$\tau_\nu = \gamma, \tag{9.4.4}$$

where $\gamma \geq 1$.

Stenholm (1977) applied Rybicki's core-saturation method to multi-dimensional non-LTE transfer problems. Stenholm and Stenflo (1977, 1978) extended this method to the transfer of radiation to magnetic flux tubes. Kalkofen and Ulmschneider (1984) applied it to moving media. We shall describe this method briefly below.

The critical frequency x_c is found in such a way that

$$\tau\phi(x_c) = \gamma, \tag{9.4.5}$$

where ϕ is the profile function and x_c is the dimensionless critical frequency such that

$$\left.\begin{array}{ll} \text{for core frequencies} & |x| < x_c, \\ \text{for wing frequencies} & |x| > x_c. \end{array}\right\} \tag{9.4.6}$$

The source function S for a two-level atom with complete redistribution is

$$S = \frac{\bar{J} + \epsilon'}{1 + \epsilon'}, \tag{9.4.7}$$

where $\epsilon' = \epsilon^\star/(1 + \epsilon^\star)$, with $\epsilon^\star$ the usual collisional parameter. B is the Planck function and the integrated mean intensity is

$$J(r) = \frac{1}{2}\int_{-1}^{+1} d\mu \int_0^\infty d\nu\, \phi(\tau, \nu, \mu) I(\tau, \nu, \mu), \tag{9.4.8}$$

where $\phi(\tau, \nu, \mu)$ is the profile function given by

$$\phi(\tau, \nu, \mu) = \psi/(\pi^{\frac{1}{2}} \Delta\nu_D), \tag{9.4.9}$$

with ψ the Voigt function,

$$\psi(\tau, \nu, \mu) = H(a, v), \tag{9.4.10}$$

where a is given in terms of the damping constant Γ as

$$a(r) = \Gamma/4\pi\Delta\nu_D \tag{9.4.11}$$

and v is given by

$$v(\tau, \nu, \mu) = \frac{\nu - \nu_0 - (\nu_0/c)\mu V}{\Delta\nu_D}. \tag{9.4.12}$$

Also $a = 0$ for pure Doppler broadening and $\psi = 1$ at the displaced line centre $\nu = \nu_0[1 + (V/c)\mu]$, and in a static atmosphere $V = 0$, the reference optical depth coincides with the monochromatic optical depth at the line centre. The profile function ϕ satisfies the normalization condition:

$$\int_0^\infty \phi(\tau, \nu, \mu)\, d\nu = 1. \tag{9.4.13}$$

The integral in equation (9.4.8) can be written as the sum of integrals over the line and wing frequencies as

$$J(\tau) = \frac{1}{2}\int_{\nu,\mu\in core} d\nu \int d\mu\, \phi I + \frac{1}{2}\int_{\nu,\mu\in wings} d\nu \int d\mu\, \phi I, \tag{9.4.14}$$

which is Rybicki's (1972) equation for the separation of the core and wing frequencies in the line. Similarly the normalization integral may also be split into core and wing frequencies. Thus, we define the wing part as

$$\Omega(\tau) = \frac{1}{2}\int_{\nu,\mu\in wings} \int d\mu\, \phi(\tau, \nu, \mu) \tag{9.4.15}$$

and the integral of the profile function over the line core is written as

$$\frac{1}{2}\int_{\nu,\mu\in core} \int d\mu\, \phi(\tau, \nu, \mu) = 1 - \Omega(\tau). \tag{9.4.16}$$

Using equation (9.4.2), we get

$$I(\tau, \nu, \mu) = S(\tau), \qquad \nu, \mu \in core. \tag{9.4.17}$$

The core-saturation approximation can now be used to write equation (9.4.14) as

$$J(\tau) = S(1 - \Omega) + \frac{1}{2}\int_{\nu,\mu\in wings} d\nu \int d\mu\, \phi I. \tag{9.4.18}$$

Using equations (9.4.7) and (9.4.18), we get

$$S = (\Omega + \epsilon)^{-1}\left[\epsilon B + \frac{1}{2}\int_{\nu,\mu\in wings} d\nu \int d\mu\, \phi I\right]. \tag{9.4.19}$$

In the above equation, there is no reference to the core photons.

The monochromatic specific intensity $I(\tau, \nu, \mu)$ depends on $S(\tau)$ in a layer whose thickness is of the order of a photon mean free path. Equation (9.4.19) can be of the form

$$S(\tau) = X(\tau, \tau')S(\tau') + Y(\tau), \tag{9.4.20}$$

where X is the angle and frequency integrated matrix operator. The above equation can be solved iteratively as

$$S^{(n)}(\tau) = X(\tau, \tau')S^{(n-1)}(\tau') + Y(\tau), \tag{9.4.21}$$

where $S^{(n)}$ is the source function in the nth approximation.

In the standard core-saturation technique of Rybicki (1972), extended by Stenholm (1977) and Scharmer (1981), it is the optical depth at a given point that decides the separation of the wing photons from those of the core at a given point. This method treats the escape of photons from the surface of the medium but neglects transfer within the medium. It cannot treat structure, such as shocks within the medium. Kalkofen and Ulmschneider (1984) treated this problem. They used the usual convention where $\mu = -1$ points in the positive direction of τ. The monochromatic optical distance δ for outward travelling photons between the spatial grid points τ_k and τ_{k+1} is given by

$$\delta = \frac{1}{\mu_m}\bar{\psi}\,(\tau_{k+1} - \tau_k), \qquad \mu_m > 0, \tag{9.4.22}$$

where $\bar{\psi}$ is the average profile in the interval $(k, k+1)$, which is

$$\bar{\psi} = \frac{1}{2}\left[\psi(\tau_k, \nu_n, \mu_m) + \psi(\tau_{k+1}, \nu_n, \mu_m)\right] \tag{9.4.23}$$

for photons of the ray (ν_n, μ_m). Similarly for the inward travelling ray, we have

$$\delta = \frac{1}{\mu_m}\bar{\psi}\,(\tau_k - \tau_{k-1}), \qquad \mu_m > 0, \tag{9.4.24}$$

for the optical distance between the grid points τ_k and τ_{k-1} along the ray $(-\nu_n, -\mu_m)$ and $\bar{\psi}$ is

$$\bar{\psi} = \frac{1}{2}\left[\psi(\tau_k, \nu_n, \mu_m) + \psi(\tau_{k-1}, \nu_n, \mu_m)\right] \tag{9.4.25}$$

in the interval $(k, k-1)$. Here ν_n is the discretized frequency displacement $(\nu - \nu_0)$ from the rest frequency ν_0 of the line. The profile function is symmetric about the displaced line centre or

$$\phi(\tau, \nu_n, \mu_m) = \phi(\tau, -\nu_n, -\mu_m). \tag{9.4.26}$$

This property is used in computing equation (9.4.25) and reduces the number of computations by half. If

$$\delta > \gamma, \tag{9.4.27}$$

the photons at τ_k of the outward beam (ν_n, μ_m) are supposed to be in the line core and if relation (9.4.27) is satisfied for δ in the interval $(k, k-1)$, the corresponding photons in the inward beam $(-\nu_n, -\mu_m)$ will be in the core. The above relation is not satisfied by the photons in the far wings (ν_n >< and $\mu = \pm\mu_m$). To obtain the intensity assuming that the source function depends linearly on monochromatic optical depth and that the distance to the boundary is large, a generalized Eddington–Barbier relation has been used. Thus (see chapter 2)

$$I(\tau_k, \nu, \mu) = S\left[\tau_k + \frac{\mu}{\psi(\tau_k, \nu, \mu)}\right]. \quad (9.4.28)$$

Since $\tau = \tau_k \pm \mu_m/\psi(\tau_k, \pm\nu_n, \pm\mu_m)$ does not coincide with any grid point, we need an interpolated value of the source function in equation (9.4.28). We write the intensity as

$$I^+(\tau_k, \nu_n, \mu_m) = W S_j + (1 - W)S_{j+1}, \quad (9.4.29)$$

where

$$W = (\delta_k^{j+1} - 1)/\delta_j^{j+1}, \quad (9.4.30)$$

δ_k^j is the optical distance along the rays (ν_n, μ_m) and $(-\nu_n, -\mu_m)$ between the spatial grid points τ_j and τ_k, and $(j, j+1)$ is the depth interval in which the source point for the outward intensity at τ_k is located. The inward intensity is

$$I^-(\tau_k, -\nu_n, -\mu_m) = W S_i + (1 - W)S_{i-1}, \quad (9.4.31)$$

with

$$W = (\delta_{i-1}^k - 1)/\delta_{i-1}^i, \quad (9.4.32)$$

where $(i, i-1)$ is the depth interval containing the source point for I^-. The intervals $(j, j+1)$ and $(i, i-1)$ are chosen so that the source points are at unit optical distance along the corresponding rays from field points τ_k at which the intensity is to be determined.

The optical distance $\bar{\tau}$ along the rays (ν_n, μ_m) and $(-\nu_n, -\mu_m)$ is

$$\bar{\tau} = \frac{1}{\mu}\int_0^{\tau_k} dt\, \psi(t, \nu_n, \mu_m). \quad (9.4.33)$$

At the grid points τ_i, the intensity I_i^+ in the direction (ν_n, μ_m), across the layer $(i, i+1)$ is

$$I_i^+ = aI_{i+1}^+ + bS_i + cS_{i+1}, \quad (9.4.34)$$

where the coefficients a, b and c depend on the optical path length over the layer $(i, i+1)$ and

$$\delta = \bar{\tau}_{i+1} - \bar{\tau}_i. \quad (9.4.35)$$

The inward intensity along $(-\nu_n, -\mu_m)$ is obtained by integrating across the layer $(i-1, i)$,

$$I_i^- = aI_{i-1}^- + bS_i + cS_{i-1}, \quad (9.4.36)$$

where the quantities a, b and c depend on δ, which is given by

$$\delta = \bar{\tau}_i - \bar{\tau}_{i-1}. \tag{9.4.37}$$

If δ is small (half implicit differencing), then the coefficients in equations (9.4.36) and (9.4.37) are

$$a = \frac{2-\delta}{2+\delta}, \quad b = c = \frac{\delta}{2+\delta}, \quad \delta \leq 1. \tag{9.4.38}$$

If the mesh is written, then

$$a = c = \frac{1}{2\delta + 1}, \quad b = (2\delta - 1)/(2\delta + 1), \quad \delta > 1. \tag{9.4.39}$$

The integration weights are obtained from the formal solution of the equation, thus

$$I_i^+ = \int_{\bar{\tau}_i}^{\bar{\tau}_{i+1}} dt \exp[-(t - \bar{\tau}_i)]S(t) + I_{i+1}^+ \exp[-(\bar{\tau}_{i+1} - \bar{\tau}_i)]. \tag{9.4.40}$$

If the source function is assumed to be a linear function of monochromatic optical depth in $(i, i+1)$, we can obtain an expression of the form (9.4.34) for the outward intensity with the following coefficients:

$$\left.\begin{aligned} a &= \exp(-b), \quad b = 1 - \frac{1 - \exp(-\delta)}{\delta}, \\ c &= \exp(-\delta)\left[\frac{\exp(\delta) - 1}{\delta} - 1\right], \end{aligned}\right\} \tag{9.4.41}$$

in terms of equation (9.4.35) for the path length δ. For short path lengths, we find

$$\left.\begin{aligned} b &= \frac{\delta}{2}\left[1 - \frac{\delta}{3}\left(1 - \frac{\delta}{4}\cdots\right)\right], \quad \delta \ll 1, \\ c &= \exp(-\delta)\frac{\delta}{2}\left[1 + \frac{\delta}{3}\left(1 + \frac{\delta}{4}\cdots\right)\right]. \end{aligned}\right\} \tag{9.4.42}$$

The coefficients for I_i are obtained in the same way using equation (9.4.37).

9.5 Scharmer's method

In this method an approximate relationship is developed between the monochromatic intensity and the source function by evaluating an approximate solution of the transfer equation along a ray. This method is sufficiently accurate even with large velocity gradients and it can be applied easily to multi-level non-LTE and multi-dimensional problems. One can choose an approximate or accurate solution to compute the models of radiation hydrodynamics (Scharmer 1981, 1983, 1984).

An integral equation for the line source function S_l is derived using the core-saturation approximation, the Eddington–Barbier relation and the upper boundary condition. The transfer equation is given by,

$$\pm \frac{dI_\nu}{d\tau_\nu} = I_\nu - S_\nu, \tag{9.5.1}$$

where the + and − signs indicate the outgoing and incoming rays respectively. τ_ν is the monochromatic optical depth along the outgoing ray and is given by

$$d\tau_\nu = \left(\frac{\phi_\nu}{r} + 1\right)\frac{d\tau_c}{|\mu|}, \tag{9.5.2}$$

where $d\tau_c$ is the continuum optical depth and r is the ratio of the continuum to line opacities.

An approximate solution, which lies between I_ν and S_ν, can be obtained when essentially no transfer of radiation occurs in the optically thick core of the line (core-saturation (Rybicki 1972) or on-the-spot approximation (Osterbrock 1962)). Therefore, for the frequencies which are optically thick, we have

$$I_\nu(\tau_\nu) \approx S_\nu(\tau_\nu), \quad \tau_\nu > \gamma,\ \mu < 0,\ \mu > 0. \tag{9.5.3}$$

When the optical depth is much smaller than unity the Eddington–Barbier relation is valid for an outgoing ray (which is exact only at the surface for a source function which varies linearly with the optical depth) and is stated as

$$I_\nu(\tau_\nu) \approx S_\nu(\tau_\nu = 1), \quad \tau_\nu < \gamma,\ \mu > 0. \tag{9.5.4}$$

For a semi-infinite medium, we need to include the upper boundary and

$$I_\nu(\tau_\nu) \approx I_\nu(\tau_\nu = 0),\ \tau_\nu < \gamma,\ \mu < 0. \tag{9.5.5}$$

Let us consider a two-level atom. The line source function S_l is given by

$$S_l = \frac{J + \epsilon^\star B}{1 + \epsilon^\dagger}, \tag{9.5.6}$$

where J is the integrated mean intensity and $\epsilon^\star$ and $\epsilon^\dagger$ refer to different creation and destruction processes of photons (Athay 1972a, page 16). J is given by

$$J = \int_{-\infty}^{+\infty} \phi_\nu J_\nu \, d\nu \tag{9.5.7}$$

where J_ν is the average of I_ν over μ. We divide J between an optically thick core J_c and the optically thin wings J_W (see section 9.4 on core saturation). Thus, we have

$$J = J_c + J_W = \int_{core} \phi_\nu J_\nu \, d\nu + \int_{wings} \phi_\nu J_\nu \, d\nu. \tag{9.5.8}$$

For the core part of the line we can use equation (9.5.3). As S_l is independent of μ equation (9.5.3) can be integrated to give

$$J_\nu \approx S_l. \tag{9.5.9}$$

In the case of wings, J_ν can be written for the rays (incoming and outgoing) as

$$J_\nu = \frac{1}{2}\int_{-1}^{0} I_\nu \, d\mu + \frac{1}{2}\int_{0}^{1} I_\nu \, d\mu. \tag{9.5.10}$$

The first term can be evaluated using the upper boundary condition in equation (9.5.5). The second term is evaluated by using the angle-averaged Eddington–Barbier relation by assuming that S_l can be approximated by a linear function of τ_ν, that continuum opacity is negligible, and that ϕ_ν is depth independent. Then S_l is given as a linear function of the line centre optical depth τ_0 by

$$S_l = a_\nu + b_\nu \tau_0 = a_\nu + b_\nu \mu \tau_\nu / \phi_\nu. \tag{9.5.11}$$

The emergent intensity from the above source function is

$$I_\nu = a_\nu + b_\nu / \phi_\nu. \tag{9.5.12}$$

Using equation (9.5.12), we get

$$\frac{1}{2}\int_{0}^{1} I_\nu \, d\mu = \frac{1}{2}\left[a_\nu + \frac{1}{2} b_\nu / \phi_\nu\right] = \frac{1}{2} S_l(\tau_0'), \tag{9.5.13}$$

where the quadrature point τ_0' depends on ν, and not on τ_0, and

$$\tau_0'(\nu) = (2\phi_\nu)^{-1}. \tag{9.5.14}$$

Therefore we can write J as the combination of contributions from: (1) the core-saturation approximation; (2) the Eddington–Barbier relation; and (3) the outer boundary condition for the incoming intensity. Thus,

$$J = 2S_l \int_{0}^{\nu_c} \phi_\nu \, d\nu + \int_{\nu_c}^{\infty} S_l(\tau_0') \phi_\nu \, d\nu + \int_{\nu_c}^{\infty} J_\nu^- \phi_\nu \, d\nu, \tag{9.5.15}$$

where J_ν^- is the mean intensity incident at the top of the atmosphere and $\nu_c(\tau_0)$ is the critical frequency which separates the saturated core ($\tau_\nu > \gamma$) from the transparent wing ($\tau_\nu < \gamma$). This frequency is defined such that τ_0' is continuous at ν_c, that is $\tau_0' = \tau_0$ when $\nu = \nu_c$. Using equation (9.5.14) we find that ν_c is given by

$$\phi_{\nu_c} = (2\tau_0)^{-1} \quad \text{for } \gamma = \frac{1}{2}. \tag{9.5.16}$$

Using equations (9.5.6), (9.5.15) and (9.5.16) and the normalization condition that

$$\int_{\nu_0}^{\infty} \phi_\nu \, d\nu + \int_{0}^{\nu_c} \phi_\nu \, d\nu = \frac{1}{2}, \tag{9.5.17}$$

we get

$$\epsilon^\dagger S_l - \epsilon^\star B = -2S_l \int_{\nu_c}^{\infty} \phi_\nu \, d\nu + \int_{\nu_c}^{\infty} S_l(\tau_0') \phi_\nu \, d\nu + \int_{\nu_c}^{\infty} J_\nu^- \phi_\nu \, d\nu. \tag{9.5.18}$$

One can use the probability of escape of photons p_e (Osterbrock, 1962) as the depth variable defined as,

$$p_e(\tau_0) = \int_{\nu(\tau_\nu)}^{\infty} \phi_\nu \, d\nu. \tag{9.5.19}$$

If p_e is the probability of a wing photon escaping in the direction $\mu > 0$, then equation (9.5.18) can be written as

$$\epsilon^\dagger S_l - \epsilon^\star B = -2S_l p_e + \int_0^{p_e} S_l \, dp_e' + \int_0^{p_e} J^- \, dp_e'. \tag{9.5.20}$$

This equation is not derived from probabilistic considerations but probabilistic interpretations can be given. It is derived from the core-saturation approximation, the Eddington–Barbier relation and the upper boundary condition. Equation (9.5.20) can be written as a first order differential equation as

$$\left(\epsilon^\dagger + 2p_e\right) \frac{dS_l}{dp_e} = -S_l + \epsilon^\star \frac{dB}{dp_e} - Sl \frac{d\epsilon^\dagger}{dp_e} + B \frac{d\epsilon^\star}{dp_e} + B \frac{d\epsilon^\star}{dp_e} + J^-. \tag{9.5.21}$$

If $\epsilon^\dagger = \epsilon^\star = \epsilon'$ and if J^-, ϵ', B are not functions of p_e, then

$$(\epsilon' + 2p_e) \frac{dS_l}{dp_e} = J^- - S_l. \tag{9.5.22}$$

In a semi-infinite medium, we have

$$S_l = J^- + (B - J^-) \left[\epsilon'/(\epsilon' + 2p_e)\right]^{\frac{1}{2}}, \tag{9.5.23}$$

as $S_l \to B$ and $p_e \to 0$. If $p_e = \frac{1}{2}$, then

$$\begin{aligned} S_l(\tau = 0) &= J^- + (B - J^-)[\epsilon'/(1 + \epsilon')]^{\frac{1}{2}} \\ &= \sqrt{\epsilon} B + (1 - \sqrt{\epsilon}) J^-, \end{aligned} \tag{9.5.24}$$

where $S_l(\tau = 0)$ is the surface value of the source function S_l and $\epsilon = \epsilon'/(1 + \epsilon')$. The solutions of equation (9.5.21) have been discussed by Athay (1972b, 1976) and Frisch and Frisch (1975).

The above analysis can be used to obtain the source function in the presence of large velocity gradients as in Sobolev's theory, neglecting the incident radiation. In the limit of large velocity gradients, it is understood that the radiation coming from below in the wings is equal to the continuum intensity which is independent of frequency. This gives us

$$S_l(\tau_0') = I_c = \text{constant}. \tag{9.5.25}$$

Substituting this into equation (9.5.20) and setting $\epsilon^\dagger = \epsilon^\star = \epsilon'$, we obtain

$$\epsilon'(S_l - B) = -2p_e S_l + p_e I_c, \tag{9.5.26}$$

or

$$S_l = [(1-\epsilon)p_e I_c + \epsilon B][\epsilon + (1-\epsilon)2p_e]^{-1}, \tag{9.5.27}$$

where $\epsilon = \epsilon'/(1+\epsilon')$. This is the same as equation (9.2.34), with $\beta = 2\beta_c = 2p_e$.

This method has been found applicable for intermediate velocity gradients also. There are discontinuities at the transition point between the core frequencies and the wing frequencies because of the fact that the non-local radiation is ignored. A more accurate lambda operator method in a semi-infinite medium is described below; this method corrects the above disadvantages.

The formal solution for $\mu > 0$ is written as

$$I_\nu(\tau_\nu) = \exp(\tau_\nu)\int_{\tau_\nu}^{\infty} S_\nu e^{-\tau'_\nu}\, d\tau'_\nu. \tag{9.5.28}$$

S_ν and I_ν are related through the quadrature points $\tau_\nu^\pm$ and the corresponding weights $w^\pm$ for outgoing (+) and incoming (−) rays. Thus

$$I_\nu^\pm(\tau_\nu) = w^\pm(\tau_\nu)S_\nu(\tau_\nu^\pm). \tag{9.5.29}$$

To evaluate $\tau_\nu^\pm$ and $w^\pm$, we assume that S_ν is a linear function of τ_ν, then

$$S_\nu = a_\nu + b_\nu \tau_\nu. \tag{9.5.30}$$

By substituting equation (9.5.30) into equation (9.5.28), we obtain

$$I_\nu(\tau_\nu) = a_\nu + b_\nu(\tau_\nu + 1). \tag{9.5.31}$$

A comparison of equations (9.5.31) and (9.5.29) gives us

$$w^+ = 1 \quad \text{and} \quad \tau_\nu^+ = \tau_\nu + 1. \tag{9.5.32}$$

By writing the formal solution for $\mu < 0$,

$$I_\nu = -\exp(-\tau_\nu)\int_0^{\tau_\nu} S_\nu \exp(-\tau'_\nu)\, d\tau'_\nu, \tag{9.5.33}$$

we obtain

$$w^- = 1 - \exp(-\tau) \quad \text{and} \quad \tau_\nu^- = \frac{\tau_\nu}{w} - 1. \tag{9.5.34}$$

These expressions contain the upper boundary condition that $I_\nu(0) = 0$. Now when $\tau_\nu \to \infty$, we have

$$w^+ = w^- = 1 \quad \text{and} \quad \tau_\nu^+ = \tau_\nu^+ = \tau_\nu \tag{9.5.35}$$

and if $\tau_\nu \to 0$, we get $\tau^+ = 1$ (Eddington–Barbier relation) and $w^- = 0$, which corresponds to $I_\nu(0) = 0$ for the incoming rays. The advantage of these quadrature formulae is that they facilitate a continuous transition from the core to the wing. This improves the accuracy.

The sources in the upper layers cause some of the non-localness of the radiation field where the optical depth is less than τ_ν. If we expand w^- and τ_ν^-, we get

$$w^- \to \tau_\nu \quad \text{and} \quad \tau_\nu^- \to \frac{1}{2}\tau_\nu. \tag{9.5.36}$$

In the case of a plane parallel slab, the formal solution is

$$I_\nu(\tau_\nu) = \exp(\tau_\nu) \int_0^{\tau_\nu} S_\nu \exp(-\tau_\nu')\, d\tau_\nu'. \tag{9.5.37}$$

The τ_ν^+s and w^+s for the outgoing rays are calculated as previously assuming S to be a linear function of optical depth as in equation (9.5.30). Then, we obtain for $I_\nu(\tau_\nu)$:

$$I_\nu(\tau_\nu) = [1 - \exp(-\delta\tau_\nu)]\,(a_\nu + b_\nu\{1 + T_\nu - \delta\tau_\nu/[1 - \exp(-\delta\tau_\nu)]\}), \tag{9.5.38}$$

where $\delta\tau_\nu = T_\nu - \tau_\nu$. From equations (9.5.29) and (9.5.38), we obtain

$$w^+ = 1 - \exp(-\delta\tau_\nu); \quad \tau_\nu^+ = 1 + T_\nu - \delta\tau_\nu/w^+. \tag{9.5.39}$$

One can obtain the result for a semi-infinite medium if $\delta\tau_\nu \gg 1$. And if T_ν, $\tau_\nu \ll 1$, then

$$w^+ \to \delta\tau_\nu, \quad \tau_\nu^+ \to T_\nu - \delta\tau_\nu/2 = \frac{1}{2}(T_\nu + \tau_\nu), \tag{9.5.40}$$

and when $\tau_\nu \ll T_\nu$, then $\tau_\nu^+ \approx \frac{1}{2}T_\nu$. This relation can be used to obtain the solution of non-LTE problems in plane parallel slabs and extended atmospheres but not in semi-infinite media.

This theory can be applied to obtain the solution of non-LTE two-level atom problems. The source function in this case is given by

$$S_\nu = \frac{\phi_\nu}{\phi_\nu + \beta} S_l + \frac{\beta}{\phi_\nu + \beta} B, \tag{9.5.41}$$

where all the symbols have their usual meanings (see chapter 1). Let us write the intensity in terms of an integral operator $\Lambda^\dagger_{\nu\mu}$ as

$$I_\nu \approx \Lambda^\dagger_{\nu\mu}[S_\nu]. \tag{9.5.42}$$

From equations (9.5.6), (9.5.41) and (9.5.42) we can write the integral equation for S_l as

$$(1 + \epsilon^\dagger)S_l - \frac{1}{2}\int_{-\infty}^{+\infty}\int_0^\infty \phi_\nu \Lambda^\dagger_{\nu\mu}\left[\frac{\phi_\nu}{\phi_\nu + \beta} S_l\right] d\mu\, d\nu$$
$$\approx \epsilon^\star B + \frac{1}{2}\int_{-1}^{+1}\int_0^\infty \phi_\nu \Lambda^\dagger_{\nu\mu}\left[\frac{\beta}{\phi_\nu + \beta} B\right] d\mu\, d\nu, \tag{9.5.43}$$

where β is very small and at great depths the second term of the above equation is of the order $(1 - p_e)S_l$. Thus the two terms on the LHS cancel at large depths, where p_e is small. This situation gives rise to numerical problems at large optical depths and if ϵ is of the same order as the accuracy of the computer. To avoid this the following procedure is used.

We introduce a new operator $\delta\Lambda^{\dagger}_{\nu\mu}$ which is the difference between $\Lambda^{\dagger}_{\nu\mu}$ and the operator corresponding to complete saturation $I_\nu = S_\nu$. Then

$$\delta\Lambda^{\dagger}_{\nu\mu} = \Lambda^{\dagger}_{\nu\mu} - 1. \tag{9.5.44}$$

From this equation and equation (9.5.43) we obtain

$$(1 + \epsilon^{\dagger})S_l = \left(\frac{1}{2}\int_{-1}^{+1}\int_0^{\infty} \frac{\phi_\nu + \beta}{\phi_\nu + \beta}\phi_\nu \, d\mu \, d\nu + \epsilon^{\dagger}\right) S_l \tag{9.5.45}$$

and

$$\begin{aligned}\left(e^{\dagger} + \delta\right) S_l - \frac{1}{2}\int_{-1}^{+1}\int_0^{\infty} \phi_\nu \delta\Lambda^{\dagger}_{\nu\mu}\left(\frac{\phi_\nu}{\phi_\nu + \beta} S_l\right) d\mu \, d\nu \\ \approx (\epsilon^{\star} + \delta)B + \frac{1}{2}\int_{-1}^{+1}\int_0^{\infty} \phi_\nu \delta\Lambda^{\dagger}_{\nu\mu}\left(\frac{\beta}{\phi_\nu + \beta} B\right) d\mu \, d\nu,\end{aligned} \tag{9.5.46}$$

where

$$\delta = \frac{1}{2}\int_{-1}^{+1}\int_0^{\infty} \frac{\beta}{\phi_\nu + \beta}\phi_\nu \, d\mu \, d\nu \tag{9.5.47}$$

is the probability that a line photon is absorbed in a continuum transition.

At great depths where $\delta\Lambda^{\dagger}_{\nu\mu}$ vanishes, we get

$$(\epsilon^{\dagger} + \delta)S_l \approx (\epsilon^{\dagger} + \delta)B, \tag{9.5.48}$$

which gives fairly accurate results for small $\epsilon^{\star}$, $\epsilon^{\dagger}$ and δ.

We shall now describe a numerical procedure for solving equation (9.5.46). We shall write

$$\mathbf{w} \cdot \mathbf{S}_l = -\mathbf{E}, \tag{9.5.49}$$

where $\mathbf{w}$ is a matrix and $\mathbf{S}_l$ is the vector of the source function $S_l(1), S_l(2), \ldots, S_l(n_r)$, where n_r is the number of depth points, and $\mathbf{E}$ is the vector of the components of the RHS of equation (9.5.46). By using the symmetry relation

$$\left.\begin{aligned}\phi_\nu(-\mu, \nu, \tau_c) &= \phi_\nu(\mu, -\nu, \tau_c), \\ \tau_\nu(-\mu, \nu, \tau_c) &= \tau_\nu(\mu, -\nu, \tau_c),\end{aligned}\right\} \tag{9.5.50}$$

we can treat μ as a positive variable. Then for each $+\mu$, we have $\tau_\nu^{\pm}$ and weights $w^{\pm}$. The integral on the LHS of equation (9.5.46) is written as

$$\frac{1}{2}\int_{-1}^{+1}\int_{0}^{\infty}\phi_\nu \delta\Lambda^{\dagger}_{\nu\mu}\left(\frac{\phi_\nu}{\phi_\nu+\beta}S_l\right)d\mu\, d\nu$$
$$=\sum_{i=1}^{n_\mu}\sum_{j=1}^{n_\nu}\phi_{ijk}\left(\delta\Lambda^{\dagger}_{-ijk}+\delta\Lambda^{\dagger}_{ijk}\right)\left(\frac{\phi_{ijk}}{\phi_{ijk}+\beta}S_{lk}\right)w_i^\mu w_j^\nu w_k^\tau, \tag{9.5.51}$$

where the operators $\delta\Lambda^{\dagger}_{-ijk}$ and $\delta\Lambda^{\dagger}_{ijk}$ correspond to the incoming and outgoing rays, $i = 1,\ldots,n_r$ is the angular index, $j = 1,\ldots,n_\nu$ is the frequency variable and $k = 1,\ldots,n_\tau$ is the depth index. The angular weights w_i are calculated using the Gaussian quadrature and the frequency weights w_j^ν are calculated from the trapezoidal rule. The weights w_k^τ are chosen so that correct normalization of the line profile at all depths is ensured: or

$$w_k^\tau=\left(2\sum_{i=1}^{n_\mu}\sum_{j=1}^{n_\nu}\phi_{ijk}w_i^\mu w_j^\nu\right)^{-1} \tag{9.5.52}$$

at each depth point k. The operator $\delta\Lambda^{\dagger}_{\nu\mu}$ demands that the product $[\phi_{ijk}/(\phi_{ijk}+\beta_k)]S_l$ should be evaluated at τ_ν^+ and not at τ_ν. We assume that this product varies linearly with depth between k^+ and $k^+ + 1$, where k^+ is defined such that

$$\tau_{ijk^+}\le\tau_\nu^+<\tau_{ijk^++1}, \tag{9.5.53}$$

with τ_{ijk^+} the monochromatic optical depth at the depth point k^+. Therefore we have

$$c_{ijk}\left(\tau_\nu^+\right)S_l(\tau_\nu^+)\approx a_0^+c_{ijk^+}S_{lk}+\left(1-a_0^+\right)c_{ijk^++1}S_{lk^++1}, \tag{9.5.54}$$

where

$$c_{ijk}=\phi_{ijk}/(\phi_{ijk}+\beta_k) \tag{9.5.55}$$

and a_0^+ is a linear interpolation coefficient that is given by

$$a_0^+=\left(\tau_{ijk^++1}-\tau_\nu^+\right)/\left(\tau_{ijk^++1}-\tau_{ijk^+}\right). \tag{9.5.56}$$

Therefore, we can now write,

$$\delta\Lambda^{\dagger}_{ijk}\left[C_{ijk}S_{lk}\right]=w^+a_0^+c_{ijk^+}S_{lk^+}+w^+(1-a_0^+)c_{ijk^++1}S_{ik^++1}-c_{ijk}S_{lk}. \tag{9.5.57}$$

For a finite slab w^+ and τ_ν^+ are given in equation (9.5.39) and for a semi-infinite medium $w^+ = 1$ and $\tau_\nu^+ = \tau_{ijk} + 1$. From equations (9.5.57) and (9.5.51), we can derive an explicit expression for the double integral which contains S_l. The coefficients of S_{lk}, S_{lk^+} and S_{lk^++1} give the contributions for the angle i, frequency j and depth k to the matrix elements of w_{kk}, w_{kk^+} and w_{kk^++1}. The summation over i and j for each k will give the contribution to w for the outgoing rays and a similar treatment will give the contribution to the incoming rays. The vector **E** is calculated in a similar way. For several physical aspects of this method's applications, one can refer to Scharmer (1984).

We shall now describe a linearization procedure for solving a non-LTE problem exactly. It is similar to the perturbation technique of Cannon (1973a,b) although this is simpler to formulate. Equation (9.5.46) is written exactly as

$$(\epsilon^{\dagger} + \delta) - \frac{1}{2}\int_{-1}^{+1}\int_{0}^{\infty} \phi_\nu \delta\Lambda_{\nu\mu}\left(\frac{\phi_\nu}{\phi_\nu + \beta} S_l\right) d\mu\, d\nu$$
$$= (\epsilon^{\star} + \delta)B + \frac{1}{2}\int_{-1}^{+1}\int_{0}^{\infty} \phi_\nu \delta\Lambda_{\nu\mu}\left(\frac{\beta}{\phi_\nu + \beta} B\right) d\mu\, d\nu, \qquad (9.5.58)$$

where $\delta\Lambda_{\nu\mu}$ is the exact operator given by

$$I_\nu - S_\nu = \delta\Lambda_{\nu\mu}\,[S_\nu]. \qquad (9.5.59)$$

If $S_l(n)$ is an estimate of S_l and if this estimate does not satisfy equation (9.5.55), then

$$\left(\epsilon^{\dagger} + \delta\right) S_l^{(n)} - \frac{1}{2}\int_{-1}^{+1} \in t_0^{\infty} \phi_\nu \delta\Lambda_{\nu\mu}\left(\frac{\phi_\nu}{\phi_\nu + \beta} S_l^{(n)}\right) d\mu\, d\nu$$
$$= \left(\epsilon^{\star} + \delta\right) B + \frac{1}{2}\int_{-1}^{+1}\int_{0}^{\infty} \phi_\nu \delta\Lambda_{\nu\mu}\left(\frac{\beta}{\phi_\nu + \beta} B\right) d\mu\, d\nu + E^{(n)},$$
$$(9.5.60)$$

where $E^{(n)}$ is a correction term. This vanishes if $S_l^{(n)}$ is the correct solution of equation (9.5.58). In order to get the correct solution, we set

$$S_l^{(n+1)} = S_l^{(n)} + \delta S_l^{(n)}. \qquad (9.5.61)$$

We need equation (9.5.60) to be satisfied when $S_l^{(n)}$ is replaced by $S_l^{(n+1)}$ where $E^{(n)} \to 0$. Substituting equation (9.5.61) into equation (9.5.60), we obtain for $\delta S_l^{(n)}$

$$(\epsilon^{\dagger} + \delta)\delta S_l^{(n)} - \frac{1}{2}\int_{-1}^{+1}\int_{0}^{\infty} \phi_\nu \delta\Lambda_{\nu\mu}\left(\frac{\phi_\nu}{\phi_\nu + \beta}\delta S_l^{(n)}\right) d\mu\, d\nu = -E^{(n)},$$
$$(9.5.62)$$

which gives the exact values for the correction term to the line source function. However, this does not save much computer time. A better estimate of $\delta S_l^{(n)}$ can be obtained if $\delta\Lambda_{\nu\mu}$ is replaced by $\delta\Lambda^{\dagger}_{\nu\mu}$. Therefore we can estimate $\delta s_l^{(n)}$ from the equation:

$$\left(c^{\dagger} + \delta\right)\delta S_l^{(n)} - \frac{1}{2}\int_{-1}^{+1}\int_{0}^{\infty} \phi_\nu \delta\Lambda^{\dagger}_{\nu\mu}\left(\frac{\phi_\nu}{\phi_\nu + \beta}\delta S_l^{(n)}\right) d\mu\, d\nu \approx -E^{(n)}.$$
$$(9.5.63)$$

This equation is of similar structure to equation (9.5.46) and can be solved similarly.

The following steps can be used to obtain an accurate solution of the equivalent two-level atom problem:

1. assume $S_l^{(0)} = 0$ unless a better value of S_l is known;
2. calculate $E^{(n)}$ from equation (9.5.60);
3. calculate $\delta S_l^{(n)}$ by solving equation (9.5.63) and then compute $S_l^{(n+1)} = S_l^{(n)} + \delta S_l^{(n)}$;
4. repeat the above three steps until convergence is obtained to the accuracy of $\max\left|\delta S_l^{(n)}/S_l^{(n)}\right| < \eta$, where η is small enough to satisfy the desired accuracy.

In the case of large optical depths, the final solution can be obtained as follows. The error term is written as

$$E^{(n)} = \left(e^{\dagger} + \delta\right) S_l^{(n)} - \left(e^{\star} + \delta\right) B - \delta I^{(n)}, \tag{9.5.64}$$

where $\delta J^{(n)}$ is given by

$$\delta J^{(n)} = \frac{1}{2}\int_{-1}^{+1}\int_0^{\infty} \phi_\nu \left(I_\nu^{(n)} - S_\nu^{(n)}\right) d\mu\, d\nu, \tag{9.5.65}$$

and $I_\nu^{(n)}$ is calculated from the source function $S_\nu^{(n)}$:

$$S_\nu^{(n)} = \frac{\phi_\nu}{\phi_\nu + \beta} S_l^{(n)} + \frac{\beta}{\phi_\nu + \beta} B. \tag{9.5.66}$$

$I_\nu^{(n)}$ approaches $S_\nu^{(n)}$ at large optical depths, therefore $\delta J^{(n)} \rightarrow 0$ in equation (9.5.65) and cannot be used to evaluate $\delta J^{(n)}$. Instead, whenever $\tau_\nu \geq 1$, we can use the second order form of the transfer equation:

$$j'_\nu - S_\nu = \frac{d^2 j'_\nu}{d^2\tau_\nu^2}, \tag{9.5.67}$$

where

$$j'_\nu = \frac{1}{2}\left[I_\nu(\mu, \nu, \tau_c) + I_\nu(-\mu, -\nu, \tau_c)\right]. \tag{9.5.68}$$

This gives a relatively high accuracy for $\delta I^{(n)}$ even for large optical depths. This procedure has been applied to the non-LTE problem with $\epsilon = 10^{-10}$ and optical depths as high as 10^{14}.

This method of linearization has been applied to partial redistribution functions (see Scharmer (1983)). It eliminates the arbitrary distinction between the core and the wing which helps to simplify the book keeping. The intensities and the source functions are related by the Eddington–Barbier relations in the core and wing of the line and the problem is reformulated in terms of lambda operators using the single-point quadrature formulae for all frequencies.

The lambda operator is replaced by a numerical matrix operator that is strongly upper triangular with few elements appearing closely below the diagonal elements. This is a general method that can be applied in media with differential velocity fields, partial frequency redistribution and in multi-dimensional geometries which very few other second order methods are capable of. This allows saturation at all points in the medium – at boundaries as well as at interior points. This is most useful in dealing with radiation hydrodynamical phenomena, shocks etc. This kind of situation is dealt with by the core-saturation technique extended by Kalkofen and Ulmschneider (1984) (see the previous section).

9.6 Probabilistic equations for line source function

In this section we shall study the probabilistic equations following Athay (1972b, 1984).

9.6.1 Empirical basis for probabilistic formulations

For a two-level atom, the transfer equation for a monochromatic intensity in a plane parallel slab is written as

$$\mu \frac{dI_\nu}{d\tau_\nu} = I_\nu - S_\nu. \tag{9.6.1}$$

If we apply $\int \cdots d\mu$ over the above equation, we obtain

$$\frac{dH_\nu}{d\tau_\nu} = J_\nu - S_\nu, \tag{9.6.2}$$

where as usual H_ν and J_ν are the net outward flux and mean intensity respectively. The source function S_ν is

$$S_\nu = \frac{\phi_\nu}{\phi_\nu + \beta} S + \frac{\beta}{\phi_\nu + \beta} B, \tag{9.6.3}$$

where S is the line source function (which is independent of frequency), ϕ_ν is the line profile function, β is the ratio of opacities in the continuum and line and B is the Planck function. If we multiply equation (9.6.2) by $\phi_\nu \, d\nu$ and integrate, we get

$$\int_0^\infty \phi_\nu \frac{dH_\nu}{d\tau_\nu} d\nu = M \frac{dH}{d\tau_0} = \int_0^\infty J_\nu \phi_\nu \, d\nu - (1 - \delta) S - \delta B, \tag{9.6.4}$$

where τ_0 is the line centre optical depth and

$$M \, dH = \int_0^\infty \frac{\phi_\nu}{\phi_\nu + \beta} dH_\nu \, d\nu \tag{9.6.5}$$

and

$$\delta = \int_0^\infty \frac{\phi_\nu \beta}{\phi_\nu + \beta} d\nu. \tag{9.6.6}$$

The profile function ϕ_ν is normalized to unity

$$\int \phi_\nu \, d\nu = 1. \tag{9.6.7}$$

Here δ gives the probability of continuum absorption of the photon in the band width of the line. The line source function S is written as

$$S = \left[\int_0^\infty J_\nu \phi_\nu \, d\nu + (\epsilon + \eta) \, B \right] [1 + \epsilon + \sigma]^{-1}, \tag{9.6.8}$$

where ϵ is the ratio of the collisional de-excitation rate (from the upper to the lower level) to the spontaneous transition probability. The quantities σ and η represent losses and gains of photons due to atomic transitions of the line under consideration.

If we define ρ as

$$\rho = 1 - S^{-1} \int_0^\infty J_\nu \phi_\nu \, d\nu, \tag{9.6.9}$$

equations (9.6.4) and (9.6.8) become

$$M \frac{dH}{dt} = -\rho S + \delta(S - B) \tag{9.6.10}$$

and

$$\rho S = (\epsilon + \eta) B - (\epsilon + \sigma) S \tag{9.6.11}$$

respectively.

Combining equations (9.6.10) and (9.6.11), we get

$$M \frac{dH}{d\tau} = (\epsilon + \sigma + \delta) S - (\epsilon + \eta + \delta) B. \tag{9.6.12}$$

The quantities ϵ, δ and η give the probabilities for photon creation and destruction and influence the ratio S/B.

We need to make a comparison between ρ, which is the flux divergent coefficient determined non-locally, and p_e, the locally determined escape probability. At $\tau_0 = 0$, we can write

$$H_\nu(0) = \frac{1}{2} \int_0^\infty S_\nu E_2(\tau_\nu) \, d\tau_\nu, \tag{9.6.13}$$

where $E_2(\tau_\nu)$ is the second exponential integral. The contribution $\delta H_\nu(0)$ in the interval $d\tau_\nu$ is given by

$$\frac{\delta H_\nu(0)}{d\tau_\nu} = \frac{1}{2} S_\nu E_2(\tau_\nu). \tag{9.6.14}$$

If we multiply by ϕ_ν and integrate over frequency, equation (9.6.14) gives (by using equation (9.6.3))

$$M_0 \frac{\delta H(0)}{d\tau_0} = \frac{1}{2}\int_0^\infty \phi_\nu S_\nu E_2(\tau_\nu)\, d\nu$$
$$= p_e S - \frac{1}{2}\int_0^\infty \frac{\phi_\nu}{\phi_\nu + \beta}\beta(S - B)E_2(\tau_\nu)\, d\nu, \quad (9.6.15)$$

where

$$p_e = \frac{1}{2}\int_0^\infty \phi_\nu E_2(\tau_\nu)\, d\nu. \quad (9.6.16)$$

Equations (9.6.15) and (9.6.10) are very similar with p_e replacing ρ and

$$\frac{1}{2}\int \phi_\nu \beta/(\phi_\nu + \beta)E_2(\tau_\nu)\, d\nu$$

replacing δ. But the interpretations of these two equations are different. The quantity dH in equation (9.6.10) is the change in the flux at depth τ_0, while $\delta H(0)$ in equation (9.6.15) is the change in flux at the surface $\tau_0 = 0$. Thus, ρ corresponds to the local probability of photon emission, while p_e corresponds to that at the surface or the escape probability. An empirical relation exists between ρ and p_e, which is given by

$$\frac{1}{\rho^2} \approx \frac{1}{p_e^2} + \frac{1}{2\epsilon p_e}. \quad (9.6.17)$$

Frisch and Frisch (1975) derived the probabilistic equation for the two-level atom with $\delta = 0$ as

$$2p\frac{dS}{dp} = -S + \epsilon\frac{dB}{dp}, \quad (9.6.18)$$

where

$$p = \epsilon + (1 - \epsilon)p_e. \quad (9.6.19)$$

Frisch and Frisch (1975) and Athay (1972b) derived the equation

$$2p_e\frac{dS}{dp_e} = -\left(1 + \frac{\epsilon}{p_e}\right)S + \frac{\epsilon}{p_e}B. \quad (9.6.20)$$

An equation equivalent to (9.6.18) was derived by Scharmer (1981), (see section 9.5). Equation (9.6.20) is not as accurate as equation (9.6.18), although both equations are approximations of the exact transfer equation. For constant ϵ and B and $\epsilon \ll 1$, equation (9.6.18) has the solution

$$\frac{S}{B} = \epsilon(\epsilon^2 + 2\epsilon p_e)^{\frac{1}{2}} \quad (9.6.21)$$

and equation (9.6.20) has the solution

$$\frac{S}{B} = \frac{\pi^{\frac{1}{2}}\epsilon}{(2\epsilon p_e)^{\frac{1}{2}}}, \quad p_e \gg \epsilon, \quad (9.6.22)$$

while the exact relation is

$$\frac{S}{B} = \frac{\epsilon}{\rho + \epsilon}, \quad \epsilon \ll p_e. \tag{9.6.23}$$

Equation (9.6.21) appears to be nearer to the exact solution (9.6.23) than equation (9.6.22) with the factor $\pi^{\frac{1}{2}}$ in it. However, the presence of square roots defies any physical interpretations in probability theory.

9.6.2 Exact equation for S/B

From equations (9.6.4) and (9.6.12), we obtain,

$$\int \phi_\nu (J_\nu - S_\nu)\, d\nu = (\epsilon + \sigma + \delta) S - (\epsilon + \eta + \delta) B. \tag{9.6.24}$$

We need to evaluate the integral on the LHS. For this we use the relation

$$J_\nu - S_\nu = \Lambda_\nu S_\nu - S_\nu = \frac{1}{2}\int_0^\infty S_\nu E_1 \{|t_\nu - \tau_\nu|\}\, dt_\nu - S_\nu, \tag{9.6.25}$$

where E_1 is the first exponential integral. We assume a depth interval (t_j, t_{j+1}) in which S, B, ϕ, β are constant and at the optical depth τ_i (i denotes the local depth), the constant values being specified with the mean optical depth τ_j as given by

$$\tau_j = (t_j t_{j+1})^{\frac{1}{2}} \tag{9.6.26}$$

(Finn and Jefferies 1968, Athay 1976). Equations (9.6.24) and (9.6.25) can then be written as,

$$\begin{aligned}
&\left(\epsilon_i + \sigma_i + \delta_i + p_i^i + p_{i+1}^i\right) S_i - \sum_1^{i-1} S_j \left(p_{j+1}^i - p_j^i\right) - \sum_{i+1}^{n} S_j \left(p_j^i - p_{j+1}^i\right) \\
&\quad = \left(\epsilon_i + \eta_i + \delta_i - Q_i^i - Q_{i+1}^i\right) B_i \\
&\qquad + \sum_1^{i-1} B_j \left(Q_{j+1}^i - Q_j^i\right) + \sum_{i+1}^{\eta} B_j \left(Q_j^i - Q_{j+1}^i\right),
\end{aligned} \tag{9.6.27}$$

where

$$p_j^i = R_j^i - Q_j^i, \tag{9.6.28}$$

$$R_j^i = \frac{1}{2}\int_0^\infty \phi_i E_2 \left(|t_j - \tau_i|\right) d\nu, \tag{9.6.29}$$

$$Q_j^i = \frac{1}{2}\int_0^\infty \phi_i \frac{\beta_j}{\phi_j + \beta_i} E_2 \left(|t_j - \tau_i|\right) d\nu. \tag{9.6.30}$$

Equation (9.6.27) represents the fact that the sinks are balanced by the sources.

If we set $\phi_i = \phi_j$ and consider a two-level atom with no continuum absorption, equation (9.6.27) can be written for $\tau_0 = 0$ as

$$(\epsilon + 2p_e)\, S = \epsilon B + \int_0^{p_e} S\, dp_e. \tag{9.6.31}$$

It should be noted that this equation is exact at $\tau_0 = 0$ only. The numerical evaluation of the solutions for S/B for a few special cases have been given in Athay (1984).

9.6.3 Approximate probabilistic equations

Sometimes approximate probabilistic equations are useful in situations where accurate solutions are not necessary, particularly when it comes to saving a lot of computing time. As an illustration, we consider a two-level atom without continuum absorption. From equation (9.6.24), we have

$$\int J_\nu \phi_\nu\, d\nu - S_i = \epsilon_i S_i - \epsilon_i B_i. \tag{9.6.32}$$

We expand S_i in a Taylor series retaining the first three terms, and write

$$S_j = S_I + \frac{dS_i}{d\tau_{0i}} \left(t_{0j} - \tau_{0i}\right) + \frac{1}{2} \left(t_{0j} - \tau_{0i}\right)^2, \tag{9.6.33}$$

or

$$S_j = S_i + \frac{1}{\phi_i} \frac{dS_i}{d\tau_{0i}} \left(t_j - \tau_i\right) + \frac{1}{2\phi_i^2} \frac{d^2 S_i}{d\tau_{0i}^2} \left(t_j - \tau_i\right)^2, \tag{9.6.34}$$

where t_j and τ_i are the frequency dependent optical depths and ϕ_i is kept constant. Applying the lambda transform on S_j then gives us

$$\begin{aligned} J_i = S_i &\left[1 - \frac{1}{2} E_2(\tau_i)\right] + \frac{dS_i}{d\tau_{0i}} \left[\frac{\tau_{0i}}{2} E_2(\tau_i) + \frac{1}{2\phi_i} E_3(\tau_i)\right] \\ &+ \frac{d^2 S_i}{d\tau_{0i}^2} \left[\frac{1}{3\phi_i^2} - \frac{\tau_{0i}^2}{4} E_2(\tau_i) - \frac{\tau_{0i}}{2\phi_i} E_3(\tau_i) - \frac{1}{2\phi_i^2} E_4(\tau_i)\right]. \end{aligned} \tag{9.6.35}$$

Multiplying the above equation by $\phi_i\, dx$, where x is measured in Doppler width, we obtain

$$\begin{aligned} \int J_i \phi_i\, dx = S_i (1 - p_e) &+ \frac{dS_i}{d\tau_{0i}} \left[\tau_{0i}\, p_e + \frac{1}{2\pi^{1/2}} \int E_3(\tau_i)\, dx\right] \\ &+ \frac{d^2 S_i}{d\tau_{0i}^2} \left\{\frac{1}{2\pi^{1/2}} \int \left[\frac{2}{3} - E_4(\tau_i)\right] \frac{dx}{\phi_i} \right. \\ &\left. - \frac{\tau_{0i}^2}{2} p_e - \frac{\tau_{0i}}{2\sqrt{\pi}} \int E_3(\tau_l)\, dx \right\}. \end{aligned} \tag{9.6.36}$$

The integrals in the above equation diverge for large x. This can be avoided by adding the continuum opacity, but this does not solve the problem. The problem lies in the Taylor expansion which includes the derivatives $dS/d\tau_{0i}$ and $d^2S/d\tau_{0i}^2$ which increase without limit as t_j increases. We need to truncate the optical depth as is done in the Taylor expansion.

Let τ_0 be the range over which the expansion in equation (9.6.33) is valid and define

$$L_1 = \frac{t_{0j}}{\tau_{0i}}, \tag{9.6.37}$$

with the corresponding limit on frequency integrals in equation (9.6.36) being

$$L_1\phi(x_1) = 1. \tag{9.6.38}$$

For a Doppler core, we get

$$x_1 = (\ln L_1)^{\frac{1}{2}}. \tag{9.6.39}$$

From equations (9.6.32) and (9.6.36) we get

$$\left(-\frac{\tau_0^2}{2}p_e - \tau_0 p_3 + p_4\right)\frac{d^2S}{d\tau_0^2} + (\tau_0 p_e + p_3)\frac{dS}{d\tau_0} = (p_e + \epsilon)S - \epsilon B, \tag{9.6.40}$$

where

$$p_3 = \frac{1}{\pi^{1/2}}\int_0^{x_1} E_3(\tau_x)\,dx \tag{9.6.41}$$

and

$$p_4 = \frac{1}{\pi^{1/2}}\int_0^{x_1} \frac{1}{\phi_x}\left[\frac{2}{3} - E_4(\tau_x)\right]dx. \tag{9.6.42}$$

The integration limits in equations (9.6.41) and (9.6.42) are arbitrary. This converts equation (9.6.40) to a form similar to equation (9.4.18). Differentiating with respect τ_0 and using the relations

$$\frac{dp_4}{d\tau_0} = p_3, \quad \frac{dp_3}{d\tau_0} = -p_e, \quad \frac{dp_e}{d\tau_0} = -p_1, \quad p_1 = \frac{1}{\pi^{1/2}}\int_0^{\infty} \phi_x^2 E_1(\tau_x)\,dx, \tag{9.6.43}$$

we obtain

$$\left(-\frac{\tau_0^2}{2}p_1 + \tau_0 p_e + p_3\right)\frac{d^2S}{d\tau_0^2} + (\tau_0 p_1 + p_e + \epsilon)\frac{dS}{d\tau_0}$$
$$= \left(p_1 - \frac{d\epsilon}{d\tau_0}\right)S + \epsilon\frac{dB}{d\tau_0} + B\frac{d\epsilon}{d\tau_0}, \tag{9.6.44}$$

to terms of order of $d^2S/d\tau_0^2$.

Equations (9.6.40) and (9.6.44) are obtained on the assumption that ϕ_ν is kept constant; these equations cannot be applied when ϕ_ν varies with depth. Equation (9.6.44) gives an improved solution if the second order term is included. The accuracies of various approximations are discussed in Athay (1984).

9.7 Probabilistic radiative transfer

In this section a probabilistic radiative transfer equation is derived in plane parallel, finite and semi-infinite atmospheres in lines and continua. The boundary conditions, escape probabilities and complete linearization are discussed. We shall study the one-stream probabilistic equation (Canfield *et al.* 1984).

The radiative transfer equation in a plane parallel medium is written as

$$\mu \frac{\partial I_\nu(\mu)}{\partial \tau} = -\phi_\nu \left[I_\nu(\mu) - S_\nu \right], \tag{9.7.1}$$

where all the symbols have their usual meanings and

$$d\tau = \kappa_0 \, dx \quad \text{and} \quad d\tau_\nu = \phi_\nu \, d\tau. \tag{9.7.2}$$

We can write the solution of equation (9.7.1) (with no incident radiation (see chapter 2)) as

$$I_\nu^+(\mu, \tau_\nu) = \int^{\tau_\nu} S_\nu \exp[-(\tau_\nu - t)/\mu] \frac{dt}{\mu}, \quad \mu > 0 \tag{9.7.3}$$

and

$$I_\nu^-(\mu, \tau_\nu) = -\int_{\tau_\nu}^{T_\nu} S_\nu \exp[-(\tau_\nu - t)] \frac{dt}{\mu}, \quad \mu < 0, \tag{9.7.4}$$

where I_ν^+ and I_ν^- are the intensities flowing into and out of the atmosphere respectively. τ, τ_ν and T_ν are the optical depths at the line centre, at frequency ν, and of the slab at frequency ν respectively. The mean intensity of the diffuse radiation field is written as

$$J_\nu^d(\tau_\nu) = \frac{1}{2} \int_0^{T_\nu} S_\nu E_1(|t - \tau_\nu|) \, dt. \tag{9.7.5}$$

The frequency integrated mean intensity is

$$J^d(\tau) = \int_0^\infty J_\nu \Phi_\nu \, d\nu, \tag{9.7.6}$$

where

$$\Phi_\nu = M \phi_\nu, \tag{9.7.7}$$

with M the normalization constant. Using equation (9.7.5) we can rewrite equation (9.7.6) as

$$J^d(\tau) = \frac{1}{2} \int_0^\infty d\nu \, \Phi_\nu \int_0^\tau \phi_\nu S_\nu E_1(|t - \tau| \, \phi_\nu) \, d\nu. \tag{9.7.8}$$

If Ψ_ν and Φ_ν are the emission and absorption profiles, we have

$$S_\nu = S_0 \Psi_\nu / \Phi_\nu. \tag{9.7.9}$$

Then $J^d(\tau)$ can be written as

$$J^d(\tau) = \frac{1}{2}\int_0^T S_0(t)\,dt \int_0^\infty \phi_\nu \Psi_\nu E_1(|t|\phi_\nu)\,d\nu = \int_0^T S_0(t) K_1(t-\tau)\,dt, \tag{9.7.10}$$

where

$$K_1(\tau) = \frac{1}{2}\int_0^\infty \phi_\nu \Psi_\nu E_1(|t|\phi_\nu)\,d\nu, \tag{9.7.11}$$

with τ and T the line centre optical depths. The quantity $J^d(\tau)$ given in equation (9.7.10) represents the diffuse component of the radiation field. We must add the incident radiation to obtain the total mean intensity, $J(\tau)$. Therefore

$$J(\tau) = J^d(\tau) + J^{inc}(\tau), \tag{9.7.12}$$

where

$$J^{inc}(\tau) = \frac{1}{2}\int_0^\infty \Phi_\nu\,d\nu \int_0^1 I_\nu^-(\mu, \tau = 0)\exp(-\tau\phi_\nu/\mu)\,d\mu$$
$$+ \frac{1}{2}\int_0^\infty \Phi_\nu\,d\nu \int_{-1}^0 I_\nu^-(\mu, \tau = T)\exp[(T - \tau\phi_\nu)/\mu]\,d\mu. \tag{9.7.13}$$

If the boundary intensities I_ν^+ and I_ν^- in the above equation are given, then $J^{inc}(\tau)$ can be calculated easily. Therefore, we need to calculate $J^d(\tau)$, the diffuse field.

Applying the operator $\int_0^\sigma d\tau\, S_0(\tau)\partial/\partial\tau$ to both sides of equation (9.7.10), we obtain

$$\int_0^\sigma d\tau\, S_0(\tau)\frac{\partial J^d(\tau)}{\partial\tau} = \int_0^\sigma d\tau\, S_0(\tau)\frac{\partial}{\partial\tau}\int_0^\sigma dt\, S_0(t)K_1(t-\tau)$$
$$+\int_0^\sigma d\tau\, S_0(\tau)\frac{\partial}{\partial\tau}\int_0^T dt\, S_0(t)K_1(t-\tau). \tag{9.7.14}$$

The first term on the RHS of equation (9.7.14) vanishes as $K_1(\tau - t) = K(t-\tau)$. In the case of the second term, since $0 < \tau < \sigma < t < T$ and the integrand vanishes for $\tau - t \gg 1$, $S(\tau) \approx S(t) \approx S(\sigma)$ and if S is slowly varying or $(d\ln S_0/d\ln\tau) \ll 1$, we can write

$$\int_0^\sigma d\tau\, S_0(\tau)\frac{\partial J^d(\tau)}{\partial\tau} = S_0^2(\sigma)\int_0^\sigma d\tau\frac{\partial}{\partial\tau}\int dt\, K_1(t-\tau). \tag{9.7.15}$$

If $p_e(\tau)$ is the probability that a photon will escape the medium in a single flight, then,

$$p_e(\tau) = \frac{1}{2}\int_0^\infty d\nu \int_0^1 d\mu\, \Psi_\nu \exp(-\tau_\nu/\mu) = \frac{1}{2}\int_0^\infty d\nu\, \Psi_\nu E_2(\tau_\nu), \tag{9.7.16}$$

where

$$\tau_\nu = \tau\phi_\nu. \tag{9.7.17}$$

From the definition of $p_e(\tau)$ and the relationship between $E_1(x)$ and $E_2(x)$ (see Abromowitz and Stegun (1964)) we can write

$$\int_0^\tau dt\, K_1(t) = p_e(0) - p_e(\tau) = \frac{1}{2} - p_e(\tau) \tag{9.7.18}$$

and

$$\int_\sigma^T dt\, K_1(t-\tau) = p_e(\sigma-\tau) - p_e(T-\tau). \tag{9.7.19}$$

Therefore,

$$\int_0^\sigma d\tau \frac{\partial}{\partial\tau}\int_\sigma^T dt\, K_1(t-\tau) = p_e(0) + p_e(T) - p_e(\sigma) - p_e(T-\sigma). \tag{9.7.20}$$

Substituting the above equation into equation (9.7.15), applying the operator $S^{-1}(\sigma)\partial/\partial\tau$ to the result and then replacing σ with τ, we obtain

$$\frac{\partial J^d}{\partial\tau} = 2\,[p_e(0) + p_e(\tau) - p_e(T-\tau)]\,\frac{\partial S_0}{\partial\tau} - S_0\frac{\partial}{\partial\tau}\,[p_e(\tau) + p_e(T-\tau)]. \tag{9.7.21}$$

Since $p_e(0) = \frac{1}{2}$, this equation can be written as

$$\frac{\partial\rho}{\partial\tau} = \frac{\partial p_e^\star}{\partial\tau} + (2p_e^\star - \rho)\frac{\partial \ln S_0}{\partial\tau}, \tag{9.7.22}$$

where

$$\rho = 1 - \frac{J^d}{S_0}, \tag{9.7.23}$$

$$p_e^\star(\tau) = p_e(\tau) + p_e(T-\tau) - p_e(T). \tag{9.7.24}$$

Equation (9.7.22) is the probabilistic equation for the diffuse radiation field. The radiation field due to incident radiation must be added to obtain the total radiation field. The above relation is obtained with the assumption that

$$\left|\frac{d\ln S_0}{d\ln\tau}\right| \ll 1. \tag{9.7.25}$$

So far we have treated the radiation field in a one-stream approach. Two-stream treatments of the radiation field in a finite slab have the advantage of allowing the specification of the mean intensities at each of the boundaries. No diffuse radiation will propagate into the medium at these boundaries.

The mean intensity J_ν can be written in two parts, J_ν^+ and J_ν^-, corresponding to I_ν^+ and I_ν^-. We can write these (with frequency integration) as

$$J^+(\tau) = \int_0^\tau dt\, S_0(t) K_1(t-\tau) \tag{9.7.26}$$

and

$$J^-(\tau) = \int_\tau^T dt\, S_0(t) K_1(t-\tau). \tag{9.7.27}$$

Following a procedure similar to that used for one-dimensional stream transfer, we obtain for the two-stream probabilistic equation:

$$\frac{\partial J(\tau)}{\partial \tau} = \frac{\partial}{\partial \tau}\left[J^+(\tau) + J^-(\tau)\right] = 2\left[p_e(0) - p_e(\tau) - p_e(T-\tau)\right]\frac{\partial S_0}{\partial \tau} - S_0 \frac{\partial}{\partial \tau}\left[p_e(\tau) + p_e(T-\tau)\right]. \tag{9.7.28}$$

Equation (9.7.28) is almost the same as equation (9.7.27) except for the term $2p_e(T)\partial S_0/\partial\tau$ in equation (9.7.28) which will disappear when T is large. We define

$$p_e^+ = p_e(\tau) - p_e(T)/2 \tag{9.7.29}$$

and

$$p_e^- = p_e(T-\tau) - p_e(T)/2, \tag{9.7.30}$$

then

$$\frac{\partial J^+(\tau)}{\partial \tau} = \left[1 - 2p_e^+(\tau)\right]\frac{\partial S_0(\tau)}{\partial \tau} - S_0 \frac{\partial p_e^+(\tau)}{\partial \tau} \tag{9.7.31}$$

and

$$\frac{\partial J^-(\tau)}{\partial \tau} = \left[1 - 2p_e^-(\tau)\right]\frac{\partial S_0(\tau)}{\partial \tau} - S_0 \frac{\partial p_e^-(\tau)}{\partial \tau}. \tag{9.7.32}$$

The two-stream approach is similar to the one-stream approach but the former has the advantage of allowing two boundary conditions.

In the case of semi-infinite atmospheres the probability of photon escape tends to zero in the deeper layers of the atmosphere. The source function then approaches the Planck function, or

$$J(\tau \to \infty) = S_0(\tau \to \infty) = B(\tau \to \infty), \tag{9.7.33}$$

where

$$B = \int_0^\infty \Phi_\nu B_\nu\, d\nu. \tag{9.7.34}$$

One-stream transfer is of sufficient accuracy. In the two-stream case, the boundary condition in a semi-infinite atmosphere is given by

$$J^{+}(\tau \to \infty) = J^{-}(\tau \to \infty) = B(\tau \to \infty)/2. \tag{9.7.35}$$

In the case of a finite atmosphere, two-stream transfer is better with no diffuse radiation entering from either side, that is

$$J^{+}(\tau = 0) = J^{-}(\tau = T) = 0. \tag{9.7.36}$$

If we apply one-stream transfer, the following boundary condition can be used:

$$J(\tau_0) = \int_0^T dt\, S(t)(K_1(\tau_0 - t)), \tag{9.7.37}$$

where τ_0 is the optical depth at which the boundary condition is imposed. We must ensure the global energy balance condition (see Hummer and Rybicki (1982)) given by

$$\int_0^T \rho S\, dT = \int_0^T p_e S\, dt, \tag{9.7.38}$$

which means that the number of emissions – number of absorptions = number of escaping photons integrated over the atmosphere from $\tau = 0$ to $\tau = T$.

The probability that a photon will escape the medium in a single flight is given by

$$p_e(\tau) = \frac{1}{2}\int_0^\infty d\nu \int_0^1 d\mu\, \Psi_\nu \exp(-\tau_\nu/\mu), \tag{9.7.39}$$

or

$$p_e(\tau) = \frac{1}{2}\int_0^\infty d\nu\, \Psi_\nu E_2[\tau_\nu]. \tag{9.7.40}$$

In the spectral lines we measure the displacement in the form of $x = (\nu - \nu_0)/\Delta\nu_D$, $\Delta\nu_D$ being the Doppler width, and assume that the emission profiles and absorption profiles are the same. Then

$$p_e(\tau) = \int_0^\infty dx\, \Psi(x) E_2[\tau\phi(x)], \tag{9.7.41}$$

where the displacement is measured from the centre of the line (this is true for a static atmosphere) or

$$p_e(\tau) = M \int_0^\infty dx\, \phi(x) E_2[\tau\phi(x)], \tag{9.7.42}$$

where

$$\Phi_\nu = M\phi_\nu. \tag{9.7.43}$$

We shall derive some asymptotic expressions for the p_es which are fairly accurate even at high optical depths and denote them by $\bar{p}_e$. Osterbrock (1962) assumed that photons with an optical depth less than unity escaped the atmosphere. Then $\bar{p}_e$ is given by

$$\bar{p}_e(\tau) \approx \int_{x_1}^{\infty} \Phi(x)\,dx, \tag{9.7.44}$$

where

$$\tau\Phi(x) = 1. \tag{9.7.45}$$

For a Doppler profile,

$$\Phi(x) = \frac{1}{\pi^{1/2}} \exp(-x^2), \tag{9.7.46}$$

$x_1 \to \infty$ as $\tau \to \infty$, therefore

$$\bar{p}_e = \frac{\pi^{1/2}}{2} x_1 \exp(-x_1^2) \tag{9.7.47}$$

or

$$\bar{p}_e(\tau) = \left[2\pi^{\frac{1}{2}}\tau(\ln\tau)^{\frac{1}{2}}\right]^{-1}. \tag{9.7.48}$$

Following Ivanov (1973), we change the variable of integration in equation (9.7.42) to $z = 1/\phi(x)$, the probability is then $k_2/2$ and we get

$$p_e(\tau) = M \int_1^{\infty} dz\, z^{-1} X'(z) E_2(\tau/z), \tag{9.7.49}$$

where $X'(z)$ is the derivative of x with respect to z. If we set $y = \tau/z$, we obtain

$$p_e(\tau) = M X'(\tau) \int_0^{\tau} dy\, y^{-1} E_2(y) X'(\tau/y)/X'(\tau). \tag{9.7.50}$$

We will now derive the asymptotic form for $p_e(\tau)$. If $\tau \to \infty$ in equation (9.7.50), we have

$$\lim_{\tau\to\infty} \left[X'(\tau/y)/X'(\tau)\right] = y^{2\delta}. \tag{9.7.51}$$

Therefore the asymptotic form of p_e is

$$\bar{p}_e = M X'(\tau) \int_0^{\infty} y^{2\delta-1} E_2(y)\,dy. \tag{9.7.52}$$

After some algebra, we obtain $\bar{p}_e$ as

$$\bar{p}_e(\tau) = M\Gamma(2\delta) X'(\tau)/(2\delta + 1) \tag{9.7.53}$$

and the asymmetric form of K_1 is given by

$$K_1(\tau \to \infty) = \frac{-M\Gamma(2\delta)X''(\tau)}{(2\delta + 1)}, \tag{9.7.54}$$

where Γ is the gamma function. For Doppler broadening, equation (9.7.53) gives

$$\bar{p}_e(\tau) = \left[\pi^{\frac{1}{2}}\tau(\ln\tau)^{\frac{1}{2}}\right]^{-1}. \tag{9.7.55}$$

If the absorption profile has the form

$$\Phi(x) = \alpha\,|x|^{-\beta}\,,\ \beta > 1, \tag{9.7.56}$$

equation (9.7.54) will give the exact asymptotic probability as

$$\bar{p}_e(\tau) = \alpha^{\frac{1}{\beta}} M^{\frac{(\beta-1)}{\beta}} \Gamma\left[\frac{(\beta-1)}{\beta}\right] \tau^{-\frac{(\beta-1)}{\beta}} / (2\beta - 1), \tag{9.7.57}$$

which has been evaluated for several physical cases by Puetter (1981).

A strongly interlocked resonance line with radiative and collisional wings, which has a Lorentz profile, has its $\bar{p}_e$ given by

$$\bar{p}_e(\tau) = \frac{1}{3}\left(\frac{a}{\pi^{\frac{1}{2}}\tau}\right)^{\frac{1}{2}}, \tag{9.7.58}$$

where $a = \Gamma/4\pi\,\Delta\nu_D$ and Γ is the total damping width of the transition.

In the case of linear Stark broadening (with the Holtsmark profile $\Phi(x) = \alpha_H x^{-\frac{5}{2}}$), which dominates the absorption profile, $\bar{p}_e$ for hydrogen ions is given by

$$\bar{p}_e(\tau) = \frac{1}{4}\alpha_H^{2/5}\pi^{-3/10}\Gamma\left(\frac{3}{5}\right)\tau^{-3/5}, \tag{9.7.59}$$

where

$$\alpha_H = 6.9 \times 10^{-6} z^{-9/2} (n_u n_i)^3 \chi_i T_4^{3/4} n_{12}. \tag{9.7.60}$$

Here n_u and n_i are the principal quantum numbers of the upper and lower levels, $\chi_i = 0.5$ if $n_u = n_i + 1$ and $\chi_i = 1$ otherwise, $T_4 = T_e/10^4$ K, $n_{12} = n_e/10^{12}$ cm^{-2}, z is the charge of the ion and $n_p = n_e$.

Canfield and Ricchiazzi (1980) derived an approximate form of p_e in bound free continua given by

$$p_e(\tau) = (2\beta)^{-1} \exp\left[-\tau\beta^3 - \alpha(\beta - 1)\right], \tag{9.7.61}$$

where

$$\beta = \max\left\{(3\tau\alpha^{-1})^{\frac{1}{4}}, 1\right\}, \tag{9.7.62}$$

and

$$\alpha = \frac{h\nu_0}{kT}. \tag{9.7.63}$$

Equation (9.7.61) is reasonably accurate as long as α is significantly larger than 1 and β is not much larger than 1. This equation gives good values for photon escape probabilities between $\tau = 0$ and fairly large τs.

9.8 Mean escape probability for resonance lines

An approximate procedure is described below for a two-level atom in a plane parallel atmosphere of thickness T in which photons have a small probability ϵ of destruction for each scattering (see Frisch (1982, 1984)).

We obtain first a relationship between the mean escape probability $\langle p \rangle$ and the mean number of scatterings $\langle N \rangle_T$, using the principle of global conservation of energy. Then an approximate relationship of $\langle N \rangle_T$ with ϵ and $\langle N_0 \rangle_T$ (the mean number of scatterings when $\epsilon = 0$) is derived from scaling arguments. In addition, the mean net radiative bracket $\langle \rho \rangle_T$ is derived for a given slab of optical thickness.

The source function for a two-level atom is written as

$$S(\tau, x, \mathbf{n}) = \frac{1-\epsilon}{\phi(x)} \int \frac{d\mathbf{n}'}{4\pi} \int_{-\infty}^{+\infty} dx' R(x, \mathbf{n}, x', \mathbf{n}') I(\tau, x', \mathbf{n}) + G(\tau), \tag{9.8.1}$$

where I is the intensity, τ is the mean line optical depth, $\mathbf{n}$ is the direction of propagation, $\phi(x)$ is the profile function, x is the frequency shift measured in Doppler widths, R is the redistribution function, ϵ is the probability per scattering that a photon leaves the line and $G(\tau)$ is distribution of line photons. Multiplying by $\phi(x)$, integrating over frequencies and angles and normalizing, equation (9.8.1) can be written as

$$\bar{S}(\tau) = (1-\epsilon)\bar{J} + G(\tau), \tag{9.8.2}$$

where $\bar{S}(\tau)$ is the frequency integrated source function given by

$$\bar{S}(\tau) = \int \frac{d\mathbf{n}}{4\pi} \int_{-\infty}^{+\infty} \phi(x) S(\tau, x, \mathbf{n})\, dx, \tag{9.8.3}$$

and $\bar{J}(\tau)$ is similarly defined with I replacing S.

The mean escape probability $\langle p \rangle_T$ is then defined as the ratio of the total emergent flux to the rate of emission integrated over space, frequency and directions. Therefore for a plane parallel case with optical depth T, we can write

$$\langle p \rangle_T = \text{Total emergent flux}/4\pi \int_0^T \bar{S}(\tau)\, d\tau. \tag{9.8.4}$$

From the global conservation of photons in the line, $\langle p \rangle_T$ can be written as

$$\langle p \rangle_T = \int_0^T \left[\bar{S}(\tau) - \bar{J}(\tau)\right] d\tau \Big/ \int_0^T \bar{S}(\tau)\, d\tau. \tag{9.8.5}$$

By using equation (9.8.2), we get

$$\langle p \rangle_T = \frac{1}{1-\epsilon}\left[\int_0^T G(\tau)\, d\tau \Big/ \int_0^T \bar{S}(\tau)\, d\tau - \epsilon\right]. \tag{9.8.6}$$

The first term in the square brackets on the RHS is the inverse of the mean number of scatterings (Hummer 1964) $\langle N \rangle_T$. Therefore equation (9.8.6) can be written as

$$\langle p \rangle_T = \frac{1}{1-\epsilon}\left[\frac{1}{\langle N \rangle_T} - \epsilon\right]. \tag{9.8.7}$$

The mean number of scatterings $\langle N \rangle_T$ accounts for the escape of photons through the boundaries and the destruction of the photons. As can be seen from equation (9.8.7), as $\epsilon \to 0$, the inverse of the mean number of scatterings tends to the mean escape probability. Therefore from equation (9.8.7) one gets

$$\langle p \rangle_T = (\langle N_0 \rangle_T)^{-1}, \tag{9.8.8}$$

where $\langle N_0 \rangle_T$ is the mean number of scatterings when $\epsilon = 0$ (Hummer 1964).

We shall now find the relationship between $\bar{S}$ and $\bar{J}$ from $\langle p \rangle_T$. We introduce the net radiative bracket

$$\rho(\tau) = 1 - \bar{J}(\tau)/\bar{S}(\tau). \tag{9.8.9}$$

Iron's (1978) definition of net radiative bracket $\langle p \rangle_T$ is

$$\langle p \rangle_T = \int_0^T \rho(\tau)\bar{S}(\tau)\,d\tau \bigg/ \int_0^T \bar{S}(\tau)\,d\tau. \tag{9.8.10}$$

From equations (9.8.5) and (9.8.9), we obtain

$$\langle \rho \rangle_T = \langle p \rangle_T\,. \tag{9.8.11}$$

If $\tau_{mn} = \min(\tau, T - \tau)$ and we set

$$\rho(\tau) = \langle \rho \rangle_{\tau_{mn}}\,, \tag{9.8.12}$$

then from equations (9.8.8), (9.8.9) and (9.8.12) we obtain

$$\bar{J}(\tau) = \left[1 - 1/\langle N_0 \rangle_{\tau_{mn}}\right]\bar{S}(\tau). \tag{9.8.13}$$

The above approximation is reasonably accurate far away from the boundaries. If an appropriate scaling law is used for $\langle N_0 \rangle_T$, equation (9.8.13) can be used for any kind of frequency redistribution. This scaling law enters the large scale behaviour of the radiation field. In the case of complete redistribution the above approximation can be compared with

$$\bar{J}(\tau) = \{1 - [p_e(\tau) + p_e(T - \tau)]\}\,S(\tau), \tag{9.8.14}$$

where

$$p_e(\tau) = \frac{1}{2} - \int_0^T K_1(t)\,dt \tag{9.8.15}$$

is the one-sided escape probability. For resonance lines, equation (9.8.13) becomes

$$\bar{J}(\tau) = [1 - \alpha/\tau_{mn}]\,\bar{S}(\tau),\ 1 \ll \tau \le \frac{T}{2}, \tag{9.8.16}$$

where α is a constant of the order of unity. If α is the chosen to be between 1 and 0.5, the errors on the net radiative bracket for large depths and on emergent fluxes will be about a factor 2 (see figure 9.8).

9.9 Probability of quantum exit

In this section we shall study the concept of the probability of quantum exit from the medium (Sobolev 1951, 1963) and derive some useful relations in scattering media. When a quantum is absorbed at an optical depth τ, it will have certain probability of leaving the medium in a given direction with an altered or unaltered frequency depending upon the optical properties of the medium. This helps us to find the radiation field at any point in the medium. Once the probability is known at any given point in the medium, the radiation field can be calculated by multiplying the amount of energy coming directly from the source and absorbed at a given point by the probability of quantum exit at the same depth and integrating all such products at different depths. This probability can correctly represent the source function.

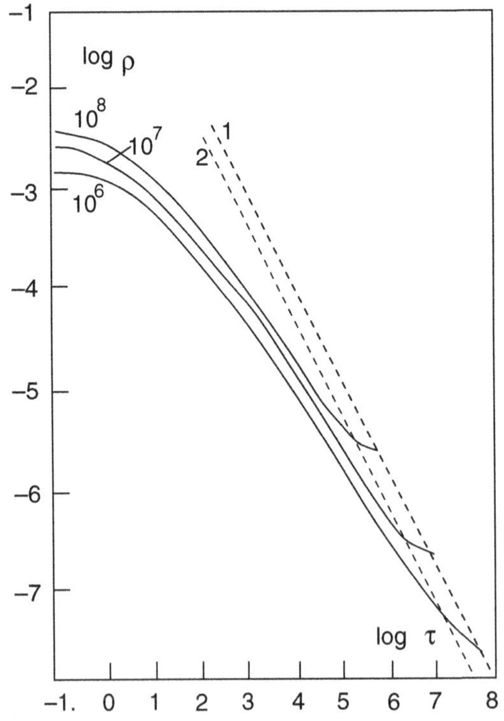

Figure 9.8 Net radiative bracket versus optical depth for $\epsilon = 0$ and a uniformly distributed source ($G(\tau) = 1$). The Voigt parameter for the line $a = 4.7 \times 10^{-2}$. The full lines are the large numerical solution curves and are labelled with total optical thickness T. The dashed lines represent the analytical fitting formulae (1) $\rho = 1/\tau$ (2) $\rho = 1/2\rho$. Only half the slab is shown because of the symmetry of the problem (from Frisch (1984), with permission).

We consider a one-dimensional medium with an optical depth τ_0 (rod model, see chapter 6), isotropic scattering ($p = 1/2$) and the albedo for single scattering ω. If the quantum is absorbed at the optical depth τ, let $P(\tau)$ denote the probability of quantum exit at the boundary $\tau = 0$ directly from τ after some scattering in the medium. If $f(\tau)$ is the energy coming out of the source that is absorbed between τ and $\tau + d\tau$, then the emergent radiation is given by

$$I(\tau = 0) = \int_0^{\tau_0} P(\tau) f(\tau)\, d\tau. \tag{9.9.1}$$

On the other side of the medium where $\tau = \tau_0$, the intensity is

$$I(\tau = T_0) = \int_0^{\tau_0} P(\tau_0 - \tau) f(\tau)\, d\tau. \tag{9.9.2}$$

If a quantum is absorbed at τ, its probability of exit from the medium without scattering is $(\omega/2)\exp(-\tau)$. If the quantum undergoes a series of scatterings, the probability is

$$\frac{\omega}{2} \int_0^{\tau_0} P(\tau') \exp\left[-\left(\tau - \tau'\right)\right] d\tau'. \tag{9.9.3}$$

Therefore, the total probability is

$$P(\tau) = \frac{\omega}{2} \exp(-\tau) + \frac{\omega}{2} \int_0^{\tau_0} P(\tau') \exp\left[-(\tau - \tau')\right] d\tau'. \tag{9.9.4}$$

In figure 9.9, the medium has thickness τ_0. The probability of quantum exit at the depth $\tau + \Delta\tau$ is equal to the probability of quantum exit from the depth τ with thickness $\tau_0 - \Delta\tau$ with the immediate passage through the additional layer $\Delta\tau$. We denote this probability by $P(\tau)$. The probability of quantum exit from the depth τ and subsequent passage through additional layers without absorption is equal to $P_1(\tau)(1 - \Delta\tau)$. The probability of exit from τ followed by absorption in the adjacent layer and leaving the medium after a series of scatterings is $P_1(\tau)\Delta\tau P(0)$. Therefore,

$$P(\tau + \Delta\tau) = P(\tau)(1 - \Delta\tau) = P_1(\tau)\Delta\tau P(0), \tag{9.9.5}$$

$$P_1(\tau) = P(\tau) - P(\tau_0 - \tau)\Delta\tau P(\tau_0). \tag{9.9.6}$$

Inserting equation (9.9.6) into equation (9.9.5), and letting $\Delta\tau \to 0$, we get

$$\frac{dP(\tau)}{d\tau} = -P(\tau) + P(\tau)P(0) - P(\tau_0 - \tau)P(\tau_0). \tag{9.9.7}$$

Therefore, we obtain

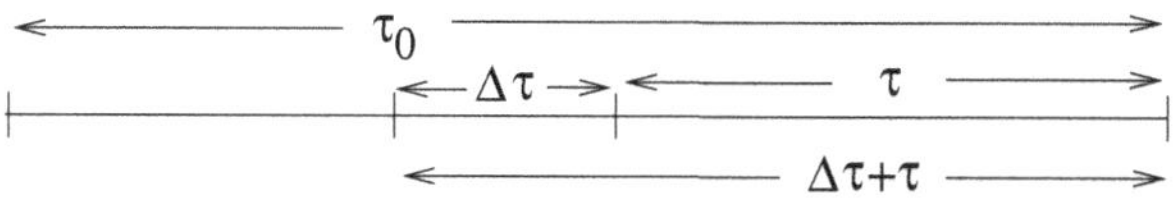

Figure 9.9 Schematic diagram of the rod model.

$$P(0) = \frac{\omega}{2}(1 + r) \tag{9.9.8}$$

and

$$P(\tau_0) = \frac{\omega}{2}(\exp(-\tau_0) + t), \tag{9.9.9}$$

where

$$r = \int_0^{r_0} P(\tau)\exp(-\tau)\,d\tau, \tag{9.9.10}$$

and

$$t = \int_0^{\tau_0} P(\tau_0 - \tau)\exp(-\tau)\,d\tau, \tag{9.9.11}$$

r and t being the diffuse reflection and transmission respectively of the quantum exit from the medium. r and t are connected by the relation

$$2r = \frac{\omega}{2}\left\{(1 + r)^2 - \left[t + \exp(-\tau_0)\right]^2\right\}. \tag{9.9.12}$$

The solution of equation (9.9.7) can be of the form

$$p(\tau) = C\exp(-k\tau) + D\exp(k\tau), \tag{9.9.13}$$

where

$$[1 - P(0) - k]\,C + P(\tau_0)e^{k\tau_0}D = 0 \tag{9.9.14}$$

and

$$P(\tau_0)\exp(-k\tau_0)C + [1 - P(0) + k]\,D = 0. \tag{9.9.15}$$

By using appropriate conditions, setting $\tau = 0$ in equation (9.9.13) and then $\tau = \tau_0$, we obtain values for C and D:

$$C = \frac{P(0)\exp k\tau - P(\tau_0)}{\exp(k\tau_0) - \exp(-k\tau_0)} \tag{9.9.16}$$

and

$$D = \frac{P(\tau_0) - P(0) - \exp(-k\tau_0)}{\exp(k\tau_0) - \exp(-k\tau_0)}. \tag{9.9.17}$$

After inserting the values of C and D into equations (9.9.13)–(9.9.15) we obtain,

$$P^2(0) - P^2(\tau_0) = (1 + k)\left[P(0) - P(\tau_0) - P(\tau_0)\exp(k\tau_0)\right] \tag{9.9.18}$$

and

$$k^2 = [P(0) - 1]^2 - P^2(\tau_0). \tag{9.9.19}$$

Solving the above two equations, we obtain

$$P(0) = (1 - k^2)\frac{(1+k)\exp(k\tau_0) - (1-k)\exp(-k\tau_0)}{(1+k)^2\exp(k\tau_0) - (1-k)^2\exp(-k\tau_0)}. \tag{9.9.20}$$

From equations (9.9.16)–(9.9.20) we obtain for $P(\tau)$:

$$P(\tau) = (1 - k^2)\frac{(1+k)\exp[k(\tau_0 - \tau)] - (1-k)\exp[-k(\tau_0 - \tau)]}{(1+k)^2\exp(k\tau_0) - (1-k)^2\exp(-k\tau_0)}. \tag{9.9.21}$$

Inserting equations (9.9.8) and (9.9.9) into equation (9.9.19), we obtain

$$k^2 = 1 - \omega(1+r) + \frac{\omega^2}{4}(1+r)^2 - \frac{\omega^2}{4}[t + \exp(-\tau_0)]^2. \tag{9.9.22}$$

The quantity $P(\tau)$ helps us to calculate the emergent radiation. We shall connect $P(\tau)$ to the emergent radiation field through the resolvent kernel for a one-dimensional rod following Sobolev (1963) and Grant (1968). If $I^+(\tau)$ and $I^-(\tau)$ denote the intensities in the forward and backward directions, the transfer equations are written as

$$\frac{dI^+}{d\tau} + I^+ = S^+ \tag{9.9.23}$$

and

$$-\frac{dI^-}{d\tau} + I^- = S^- \tag{9.9.24}$$

($0 < \tau < T$ = total optical depth), where the source function S^+ and S^- are given by

$$S^+(\tau) = B^+(\tau) + \omega(\tau)\left[p(\tau)I^+(\tau) + (1 - p(\tau))I^-(\tau)\right] \tag{9.9.25}$$

and

$$S^-(\tau) = B^-(\tau) + \omega(\tau)\left[(1 - p(\tau))I^+(\tau) + p(\tau)I^-(\tau)\right]. \tag{9.9.26}$$

Here $B^{\pm}$ are the Planck functions, the + and − signs representing the forward and backward directions (useful in moving media). $\omega(\tau)$ is the albedo for single scattering and $p(\tau)$ is the phase function which is equal to $\frac{1}{2}$ for isotropic scattering. The boundary conditions are given at $\tau = 0$ and $\tau = T$, or $I^+(0) = I_1$ and $I^-(T) = I_2$. We can use these separately if we wish to interpret I^+ and I^- as corresponding to the total field of radiation. If we separate the incident and diffuse radiation fields, we should replace the source functions S^+, S^- by S_d^+ and S_d^- respectively which gives us

$$S_d^+(\tau) = S^+(\tau) + \omega(\tau)\left[p(\tau)I_1\exp(-\tau) + (1 - p(\tau))I_2\exp[-(T - \tau)]\right] \tag{9.9.27}$$

and

$$S_d^-(\tau) = S^-(\tau) + \omega(\tau)\{(1 - p(\tau))I_1 \exp(-\tau) + p(\tau)I_2 \exp[-(T - \tau)]\}. \tag{9.9.28}$$

The boundary values of the Is are

$$I^+(0) = I^-(T) = 0. \tag{9.9.29}$$

The solution of equations (9.9.23) and (9.9.24) can be written (see chapter 2) as

$$I^+(\tau) = I_1 \exp(-\tau) + \int_0^\tau S^+(t) \exp[-(\tau - t)]\, dt \tag{9.9.30}$$

and

$$I^-(\tau) = I_2 \exp[-(T - \tau)] + \int_0^\tau S^-(t) \exp[-(\tau - t)]\, dt. \tag{9.9.31}$$

Substituting equations (9.9.30) and (9.9.31) into equations (9.9.25)–(9.9.28), we get

$$S = g + \omega\xi[S], \tag{9.9.32}$$

where

$$S = \begin{pmatrix} S^+ \\ S^- \end{pmatrix}, \tag{9.9.33}$$

$$g = \begin{pmatrix} B^+ \\ B^- \end{pmatrix} + \omega I_1 \begin{pmatrix} p \\ 1-p \end{pmatrix} + \omega I_2 \exp[-(t - \tau)] \begin{pmatrix} 1-p \\ p \end{pmatrix} \tag{9.9.34}$$

and

$$\xi[S(t)] = \begin{pmatrix} p\int_0^\tau \exp[-(\tau - t)]S^+(t)\, dt + (1-p)\int_\tau^T \exp[-(\tau - t)]S^-(t)\, dt \\ (1-p)\int_0^\tau \exp[-(\tau - t)]S^+(t)\, dt + p\int_\tau^T \exp[-(\tau - t)]S^-(t)\, dt \end{pmatrix}. \tag{9.9.35}$$

Or,

$$\xi[S(t)] = \begin{pmatrix} p & (1-p) \\ (1-p) & p \end{pmatrix} \begin{pmatrix} \int_0^\tau dt\, S^+(t) \exp[-(\tau - t)] \\ \int_\tau^T dt\, S^-(t) \exp[-(t - \tau)] \end{pmatrix}. \tag{9.9.36}$$

We set $p = \frac{1}{2}$ for isotropic scattering and $B^+ = B^-$, $S^+ = S^- = S^0$, then equation (9.9.32) becomes

$$S^0 = g^0 + \omega\xi^0[S^0], \tag{9.9.37}$$

where

$$\xi^0[S^0(\xi)] = \frac{1}{2}\int_0^T dt\, S^0(t) \exp[-(t-\tau)]. \tag{9.9.38}$$

In the case of slab geometry in which the angular distribution enters, the exponential in the above equation is written as

$$\int_0^1 \frac{dt}{\mu} \exp[-(t-\tau)/\mu] = E_i\,(|t-\tau|). \tag{9.9.39}$$

Equation (9.9.37) can be solved by successive approximation. We can write $S_n,\ n = 0, 1, 2, \ldots$ and

$$S_0 = g, \tag{9.9.40}$$

$$S_{n+1} = g + \omega\xi\,[S_n]. \tag{9.9.41}$$

We easily see that

$$S = \left[I + \omega\xi + (\omega\xi)^2 + \cdots\right] g = [1 - \omega\xi]^{-1}\, g, \tag{9.9.42}$$

where I is unity and unit matrix in slab geometry, and

$$\xi^n[g] = \xi\left[\xi^{n-1}[g]\right], \quad n = 1, 2, \ldots, \tag{9.9.43}$$

the nth iterate is Neumann series (or N series) (Busbridge 1960).

We shall now examine the existence of the solution to this problem. We present Grant's (1968) analysis. We assume S is bound on [0,1]. We introduce norms on Banach space of the following forms:

$$\sigma^+ = \max\left[\left|S^+(t)\right|;\ t \in (0, T)\right], \tag{9.9.44}$$

$$\sigma^- = \max\left[\left|S^-(t)\right|;\ t \in (0, T)\right] \tag{9.9.45}$$

and

$$\|S\| = \sigma^+ + \sigma^- \tag{9.9.46}$$

S is a number on Banach space (complete vector normed space), say B, and

$$\begin{aligned}\|\xi[S]\| &\le \sigma^+ \int_0^\tau \exp[-(\tau - t)]\,dt + \sigma^- \int_\tau^T \exp[-(\tau - t)]\,dt \\ &\le (\sigma^+ - \sigma^-)[1 - \exp(-T)]\end{aligned}$$

$$= (1 - \exp(-T))\|S\|. \tag{9.9.47}$$

If $g \in B$, so are successive terms $S_1, S_2, \ldots$. Therefore from equations (9.9.47) and (9.9.42), we have

$$\|S_{n+1}\| \leq \|g\| + \omega_0[1 - \exp(-T)]\|S\|, \tag{9.9.48}$$

where

$$\omega_0 = \max[\omega(\tau);\ \tau \in (0, T)]. \tag{9.9.49}$$

Therefore,

$$\|S_{n+1}\| \leq (1 + \lambda + \cdots + \lambda^n)\|g\| = \frac{1 - \lambda^{n+1}}{1 - \lambda}\|g\|, \tag{9.9.50}$$

where

$$\lambda = \omega_0[1 - \exp(-T)] \tag{9.9.51}$$

and

$$\|S_{n+p} - S_n\| \leq \left(\lambda^{n+1} + \cdots + \lambda^{n+p}\right)\|g\| \to 0 \tag{9.9.52}$$

$$\text{as } n, p \to \infty \text{ if } \lambda < 1. \tag{9.9.53}$$

Therefore S_n is a Cauchy sequence provided $\lambda < 1$ and the solution of equation (9.9.32) exists. We notice that when $T \to \infty$ and $\omega_0 = 1$ (a purely scattering medium) $\lambda \to 1$, in which case the series (9.9.52) converges slowly or diverges and when $\lambda = 1$, the solution does not exist.

We shall now prove the uniqueness of the solution. If there are two solutions S_1 and S_2 and ξ is a linear operator then $S = S_1 - S_2$ should satisfy the homogeneous equation

$$S = \omega\xi[S]. \tag{9.9.54}$$

Therefore,

$$\|S\| = \|\omega\xi[S]\| \leq \|S\|, \tag{9.9.55}$$

so that

$$(1 - \lambda)\|S\| \leq 0. \tag{9.9.56}$$

If $\lambda \leq 1$, we have a contradiction unless $||S|| = 0$, which implies $S_1 = S_2$. For large T and $\sigma_0 = 1$, the computation is no longer practical as the Neumann series will not converge. We note that $\lambda < 1$. Therefore if

$$\lambda = \omega_0[1 - \exp(-T)] < 1$$

we have $\omega_0 > 1$ provided that

$$T < T_e = \ln\left(\frac{\omega_0}{\omega_0 - 1}\right), \tag{9.9.57}$$

which is important in the design of reactors. A similar analysis is given by Anselone (1960, 1961) for the case of diffuse reflection from a homogeneous slab and for the general case.

9.9.1 The resolvents and Milne equations

We start with equation (9.9.32):

$$\omega\xi[S(\tau)] = \int_0^T dt\, S(t)K(|t-\tau|), \tag{9.9.58}$$

where

$$K(|t-\tau|) = \frac{1}{2}\omega \exp(-|t-\tau|). \tag{9.9.59}$$

The kernel K depends only on $|t-\tau|$. The solution of equation (9.9.32) can be written in terms of resolvent operator L, which is defined as

$$S = g + L[g], \tag{9.9.60}$$

where

$$L[g(\tau)] = \int_0^T \Gamma(\tau, \tau')g(\tau')\, d\tau'. \tag{9.9.61}$$

Therefore, the resolvent kernel is given by

$$\Gamma(\tau, \tau') = K(|\tau - \tau'|) + \int_0^T K(|\tau'' - \tau|)\Gamma(\tau'', \tau)\, d\tau''. \tag{9.9.62}$$

It can be shown that $\Gamma(\tau, \tau') = \Gamma(\tau', \tau)$. Equation (9.9.62) can be written, using this symmetry property, as

$$\Gamma(\tau', \tau) = K(|\tau - \tau'|) + \int_0^\tau K(x)\Gamma(\tau - x, \tau')\, dx$$
$$+ \int_0^{T-\tau} K(x)\Gamma(\tau + x, \tau')\, dx. \tag{9.9.63}$$

We shall write $\partial\Gamma/\partial\tau'$ and $\partial\Gamma/\partial\tau$ as follows:

$$\frac{\partial\Gamma}{\partial\tau'} = \frac{\partial K(|\tau - \tau'|)}{\partial\tau'} + \int_0^T K(|\tau'' - \tau|)\frac{\partial\Gamma(\tau'', \tau')}{\partial\tau'}d\tau'', \tag{9.9.64}$$

$$\frac{\partial\Gamma}{\partial\tau} = \frac{\partial K(|\tau - \tau'|)}{\partial\tau} + K(\tau)\Gamma(0, \tau') - K(T - \tau)\Gamma(T, \tau')$$

$$+\int_0^T K(|\tau''-\tau|)\frac{\partial\Gamma(\tau''-\tau')}{\partial\tau''}d\tau''. \tag{9.9.65}$$

Therefore,

$$\frac{\partial\Gamma}{\partial\tau}+\frac{\partial\Gamma}{\partial\tau'}=K(\tau)\Gamma(0,\tau')-K(T-\tau)\Gamma(T-\tau')$$
$$+\int_0^T K(|\tau''-\tau|)\left(\frac{\partial\Gamma}{\partial\tau'}+\frac{\partial\Gamma}{\partial\tau''}\right)d\tau''. \tag{9.9.66}$$

Furthermore, we can write

$$\Phi(\tau)=\Gamma(0,\tau)=\Gamma(\tau,0)=K(\tau)+\int_0^T K(|x-\tau|)\Phi(x)\,dx, \tag{9.9.67}$$

$$\Gamma(\tau,T)=K(T-\tau)+\int_0^T K(|y-t|)\Gamma(y,T)\,dy. \tag{9.9.68}$$

We put $x=T-\tau$, $z=T-y$, then

$$\Gamma(T-x_1,T)=K(x)+\int_0^T K(|x-z|)\Gamma(T-z,T)\,dz. \tag{9.9.69}$$

If we compare equations (9.9.68) and (9.9.69), we get

$$\Gamma(\tau,T)=\Gamma(T,\tau)=\Phi(T-\tau). \tag{9.9.70}$$

Inserting this into equation (9.9.67), we get

$$\Psi(\tau,\tau')=\int_0^T K(|x-\tau|)\Psi(x,\tau')\,dx, \tag{9.9.71}$$

where

$$\Psi(\tau,\tau')=\frac{\partial\Gamma}{\partial\tau}+\frac{\partial\Gamma}{\partial\tau'}-\Phi(\tau)\Phi(\tau')+\Phi(T-\tau)\Phi(T-\tau'). \tag{9.9.72}$$

Equation (9.9.7) has a non-trivial solution for $\omega_0\le 1$. Therefore for all τ, τ'

$$\frac{\partial\Gamma}{\partial\tau}+\frac{\partial\Gamma}{\partial\tau'}=\Phi(\tau)\Phi(\tau')-\Phi(T-\tau)\Phi(T-\tau'), \tag{9.9.73}$$

which is clearly symmetric in τ and τ'.

We note that the quantity $(\omega/4\pi)\,\Gamma(\tau,\tau')$ represents the probability that the quantum absorbed at τ will be emitted after several scatterings between depths τ'

and $\tau' + d\tau'$ per unit solid angle. Therefore, we get the following probability of quantum exit in slab geometry:

$$P(\tau, \mu) = \frac{\omega}{4\pi} \exp(-\tau/\mu) + \frac{\omega}{4\pi} \int_0^\infty \Gamma(\tau, \tau') \exp(\tau'/\mu)\, dt'. \tag{9.9.74}$$

Equation (9.9.62) can be written for the slab as

$$\Gamma(\tau, \tau')\, d\tau' = \frac{\omega}{2} E_i|\tau - \tau'|d\tau' + \frac{\omega^2}{4} d\tau' \int_0^\infty E_i|\tau - \tau'|E_i|\tau'' - \tau'| + \cdots. \tag{9.9.75}$$

The first term represents the probability of quantum exit at τ in which the quantum will be emitted between depths τ' and $\tau' + d\tau'$ after one scattering. The second term represents the emission after two scatterings and so on.

We shall derive the Chandrasekhar's X- and Y-functions in terms of resolvent kernels. We start with equation (9.9.58). We have $g(\tau) = \exp(-\tau)$ (for the rod case) and

$$\begin{aligned} S(\tau) &= \exp(-\tau) + \int_0^T K(|t - \tau|) S(t)\, dt \\ &= \exp(-\tau) + \int_0^T \Gamma(\tau, \tau') \exp(-\tau')\, d\tau', \end{aligned} \tag{9.9.76}$$

from equation (9.9.61).

By differentiating, we obtain

$$\begin{aligned} \frac{dS}{d\tau} + S &= \Gamma(\tau, 0) - \exp(-\tau)\Gamma(\tau, T) + \int_0^T \left(\frac{\partial \Gamma}{\partial \tau} + \frac{\partial \Gamma}{\partial \tau'} \right) \exp(-\tau')\, d\tau' \\ &= \Phi(\tau) X(\tau) - \Phi(T - \tau) Y(T), \end{aligned} \tag{9.9.77}$$

where

$$\begin{aligned} X(T) &= 1 + \int_0^T \Phi(t) \exp(-t)\, dt = S(0), \\ Y(t) &= \exp(-T) + \int_0^T \Phi(T - \tau) \exp(-t)\, dt = S(t). \end{aligned} \tag{9.9.78}$$

These are Chandrasekhar's X- and Y-functions for the rod. For the slab they are given by

$$X\left(T, \frac{1}{\mu}\right) = 1 + \int_0^T \Phi(t) \exp(-\mu t)\, dt \tag{9.9.79}$$

and

$$Y\left(T, \frac{1}{\mu}\right) = \exp(-\mu T) + \int_0^T \Phi(T - \tau) \exp(-\mu t)\, dt. \tag{9.9.80}$$

We can determine $S(\tau),\ X(T),\ Y(T)$, once we know $\Phi(\tau)$ from equation (9.9.67). We can determine the reflection and transmission factors in terms of X- and Y-functions. They are defined as

$$r(T) = \int_0^T S(\tau)\exp(-\tau)\,d\tau, \tag{9.9.81}$$

$$t(T) = \int_0^T S(T)\exp[-(T-\tau)]\,d\tau. \tag{9.9.82}$$

These can be expressed in terms of X- and Y-functions by first integrating by parts and then using equation (9.9.76), giving

$$r(T) = \frac{1}{2}\left[X^2(T) - Y^2(T)\right] \tag{9.9.83}$$

and

$$t(T) = \lim_{\mu\to 1} t\left(T, \frac{1}{\mu}\right), \tag{9.9.84}$$

where

$$t\left(T, \frac{1}{\mu}\right) = \int_0^T S(\tau)\exp[-\mu(T-\tau)]\,d\tau \tag{9.9.85}$$

or

$$t\left(T, \frac{1}{\mu}\right) = \frac{Y(T)X\left(T, \frac{1}{\mu}\right) - X(T)Y\left(T, \frac{1}{\mu}\right)}{\mu - 1}, \tag{9.9.86}$$

so that

$$t(T) = Y(t)X'(T) - X(T)Y'(T), \tag{9.9.87}$$

where

$$X'(T) = \lim_{\mu\to 1}\frac{d}{d\mu}\left[X\left(T, \frac{1}{\mu}\right)\right], \tag{9.9.88}$$

$$Y'(T) = \lim_{\mu\to 1}\frac{d}{d\mu}\left[Y\left(T, \frac{1}{\mu}\right)\right]. \tag{9.9.89}$$

We have still to solve equation (9.9.67) for $\Phi(\tau)$. We write

$$\Phi(\tau) = A\exp(k\tau) + B\exp(-k\tau). \tag{9.9.90}$$

This is subjected to the condition that

$$k^2 = 1 - \omega. \tag{9.9.91}$$

This gives us

$$\Phi(\tau) = \frac{r\exp(-kT)}{1 - r^2\exp(-2kT)} \times \{(1+k)\exp[k(T-\tau)] - (1-k)\exp[-k(T-\tau)]\}, \quad (9.9.92)$$

where

$$r = \frac{1-k}{1+k}. \quad (9.9.93)$$

Now, we can determine $\Gamma(\tau, \tau')$, which is given by

$$\Gamma(\tau, \tau') = \frac{1-k^2}{2k}\frac{1}{1 - r^2\exp(-2kT)} \times \Big(\exp(-k|\tau - \tau'|) + r^2\exp(-2kT)\exp(k|T - \tau'|) - r\left\{\exp[-k(\tau+\tau')] + \exp(-2kT)\exp[k(\tau+\tau')]\right\}\Big). \quad (9.9.94)$$

We see that

$$\Gamma(\tau, 0) = \Phi(\tau). \quad (9.9.95)$$

Inserting equation (9.9.94) into equation (9.9.61), we get

$$S(\tau) = \frac{(1-r^2)\exp(-kt)}{2k(1 - r^2\exp(-2kt))} \times \{(1+k)\exp[k(T-\tau)] - (1-k)\exp[-k(t-\tau)]\} = \frac{2}{\omega}\Phi(\tau). \quad (9.9.96)$$

Therefore $X(T)$ and $Y(T)$ can be written as

$$X(T) = S(0) = \frac{(1+r)[1 - r\exp(-2kT)]}{1 - r^2\exp(-2kT)} \quad (9.9.97)$$

and

$$Y(T) = S(T) = \frac{(1-r^2)\exp(-kT)}{1 - r^2\exp(2kT)}. \quad (9.9.98)$$

Substituting equation (9.9.97) into equations (9.9.81) and (9.9.82) and multiplying by a scaling factor $\frac{1}{2}\omega$, as we need to get $g = \frac{1}{2}\omega\exp(-\tau)$, we obtain

$$r(T) = \frac{1 - \exp(2kT)}{1 - r^2\exp(2kT)} \quad (9.9.99)$$

and

$$t(T) = \frac{(1-r^2)\exp(-kT)}{1 - r^2\exp(-kT)} - \exp(-T). \quad (9.9.100)$$

Thus we are able to connect the X- and Y-functions of Chandrasekhar to the probability of quantum exit through resolvent kernels.

A scheme for obtaining resolvent kernels of the integral equation for transfer problems in isotropically scattering media was given by Wilson and Sen (1973).

Exercises

9.1 Draw surfaces of constant radial velocity when $V_r = c$ and $V(r) = r$, where c is a constant and show that these are the cones $\theta = \cos^{-1}\mu =$ constant and the planes $z =$ constant respectively in (r, θ) and (p, z) coordinates.

9.2 Derive equations (9.2.38) and (9.2.39).

9.3 If $\beta(T, \sigma)$ and $\beta_c(\tau, \sigma, z)$, where $\sigma = d\ln V/d\ln r - 1$, are defined as $\beta'(\tau, \sigma) = \int_\xi^1 A_x A_x\, dx$ where

$$A_x = \tau^{-1}(1 + \sigma X^2)\left\{1 - \exp\left[-\tau/(1 + \sigma X^2)\right]\right\},$$

and

$$\beta'(\tau, \sigma) = \beta(\tau, \sigma) \quad \text{when } \xi = 0$$

and

$$\beta'(\tau, \sigma) = \beta_c(\tau, \sigma, z) \quad \text{when } \xi = (1 - z^2)^{\frac{1}{2}},$$

with the condition that $\sigma > -1$, calculate β and β_c for large τ, that is $\tau \gg \max(1, 1 + \sigma)$. Similarly calculate β and β_c for small τ, that is $\tau \ll \min(1, 1 + \sigma)$ for $+\sigma$ and $-\sigma$. Find also $\beta(\tau, 0)$ and $\beta_c(\tau, 0, z)$.

9.4 Show that at resonance points (see equation (9.3.9)), $x = 1$.

9.5 If there are N resonance points, show that I_ν^{emg} is given by

$$I_\nu^{emg} = I_\nu^{inc} \exp\left(-\sum_{i=1}^{N} \tau_i\right) + \sum_{j=1}^{N} S_j \left[1 - \exp(-\tau_j)\right] \exp\left(-\sum_{i=1}^{j-1} \tau_i\right)$$

and the intensity $I(\mathbf{r}, \mathbf{n})$ is given by

$$I(\mathbf{r}, \mathbf{n}) = I_\nu^{inc} \frac{1 - \exp(-\tau_1)}{\tau_1} \exp\left(-\sum_{i=2}^{N} \tau_i\right) + \left[1 - \frac{1 - \exp(-\tau_1)}{\tau_1}\right] S_1$$
$$+ \frac{1 - \exp(-\tau_1)}{\tau_1} \sum_{j=2}^{N} S_j \left[1 - \exp(-\tau_j)\right] \exp\left(-\sum_{i=2}^{j-1} \tau_i\right).$$

9.6 If

$$\beta = \frac{1}{4\pi} \int d\Omega \frac{1 - \exp(-\tau)}{\tau}$$

and

$$\beta_c = \frac{1}{4\pi} \int_{\Omega_c} d\Omega \frac{1 - \exp(-\tau)}{\tau},$$

by integrating over all solid angles show that

$$\bar{J}(\mathbf{r}_1) = \bar{\beta}_c I_c + (1-\beta)S(r_1) + \frac{1}{4\pi}\int d\Omega \frac{1-\exp(-\tau_1)}{\tau_1}$$
$$\times \sum_{j=2}^{N}\left[1-\exp(-\tau_j)\right]\exp\left(-\sum_{i=2}^{j-1}\tau_i\right)S(r_j),$$

where

$$\bar{\beta}_c = \frac{1}{4\pi}\int_{\Omega_c} d\Omega \exp\left(-\sum_{i=2}^{N}\tau_i\right)\frac{1-\exp(\tau_1)}{\tau_1}.$$

9.7 If ν_c is the critical frequency separating the core and wing frequencies and is determined by

$$\phi(\nu_c)\tau = \gamma,$$

show that for a Doppler profile ν_c is given by

$$\nu_c = \left(\log\frac{\tau}{\gamma\pi^{1/2}}\right)^{\frac{1}{2}}.$$

If

$$N_W = \frac{2}{\sqrt{\pi}}\int_{\nu_c}^{\infty}\exp(-\nu^2)\,d\nu \sim \frac{\phi(\nu_c)}{\nu_c}$$

show that $N_W = c\gamma\tau^{-1}$, where c is a slowly varying function of τ. If the Voigt profile is given by

$$\phi(\nu) = \frac{a}{\pi\nu^2},$$

show that $\nu_c \sim (a\tau/\pi\gamma)^{\frac{1}{2}}$ and that $N_w \approx 2a/\pi\nu_c \approx (4a\gamma/\pi\tau)^{1/2}$.

9.8 Derive equations (9.5.20) and (9.5.21).

9.9 Derive equation (9.5.34).

9.10 Derive equations (9.6.36) and (9.6.40).

9.11 Using equation (9.7.21) show that the probabilistic radiative transfer equation for media in a steady state and with a two-level atom is of the form

$$S_0\frac{\partial}{\partial\tau}\left(p_e^\star+\epsilon\right) + \left(2p_e^\star+\epsilon\right)\frac{\partial S_0}{\partial\tau} = \frac{\partial}{\partial\tau}(\epsilon B),$$

where the source function S_0 for the two-level atom is

$$S_0 = (J^d + \epsilon B)/(1+\epsilon).$$

9.12 Derive equation (9.7.27).

9.13 If $\rho^+ = \frac{1}{2} - J^+/S_0$ and $\rho^- = \frac{1}{2} - J^-/S_0$, show that equations (9.7.3) and (9.7.31) can be written as

$$\frac{\partial}{\partial \tau}(S_0 - J^+) = 2(p_e^+)^{\frac{1}{2}} \frac{\partial}{\partial \tau}\left[(p_e^+)^{\frac{1}{2}} S_0\right]$$

and

$$\frac{\partial}{\partial \tau}\left(S_0 - J^-\right) = 2(p_e^-)^{\frac{1}{2}} \frac{\partial}{\partial \tau}\left[(P_e^-)^{\frac{1}{2}} S_0\right].$$

Further, show that

$$\frac{\partial \rho^+}{\partial \tau} = \frac{\partial p_e^+}{\partial \tau} + (2p_e^+ - \rho^+)\frac{\partial \ln S_0}{\partial \tau}$$

and

$$\frac{\partial \rho^-}{\partial \tau} = \frac{\partial p_e^-}{\partial \tau} + (2p_e^- - \rho^-)\frac{\partial \ln S_0}{\partial \tau}.$$

9.14 If the optical depth is unity at frequency x ($x = (\nu - \nu_0)/\Delta\nu_D$), show that for a Doppler profile $x_1 = (\ln \tau_0)^{\frac{1}{2}}$, where τ_0 is the optical depth at the line centre.

9.15 Show that for a Doppler profile

$$\langle N_0 \rangle_T \approx T(\ln T)^{1/2},$$

$$p_e(\tau) = \frac{1}{4\tau(\ln \tau)^{1/2}} \quad \tau \to \infty$$

and for the Voigt profile

$$\langle N_0 \rangle_t = (T/a)^{\frac{1}{2}}, \quad T \gg \frac{1}{a}$$

and

$$p_e(\tau) = \frac{1}{3}(a/\tau)^{\frac{1}{2}}, \quad \tau \to \infty.$$

9.16 Supply all the steps leading to equation (9.9.21).

9.17 Derive equations (9.9.83) and (9.9.84).

REFERENCES

Abromowitz, M., Stegun, I.A., 1964, *Handbook of Mathematical Functions*, US Government Printing Office, Washington, DC.

Anselone, P.M., 1960, *MNRAS*, **120**, 498.

Anselone, P.M., 1961, *J. Math. and Mech.*, **1**, 537.

Athay, R.G., 1972a, *Radiation Transport in Spectral Lines*, Reidel, Dordrecht.

Athay, R.G., 1972b, *ApJ*, **176**, 659.

Athay, R.G., 1976, *ApJ*, **204**, 160.

Athay, R.G., 1984, in *Methods in Radiative Transfer*, ed. W. Kalkofen, Cambridge University Press, Cambridge, page 79.

Beals, C., 1950, *Publ. Dominion Astrophys. Obs., Victoria*, **9**, 1.

Busbridge, I.W., 1960, *The Mathematics of Radiative Transfer*, Cambridge University Press, Cambridge.

Canfield, R.C., McClymont, Puetter, R.C., 1984, in *Methods in Radiative Transfer*, ed. W. Kalkofen, Cambridge University Press, Cambridge, page 101.

Canfield, R.C., Puetter, R.C., Ricchiazzi, P.J., 1981, *ApJ*, **248**, 82.

Canfield, R.C., Ricchiazzi, P.J., 1980, *ApJ*, **239**, 1036.

Cannon, C.J., 1973a, *JQSRT*, **13**, 627.

Cannon, C.J., 1973b, *ApJ*, **185**, 621.

Cannon, C.J., 1984, in *Methods in Radiative Transfer*, ed. W. Kalkofen, Cambridge University Press, Cambridge, page 157.

Castor, J.I., 1970, *MNRAS*, **149**, 111.

Castor, J.I., 1972, *ApJ*, **178**, 779.

Castor, J.I., Nussbaumer, H., 1972, *MNRAS*, **155**, 293.

Castor, J.I., Van Blerkom, D., 1970, *ApJ*, **161**, 485.

Elitzur, M., 1984, *Ap.J*, **280**, 653.

Finn, G.D., 1971, *JQSRT*, **11**, 203.

Finn, G.D., 1972, *JQSRT*, **12**, 35.

Finn, G.D., Jefferies, J.T., 1968, *JQSRT*, **8**, 1675.

Frisch, H., 1982, *A&A*, **114**, 119.

Frisch, H., 1984, in *Methods in Radiative Transfer*, ed. W. Kalkofen, Cambridge University Press, Cambridge, page 65.

Frisch, U., Frisch, H., 1975, *MNRAS*, **173**, 167.

Grachev, S.L., Grinin, V.P., 1975, *Astrophysics*, **11**, 20.

Grant, I.P., 1968, in Lectures Notes on New Methods in Radiative Transfer, Pembroke College, Oxford University, Oxford, unpublished.

Hummer, D.G., 1964, *ApJ*, **140**, 276.

Hummer, D.G., Rybicki, G.B., 1982, *ApJ*, **263**, 925.

Irons, F.E., 1978, *MNRAS*, **182**, 705.

Irons, F.E., 1979a, *JQSRT*, **22**, 1.

Irons, F.E., 1979b, *JQSRT*, **22**, 21.

Ivanov, V.V., 1973, in *Transfer of Radiation in Spectral Lines*, NBS. Spl. Publ., 385, US Government Printing Office, Washington, DC.

Kalkofen, W., 1966, *JQSRT*, **6**, 633.

Kalkofen, W. (ed.), 1987, *Numerical Radiative Transfer*, Cambridge University Press, Cambridge.

Kalkofen, W., Ulmschneider, P., 1984, in *Methods in Radiative Transfer*, ed. W. Kalkofen, Cambridge University Press, Cambridge, page 131.

Kogure, T., 1959, *Publ. Astro. Soc. Japan*, **11**, 127.

Kogure, T., 1959, *Publ. Astro. Soc. Japan*, **11**, 278.

Kogure, T., 1961, *Publ. Astro. Soc. Japan*, **13**, 335.

Kogure, T., 1967, *Publ. Astro. Soc. Japan*, **19**, 30.

Kuan, P.K., Kuhi, L.V., 1975, *ApJ*, **199**, 148.

Magnan, C., 1970, *JQSRT*, **10**, 1.

Marti, F., Noerdlinger, P.D., 1977, *ApJ*, **215**, 247.

Mihalas, D., 1978, *Stellar Atmospheres*, 2nd Edition, Freeman and Company, San Francisco.

Miyamoto, S., 1949, *Jap. J. Astron.*, **1**, 17.

Miyamoto, S., 1952, *Publ. Astron. Soc. Japan*, **4**, 1.

Miyamoto, S., 1952, *Publ. Astron. Soc. Japan*, **4**, 28.

Nordlund, Å., 1984, *Numerical Methods in Radiative Transfer*, ed. W. Kalkofen, Cambridge University Press, Cambridge, page 211.

Osterbrock, D.E., 1962, *ApJ*, **135**, 195.

Panagia, N., Ranieri, M., 1973, *A&A*, **24**, 219.

Pozdnyakov, L.A., Soboĺ, I.M., Syunyaev, R.A., 1976, *Sov. Astron. Lett.*, **2**, 55.

Puetter, R.C., 1981, *ApJ*, **251**, 446.

Rybicki, G., 1972, in *Line Formation in the Presence of Magnetic Fields*, eds. R.G. Athay, L.L. House and G. Newkirk Jr, Boulder High Altitude Observatory, Colorado, page 145.

Rybicki, G., 1984, in *Methods in Radiative Transfer*, ed. W. Kalkofen, Cambridge University Press, Cambridge, page 21.

Rybicki, G.B., Hummer, D.G., 1969, *MNRAS*, **144**, 313.

Rybicki, G.B., Hummer, D.G., 1978, *ApJ*, **219**, 654.

Scharmer, G.B., 1981, *ApJ*, **249**, 720.

Scharmer, G.B., 1983, *A&A*, **117**, 83.

Scharmer, G.B., 1984, in *Methods in Radiative Transfer*, ed. W. Kalkofen, Cambridge University Press, Cambridge, page 173.

Sen, K.K., Wilson, S.J, 1998, in *Radiative Transfer in Moving Media*, Springer, Singapore.

Sobolev, V.V., 1947, *Moving Atmospheres of Stars*, (English translation, S. Gaposchkin, 1960, Harvard University Press, Cambridge, MA).

Sobolev, V.V., 1951, *A.J.*, **28**, (Bulletin 5, A New Method in the Theory of Light Scattering).

Sobolev, V.V., 1957, *Soviet Astr - A.J.*, **1**, 678

Sobolev, V.V., 1963, *A Treatise on Radiative Transfer*, translated by S.I. Gaposchkin, Van Nostrand Company Inc., New York.

Stenholm, L.G., 1977, *A&A*, **54**, 577.

Stenholm, L.G., Stenflo, J.O., 1977, *A&A*, **58**, 273.

Stenholm, L.G., Stenflo, J.O., 1978, *A&A*, **67**, 33.

Ueno, S., 1965, *J. Math. Annal. and Applcs.*, **11**, 11.

Williams, A.C., Elsner, R.F., Weisskopf, M.C., Darbro, W., 1984, *ApJ*, **276**, 691.

Wilson, S.J., Sen, K.K., 1973, *JQSRT*, **13**, 83.

Chapter 10

Operator perturbation methods

10.1 Introduction

The complete linearization method of Auer and Mihalas (1969) was a significant advance in solving complex problems of radiative transfer and was followed by the work of Rybicki (1971), Kalkofen (1974) and others. These are basically Newton–Raphson linearization methods which are highly efficient but are not favourably oriented towards computer time and storage. Certain problems such as those which involve radiation hydrodynamics require faster methods with sometimes a little loss of accuracy. Operator perturbation techniques were developed to meet the needs of these problems. An excellent survey of these methods is given in Kalkofen (1987).

Wu (1992) developed a method that can deal with complex models with a high rate of convergence in multi-level non-LTE line formation calculations. It essentially consists of linearization of the transfer equation and constraints, then solving them separately. It overcomes the disadvantage of requiring the simultaneous solution of the corresponding equations by the complete linearization method and the poor convergence rate. Hubený and Lanz (1992) suggested two approaches to accelerate the method of complete linearization. The first one is the so called Kantorovich variant of the Newton–Raphson method by which the Jacobi matrix of the system is fixed. This reduces the calculation of the number of matrix inversions considerably and retains them fixed during the subsequent computations. The second approach is the application of Ng acceleration. These approaches reduce the computer time by about 2–5 times. Heinzel (1995) developed a new multi-level transfer code for isolated solar atmospheric structures, based upon the approach of Rybicki and Hummer (1991), called MALI (multi-level accelerated lambda iteration). This is faster than the linearization method of Auer and Mihalas

(1969) by an order of magnitude and the solution is of sufficient accuracy in slab geometry.

10.2 Non-local perturbation technique of Cannon

Cannon (1973a,b) introduced a novel and significant departure in numerical radiative transfer. He reduced the order of the system of equations while retaining the accuracy, which reduces the computer storage and time although with a slower convergence rate. This technique of operator perturbation consists of dividing the transfer problem into two parts: (1) corrections to a solution calculated with the help of an approximate (differential or integral) operator; and (2) the error made by a solution in satisfying the conservation equation. Linear problems also require iterations but have the advantage of favourable computer time. We describe this procedure below following Cannon (1984).

The time dependent transfer equation is

$$(\mathbf{\Omega} \cdot \nabla) I = -\kappa (I - S), \tag{10.2.1}$$

where, with the usual notation,

$$\kappa = \frac{h\nu}{4\pi} (N_L B_{LU} - N_U B_{UL}) \phi(\Delta\nu) \tag{10.2.2}$$

and

$$S = \frac{1-\epsilon}{4\pi\phi(\Delta\nu)} \int_{-\infty}^{+\infty} d(\Delta\nu') \int_{4\pi} d\Omega' \, R(\Delta\nu', \mathbf{\Omega}'; \Delta\nu, \mathbf{\Omega}) I(\mathbf{r}, \Delta\nu', \mathbf{\Omega}') + \epsilon B_\nu(T). \tag{10.2.3}$$

Here R is the redistribution function.

We define the operator

$$\Lambda = \frac{1}{4\pi\phi(\Delta\nu)} \int_{-\infty}^{+\infty} d(\Delta\nu') \int_{4\pi} d\Omega' \, R(\Delta\nu', \mathbf{\Omega}'; \Delta\nu, \mathbf{\Omega}). \tag{10.2.4}$$

Now using the operator Λ in equation (10.2.4), equation (10.2.1) can be written as

$$\frac{1}{\kappa} (\mathbf{\Omega} \cdot \nabla) I = -I + (1-\epsilon) \Lambda I + \epsilon B_\nu(T). \tag{10.2.5}$$

The double integral in equation (10.2.4) can be replaced by the quadrature:

$$\Lambda I \to \frac{1}{4\pi\phi(\Delta\nu)} \sum_{i=1}^{N_{\nu'}} \sum_{k=1}^{N_{\Omega'}} \omega_{ik} R(\Delta\nu_i', \mathbf{\Omega}_k'; \Delta\nu, \mathbf{\Omega}) I(\mathbf{r}, \Delta\nu_i', \mathbf{\Omega}_k'), \tag{10.2.6}$$

where N_ν and $N_{\Omega'}$ are the grid points in $\Delta\nu'$ and $\mathbf{\Omega}'$ respectively and ω_{ik} are the appropriate weights. Therefore the numerical equivalent of Λ is

$$\Lambda \to \frac{1}{4\pi\phi(\Delta\nu)} \sum_{i=1}^{N_{\nu'}} \sum_{k=1}^{N^\star_{\Omega'}} \omega_{ik} R(\Delta\nu'_i, \mathbf{\Omega}'_k; \Delta\nu, \mathbf{\Omega}). \tag{10.2.7}$$

The perturbation method needs a less accurate operator, say $\Lambda^\star$:

$$\Lambda^\star \to \frac{1}{4\pi\phi(\Delta\nu)} \sum_{i=1}^{N^\star_{\nu'}} \sum_{k=1}^{N^\star_{\Omega'}} \omega^\star_{ik} R(\Delta\nu_i'^\star, \mathbf{\Omega}_k'^\star; \Delta\nu, \mathbf{\Omega}), \tag{10.2.8}$$

where $N^\star_{\nu'} < N_{\nu'}$, $N^\star_{\mathbf{\Omega}'} < N_{\Omega'}$ and $\omega^\star_{ik}$ are the correctly normalized weights on the coarse grid $\{\Delta\nu_i'^\star\}$ and $\{\mathbf{\Omega}_k'^\star\}$. R_0 is the complete redistribution which satisfies the relation

$$R_0(\Delta\nu'_1, \mathbf{\Omega}'; \Delta\nu, \mathbf{\Omega}) = \phi(\Delta\nu')\phi(\Delta\nu). \tag{10.2.9}$$

Equation (10.2.8) involves the angle quadrature, frequency quadrature and redistribution perturbation technique (AQPT, FQPT and RPT) (see Cannon (1975a,b) and Cannon *et al.* (1975)). Equation (10.2.5) may be written in the form

$$\frac{1}{\kappa}(\mathbf{\Omega}\cdot\nabla)\, I = -I + (1-\epsilon)\Lambda^\star I + (1-\epsilon)(\Lambda - \Lambda^\star)I, \tag{10.2.10}$$

where $(\Lambda - \Lambda^\star)$ is the perturbation operator. We write that,

$$I = \sum_{l=0}^{\infty} \lambda^l I^{(l)}, \tag{10.2.11}$$

where the parameter λ may be set equal to unity. By substituting equation (10.2.11) into equation (10.2.10), we obtain for the zeroth order

$$\frac{1}{\kappa}(\mathbf{\Omega}\cdot\nabla)\, I^{(0)} = -I^{(0)} + (1-\epsilon)\Lambda^\star I^{(0)} + \epsilon B_\nu(T), \tag{10.2.12}$$

and higher order expressions are given by

$$\frac{1}{\kappa}(\mathbf{\Omega}\cdot\nabla)\, I^l = -I^{(l)} + (1-\epsilon)\Lambda^\star I^{(l)} + (1-\epsilon)(\Lambda - \Lambda^\star)I^{(l-1)}, \tag{10.2.13}$$

for all $l \geq 1$.

If we write

$$\left.\begin{aligned} \mathcal{E}^{(0)} &= \epsilon B_\nu(T), \\ \mathcal{E}^{(l)} &= (1-\epsilon)(\Lambda - \Lambda^\star)I^{(l-1)}, \quad l \geq 1, \end{aligned}\right\} \tag{10.2.14}$$

then equations (10.2.12) and (10.2.13) will have a general form given by $(l \geq 0)$

$$\frac{1}{\kappa}(\mathbf{\Omega}\cdot\nabla)\, I^{(l)} = -I^{(l)} + (-\epsilon)\Lambda^\star I^{(l)} + \mathcal{E}^{(l)}. \tag{10.2.15}$$

The solution of equation (10.2.15) can be obtained by first solving for $l = 0$:

$$\frac{1}{\kappa}\left(\boldsymbol{\Omega}\cdot\nabla\right)I^{(0)} = -I^{(0)} + \frac{1-\epsilon}{2}\sum_{i=1}^{N^\star_{\nu'}}\sum_{k=1}^{N^\star_{\Omega'}}\omega^\star_{ik}\phi(\Delta\nu_i'^\star)I^{(0)}(r,\Delta\nu_i'^\star,\boldsymbol{\Omega}_k'^\star) + \epsilon B_\nu(T). \tag{10.2.16}$$

For a plane parallel geometry, $\boldsymbol{\Omega}\cdot\nabla = \mu\partial/\partial z$, where $\mu = \cos\theta$, that is equation (10.2.16) becomes

$$\frac{\mu}{\kappa}\frac{\partial I^{(0)}}{\partial z} = -I^{(0)} + \frac{1-\epsilon}{2}\sum_{i=1}^{N^\star_{\nu'}}\sum_{k=1}^{N^\star_{\nu'}}\omega^\star_{ik}\phi(\Delta\nu_i'^\star)I^{(0)}(z,\Delta\nu_i'^\star,\mu_k'^\star) + \epsilon B_\nu(T). \tag{10.2.17}$$

We write

$$\Phi(z,\Delta\nu,\mu) = \frac{1}{2}\left[I(z,\Delta\nu,\mu>0) + I(z,\Delta\nu,\mu<0)\right]. \tag{10.2.18}$$

Then using equations (10.2.17) and (10.2.18), we get the Feautrier second order differential equation:

$$\left[\frac{\mu}{\kappa}\frac{\partial}{\partial z}\right]^2\Phi^{(0)} = \Phi^{(0)} - (1-\epsilon)\sum_{i=1}^{N^\star_{\nu'}}\sum_{k=1}^{N^\star_{\mu'}}\omega^\star_{ik}\phi(\Delta\nu_i'^\star)\Phi^{(0)}(z,\Delta\nu_i'^\star,\mu_k'^\star) - \epsilon B_\nu(T). \tag{10.2.19}$$

The double derivative in this equation can be replaced by a difference equation on a depth grid $\{z_j\}$ with N_z points. We then have

$$\left\{\left[\frac{1}{\kappa}\frac{\partial}{\partial z}\right]^2\Phi^{(0)}\right\}_j = \frac{\dfrac{\Phi^{(0)}_{j+1}-\Phi^{(0)}_j}{\Delta j} - \dfrac{\Phi^{(0)}_j-\Phi^{(0)}_{j-1}}{\nabla_j}}{\frac{1}{2}\left(\Delta_j+\nabla_j\right)}, \tag{10.2.20}$$

where

$$\left.\begin{aligned}\Delta_j &= \frac{1}{2}\left(\kappa_{j+1}+\kappa_j\right)\left(z_{j+1}-z_j\right),\\ \nabla_j &= \frac{1}{2}\left(\kappa_j+\kappa_{j-1}\right)\left(z_j-z_{j-1}\right).\end{aligned}\right\} \tag{10.2.21}$$

From equations (10.2.19)–(10.2.21), we obtain the matrix system of equations,

$$-\mathbf{A}^\star_j\Phi^{(0)}_{j-1} + \mathbf{B}^\star_j\Phi^{(0)}_j - \mathbf{C}^\star_j\Phi^{(0)}_{j+1} = [\epsilon B_\nu]_j\,\boldsymbol{\Pi}^\star, \tag{10.2.22}$$

for $j = 1,\ldots,N_2$. Here $\Phi^{(0)}_j$ and $\boldsymbol{\Pi}^\star$ (unit vector) are vectors of dimension $N^\star_{\nu'}$ $N^\star_{\mu'}$, the matrix $\mathbf{B}^\star_j$ is full with the quadrature weights $\omega^\star_{ik}$, the boundary conditions are

$$\left.\begin{aligned} \mathbf{A}_1^\star &\equiv 0, \\ \mathbf{C}_{N_z}^\star &= 0. \end{aligned}\right\} \tag{10.2.23}$$

The solution of equation (10.2.22) can now be obtained by the following recursive scheme:

$$\mathbf{\Phi}_j^{(0)} = U_j^{(0)} + \mathbf{V}_j^{(0)} \mathbf{\Phi}_{j+1}^{(0)}, \tag{10.2.24}$$

where

$$\mathbf{U}_j^{(0)} = \left[\mathbf{B}_j^\star - A_j^\star \mathbf{V}_{j-1}^0\right]^{-1} \left\{[\epsilon B_\nu]\, \mathbf{\Pi}^\star + \mathbf{A}_j^\star \mathbf{U}_{j-1}^{(0)}\right\} \tag{10.2.25}$$

and

$$\mathbf{V}_j^{(0)} = \left[\mathbf{B}_j^\star - \mathbf{A}_j^\star \mathbf{V}_{j-1}^{(0)}\right]^{-1} \mathbf{C}_j^\star. \tag{10.2.26}$$

(Note that $\mathbf{U}$ is a vector while $\mathbf{V}$ is a matrix.) $\mathbf{\Phi}_j^{(0)}$ has been obtained from the above equations for all $j = 1, \ldots, N_z$, and then $\mathcal{E}^{(1)}$ can be obtained from the following equation:

$$\begin{aligned} \mathcal{E}^{(1)} &= (1-\epsilon)\left(\Lambda - \Lambda^\star\right) I^{(0)} \\ &\rightarrow \frac{1-\epsilon}{4\pi\phi(\Delta\nu)} \sum_{i=1}^{N_{\nu'}} \sum_{k=1}^{N_{\Omega'}} \omega_{ik} R(\Delta\nu_i', \mathbf{\Omega}_k', \Delta\nu, \mathbf{\Omega}) I^{(0)}\left(r, \Delta\nu_i', \mathbf{\Omega}_i'\right) \\ &\quad - \frac{1-\epsilon}{2} \sum_{i=1}^{N_{\nu'}} \sum_{k=1}^{N_{\mu'}} \omega_{ik}^\star \phi(\Delta\nu_i'^\star) I^{(0)}(z, \Delta\nu_i'^\star, \mu_k'^\star) \\ &= \frac{1-\epsilon}{4\pi\phi(\Delta\nu)} \sum_{i=1}^{N_{\nu'}} \sum_{k=1}^{N_{\Omega'}} \omega_{ik} R(\Delta\nu_i', \mathbf{\Omega}_k', \Delta\nu, \mathbf{\Omega}) I^{(0)}(r, \Delta\nu_i', \mathbf{\Omega}_i') \\ &\quad - (1-\epsilon) \sum_{i=1}^{N_{\nu'}} \sum_{k=1}^{N_{\mu'}} \omega_{ik}^\star \phi(\Delta\nu_i'^\star) \Phi^{(0)}(z, \Delta\nu_i'^\star, \mu_k'^\star). \end{aligned} \tag{10.2.27}$$

The second term in equation (10.2.27) is known because $\Phi^{(0)}$ has already been computed on the grid $\left[\Delta\nu_i^\star, \mu_k^\star\right]$ at the points z_i. We need to evaluate the first term from $I^{(0)}$ on the grid $[\Delta\nu_i, \mu_k]$. We can obtain this as follows.

We write equation (10.2.1) for the zeroth order :

$$\frac{\mu}{\kappa} \frac{\partial I^{(0)}}{\partial z} = -I^{(0)} + S^{(0)}, \tag{10.2.28}$$

where $S^{(0)}$ is given by

$$S^{(0)} = (1-\epsilon) \sum_{i=1}^{N_{\nu'}^\star} \sum_{k=1}^{N_{\mu'}^\star} \omega_{ik}^\star \phi(\Delta\nu_i'^\star) \Phi^{(0)}(z, \Delta\nu_i'^\star, \mu_k'^\star) + \epsilon B_\nu(T), \tag{10.2.29}$$

which is already known. We can then compute $I^{(0)}$ from the formal solution:

$$I^{(0)}(z, \Delta\nu_i, \mu_k > 0) = \int_{z_\infty}^{z} \kappa(z', \Delta\nu_i) S^{(0)}(z') \times \exp\left[-\frac{1}{\mu_k}\int_{z'}^{z} \kappa(z'', \Delta\nu_i)\, dz''\right] \frac{dz'}{\mu_k} \tag{10.2.30}$$

and

$$I^{(0)}(z, \Delta\nu_i, \mu_k < 0) = \int_{z_\infty}^{z} \kappa(z', \Delta\nu_i) S^{(0)}(z') \times \exp\left[-\frac{1}{\mu_k}\int_{z'}^{z} \kappa(z'', \Delta\nu_i)\, dz''\right] \frac{dz'}{\mu_k}, \tag{10.2.31}$$

on the fine grid $[\Delta\nu_i, \mathbf{\Omega}_k]$.

After obtaining the first order correction $\mathcal{E}^{(1)}$ to the zeroth order problem, we now consider equation (10.2.15) for $l = 1$. The same procedure is repeated using the Feautrier technique to give the recursive scheme for all $l \geq 0$:

$$\mathbf{\Phi}_j^{(l)} = \mathbf{U}_j^{(l)} + \mathbf{V}_j^{\star}\mathbf{\Phi}_{j+1}^{(l)}, \tag{10.2.32}$$

where

$$\mathbf{U}_j^{(l)} = \mathbf{O}_j\left[\mathcal{E}^{(l)} + \mathbf{A}_j^{\star}\mathbf{U}_{j-1}^{(l)}\right], \tag{10.2.33}$$

$$\mathbf{V}_j^{\star} = \mathbf{O}_j \mathbf{C}_j^{\star} \tag{10.2.34}$$

and

$$\mathbf{O}_j = \left[\mathbf{B}_j^{\star} - \mathbf{A}_j^{\star}\mathbf{V}_{j-1}^{\star}\right]^{-1}. \tag{10.2.35}$$

If the series (10.2.11) converges, then it will converge to the exact value as the transfer equation is linear. The convergence properties were discussed by Cannon (1973a,b) and Cannon *et al.* (1975). Local perturbation theory would yield strong divergence while non-local perturbation gives good convergence even for severe situations. We need to stress two points here. (1) When we invert the Feautrier matrices, they are of the size $N_\nu N_\mu$ corresponding to the exact operator Λ (see equation (10.2.7)). However, in the non-local perturbation technique, the matrices to be inverted are of the size $N_\nu^\star N_\mu^\star \times N_\nu^\star N_\mu^\star$, where $N_\nu^\star < N_\nu$ and $N_\mu^\star < N_\mu$. So there is a saving of computer time of the order $(N_\nu N_\mu / N_\nu^\star N_\mu^\star)^3$. Sample calculations with $N_\nu^\star = 3$ and $N_\mu^\star = 1$ give convergence to several significant figures after approximately 10 terms in the perturbation series (10.2.11), whereas the more accurate Λ operator representation requires at least $N_\nu \geq 10$, $N_\mu \sim 3$ even for simple line calculations. We can see that there is a considerable saving in computer time. (2) In computing the more numerically accessible redistribution functions R_0 instead of R_{II} and R_{III}, which require a large number of frequency and angle points, the inversions in the Feautrier technique require large matrices that create amplifications of round-off errors which reduces the accuracy of the inversions. This perturbation technique uses the redistribution R_0 which is well behaved and requires fewer frequencies and angles. This contributes to the saving of computer time.

This method can be applied to the perturbed integral equation. The one-dimensional integral form of equation (10.2.1) is

$$S = (1-\epsilon)\Lambda' S + \epsilon B_\nu(T), \tag{10.2.36}$$

where Λ' is the operator given by

$$\Lambda' = \frac{1}{2}\int_{-\infty}^{+\infty}\int_0^1 \frac{R(\Delta\nu', \mu'; \Delta\nu, \mu)}{\phi(\Delta\nu)} d(\Delta\nu')\frac{d\mu'}{\mu'} \times \int_{z_\infty}^{0} \exp\left[-\frac{1}{\mu'}\left|\int_{z'}^{z}\kappa(z'', \Delta\nu')\,dz''\right|\right]\kappa(z', \Delta\nu')\,dz'. \tag{10.2.37}$$

Direct numerical solution requires that equation (10.2.37) is discretized on a grid of $[\Delta\nu, \mu]$ by replacing the integrals by approximate quadrature sums. Equation (10.2.36) is written as

$$S = (1-\epsilon)\Lambda'^{\star} S + \epsilon B_\nu(T) + (1-\epsilon)(\Lambda' - \Lambda'^{\star})S. \tag{10.2.38}$$

Here AQPT, FQPT and RPT are incorporated as earlier. We use perturbation of the type (10.2.11), that is

$$S = \sum_{l=0}^{\infty} \lambda^l S^{(l)}, \tag{10.2.39}$$

from which the following recursive scheme is developed:

$$S^{(l)} = (1-\epsilon)\Lambda'^{\star} S^{(l)} + E^{(l)}, \tag{10.2.40}$$

where

$$\left.\begin{aligned} E^{(0)} &= \epsilon B_\nu(T), \\ E^{(l)} &= (1-\epsilon)(\Lambda' - \Lambda'^{\star})S^{(l-1)} \quad \text{for all } l \geq 1. \end{aligned}\right\} \tag{10.2.41}$$

The rest of the calculations follow as described earlier.

This method has been adapted by Scharmer by combining it with the core-saturation method of Rybicki (see chapter 9). Cram and Lopert (1976) and Cram (1977) applied this technique to solve the velocity dependent transfer equation, which is given by

$$\frac{\mu}{\kappa}\frac{\partial I}{\partial z} = -\phi\left(\Delta\nu - \frac{\nu}{c}V_\mu\right)(I - S), \tag{10.2.42}$$

where

$$S = \frac{1-\epsilon}{2} \int_{-1}^{+1} d\mu \int_{-\infty}^{+\infty} d(\Delta\nu)\phi\left(\Delta\nu - \frac{\nu}{c}V\mu\right) I(z, \Delta\nu, \mu) + \epsilon B_\nu(T) \tag{10.2.43}$$

and

$$\kappa(z, \Delta\nu, \mu) = \kappa_0(z)\phi\left(\Delta\nu - \frac{\nu}{c}V\mu\right). \tag{10.2.44}$$

Here only complete redistribution is treated. Changing the independent variable $\Delta\nu$ measured in the observer's frame of reference to $\Delta\xi$ measured in the local rest frame of the moving gas or $\Delta\xi = \Delta\nu - (\nu/c)V\mu$, equation (10.2.42) becomes

$$\frac{\mu}{\kappa_0}\frac{\partial I}{\partial z} - \frac{\mu^2}{\kappa_0}\frac{\nu}{c}\frac{dV}{dz}\frac{\partial I'}{\partial(\Delta\xi)} = -\phi(\Delta\xi)(I' - S), \tag{10.2.45}$$

where

$$S = \frac{1-\epsilon}{2} \int_{-1}^{+1} d\mu \int_{-\infty}^{+\infty} d(\Delta\xi)\phi(\Delta\xi)I'(z, \Delta\xi, \mu) + \epsilon B_\nu(T). \tag{10.2.46}$$

Note that $I(= I')$ is invariant under the above transformation. One should take $I/\nu^3(= I'/\xi^3)$ as the invariant. We write

$$I' = \sum_{l=0}^{\infty} \lambda^l I'_l \tag{10.2.47}$$

and take the term with $\partial I'/\partial(\Delta\xi)$ in equation (10.2.45) to be of the order of λ. We have the recursive relation

$$\begin{aligned} \frac{\mu}{\kappa_0}\frac{\partial I'_l}{\partial z} = & -\phi(\Delta\xi)\left[I'_l - \frac{1-\epsilon}{2}\int_{-1}^{+1} d\mu' \right. \\ & \left. \times \int_{-\infty}^{+\infty} d(\Delta\xi')\,\phi(\Delta\xi')I'_l(z, \Delta\xi', \mu') - \mathcal{E}_l\right], \end{aligned} \tag{10.2.48}$$

where

$$\left.\begin{aligned} \mathcal{E}_0 &= \epsilon B_\nu(T), \\ \mathcal{E}_l &= \frac{\mu^2\nu}{\phi(\Delta\xi)\kappa_0^c}\frac{dV}{dz}\frac{\partial I'_{l-1}}{\partial(\Delta\xi)} \quad \text{for all } l \geq 1. \end{aligned}\right\} \tag{10.2.49}$$

The zeroth order solution from equation (10.2.48) corresponds to the zero velocity problem or the static case. The higher order terms are obtained by correcting this solution using the error source/sink terms $\mathcal{E}_l$ given in equations (10.2.49), which contain the velocity gradient.

10.3 Multi-level calculations using the approximate lambda operator

We describe this method following Hamann (1985,a,b, 1986, 1987) and Sen and Wilson (1998).

The angle-averaged intensity $\mathbf{J}_\nu$ at frequency ν is calculated from a given source function $\mathbf{S}_\nu$ by the relation

$$\mathbf{J}_\nu = \mathbf{\Lambda}_\nu \mathbf{S}_\nu, \tag{10.3.1}$$

where $\mathbf{\Lambda}_\nu$ is a linear operator called the lambda operator. In the case of line scattering, we get the averaged intensity $\mathbf{J}$:

$$\mathbf{J} = \int_{-\infty}^{+\infty} \mathbf{J}_\nu \varphi_\nu \, d\nu, \tag{10.3.2}$$

where φ_ν is the normalized profile. We now define $\mathbf{\Lambda}$ as

$$\mathbf{J} = \mathbf{\Lambda S}. \tag{10.3.3}$$

In the case of a two-level atom, $\mathbf{S}$ is given by

$$\mathbf{S} = (1 - \epsilon)\mathbf{J} + \epsilon \mathbf{B}. \tag{10.3.4}$$

The first term describes the scattering in the line while the second term represents the thermal sources. From equations (10.3.3) and (10.3.4) we get

$$\mathbf{J} = \mathbf{\Lambda} \left[(1 - \epsilon)\mathbf{J} + \epsilon \mathbf{B}\right]. \tag{10.3.5}$$

The solution of this is found by using the $\mathbf{\Lambda}$-iteration technique:

$$\mathbf{J}_{new} = \mathbf{\Lambda} \left[(1 - \epsilon)\mathbf{J}_{old} + \epsilon B\right]. \tag{10.3.6}$$

Cannon (1984) (see section 10.2) introduced an ‘approximate lambda operator’ $\Lambda^\star$:

$$\mathbf{\Lambda} = (\mathbf{\Lambda} - \mathbf{\Lambda}^\star) + \mathbf{\Lambda}^\star, \tag{10.3.7}$$

with the following iteration scheme:

$$\mathbf{J} = (\mathbf{\Lambda} - \mathbf{\Lambda}^\star)\mathbf{S} + \mathbf{\Lambda}^\star \mathbf{S}. \tag{10.3.8}$$

The iteration scheme is then

$$\mathbf{J}_{new} - \mathbf{\Lambda}^\star S_{new} = \left(\mathbf{\Lambda} - \mathbf{\Lambda}^\star\right) \mathbf{S}_{old} \tag{10.3.9}$$

(see Cannon (1973), Scharmer (1981)). $\mathbf{\Lambda}^\star$ needs to have the following properties: (1) $\mathbf{\Lambda}^\star$ should be easily converted so that $\mathbf{J}_{new}$ and $\mathbf{S}_{new}$ can be determined consistently and easily; and (2) $\mathbf{\Lambda} - \mathbf{\Lambda}^\star$ should not give rise to extraneous contributions from the optically thick line cores so that the usual convergence problems associated with the conventional lambda iteration are avoided in spite of the fact that this term acts on the old source function. Substituting equation (10.3.4) into equation (10.3.9) we get

$$\mathbf{J}_{new} - \boldsymbol{\Lambda}^{\star}\left[(1-\epsilon)\mathbf{J}_{new} + \epsilon\mathbf{B}\right] = (\boldsymbol{\Lambda} - \boldsymbol{\Lambda}^{\star})\left[(1-\epsilon)\mathbf{J}_{old} + \epsilon\mathbf{B}\right]$$
$$= \boldsymbol{\Lambda}\mathbf{S}_{old} - \boldsymbol{\Lambda}^{\star}(1-\epsilon)\mathbf{J}_{old} - \boldsymbol{\Lambda}^{\star}\epsilon B. \tag{10.3.10}$$

By setting $\mathbf{S}_{old}$ equal to J_{FS}, the above equation can be written as

$$\mathbf{J}_{new} = \left[1 - \boldsymbol{\Lambda}^{\star}(1-\epsilon)\right]^{-1}\left[\mathbf{J}_{FS} - \boldsymbol{\Lambda}^{\star}(1-\epsilon)\mathbf{J}_{old}\right]. \tag{10.3.11}$$

In the above scheme $\mathbf{J}_{FS}$, which is a formal solution, need not be very accurate and a rough approximation is sufficient for convergence. A simple operator $\boldsymbol{\Lambda}^{\star}$ which satisfies the above requirements may be taken from the idea of 'core saturation' of Rybicki (see chapter 9) as

$$\boldsymbol{\Lambda}^{\star}_{\nu}\mathbf{S} = \begin{cases} \mathbf{S}, & \text{in the optically thick line cores,} \\ \mathbf{O}, & \text{elsewhere.} \end{cases} \tag{10.3.12}$$

We define the (depth dependent) 'core fraction' f_c as that part of the scattering integral that falls into the optically thick domain

$$f_c = \int_{\nu_{red}}^{\nu_{blue}} \varphi_\nu \, d\nu. \tag{10.3.13}$$

Now the use of the approximate operator $\Lambda^{\star}$ becomes a simple multiplication with the assumption of complete redistribution, which gives a frequency independent source function or

$$\boldsymbol{\Lambda}^{\star}\mathbf{S} = f_c\mathbf{S}. \tag{10.3.14}$$

The inversion of the local operator is easy, and equation (10.3.11) becomes

$$\mathbf{J}_{new} = \left[1 - (1-\epsilon)f_c\right]^{-1}\left[\mathbf{J}_{FS} - (1-\epsilon)f_c\mathbf{J}_{old}\right], \tag{10.3.15}$$

or

$$\mathbf{J}_{new} - \mathbf{J}_{old} = \left[1 - (1-\epsilon)f_c\right]^{-1}(\mathbf{J}_{FS} - \mathbf{J}_{old}). \tag{10.3.16}$$

This is the 'accelerated lambda iteration' (ALI). The quantity ($\mathbf{J}_{FS} - \mathbf{J}_{old}$, obtained through a formal solution, is the source of iteration and the corrections are amplified by the factor $1/[1-(1-\epsilon)f_c]$. The amplification factor increases with the size of the core fraction.

We shall now determine the core fraction f_c in a spherically symmetric expanding atmosphere. The core fraction f_c is defined as

$$f_c = \int_{x_{red}}^{x_{blue}} \varphi(c)\,dx \Big/ \int_{-x_{max}}^{x_{max}} \varphi(x)\,dx. \tag{10.3.17}$$

Any angle averaging in moving media must be done in the fluid frame, that is for a fixed comoving frame (CMF) frequency. In equation (10.3.17), the profile function $\varphi(x)$ is isotropic and $x = (\nu - \nu_0)/\Delta\nu_D$, $\Delta\nu_D$ being the Doppler width. The numerator consists of an optically thick line core between x_{red} and x_{blue} and the denominator is the normalized profile in the frequency band width $+x_{max}$ to $-x_{max}$ (3–4 Doppler widths). A shorter notation is introduced by defining the profile function:

$$\phi(x) = \int_{-x_{max}}^{x} \varphi(x')\,dx', \tag{10.3.18}$$

then equation (10.3.17) becomes

$$f_c = \frac{[\phi(x_{blue}) - \phi(x_{red})]}{\phi(x_{max})}. \tag{10.3.19}$$

If there exists a ray starting from a given point that reaches the boundary in an optical depth shorter than γ, a parameter of order of unity, then this frequency is outside the core frequency (x_{red} and x_{blue}). Two simplifying assumptions are adopted: (1) only rays in the radial and transverse directions are considered; and (2) the velocity gradient and opacity are assumed to be constant in the neighbourhood of the point under consideration.

The core confining frequencies x_{red} and x_{blue} must be estimated separately for the two spatial directions considered. Interior points and points closer to the boundaries need different treatments.

When a photon travels along a given direction with a given impact parameter p across the spherically expanding atmosphere, it will encounter exact resonance with the line absorbing atom only at one spatial point at radius r. But the absorption profile in the fluid frame has a finite width and the interaction may therefore take place within the finite spatial range around this exact resonance. We consider the

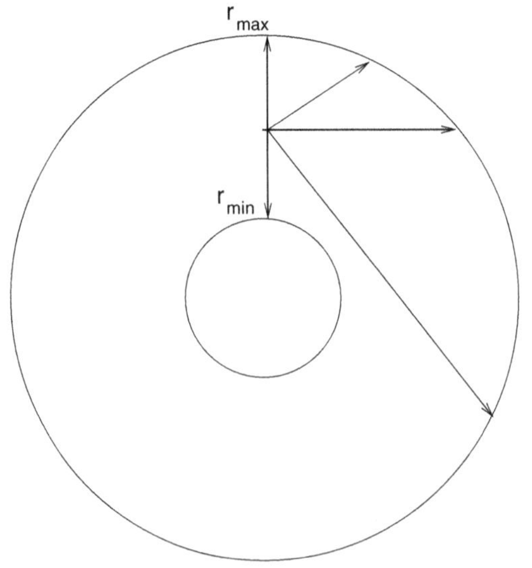

Figure 10.1 Schematic representation of the expanding atmosphere. Photons from a point at radius r may escape in all spatial directions to a boundary. When estimating f_c, only the radial and transverse directions are considered (from Hamann (1987), page 42, figure 1, with permission).

case of 'interior points', in which this 'scattering zone' lies completely within the boundaries of the atmosphere and estimate the optical depth across this zone. We will calculate the optical depth along a ray. This is nothing but the opacity integrated along a ray. In an expanding spherical atmosphere with $dv(r)/dr > 0$, any two volume elements recede from each other. Therefore a photon of constant frequency seen in the observer's frame is continuously red shifted in the comoving frame when travelling. To evaluate the optical depth, the integration along the spatial coordinate can be transformed into integration over the comoving frequency x, which involves the velocity gradient (projected on the ray considered) $v'(r, p)$, which is measured in Doppler units. Using the second assumption of constancy, we write

$$\tau(r, p) = \frac{\kappa(r)}{v'(r, p)}. \tag{10.3.20}$$

The projected velocity gradients in the two spatial directions are

$$\left.\begin{aligned} v'(r, p = 0) &= \frac{dv(r)}{dr} \quad \text{radial direction,} \\ v(r, p = r) &= \frac{v(r)}{r} \quad \text{transverse direction.} \end{aligned}\right\} \tag{10.3.21}$$

For an optically thin profile $\tau(r) < \gamma$ and we set $f_c = 0$.

We have seen that two volume elements recede from each other, or the CMF frequency of any photon is red shifted during its motion. Therefore a photon in the blue wing will have to travel through the whole line core before it can escape. Therefore if $\tau(r) > \gamma$, the whole blue side of the line belongs to the line core which means that photons cannot escape but are trapped locally (almost):

$$\text{if } \tau(r) > \gamma \text{ then } x_{blue} = x_{max}. \tag{10.3.22}$$

On the other hand, photons from the red wing will leave the scattering zone. x_{red} is obtained by transforming the optical depth integral into an integral over the CMF frequency x. Thus,

$$\phi(x_{red}) - \phi(-x_{max}) = \frac{\gamma}{\tau}. \tag{10.3.23}$$

The above arguments apply only for radii points that are well away from the boundaries (or the interior points), where we have

$$x_{red} - (-x_{max}) < \Delta v, \tag{10.3.24}$$

where Δv is the relative velocity between the radius point considered and the boundary. In the radial direction, the photon may escape through the outer or inner boundary, that is

$$\Delta v = \max\left[v(r_{max}) - v(r);\, v(r) - v(r_{min})\right]. \tag{10.3.25}$$

In the transverse direction, the velocity relative to its intersection point with the boundary (see figure 10.1) is

$$\Delta v = v(r_{max}) \left(1 - \frac{r^2}{r_{max}^2}\right)^{\frac{1}{2}}, \tag{10.3.26}$$

If condition (10.3.24) is not satisfied, we are in the so called 'boundary zone', in which case x_{red} for the core range must satisfy

$$\phi(x_{red}) - \phi(x_{red} - \Delta v) = \frac{\gamma}{\tau}. \tag{10.3.27}$$

This equation has two solutions if

$$\phi\left(\frac{\Delta v}{2}\right) - \phi\left(-\frac{\Delta v}{2}\right) = 2\phi\left(\frac{\Delta v}{2}\right) - 1 > \frac{\gamma}{\tau}. \tag{10.3.28}$$

Otherwise, the core is optically thin ($f_c = 0$). The smaller of the solutions lies between the x_{red} for the interior points (see equation (10.3.23), and $\Delta v/2$. The other solution lies between $\Delta v/2$ and x_{max}. As both the solutions lie symmetrically to $\Delta v/2$, we get

$$x_{blue} = \Delta v - x_{red}, \tag{10.3.29}$$

from which one obtains the core range. This method is used in both the radial and transverse directions at each radial point and the minimum extension of the core is taken to estimate the core range. Hamann's ALI process depends on the core fraction f_c and the quantity γ which is chosen so that best convergence rate is achieved. Hamann (1987) used the ALI scheme to solve the multi-level atom problem in an expanding spherical atmosphere. We shall now study this problem. We can write the 'formal solution' as

$$\mathbf{J}_{FS} = \mathbf{\Lambda} S_{old}, \tag{10.3.30}$$

where S_{old} is the source function calculated from the 'old' population numbers $\mathbf{N}_{old}$. The population density is dependent on the radiation field in the multi-level non-LTE statistical equilibrium equation. The population numbers can be written symbolically as

$$\mathbf{n} = \mathbf{n}(\mathbf{J}). \tag{10.3.31}$$

Equations (10.3.30) and (10.3.31) are solved simultaneously. New population numbers $\mathbf{n}_{new}$ are calculated from the formal solution, that is

$$\mathbf{n}_{new} = \mathbf{n}(J_{FS}), \tag{10.3.32}$$

as in the case of a two-level atom. Equation (10.3.30) and (10.3.32) are repeated in turn. In analogy to the two-level atom following Cannon (1984; see section 10.2) we introduce the approximate $\mathbf{\Lambda}^{\star}$ operator and obtain the 'new' radiation field, that is,

$$\mathbf{J}_{new} = \mathbf{J}_{FS} + \mathbf{\Lambda}^{\star}\left(\mathbf{S}_{new} - \mathbf{S}_{old}\right). \tag{10.3.33}$$

Now we introduce equation (10.3.33) into equation (10.3.32) to obtain

$$\mathbf{n}_{new} = \mathbf{n}(\mathbf{J}_{new}). \tag{10.3.34}$$

We let $\mathbf{n} = (n_1, \ldots, n_N)$ denote the row vector of population numbers of levels 1 to N at a given depth point, at which the rate equations are given by

$$\mathbf{n}_{new}\mathbf{P}\left(\mathbf{n}_{new}, \mathbf{n}_{old}, \mathbf{J}_{FS}\right) = \mathbf{b}, \tag{10.3.35}$$

where $\mathbf{P}$ is the rate coefficient matrix and the row vector $\mathbf{b}$ contains essentially zeros except those columns representing core or number conservation. The off-diagonal elements of P_{ij} represent the transitions from level i to level j. The diagonal elements P_{jj} represent the losses from level j to all other levels. Thus we have

$$P_{jj} = -\sum_{m \neq j} P_{jm}, \tag{10.3.36}$$

(see Mihalas (1978)). The Newton–Raphson iteration technique is then applied to obtain $\mathbf{n}^{(k+1)}$. Thus

$$\mathbf{n}^{(k+1)} = \mathbf{n}^{(k)} - \left(\mathbf{n}^{(k)}\mathbf{P} - \mathbf{b}\right)\mathbf{M}^{-1}, \tag{10.3.37}$$

where $\mathbf{P}$ and $\mathbf{M}$ are evaluated from the current solution $\mathbf{n}^{(k)}$. The matrix $\mathbf{M}$ is given by

$$M_{ij} = \frac{\partial}{\partial n_i}\left[\sum_m n_m P_{m_j}\right]. \tag{10.3.38}$$

By using equation (10.3.36), we get

$$M_{ij} = P_{ij} + \sum_{m \neq j} D_{im_j}, \tag{10.3.39}$$

where

$$D_{im_j} = n_m \frac{\partial P_{m_j}}{\partial n_i} - n_j \frac{\partial P_{j_m}}{\partial n_i}, \tag{10.3.40}$$

where $D_{im_j} = -D_{ij_m}$. The coefficient matrix $\mathbf{P}$ consists of radiative and collisional terms:

$$\mathbf{P} = -\mathbf{R} - \mathbf{C}. \tag{10.3.41}$$

The new radiation field (10.3.33) must be used when computing the radiative rates $\mathbf{R}$. The radiative rates create problems because of disproportion between the large rates in the resonance transitions and the small gains and losses which strongly influence the solutions (see Scharmer and Carlson (1985)). An appropriate analytical solution is to introduce a 'net radiative bracket' by which these disproportional upward and downward rates are cancelled (see Mihalas (1978)). Instead of equation (10.3.41), we can write

$$\mathbf{P} = -\mathbf{Z} - \mathbf{C}, \tag{10.3.42}$$

where

$$Z_{ul} = R_{ul} - \frac{n_l}{n_u} R_{lu}, \tag{10.3.43}$$

and u and l stand for the upper and lower levels respectively. Furthermore,

$$Z_{lu} = 0. \tag{10.3.44}$$

For line transitions, we find that

$$Z_{ul} = A_{ul}\left(1 - \frac{J_{new}}{S_{new}}\right), \tag{10.3.45}$$

where S_{new} is the non-LTE line source function. Now, the derivative in equation (10.3.40) becomes

$$D_{iul} = A_{ul} n_u \frac{\partial}{\partial n_i}\left(\frac{J_{new}}{S^L_{new}}\right). \tag{10.3.46}$$

One of the terms that are in $(\partial/\partial n_i)(\cdots)$ is

$$J_{new}\frac{\partial}{\partial n_i}\left(S^{-1}_{new}\right), \tag{10.3.47}$$

which contributes only when $i = 1$ or $i = u$. This reflects the 'net radiative bracket'. The other term in the derivative is given by

$$(S_{new})^{-1}\left(\frac{\partial J_{new}}{\partial n_i}\right) = (S_{new})^{-1}\int_{core}\left(\frac{\partial S_{new}}{\partial n_i}\right)\phi_x \, dx. \tag{10.3.48}$$

Equations (10.3.32) and (10.3.35) are solved simultaneously as is done in the lambda iteration procedure.

The integration over the line core is performed numerically in ALI as

$$\Lambda^\star S = \int_{x_{red}}^{x_{blue}} S_x \varphi_x \, dx. \tag{10.3.49}$$

Doppler effects can be neglected in the continuum. The approximate $\Lambda^\star$ operator at r can be defined as

$$\Lambda^\star_\nu(r) = 1 - \exp\left[-\frac{\tau_\nu(r)}{\gamma_c}\right], \tag{10.3.50}$$

where

$$\tau_\nu(r) \rightarrow \min\{\tau_\nu(r), \tau_\nu^{max} - \tau_\nu(r)\}. \tag{10.3.51}$$

τ_ν^{max} is the optical depth at the bottom of the atmosphere, γ_c is a parameter of order of unity, and $\Lambda_\nu^\star(r)$ is taken as

$$\Lambda_\nu^\star(r) = \begin{cases} 1 & \text{if } \tau_\nu(r) > \gamma_c, \\ 0 & \text{otherwise.} \end{cases} \tag{10.3.52}$$

The application of the above method to some problems is described in Hamann (1987).

An alternative method was developed by Werner and Husfeld (1985) by combining Scharmer's local operator and the perturbation technique for line formation in static, plane parallel atmospheres which can take upto 100 non-LTE levels. This was extended by Werner (1986) to compute the non-LTE plane parallel model atmospheres (see Werner (1987)).

10.4 Complete linearization method

In the non-LTE problem, the radiative transfer equation and the statistical equilibrium equation are solved simultaneously. This is a non-linear problem and therefore the complete linearization method is best suited. All the variables are globally coupled and interact with each other. Therefore no variable is considered to be more fundamental than the others. These variables are strongly coupled through the radiative transfer equation. Any change in any variable at any given point will propagate throughout the medium to the variables at other points. This system of linear equations is indeterminate and has to be solved by iteration (see Mihalas (1978), chapters 7 and 12, Werner (1987), Sen and Wilson (1998)). The system of linear constraint equations (statistical equilibrium, particle charge conservation etc.) at each depth point d is written as

$$\mathbf{M}_d \delta \mathbf{\Psi}_d = \mathbf{C}_d \quad (d = 1, 2, \ldots, D), \tag{10.4.1}$$

where $\mathbf{M}_d$ is a matrix of dimension $N_c \times N_T$ (N_c is the number of constraint equations, N_T is the number of variables describing the population and the radiation field), $\mathbf{C}_d$ is a vector of length N_c which contains the error in the constraint equations evaluated with the assumed populations and radiation field, and

$$\delta \mathbf{\Psi}_d = \left(\delta n, \ldots, \delta n_D, \delta N_e, \delta T_d, \delta J_1, \ldots, \delta \bar{J}_1, \ldots\right), \tag{10.4.2}$$

where $\bar{J}$ is the frequency integrated mean intensity, n_d represents the occupation numbers, T is the temperature and n_e is the number of electrons. Let $\mathbf{\Psi}_d$ be the solution of the complete system of constraint equations $f_d(\mathbf{\Psi}_d) = 0$. If $\mathbf{\Psi}_d^0$ is the approximate solution due to the initial approximated population, then

$$\mathbf{\Psi}_d = \mathbf{\Psi}_d^0 + \delta \mathbf{\Psi}_d, \tag{10.4.3}$$

where $\delta\mathbf{\Psi}_d$ is the perturbation to the current approximate solution $\mathbf{\Psi}_d^0$. The quantity $\delta\mathbf{\Psi}_d$ is so chosen that

$$f(\mathbf{\Psi}_d^0 + \delta\mathbf{\Psi}_d) \to 0, \tag{10.4.4}$$

due to iteration. This quantity $\delta\mathbf{\Psi}_d$ can be obtained from the linearized original system, that is

$$f_d(\mathbf{\Psi}_d^0) + \sum_j \frac{\partial f_d}{\partial \mathbf{\Psi}_{d_j}} \delta\mathbf{\Psi}_{d_j} = 0. \tag{10.4.5}$$

We need to express δ_J in terms of δn, δn_e and δT by using the linearized transfer equation. For this the transfer equation in a spherical medium is discretized through the Feautrier scheme (see chapter 8) and the difference equation is written out together with the boundary conditions. The solution is obtained by the Gaussian elimination scheme (Feautrier's method or the Rybicki scheme – see chapters 4 and 8). The system of equations can be written as

$$\mathbf{T}_n\mathbf{J}_n = \mathbf{X}_n + \mathbf{U}_n\mathbf{J}_{n-1} + \mathbf{V}_n\mathbf{H}_{n-1}, \tag{10.4.6}$$

$$\mathbf{H}_n = \mathbf{A}_n\mathbf{J}_n + \mathbf{B}_n\mathbf{H}_{n-1}, \tag{10.4.7}$$

where **J** and **H** are the mean intensity and flux vectors. Furthermore, **T** is the tri-diagonal matrix, **U** is the diagonal matrix, **V** is the lower bi-diagonal matrix, **A** is the lower bi-diagonal matrix, **B** is the upper tri-diagonal matrix and **X** is a vector. The subscript n is the frequency index. Hillier (1990) used the above complete linearization of Auer and Mihalas (1969) to solve the radiative transfer equation and the statistical equilibrium equations. He used an improved tri-diagonal Newton–Raphson operator of Hempe and Schönberg (1986) and Schönberg and Hempe (1986). He studied line intensities (Hillier 1990). The frequency integrated mean intensity $\bar{J}$ is written as

$$\bar{J} = \int_{-\infty}^{+\infty} \phi_\nu J_\nu \, d\nu = \sum_{n=1}^{n_{max}} \omega_n \phi_n J_n, \tag{10.4.8}$$

ω_n being the quadrature weights for the frequency points. If the quantities κ_c, S_c, κ_L and S_L denoting the continuum opacity, continuum source function, line opacity and line source function are represented by a single variable x_j ($j = 1, 2, 3, 4$), then at each depth point we may write

$$\delta J_l = \sum_{d=1}^{D} \sum_{s=1}^{4} \frac{\partial J_l}{\partial x_{sd}} \delta x_{sd}. \tag{10.4.9}$$

Equations (10.4.6) and (10.4.7) are linearized with respect to the independent variables and we get

$$\mathbf{T}_n\partial\mathbf{J}_n = \mathbf{U}_n\partial\mathbf{J}_{n-1} + \mathbf{V}_n\partial\mathbf{H}_{n-1} + \partial\mathbf{X}_n - \partial\mathbf{T}_n\mathbf{J}_n + \partial\mathbf{U}_n\mathbf{J}_{n-1} + \partial\mathbf{V}_n\mathbf{H}_{n-1} \tag{10.4.10}$$

and

$$\partial\mathbf{H}_n = \mathbf{A}_n\partial\mathbf{J}_n + \mathbf{B}_n\partial\mathbf{H}_{n-1} + \partial\mathbf{A}_n\mathbf{J}_n + \partial\mathbf{B}_n\mathbf{H}_{n-1}. \tag{10.4.11}$$

∂J is a three-dimensional matrix with $\partial J_l/\partial x_{sd}$ as its elements. From equations (10.4.8) and (10.4.9), we get

$$\partial\bar{J}_l = \sum_{d=1}^{D}\sum_{s=1}^{H} \partial\bar{J}_{lds}\delta x_{ds}. \tag{10.4.12}$$

Now, we introduce the assumption that the radiation field is influenced by the level conditions and neighbourhood points, ignoring the effects due to the non-local terms. Hence we obtain a local operator. The three-dimensional matrix $\partial\mathbf{J}$ can be replaced by $(\partial\mathbf{J}')$, ignoring the non-local terms in the variation matrices. The dimension of $(\partial\mathbf{J}')$ is $D \times N_B \times 4$, where N_B indicates the degree of depth coupling ($N_B = 1$ represents pure local variation and $N_B = 3$ indicates that $\bar{J}$ depends on the independent variables in its immediate neighbourhood).

Now, we linearize x in terms of the unknown population numbers and write δx_{sd} at each depth point and for each independent variable as

$$\delta x_{sd} = \sum_{j=1}^{N_c}\left(\frac{\partial x_s}{\partial n_j}\right)_d \delta n_{jd}, \tag{10.4.13}$$

where N_c is the number of constraint equations. Then from equations (10.4.9), (10.4.12) and (10.4.13) we get

$$\begin{aligned}
\delta\bar{J}_l &= \sum_{d=1}^{N_B}\sum_{s=1}^{4} \frac{\partial\bar{J}_l}{\partial x_{sr}}\delta x_{sr} \\
&= \sum_{d=1}^{N_B}\sum_{j=1}^{N_c}\sum_{s=1}^{4} \frac{\partial\bar{J}_l}{\partial x_{sr}}\left(\frac{\partial x_s}{\partial x_j}\right)\delta n_{jr} \\
&= \sum_{d=1}^{N_B}\sum_{j=1}^{N_c}\sum_{s=1}^{4} \delta J'_{lds}\left(\frac{\partial x_s}{\partial n_j}\right)_r \delta n_{jr} \\
&= \sum_{d=1}^{N_B}\sum_{j=1}^{N_c} \delta\mathbf{J}''_{ldj}\delta n_{jr}.
\end{aligned} \tag{10.4.14}$$

In equation (10.4.14), if $N_B = D$, $r = d$, otherwise $r = l - (N_0 + 1)/2 + d$. We can then replace $\delta\mathbf{J}$ by $\delta\mathbf{n}$ in equations (10.4.1) and (10.4.3) and they are therefore coupled at different depths through the radiation field $\bar{J}$. The system is solved by an iterative scheme. The convergence of the scheme depends very crucially on the starting solution of the population density. Hillier (1990) points out that a locally computed Sobolev approximation may be used as a starting model for the CMF solution. The Sobolev approximation may be continued until it has sufficiently converged and then the CMF solution can be started, although this switching is arbitrary. The linearization results in population densities that are unstable and inconsistent. To avoid these difficulties, Hillier suggests linearization and lambda iteration and the switching on and off of the temperature variation and switching off of the Newton–Raphson operation before complete convergence. One requires a lot of experience in this procedure to obtain the correct results. Furthermore, at each depth, Ng's (1974) acceleration with tri-diagonal and penta-diagonal operators and weights inversely proportional to the variables (Auer 1987) should be used.

10.5 Approximate lambda operator (ALO)

Puls (1991) noted that the lambda operator acting on a line source function is of affine type. He constructed a simple, local, optimum, parameter free ALO for line transfer in the CMF, similar to the Sobolev approximation (SA). We describe below this procedure following Puls (1991) and Sen and Wilson (1998)

We saw earlier that the transfer equation is solved by the impact parameter (see chapters 7 and 8). The transfer equations in the Feautrier scheme are written as

$$\frac{\partial u(z,p,\nu)}{\partial \tau(z,p,\nu)} + \gamma(z,p,\nu)\left[\frac{\partial v(z,p,\nu)}{\partial \nu}\right] = v(z,p,\nu) \tag{10.5.1}$$

and

$$\frac{\partial v(z,p,\nu)}{\partial \tau(z,p,\nu)} + \gamma(z,p,\nu)\left[\frac{\partial u(z,p,\nu)}{\partial \nu}\right] = u(z,p,\nu) - S(z,p,\nu), \tag{10.5.2}$$

where u and v are the mean-intensity- and flux-like variables, which are also called the Feautrier variables and

$$\gamma(z,p,\nu) = \frac{\bar{\gamma}(z,p)}{\kappa(z,p,\nu)}, \tag{10.5.3}$$

$$\bar{\gamma}(z,p) = \frac{\nu v(r)}{cr}\left[1 - \mu^2 + \mu^2\left(\frac{d\ln v}{d\ln r}\right)\right]. \tag{10.5.4}$$

We want to solve equations (10.5.1) and (10.5.2) with the given boundary condition (see chapters 7 and 8). To obtain the solution the Feautrier or Rybicki type method is used at each depth point. We have (in the notation of Puls)

$$\mathbf{T}_k\mathbf{u}_k = \mathbf{U}_k\mathbf{u}_{k-1} + \mathbf{V}_k\mathbf{v}_{k-1} + \mathbf{S}_k, \tag{10.5.5}$$

$$\mathbf{v}_k = \mathbf{G}_k\mathbf{u}_k + \mathbf{H}_k\mathbf{v}_{k-1}. \tag{10.5.6}$$

The term $\mathbf{W}_k\mathbf{J}_k$ is ignored. The index k includes the index i of the impact parameter (or angle) and the frequency index 'n'. The source function is given by

$$S_k = \frac{j_k}{\kappa_k} = \frac{j_{ck} + \kappa_{Lk} S_L}{\kappa_{ck} + \kappa_{Lk}}. \tag{10.5.7}$$

Now it is assumed that the continuum quantities with index ck are constant across the line profile. Integration over the angle and frequency of the line transfer gives the mean intensity which can be written in terms of an affine operator as

$$\bar{\mathbf{J}}_L = \boldsymbol{\Psi}\left[u_1, v_1, j_c, \kappa_c, \kappa_L\right] + \Lambda' S_L, \tag{10.5.8}$$

where $\boldsymbol{\Psi}$ is the displacement vector. The subscripts c and L in equations (10.5.7) and (10.5.8) represent the continuum and the line respectively. u_1 and v_1 are blue wing variables of the continuum problem. This affine equation is linear only when $j_c = 0$, $u_1 = v_1 = 0$. $\bar{\mathbf{J}}_L$ is the frequency integrated mean intensity which is used in the rate equation.

We solve the transfer equation $(D+1)$ times, where D is the total number of grid points, to obtain the explicit values of the displacement vector $\boldsymbol{\Psi}$and the matrix $\boldsymbol{\Lambda}'$. We have

$$\boldsymbol{\Psi} = \bar{\mathbf{J}}(S_L = 0) \tag{10.5.9}$$

and

$$\Lambda'_{ij} = \bar{J}_i(S_L = e_j) - \boldsymbol{\Psi}_i, \tag{10.5.10}$$

where e_j is the unit vector, or

$$\Lambda'_{ij} = \bar{J}_i(S_L = e_j;\ u_1 = v_1 = j_c = 0). \tag{10.5.11}$$

For an affine ALI Λ^A, the ALI scheme can be generally written as

$$\bar{J}^n = \Lambda^A\left[S^n\right] + (\Lambda - \Lambda^A)\left[S^{(n-1)}\right], \tag{10.5.12}$$

or

$$\bar{J}^n = \Lambda\left[S^{n-1}\right] + \Lambda^A\left[S^n - S^{n-1}\right], \tag{10.5.13}$$

as in Pauldrach and Herrero (1988). Puls (1991) show that it is the linear term that has to be approximated for ALI although the lambda operator is of the affine type. In the multi-level calculations he suggest that the lines should be iterated with fixed continuum before the continuum is updated again. When the population numbers stabilize and the relative corrections become small, lines and continuum may be iterated in parallel until the required convergence is reached.

An ALO of the type

$$\Lambda^{\star}_{\nu} = \mathrm{diag}\,[\mathbf{T}]^{-1}, \tag{10.5.14}$$

is the ideal choice (Olson *et al.* 1986) for a strictly local ALO in static problems, and ensures fast convergence in ALI. $\mathbf{T}$ is a tri-diagonal matrix. In CMF calculations the coupling of the angle and frequency has to be taken into consideration in the estimation of the diagonal elements. Puls (1991) estimated the contribution of $u_{k,l}$ (the Feautrier variable) to the exact diagonal of $\Lambda' = \Lambda'_{ll}$ (where k is the frequency index for a given ray). Equation (10.5.11) then takes the form

$$\Lambda'_{l,l} = \bar{J}_l(S_L = e_l;\ u_1 = v_1 = j_c = 0). \tag{10.5.15}$$

Puls showed that for frequencies $k = 1, 2,\ u_{k,l}$ is purely local and for frequencies $k \geq 3,\ u_{k,l}$ is non-local and is coupled with all other $u_{(k-1),j}$ and also with the Feautrier flux-like variables. However, this process of finding the diagonal elements of Λ' is highly involved and time consuming.

Puls attempted to obtain a proper local approximation for $u_{k,l}$. If the operator $\mathbf{V}_k$ in equation (10.5.5) is split into diagonals $\mathbf{VA}_k$ and $\mathbf{VB}_k$ (with all elements > 0), we get for each $k,\ k \geq 2$ (Puls 1991, page 484, equation 9)

$$\begin{aligned} u_{k,2} &= \sum_i \left(\mathbf{T}_k^{-1}\right)_{li} \big[\mathbf{U}_{ki} u_{(k-1),i} - \mathbf{VA}_{ki} v_{(k-1),(i-1/2)} \\ &\quad + \mathbf{VB}_{ki} v_{(k-1),(i+1/2)} + \delta_{il}\rho_{ki}\big] \\ &= \sum_i \left(\mathbf{T}_k^{-1}\right)_{li} b_{ki}, \end{aligned} \tag{10.5.16}$$

where

$$\rho_k = \left(\frac{\kappa_{kl}}{\kappa_{ck} + \kappa_{kl}}\right), \tag{10.5.17}$$

and b_{ki} is the quantity within the square brackets in equation (10.5.16). This approximation is restricted to local terms only, that is all terms except $i = l$ are neglected in b_{ki}. We define a new ALO $\Lambda'^{\star}_{l,l}$ through $U^{\star}_{kl}$, given by

$$\Lambda'^{\star}_{l,l} = \sum \omega_{p,l} \sum_{k=1}^{K} \omega_{kl} u^{\star}_{pk,l}, \tag{10.5.18}$$

where ω_{pl} are the quadrature weights for the integration. Convergence of the ALI cycle can be achieved if $\Lambda'^{\star}_{l,l} < \Lambda'_{l,l}$, where the latter is the exact diagonal operator. The exact $u_{k,l}$ is given by

$$\mathbf{u}_k = \mathbf{T}_k^{-1} b_k = \mathbf{T}_k^{-1}\left(b_{k,1}e_1 + \cdots + b_{kD}e_D\right) = \sum_i b_{ki}\mathbf{t}_k(e_i), \tag{10.5.19}$$

where $\mathbf{t}_k(e_i)$ is the solution of

$$\mathbf{T}_k \mathbf{t}_k = e_i. \tag{10.5.20}$$

The quantity $u^{\star}_{k,l}$ of the local approximation is given by

$$u^{\star}_{k,l} = b_{k,l} t_{k,l}(e_l). \tag{10.5.21}$$

$t_{k,l}(e_l)$ is positive by definition and if we require that $u^\star_{k,l} < u_{k,l}$, then we need to have the following relation

$$b_{k,l}t_{k,l}(e_l) \leq \sum b_{k,l}t_{k,l}(e_i), \tag{10.5.22}$$

satisfied, which means that

$$\sum_{i \neq l} b_{k,i} \geq 0. \tag{10.5.23}$$

Puls noted that relation (10.5.23) holds good in most numerical calculations of a number of models except when these involved a small number of red wing frequencies in the wind around the thermal point.

Now $u^\star_{k,l}$, the local approximation, is estimated from

$$u^\star_{k,l} = \mathrm{diag}\big(\mathbf{T}_k^{-1}\big)\big[\mathbf{u}_k u^\star_{k-1,0} - \mathbf{VA}_k v^\star_{k-1,-1/2} + \mathbf{VB}_k v^\star_{k-1,1/2} + \rho_{k,0}\big]. \tag{10.5.24}$$

The Feautrier variables $v_{k-1,\pm 1/2}$ are found from equation (10.5.6):

$$v^\star_{k-1,-1/2} = \mathbf{G}_{k-1,-1/2}\left[u^\star_{k-1,0} - u_{k-1,-1}\right] + \mathbf{H}_{k-1,-1/2}v^\star_{k-2,-1/2} \tag{10.5.25}$$

and

$$v^\star_{k-1,1/2} = \mathbf{G}_{k-1,1/2}\left[u_{k-1,1} - u^\star_{k-1,0}\right] + \mathbf{H}_{k-1,1/2}v^\star_{k-2,1/2}, \tag{10.5.26}$$

with the boundary conditions at the blue and red wings given by

$$u^\star_{1,0} = v^\star_{1,\pm 1/2} = 0. \tag{10.5.27}$$

From equation (10.5.16), we have

$$u_{k-1,l\pm 1} = \sum_i \big(\mathbf{T}_k^{-1}\big)_{l\pm 1,i} b_{k-1,i}. \tag{10.5.28}$$

It can be seen from equations (10.5.24)–(10.5.26) that although $u^\star_{k,0}$ is given in the local approximation non-local dependence creeps in through the $v^\star$ terms defined in equation (10.5.25) and (10.5.26). This problem can be dealt with in the same way as before by neglecting the contributions of $b_{k-1,i}$, $i \neq l$ compared to $b_{k-1,l}$. Then from equation (10.5.28), we have

$$u^\star_{k-1,l\neq 1} = \big(\mathbf{T}_{k-1}^{-1}\big)_{l\pm 1,l} b_{k-1,l}. \tag{10.5.29}$$

Puls, however, followed a different approach for the sake of computational convenience. He neglected the contribution of the off-diagonal elements $u_{k-1,l\pm 1}$ to $u^\star_{k-1,l\pm 1/2}$. Using '$\star\star$' to represent the second approximation, we have

$$v^{\star\star}_{k-1,l-1/2} = \mathbf{G}_{k-1,l-1/2}u^{\star\star}_{k-1,l} + \mathbf{H}_{k-1,l-1/2}v^{\star\star}_{k-2,l-1/2} \tag{10.5.30}$$

and

$$v^{\star\star}_{k-1,l+1/2} = \mathbf{G}_{k-1,l+1/2} u^{\star\star}_{k-1,l} + \mathbf{H}_{k-1,l+1/2} v^{\star\star}_{k-2,l+1/2}. \tag{10.5.31}$$

To obtain the second approximation in equation (10.5.26) the quantities with superscript '$\star$' are replaced by those with the superscript '$\star\star$'. As a result of these changes,

1. the diagonal operator $\Lambda'^{\star\star}_{l,l}$ depends on purely local quantities $u^{\star\star}_{k,k}$, $v^{\star\star}_{k,l\pm1/2}$;
2. $\Lambda'_{l,l} = \Lambda'^{\star\star}_{l,l}$ in the static case.

Puls (1991) constructed the approximate $\Lambda^{\star\star}$ using the following procedure:

1. start with $u^{\star\star}_{1,0} = v^{\star\star}_{1,\pm1/2} = 0$;
2. compute the term in the brackets on the RHS of equation (10.5.26) for each frequency k ($k = 1, 2, \ldots, K$) from the previous solution of $u^{\star\star}_{k-1,0}$ and $v^{\star\star}_{k-1,\pm1/2}$;
3. compute $\mathbf{T}^{-1}$ by using the algorithm of Rybicki and Hummer (1991), $u^{\star\star}_{k,0}$ obtain from the modified equation (10.5.26), and $v^{\star\star}_{k,\pm1/2}$ from equations (10.5.30) and (10.5.31);
4. add the new values of the vectors $u'^{\star\star}_{k-1,0}$ and $v^{\star\star}_{k-1,0}$ to $\Lambda'^{\star\star}$ with the corresponding spatial and frequency weights; and
5. do the computations for all frequencies and for all rays to obtain $\Lambda'^{\star\star}$ and J.

One should note that the constructed $\Lambda'^{\star\star}$ is a purely local operator without any spatial coupling which corresponds to the Sobolev approximation. The local CMF and ALO reach the static limit for small velocities. Puls (1991) examined the ALI type solutions for rate equations and found that a formal correspondence between the local CMF and SA suggests several computationally economical methods for solving the rate equations.

Puls suggested a judicious combination of CMF and SA to obtain a fast solution of the transfer and statistical equilibrium equations. The method is parameter free and works at the optimum convergence rate. The local ALO method resembles SA in terms of basic principles.

Sellmaier *et al.* (1993) studied the the non-LTE problem of multi-level radiative transfer in the total atmosphere of stars with wind in the presence of sub and supersonic velocity fields. They used the local ALO theory of Puls (1991) and an ALI scheme to solve line transfer in CMF. The scheme gave excellent convergence in the case of complex atomic models. They found that the core-saturation method is inadequate for predicting the line formation in all ranges – strong, weak and intermediate strengths. They also found that the CMF diagonal operator of Puls (1991) is the best local ALO for CMF transfer in unified models. This ALO leads to fast convergence even when 600 line transitions are included.

Hamann *et al.* (1991) developed a quasi-Newtonian iterative scheme which they used together with the Briyden update formula. This procedure was found to give considerable computational advantages. Repeated computation of derivatives and inversion of matrices is avoided.

Koter *et al.* (1993) developed a fast non-LTE technique for computing the continuum energy distribution and lines in an expanding atmosphere. They applied a semi-empirical model with a given density velocity and temperature distribution. They solved the transfer and statistical equilibrium equations in the continuum using ALI while the line transfer was solved by using SA.

The ALI approach is the contemporary method used for obtaining the solution of the transfer equation within the constraints of statistical equilibrium in expanding spherical atmospheres. A comprehensive review of the general scheme of these methods is given in Hubený (1992) and Rybicki (1991).

In the ALI method, the following procedures are used:

1. Auer and Mihalas' (1969) complete linearization of the transfer equation and the constraint equations;
2. Cannon's (1973a,b) and Scharmer's (1981, 1984) (see chapter 9) operator perturbation scheme which consists of a simple local or non-local ALO linearization of the source function about an initial value. Furthermore, an iteration scheme is developed to correct the error introduced due to ALO. The ideas of core saturation of Rybicki and Newton–Raphson linearization are widely used to obtain the solution of the non-linear problems;
3. the Feautrier impact parameter method and SA or a mixture of both;
4. ALI to give rapid convergence with Ng's (1974) or with Auer's (1987) acceleration schemes. The main aim of these techniques is to obtain a quick solution of the transfer equation with its constraint equation without adversely sacrificing accuracy.

Scharmer and Carlson (1985) used the first order transfer equation to find the change in population density δn_i. They employed Scharmer's approximate operator

$$\delta_\nu^{(n)}(\mu) = \Lambda^\star_{\mu\nu}\left[\delta S_\nu^{(n)}\right]. \qquad (10.5.32)$$

This method was coded by Carlson (1986) in a program called MULTI which is widely used for non-LTE line formation. This program is extremely fast and can handle a few hundred levels and lines (see Rutten (1999)). MULTI can be obtained from an anonymous ftp at ftp.astro.uio.no:/pub/multi.

10.6 Characteristic rays and ALO-ALI techniques

When the source function is known (or assumed), the transfer equation becomes a first order partial differential equation for the specific intensity. Using this idea the intensity is calculated along the spatial grid points defined on the spherical atmosphere. (To understand the characteristic or characteristics of surfaces, the reader should consult any good book on partial differential equations.) The characteristics method is combined with the operator perturbation technique to obtain the solution of the transfer equation.

Kunasz and Auer (1987) and Olson and Kunasz (1987) used the method of short characteristic rays to obtain the solution of the transfer equation in planar geometry. Hauschildt (1992) extended this to expanding spherical atmospheres (see Sen and Wilson (1998)).

The time independent transfer equation in CMF is

$$\begin{aligned}
&\gamma(\mu+\beta)\frac{\partial I(\nu,r,\mu)}{\partial r} \\
&\quad+\frac{\partial}{\partial\mu}\left\{\gamma(1-\mu^2)\left[\frac{1+\beta\mu}{r}-\gamma^2(\mu+\beta)\frac{\partial\beta}{\partial r}\right]I(\nu,r,\mu)\right\} \\
&\quad-\frac{\partial}{\partial\nu}\left\{\gamma\nu\left[\frac{\beta(1-\mu^2)}{r}+\gamma^2\mu(\mu+\beta)\frac{\partial\beta}{\partial r}\right]I(\nu,r,\mu)\right\} \\
&\quad+\gamma\left[\frac{2\mu+\beta(3-\mu^2)}{r}+\gamma^2(1+\mu^2+2\beta\mu)\frac{\partial\beta}{\partial r}\right]I(\nu,r,\mu) \\
&=j-\kappa I(\nu,r,\mu),
\end{aligned} \tag{10.6.1}$$

where $\beta = v(r)/c$, $v(r)$ being the velocity of the gas, c is the velocity of light, $\gamma = (1-\beta^2)^{-\frac{1}{2}}$. Equation (10.6.1) is written in the wavelength scale λ as

$$a_r\frac{\partial I}{\partial r}+a_\mu\frac{\partial I}{\partial\mu}+a_\lambda\frac{\partial(\lambda I)}{\partial\lambda}+4a_\lambda I=j-\kappa I, \tag{10.6.2}$$

where

$$\begin{aligned}
a_r &= \gamma(\mu+\beta), \\
a_\mu &= \gamma(1-\mu^2)\left[\frac{1+\beta\mu}{r}-\gamma^2(\mu+\beta)\frac{\partial\beta}{\partial r}\right], \\
a_\lambda &= \gamma\left[\frac{\beta(1-\mu^2)}{r}+r^2\mu(\mu+\beta)\frac{\partial\beta}{\partial r}\right],
\end{aligned}$$

The characteristic rays of equation (10.6.2) are defined by

$$\frac{dr}{a_r}=\frac{d\mu}{a_\mu}. \tag{10.6.3}$$

If s is the geometrical path length along the characteristic ray (see figure 10.2) equation (10.6.3) can be written as

$$\frac{dr}{ds}=a_r, \tag{10.6.4}$$

$$\frac{d\mu}{ds} = a_\mu. \tag{10.6.5}$$

Equation (10.6.2) along the characteristic ray is written as

$$\frac{\partial I}{\partial s} + a_\lambda \frac{\partial(\lambda I)}{\partial \lambda} = j - (\kappa + 4a_\lambda)I. \tag{10.6.6}$$

Integration of equation (10.6.6) along the characteristic ray requires knowledge of the solution of equation (10.6.5), that is the variation of μ along the ray to obtain a_λ. We consider only those rays that are tangent to the spherical spatial grid of the atmosphere, with the impact parameter p. At the point of tangency, the path length s is set equal to 0. Therefore from equation (10.6.4) we get

$$|\mu|_{s=0} = -\beta(p). \tag{10.6.7}$$

We assume that μ_E in the laboratory frame is known so that the variation of μ along the characteristic is determined, or

$$\mu_E(r) = \pm \left(1 - \frac{p^2}{r^2}\right)^{\frac{1}{2}}. \tag{10.6.8}$$

The corresponding μ in the CMF can be obtained from μ_E through the Lorentz transformation (see chapter 8). We get (the aberration effect)

$$\mu(r) = \frac{\mu_E(r) - \beta(r)}{1 - \beta(r)\mu_E(r)}. \tag{10.6.9}$$

From equations (10.6.8) and (10.6.9), it is clear that $\mu(r)$ is a local function of r and depends on r through $\beta(r)$ and $\mu_E(r)$. Using equation (10.6.4), we can compute

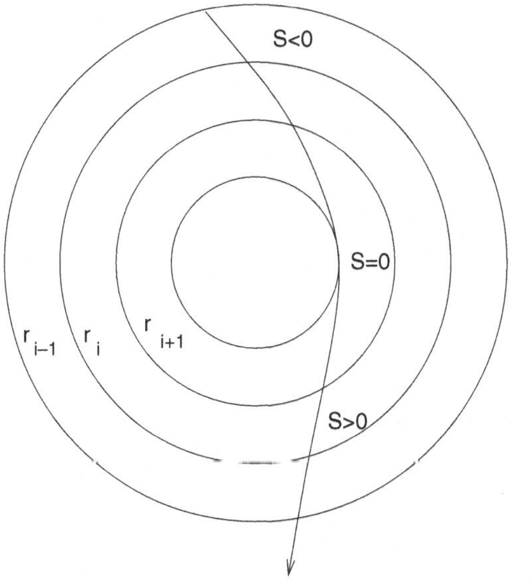

Figure 10.2 Schematic diagram showing the direction of a tangential characteristic ray in a spherical envelope.

the accurate path length s along the characteristic ray, using the Gauss quadrature formula and a linear interpolation for the velocity between the radial points. If the variation of μ along the path and the path length are known, we can integrate equation (10.6.6) and obtain the intensity. Following Hauschildt and Wehrse (1991) and Mihalas (1980) the wavelength derivative is discretized to give a static implicit scheme:

$$\left.\frac{\partial(\lambda I)}{\partial\lambda}\right|_{\lambda=\lambda_k} = \frac{\lambda_k I_{\lambda_k} - \lambda_{k-1} I_{\lambda_{k-1}}}{\lambda_k - \lambda_{k-1}}. \tag{10.6.10}$$

The substitution of equation (10.6.10) into equation (10.6.6) gives

$$\frac{dI_{\lambda k}}{ds} + a_\lambda \frac{\lambda_k I_{\lambda_k} - \lambda_{k-1} I_{\lambda_{k-1}}}{\lambda_k - \lambda_{k-1}} = j_{\lambda_k} - \left(\kappa_{\lambda_k} + 4a_\lambda\right) I_{\lambda_k}. \tag{10.6.11}$$

We shall now define a new optical depth τ along the characteristic ray as

$$-d\tau = \left[\kappa_{\lambda_k} + a_\lambda\left(4 + \frac{\lambda_k}{\lambda_k - \lambda_{k-1}}\right)\right] ds = \bar{\kappa} ds \text{ (say)}. \tag{10.6.12}$$

A new modified source function is then written as

$$\bar{S} = \frac{\kappa}{\bar{x}}\left[S + \frac{a_\lambda}{\kappa}\left(\frac{\lambda_{k-1}}{\lambda_k - \lambda_{k-1}}\right) I_{\lambda_{k-1}}\right], \tag{10.6.13}$$

with

$$S = \frac{j}{\kappa}. \tag{10.6.14}$$

Equation (10.6.11) can now be written as

$$\frac{dI}{d\tau} = I - \bar{S}. \tag{10.6.15}$$

The formal solution of equation (10.6.15) along the characteristic ray can be written as

$$I_i^- = I_{i-1}^- \exp\left(-\Delta\tau_{i-1}\right) + \int_{\tau_i}^{\tau_{i-1}} \bar{S}(t) \exp\left[-(t - \tau_i)\right] dt, \quad s < 0, \tag{10.6.16}$$

$$I_i^+ = I_{i+1}^+ \exp\left(-\Delta\tau_{i-1}\right) + \int_{\tau_i}^{\tau_{i+1}} \bar{S}(t) \exp\left[-(t - \tau_i)\right] dt, \quad s > 0. \tag{10.6.17}$$

The index i represents the discretization of the radial coordinate with $r_1 = 0$ at the outer boundary. The τ_is are the corresponding optical depths of the spatial grids along the characteristics with $\tau_1 = 0$, $\tau_i < \tau_{i+1}$ and $\Delta\tau_i = \tau_i - \tau_{i+1}$. The starting point of integration is farthest from the observer ($s < 0$) and proceeds towards the observer ($s > 0$). If the ray tangent to the shell $(j + 1)$ is labelled by the index j

and we introduce the optical thickness $\Delta\tau_{i-1,j}$ along the ray j between the spatial grid points τ_i and τ_{i-1} we have

$$\Delta\tau_{i-1,j} = \frac{1}{2}(\bar{\kappa}_{i-1} + \bar{\kappa}_i)\left|s_{i,j} - s_{i-1,j}\right|. \tag{10.6.18}$$

Now, equations (10.6.16) and (10.6.17) can be written (suppressing the ray index j for convenience)

$$I_i^- = I_{i-1}^- \exp(-\Delta\tau_{i-1}) + \Delta I_i^-, \quad s > 0 \tag{10.6.19}$$

and

$$I_i^+ = I_{i+1}^+ \exp(-\Delta\tau_{i-1}) + \Delta I_i^+, \quad s > 0. \tag{10.6.20}$$

The quantity $\Delta I_i^\pm$ can be obtained from $\bar{S}$ along a characteristic ray interpolated by either a linear or parabolic polynomial such that

$$\Delta I_i^\pm = a_i^\pm \bar{S}_{i-1} + b_i^\pm \bar{S} + c_i^\pm \bar{S}_{i+1}. \tag{10.6.21}$$

The coefficients are given by Hauschildt (1992). For the linear interpolation

$$a_i^+ = 0, \ \bar{a} = e_{0,i} - \frac{e_{1,i}}{\Delta\tau_{i-1}},$$

$$b^+ = \frac{e_{1,i+1}}{\Delta\tau_i}, \quad \bar{b} = \frac{e_{1,i}}{\Delta\tau_{i-1}},$$

$$c^+ = e_{0,i+1} - \frac{e_{1,i+1}}{\Delta\tau_i}, \quad c_i^- = 0,$$

where

$$e_{0,i} = 1 - \exp(-\Delta\tau_{i-1}),$$

$$e_{1,i} = \Delta\tau_{i-1} - e_{0,i}.$$

Once again, we shall take proper care of the disc rays and the shell rays (see chapters 7 and 8).

We shall now evaluate $\Delta I_i^\pm$ using ALO-ALI techniques. The integral operator on $\bar{S}(t)$ in equations (10.6.16) and (10.6.17) is a matrix operator acting on the vector $(\bar{S}_1, \ldots, \bar{S}_N)$, where the $\bar{S}_i$s are discretized values of $\bar{S}(t)$ at the spatial grid points r_i. We can now define the $\Lambda^\star$ of ALO which is chosen close to the Λ matrix but retaining only the diagonal terms or sometimes including a few off-diagonal matrix to give better convergence (Hauschildt 1992). We restrict our treatment to $\Lambda^\star$ with diagonal elements only. We need to distinguish between disc and shell rays in a spherical medium. We may write $\Lambda^\star$ as

$$\Lambda^\star = \Lambda^c + \Lambda^s, \tag{10.6.22}$$

where Λ^c and Λ^s represent the Λs for the core and shell rays which are tangent to the radial grid. We set $\bar{S}_i = 1$ and $\bar{S}_{i\neq j} = 0$ for the ith column of the approximate $\Lambda^\star$ intensities at the boundaries equal to zero to obtain the formal solution. This procedure will give us normalized intensities which when integrated over frequencies and characteristic rays will give the required columns of $\Lambda^\star$. This is not absolutely correct as $\bar{S}$ has contributions not only from S but also from $\partial/\partial\lambda$ term. It will, however, only save computer time. Hauschildt (1992) discussed how to include the $\partial/\partial\lambda$ term. The reader is referred to this reference for further discussion.

We consider the core intersecting rays and compute Λ^c. The inward directed normalized intensities are zero until the grid point $i-1$. Therefore

$$\left.\begin{aligned} \bar{I}^-_{i,j}(\nu) &= -b^-_{ij}, \\ \bar{I}^-_{i+1,j}(\nu) &= b^-_{i,j}\exp(\Delta\tau_{i,j}) + a^-_{i+1,j}, \\ \bar{I}^-_{k,j}(\nu) &= \bar{I}^-_{k-1,j}(\nu)\exp(-\Delta\tau_{k-1,j}), \quad k = i+2,\ldots,N. \end{aligned}\right\} \tag{10.6.23}$$

Furthermore, the outward directed normalized intensity remains zero until the grid $(i+1)$, therefore

$$\left.\begin{aligned} \bar{I}^+_{i,j}(\nu) &= b^+_{i,j}, \\ \bar{I}^+_{i-1,j}(\nu) &= b^+_{i,j}\exp(\Delta\tau_{i-1}) + c^+_{i-1,j}, \\ \bar{I}^+_{k,j}(\nu) &= \bar{I}^+_{k+1,j}\exp(-\Delta\tau_{k,j}), \quad k = i-1,\ldots,1. \end{aligned}\right\} \tag{10.6.24}$$

The diagonal elements Λ^c_{ii} are given by

$$\begin{aligned} \Lambda^c_{ii} &= \int_{-\infty}^{+\infty}\phi\,d\nu\left[\sum_j \bar{I}^-_{i,j} + \bar{I}^+_{ij}\right] \\ &= \sum_k\sum_j \omega_{kj}\left(\bar{I}^-_{ij} + \bar{I}^+_{i,j}\right), \end{aligned} \tag{10.6.25}$$

where the $\omega_{k,j}$s are the weights for the angle wavelength quadrature.

The diagonal elements of Λ can be obtained by considering multiple intersections of tangent rays with some shells. The outwardly directed intensity ($\mu > 0$) can be computed as follows. As there is a source $S_i = 1$ at r_i, there are contributions to the normalized outward intensity from the source at r_i from the points r_i with $s > 0$ and $s < 0$. For $i > j$, let

$$\left.\begin{aligned} I^{+(1)}_{i,j}(\nu) &+= \bar{I}^-_{l,j}(\nu), \quad l = i, i+1,\ldots, j+1, \\ \bar{I}^{+(2)}_{i,j}(\nu) &+= \bar{I}^+_{l,j}(\nu), \quad l = j,\ldots,i, \end{aligned}\right\} \tag{10.6.26}$$

with

$$\bar{I}^+_{i,j}(\nu) = \bar{I}^{+(1)}_{i,j}(\nu).$$

The sign $+$ denotes the logical do-loop summation in programming language. The above normalized intensity is the contribution from the source at r_i with $s < 0$. The total normalized outward directed intensity is given by

$$\bar{I}^{+(n)}_{i,j}(\nu) = \bar{I}^{+}_{i,j}(\nu) + \bar{I}^{+(2)}_{i,j}(\nu). \tag{10.6.27}$$

The inwardly directed intensity ($\mu < 0$) can be computed similarly. Thus we can obtain the elements of Λ_{ii}. To improve the accuracy we need to take more core intersecting characteristic rays. This scheme gives fast and accurate results.

Exercises

10.1 Write out the matrix Λ in equation (10.2.4).

10.2 If one used isotropic scattering, what would be the form of Λ in question 10.1.

10.3 What types of quadrature formulae (for frequency and angle) would suit the operator $\Lambda^\star$ in equation (10.2.8). Use the trapezoidal and Gauss (or Labooto) formulae for frequency and quadratures.

10.4 Expand the matrices $\mathbf{A}^\star$, $\mathbf{B}^\star$ and $\mathbf{C}^\star$ in equation (10.2.22).

10.5 Compare the two terms on the LHS of equation (10.2.27) which estimates $\mathcal{E}^{(1)}$.

10.6 If

$$\kappa_{ij}(n) = \frac{h\nu_{ij}}{4\pi} B_{ij} n_i \rho(r) \left[1 - (g_i n_j)/g_j n_i)\right]$$

and

$$S_{ij} = \frac{2h\nu_{ij}^3}{c^2} \left(\frac{n_i g_j}{n_j g_i} - 1 \right)^{-1}$$

(all the symbols have their usual meanings and $\rho(r)$ is the number density of atomic and molecular species) and assuming that no continuum opacities are present, derive $\partial \bar{J}/\partial n_i$ in terms of κ and S.

REFERENCES

Auer, L.H., 1967, *ApJ Lett.*, **150**, 53.

Auer, L.H., 1987, in *Numerical Radiative Transfer*, ed. W. Kalkofen, Cambridge University Press, Cambridge, page 101.

Auer, L.H., Mihalas, D., 1969, *ApJ*, **158**, 641.

Cannon, C.J., 1973a, *JQSRT*, **13**, 627.

Cannon, C.J., 1973b, *ApJ*, **185**, 621.

Cannon, C.J., 1984, in *Methods in Radiative Transfer*, ed. W. Kalkofen, Cambridge University Press, Cambridge, page 157.

Cannon, C.J., Lopert, P.B., Magnan, C., 1975, *A&A*, **42**, 347.

Carlson, M., 1986, *A Computer Program for Solving Multi-level Non-LTE Radiative Transfer Problems in Moving or Static Atmospheres*, Report No. 33, Uppsala Astronomical Observatory.

Cram, L.E., 1977, *A&A*, **56**, 401.

Cram, L.E., Lopert, P.B., 1976, *JQSRT*, **16**, 347.

Hamann, W.-R., 1985a, *A&A*, **145**, 443.

Hamann, W.-R., 1985b, *A&A*, **148**, 364.

Hamann, W.-R., 1986, *A&A*, **160**, 347.

Hamann, W.-R., 1987, in *Numerical Radiative Transfer*, ed. W. Kalkofen, Cambridge University Press, Cambridge, page 35.

Hamann, W.-R., Koesterke, L., Wesselowski, U., 1991, *Stellar Atmospheres Beyond Classical Models*, NATO, ASI. Ser. C. 341, eds. L. Crivellari, I. Hubený, D.G. Hummer, Kluwer Publications, Dordrecht, Holland.

Hauschildt, P.H., 1992, *JQSRT*, **47**, 433.

Hauschildt, P.H., Wehrse, R., 1991, *JQSRT*, **41**, 81.

Heinzel, P., 1995, *A&A*, **299**, 563.

Hillier, D.J., 1990, *A&A*, **231**, 116.

Hempe, K. Schönberg, K., 1986, *A&A*, **160**, 141.

Hubený, I., 1992, *The Atmosphere of Early Type Stars*, eds. V. Herber, C. Jeffrey, Lecture Notes in Physics 401, Springer, Berlin.

Hubený, I., Lanz, T., 1992, *A&A*, **262**, 501.

Kalkofen, W., 1974, *ApJ*, **188**, 105.

Kalkofen, W., 1987, in *Numerical Radiative Transfer*, ed. W. Kalkofen, Cambridge University Press, Cambridge, page 23.

Koter, de, Schmutz, W., Lamers, H.J.G.L.M., 1993, *A&A*, **277**, 561.

Kunasz, P.B., Auer, L.H., 1987, *JQSRT*, **39**, 67.

Mihalas, D., 1978, *Stellar Atmospheres*, 2nd Edition, Freeman, San Francisco.

Mihalas, D., 1980, *ApJ*, **237**, 574.

Ng, K.C., 1974, *J. Comp. Phys.*, **61**, 2680.

Olson, G.L., Auer, L.H., Buchler, J.R., 1986, *JQSRT*, **35**, 43.

Olson, G.L., Kunasz, P.B., 1987, *JQSRT*, **38**, 325.

Pauldrach, A., Herrero, A., 1988, *A&A*, **199**, 262.

Puls, J., 1991, *A&A*, **248**, 581.

Puls, J., Herrero, A., 1988, *A&A*, **204**, 219.

Rutten, R.J., 1999, *Radiative Transfer in Stellar Atmospheres*, Lecture Notes, Utrecht University, WWW edition.

Rybicki, G.B., 1971, *JQSRT*, **11**, 589.

Rybicki, G.B., 1991, in *Stellar Atmospheres Beyond Classical Models*, NATO, ASI. Ser. C, 341, eds. L. Crivellari, I. Hubený, D.G. Hummer, Kluwer Publications, Dordrecht, Holland.

Rybicki, G.B., Hummer, D.G., 1991, *A&A*, **245**, 171.

Scharmer, G.B., 1981, *ApJ*, **249**, 720.

Scharmer, G.B., 1984, in *Numerical Radiative Transfer*, ed. W. Kalkofen, Cambridge University Press, Cambridge, page 173.

Scharmer, G.B., Carlson, M., 1985, *J. Comp. Phys.*, **59**, 56.

Schönberg, K., Hempe, K., 1986, *A&A*, **163**, 151.

Sellmaier, F., Puls, J., Kudritzki, R.P., Gabler, A., Gabler, R., Vouls, S.A., 1993, *A&A*, **273**, 533.

Sen, K.K., Wilson, S.J., 1998, *Radiative Transfer in Moving Media*, Springer, Singapore.

Werner, K., 1986, *A&A*, **161**, 177.

Werner, K., 1987, in *Numerical Radiative Transfer*, ed. W. Kalkofen, Cambridge University Press, Cambridge, page 67.

Werner, K., Husfeld, D., 1985, *A&A*, **148**, 417.

Wu, G. Q., 1992, *Astrophys. Space Sci.*, **189**, 171.

Chapter 11

Polarization

We shall study the state of polarization of the radiation field in this chapter. We need to study the scattering problems exactly since light is generally polarized on scattering. A good example of this is Rayleigh scattering. An initially unpolarized beam of radiation when scattered at an angle Θ to the direction of the incident beam becomes partially plane polarized in the ratio $1 : \cos^2 \Theta$ in the directions perpendicular and parallel to the plane of scattering, which is also the plane of the direction of the incident and scattered of radiation. The diffuse radiation field arising out of the scattering of light in the atmosphere must therefore be partially polarized and we need to formulate the transfer equation correctly and conveniently, so that many important problems such as polarization in stellar (or solar) planetary atmospheres (including that of sunlit sky) are studied correctly.

There are several polarization observations of stars, for example the T Tauri stars which emit linearly and circularly polarized radiation (Bastian 1982, 1985, Nadeau and Bastian 1986). Magalhãs *et al.* (1986) obtained polarimetric observations of the semi-regular variable L_2 Puppis.

Any radiation field is described by four parameters: (i) the intensity, (ii) the degree of polarization, (iii) the plane of polarization and (iv) the ellipticity of the radiation at each point and in any given direction. However, it is very difficult to include such diverse parameters – intensity, a ratio, an angle and a pure number – in any symmetrical way in formulating the transfer equation. Therefore it is a matter of some importance to formulate the parametric representations of polarized light. In a gaseous medium, the radiative transfer equation is formulated by a set of four parameters called the Stokes parameters, which were introduced by Sir George Stokes in 1852. We shall do this following Chandrasekhar (1960).

11.1 Elliptically polarized beam

It is well known that in an elliptically polarized beam the vibrations of the electric (and the magnetic) vector in the plane transverse to the direction of propagation are such that the ratio of the amplitudes and the difference in phases of the components in any two directions at right angles to each other are absolute constants, or

$$\xi_l = \xi_l^{(0)} \sin(\omega t - \epsilon_l) \quad \text{and} \quad \xi_r = \xi_r^{(0)} \sin(\omega t - \epsilon_r), \tag{11.1.1}$$

where ξ_l and ξ_r are the components of the vibrations along the directions l and r at right angles to each other (see figure 11.1), ω is the circular frequency of the vibrations and $\xi_l^{(0)}$, $\xi_r^{(0)}$, ϵ_l and ϵ_r are constants.

Let ξ_l and ξ_r describe the principal axes of the ellipse which are in directions at angles χ and $\chi + \pi/2$ to the direction l, and then the equations representing the vibrations are

$$\xi_\chi = \xi^{(0)} \cos\beta \sin\omega t \quad \text{and} \quad \xi_{\chi+\pi/2} = \xi^{(0)} \sin\beta \cos\omega t, \tag{11.1.2}$$

where β is an angle whose tangent is the ratio of the axes of the ellipse traced by the end points of the electric vector. We have assumed that β lies between 0 and $\pi/2$ and its sign is + or − according to whether the polarization is right handed or left handed. The quantity $\xi^{(0)}$ is a quantity proportional to the mean amplitude of the electric vector and its square is equal to the intensity of the beam. Thus, we have

$$I = \left[\xi^{(0)}\right]^2 = \left[\xi_l^{(0)}\right]^2 + \left[\xi_r^{(0)}\right] = I_l + I_r. \tag{11.1.3}$$

If we have an elliptically polarized beam represented by equation (11.1.1), we can obtain the following relations. From equation (11.1.2) the vibrations in the l- and r-directions are given by:

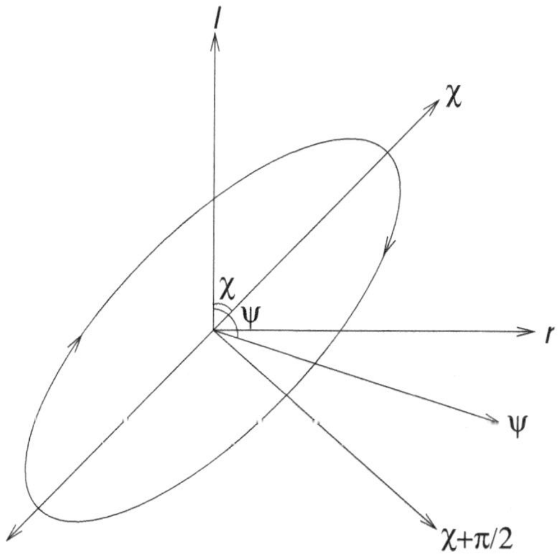

Figure 11.1 Schematic diagram showing the directions l and r.

$$\left.\begin{aligned} \xi_l &= \xi^{(0)}\left(\cos\beta\cos\chi\sin\omega t - \sin\beta\sin\chi\cos\omega t\right) \\ \xi_r &= \xi^{(0)}\left(\cos\beta\sin\chi\sin\omega t + \sin\beta\cos\chi\cos\omega t\right) \end{aligned}\right\} \quad (11.1.4)$$

Equation (11.1.4) can be reduced to the form of equation (11.1.1) by writing

$$\left.\begin{aligned} \xi_l^{(0)} &= \xi^{(0)}\left(\cos^2\beta\cos^2\chi + \sin^2\beta\sin^2\chi\right)^{\frac{1}{2}}, \\ \xi_r^{(0)} &= \xi^{(0)}\left(\cos^2\beta\sin^2\chi + \sin^2\beta\cos^2\chi\right)^{\frac{1}{2}}, \end{aligned}\right\} \quad (11.1.5)$$

$$\tan\epsilon_l = \tan\beta\tan\chi \quad \text{and} \quad \tan\epsilon_r = -\tan\beta\cot\chi. \quad (11.1.6)$$

The intensities I_l and I_r in directions l and r are given by

$$\left.\begin{aligned} I_l &= \left[\xi_l^{(0)}\right]^2 = I\left(\cos^2\beta\cos^2\chi + \sin^2\beta\sin^2\chi\right), \\ I_r &= \left[\xi_r^{(0)}\right]^2 = I\left(\cos^2\beta\sin^2\chi + \sin^2\beta\cos^2\chi\right). \end{aligned}\right\} \quad (11.1.7)$$

From equations (11.1.5) and (11.1.6) we get

$$\begin{aligned} 2\xi_l^{(0)}\xi_r^{(0)}\cos(\epsilon_l - \epsilon_r) &= 2\left[\xi^{(0)}\right]^2\left(\cos^2\beta - \sin^2\beta\right)\cos\chi\sin\chi \\ &= I\cos 2\beta\sin 2\chi. \end{aligned} \quad (11.1.8)$$

Similarly

$$2\xi_l^{(0)}\xi_r^{(0)}\sin(\epsilon_l - \epsilon_r) = I\sin 2\beta. \quad (11.1.9)$$

Whenever an elliptically polarized beam can be represented by equation (11.1.1), we can immediately write that,

$$I = \left[\xi_l^{(0)}\right]^2 + \left[\xi_r^{(0)}\right]^2 = I_l + I_r, \quad (11.1.10)$$

$$Q = \left[\xi_l^{(0)}\right]^2 - \left[\xi_r^{(0)}\right]^2 = I_l - I_r, \quad (11.1.11)$$

$$U = 2\xi_l^{(0)}\xi_r^{(0)}\cos(\epsilon_l - \epsilon_r) = I\cos 2\beta\sin 2\chi = (I_l - I_r)\tan 2\chi \quad (11.1.12)$$

and

$$V = 2\xi_l^{(0)}\xi_r^{(0)}\sin(\epsilon_l - \epsilon_r) = I\sin 2\beta = (I_l - I_r)\tan 2\beta\sec 2\chi. \quad (11.1.13)$$

I, Q, U and V are called the Stokes parameters and together represent an elliptically polarized beam. They are connected by the relations

$$I^2 = Q^2 + U^2 + V^2. \quad (11.1.14)$$

The plane of polarization and the ellipticity are given by the relation

$$\tan 2\chi = \frac{U}{Q} \quad \text{and} \quad \sin 2\beta = \frac{V}{(Q^2 + U^2 + V^2)^{\frac{1}{2}}}. \quad (11.1.15)$$

Equation (11.1.1) corresponds to vibration with constant phases and amplitudes. In practice this does not occur and even in the case of monochromatic light, the amplitude and phases vary incessantly, although they remain constant for a great number of vibrations. The relation obtained above is valid in these situations also. Further treatment of polarized light can be found in Chandrasekhar (1960).

11.2 Rayleigh scattering

In 1871, accounting for the blue colour of the sky, Lord Rayleigh discovered the physical law of the scattering of light that is called Rayleigh scattering. It was recognized by Maxwell and by Rayleigh himself that this law has a much wider scope. One of the most important applications of Rayleigh scattering is Thompson scattering by free electrons.

Rayleigh's law states that when a pencil of natural light of wavelength λ, intensity I and solid angle $d\omega$ is incident on a particle of polarizability α, energy at the rate

$$\frac{128\pi^5}{3\lambda^4}\alpha^2 I\, d\omega \times \frac{3}{4}\left(1+\cos^2\Theta\right)\frac{d\omega'}{4\pi} \tag{11.2.1}$$

is scattered in a direction at an angle Θ to the direction of incidence and in a solid angle $d\omega'$. The scattered light is partially plane polarized: the plane of polarization is at right angles to the plane of scattering which contains the directions of the incident and scattered light; and the intensities of the scattered light in directions (in the transverse plane containing the electric and the magnetic vectors) parallel and perpendicular respectively to the plane of scattering are in the ratio $\cos^2\Theta : 1$.

From equation (11.2.1), we get the scattering coefficient σ per particle as

$$\sigma = \frac{128\pi^5}{3\lambda^4}\alpha_1^2. \tag{11.2.2}$$

The Thompson scattering coefficient for electrons is

$$\sigma_e = \frac{8\pi e^4}{3m_e^2 c^4}, \tag{11.2.3}$$

where we have substituted

$$\alpha_1 = \left(\frac{\lambda}{c}\right)^2 \frac{e^2}{4\pi^2 m_e} \tag{11.2.4}$$

into equation (11.2.2). In the above equations, c is the velocity of light, e is the electronic charge and m_e is the mass of the electron.

Let us consider the incidence of an arbitrarily polarized beam on a particle. Let $\xi_\parallel$ and $\xi_\perp$ be the momentary vibrations of the incident beam resolved along directions parallel and perpendicular respectively to the plane of scattering, then

$$\xi_\parallel = \xi_\parallel^{(0)}\sin(\omega t - \epsilon_1) \quad \text{and} \quad \xi_\perp = \xi_\perp^{(0)}\sin(\omega t - \epsilon_2). \tag{11.2.5}$$

According to Rayleigh's law, the vibration of the light scattered through an angle Θ from the direction of incidence is represented by

$$\left.\begin{aligned} \xi_{\|}^{(s)} &= \left(\frac{3}{2}\sigma\right)^{\frac{1}{2}} \xi_{\|}^{(0)} \cos\Theta \sin(\omega t - \epsilon_1), \\ \xi_{\perp}^{(s)} &= \left(\frac{3}{2}\right)^{\frac{1}{2}} \xi_{\perp}^{(0)} \sin(\omega t - \epsilon_2), \end{aligned}\right\} \tag{11.2.6}$$

where the phase (ϵ_1, ϵ_2) and the amplitude, $(\xi_{\|}^{(0)}, \xi_{\perp}^{(0)})$ relations in the incident beam are maintained unaltered in the scattered beam. Therefore, the parameters representing the scattered light are proportional to (see figure 11.2)

$$\left.\begin{aligned} \frac{3}{2}\sigma\overline{\left[\xi_{\|}^{(0)}\right]^2} \cos^2\Theta &= \frac{3}{2}\sigma I_{\|} \cos^2\Theta, \\ \frac{3}{2}\sigma\overline{\left[\xi_{\perp}^{(0)}\right]^2} &= \frac{3}{2}\sigma I_{\perp}, \\ \frac{3}{2}\sigma\overline{\left[2\xi_{\|}^{(0)}\xi_{\perp}^{(0)} \cos(\epsilon_1 - \epsilon_2)\right]^2} \cos\Theta &= \frac{3}{2}\sigma U \cos\Theta, \\ \frac{3}{2}\sigma\overline{\left[2\xi_{\|}^{(0)}\xi_{\perp}^{(0)} \sin(\epsilon_1 - \epsilon_2)\right]^2} \cos\Theta &= \frac{3}{2}\sigma V \cos\Theta. \end{aligned}\right\} \tag{11.2.7}$$

If we denote the incident light by the vector

$$\mathbf{I} = (I_{\|}, I_{\perp}, U, V), \tag{11.2.8}$$

we can express the scattered intensity in the direction Θ by

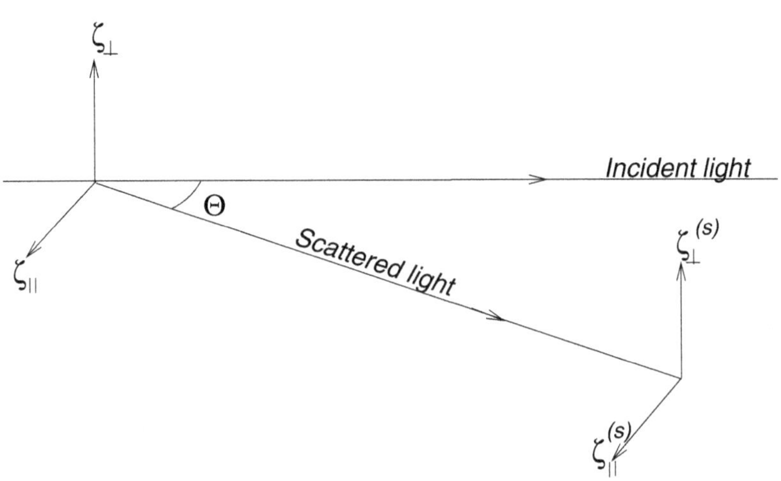

Figure 11.2 Schematic diagram of the coordinate system of the state of polarization of the incident and scattered light after single scattering. The scattering plane contains the directions of the incident and scattered light. The symbols $\|$ and $\perp$ represent directions in the planes (of scattering and incident light) transverse and perpendicular respectively to the plane of scattering.

$$\left(\sigma \frac{d\omega'}{4\pi}\right) \mathbf{RI}\, d\omega, \tag{11.2.9}$$

where

$$\mathbf{R} = \frac{3}{2} \begin{pmatrix} \cos^2\Theta & 0 & 0 & 0 \\ 0 & 1 & 0 & 0 \\ 0 & 0 & \cos^2\Theta & 0 \\ 0 & 0 & 0 & \cos^2\Theta \end{pmatrix}. \tag{11.2.10}$$

In the case of natural light, $I_{\parallel} = I_{\perp} = \frac{1}{2} I$ and $U = V = 0$. $\mathbf{R}$ is the phase matrix.

11.3 Rotation of the axes and Stokes parameters

Stokes parameters are referred to certain fixed rectangular axes. We now consider the transformation law of these parameters when the axes are rotated. The total intensity I and the parameter V are invariant under the rotation of axes, but Q and U change with the axes. Let Q' and U' be the values of Q and U when the axes are rotated through an angle ϕ in the clockwise direction, then

$$Q' = I \cos 2\beta \cos 2(\chi - \phi) \quad \text{and} \quad U' = I \cos 2\beta \sin 2(\chi - \phi), \tag{11.3.1}$$

or

$$Q' = Q \cos 2\phi - U \sin 2\phi \quad \text{and} \quad U' = -Q \sin 2\phi + U \cos 2\phi. \tag{11.3.2}$$

It is more convenient to use intensities (I_l and I_r) in two directions at right angles to each other and the parameters U and V than the original I, Q, U and V set of Stokes.

Now the transformation law of I_l, I_r, U and V for a rotation of the axes can be readily obtained from equations (11.3.2) and the invariance of I and V. Thus, we have

$$I_\phi + I_{\phi+\pi/2} = I_l + I_r, \quad V' = V, \tag{11.3.3}$$

$$I_\phi - I_{\phi+\pi/2} = (I_l - I_r) \cos 2\phi + U \sin 2\phi, \tag{11.3.4}$$

$$U' = -(I_l - I_r) \sin 2\phi + U \cos 2\phi, \tag{11.3.5}$$

or

$$\left.\begin{aligned} I_\phi &= I_l \cos^2\phi + I_r \sin^2\phi + \frac{1}{2} U \sin 2\phi, \\ I_{\phi+\pi/2} &= I_l \sin^2\phi + I_r \cos^2\phi - \frac{1}{2} U \sin 2\phi, \\ U' &= -I_l \sin 2\phi + I_r \sin 2\phi + U \cos 2\phi, \\ V' &= V. \end{aligned}\right\} \tag{11.3.6}$$

Therefore if

$$\mathbf{I} = [I_l, I_r, U, V] \tag{11.3.7}$$

is a vector whose components I_l, I_r, U and V represent an arbitrarily (partially) polarized light, the effect of a rotation of the axes through an angle ϕ in the clockwise direction is to subject $\mathbf{I}$ to the linear transformation:

$$L(\phi) = \begin{pmatrix} \cos^2\phi & \sin^2\phi & \frac{1}{2}\sin 2\phi & 0 \\ \sin^2\phi & \cos^2\phi & -\frac{1}{2}\sin 2\phi & 0 \\ -\sin 2\phi & \sin 2\phi & \cos 2\phi & 0 \\ 0 & 0 & 0 & 1 \end{pmatrix}. \tag{11.3.8}$$

It is clear that $L(\phi)$ satisfies the group relations

$$\mathbf{L}(\phi_1)\mathbf{L}(\phi_1) = \mathbf{L}(\phi_1 + \phi_2) \quad \text{and} \quad \mathbf{L}^{-1}(\phi) = \mathbf{L}(-\phi). \tag{11.3.9}$$

11.4 Transfer equation for $I(\theta, \phi)$

We consider a radiation field in an atmosphere which contains particles, such as atoms, molecules or electrons, which scatter radiation according to Rayleigh's law. We introduce a mass scattering coefficient κ, given by

$$\kappa = \frac{\sigma}{\rho} N, \tag{11.4.1}$$

where N is the number of particles per unit volume and ρ is the density. An element of mass dm scatters radiation according to the expression

$$\left(\kappa\, dm \frac{d\omega'}{4\pi}\right) \mathbf{RI}\, d\omega, \tag{11.4.2}$$

where $\mathbf{I}$ is defined in a rectangular system of coordinates in which the directions parallel and perpendicular to the plane of scattering define the axes. In the case of electrons, κ will be the Thompson scattering coefficient. For molecular scattering, Rayleigh's formula for κ can be obtained from equation (11.2.2) by setting

$$\alpha = \frac{n^2 - 1}{4\pi n}, \tag{11.4.3}$$

where n is the refractive index of the medium. Therefore κ is given by

$$\kappa = \frac{8\pi^3 (n^2 - 1)^2}{2\lambda^4 N \rho}. \tag{11.4.4}$$

We now proceed to formulate the transfer equation for polarized light. The radiation field is characterized at any point in the medium by the four intensities $I_l(\theta, \phi)$, $I_r(\theta, \phi)$, $U(\theta, \phi)$ and $V(\theta, \phi)$, where (θ, ϕ) are the polar angles corresponding to an appropriately chosen coordinate system through the point under consideration (see figure 11.3).

If l and r denote the directions in the meridian plane and at right angles to it respectively, we write

$$\mathbf{I}(\theta, \phi) = [I_l(\theta, \phi), I_r(\theta, \phi), U(\theta, \phi), V(\theta, \phi)], \tag{11.4.5}$$

then the transfer equation can be written as in the vector form

$$-\frac{d\mathbf{I}(\theta, \phi)}{\kappa\rho\, dS} = \mathbf{I}(\theta, \phi) - \mathbf{S}(\theta, \phi), \tag{11.4.6}$$

where $\mathbf{S}(\theta, \phi)$ is the vector source function for $\mathbf{I}(\theta, \phi)$. We need to evaluate the source function $\mathbf{S}(0, \phi)$. We consider the contribution $dS(\theta, \phi; \theta', \phi')$ to the source function due to scattering of a pencil of radiation of solid angle $d\omega'$ in the direction (θ', ϕ'), which is given by

$$\mathbf{RI}\frac{d\omega'}{4\pi}, \tag{11.4.7}$$

provided $I(\theta', \phi')$ is referred to the directions parallel and perpendicular to the plane of scattering. But $I(\theta', \phi')$ is referred to directions along the meridian plane OPZ and at right angles to it. According to equation (11.3.6), we can transform

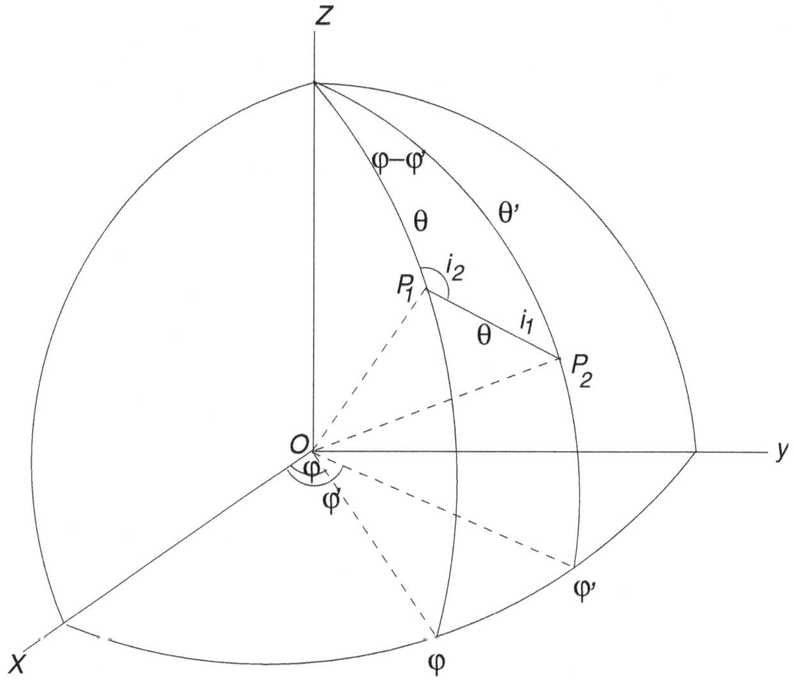

Figure 11.3 Schematic diagram of the transfer of radiation (from Chandrasekhar (1960), with permission).

$I(\theta', \phi')$ to the required directions for using equation (11.4.7) by applying the linear transformation $\mathbf{L}(-i)$ (see equation (11.3.6)) to $I(\theta', \phi')$, where i_1 denotes the angle between the meridian plane OPZ through P_1 whose coordinates are (θ', ϕ') and the plane of scattering OP_1P_2. Therefore the scattering of the pencil in the direction (θ', ϕ') contributes to the source function in the following quantity:

$$\mathbf{R}(\cos\Theta)\mathbf{L}(-i_1)\mathbf{I}(\theta', \phi')\frac{d\omega'}{4\pi}, \tag{11.4.8}$$

which refers to the Stokes parameters to directions at P_2 parallel and perpendicular to the plane of scattering. Now we can apply the linear transformation $L(\pi - i_2)$, where i_2 is the angle between the planes OP_2Z and OP_1P_2, to (11.4.8) to transform to the coordinate axes at P_2 – the directions along the great circle arc P_2Z and the direction at right angles to it. Then we obtain $d\mathbf{S}(\theta, \phi; \theta', \phi')$ as

$$d\mathbf{S}(\theta, \phi; \theta', \phi') = \mathbf{L}(\pi - i_2)\mathbf{R}(\cos\Theta)\mathbf{L}(-i_1)\mathbf{I}(\theta', \phi')\frac{d\omega}{4\pi}. \tag{11.4.9}$$

From equation (11.4.9), we can obtain the source functions by integrating over all directions (θ', ϕ'). Thus,

$$\mathbf{S}(\theta, \phi) = \frac{1}{4\pi}\int_0^{\pi}\int_0^{2\pi} \mathbf{L}(\pi - i_2)\mathbf{R}(\cos\Theta)\mathbf{L}(-i_1)\mathbf{I}(\theta'\phi')\sin\theta'\, d\theta'\, d\phi'. \tag{11.4.10}$$

We can now write the transfer equation (11.4.6) as

$$-\frac{d\mathbf{I}(\theta, \phi)}{\kappa\rho\, ds} = \mathbf{I}(\theta, \phi) - \frac{1}{4\pi}\int_0^{\pi}\int_0^{2\pi} \mathbf{P}(\theta, \phi; \theta', \phi')I(\theta', \phi')\sin\theta'\, d\theta'\, d\phi', \tag{11.4.11}$$

where the phase matrix $\mathbf{P}(\theta, \phi; \theta', \phi')$ is given by

$$\mathbf{P}(\theta, \phi; \theta', \phi') = \mathbf{L}(\pi - i_2)\mathbf{R}(\cos\Theta)\mathbf{L}(-i_1), \tag{11.4.12}$$

or

$$\mathbf{P}(\theta, \phi; \theta', \phi') = \begin{pmatrix} (l,l)^2 & (r,l)^2 & (l,l)(r,l) & 0 \\ (l,r)^2 & (r,r)^2 & (l,r)(r,r) & 0 \\ 2(l,l)(l,r) & 2(r,r)(r,l) & (l,l)(r,r) + (r,l)(l,r) & 0 \\ 0 & 0 & 0 & (l,r)(r,r) - (r,l)(l,r) \end{pmatrix}, \tag{11.4.13}$$

where from the spherical triangle ZP_1P_2 we obtain

$$\left.\begin{aligned}
(l, l) &= \sin\theta \sin\theta' + \cos\theta\cos\theta' \cos(\phi' - \phi),\\
(r, l) &= \cos\theta \sin(\phi' - \phi),\\
(l, r) &= -\cos\theta' \sin(\phi' - \phi),\\
(r, r) &= \cos(\phi' - \phi).
\end{aligned}\right\} \qquad (11.4.14)$$

The elements in the phase matrix (11.4.13) are:

$$\left.\begin{aligned}
(l, l)^2 &= \frac{1}{2}\left[2(1-\mu^2)(1-\mu'^2) + \mu^2\mu'^2\right]\\
&\quad + 2\mu\mu'(1-\mu^2)^{\frac{1}{2}}(1-\mu'^2)^{\frac{1}{2}} \cos(\phi' - \phi)\\
&\quad + \frac{1}{2}\mu^2\mu'^2 \cos 2(\phi' - \phi),\\
(r, l)^2 &= \frac{1}{2}\mu^2\left[1 - \cos 2(\phi' - \phi)\right],\\
(l, r)^2 &= \frac{1}{2}\mu'^2\left[1 - \cos 2(\phi' - \phi)\right],\\
(r, r)^2 &= \frac{1}{2}\left[1 + \cos 2(\phi' - \phi)\right],\\
(l, l)(r, l) &= \mu(1-\mu^2)^{\frac{1}{2}}(1-\mu'^2)^{\frac{1}{2}} \sin(\phi' - \phi)\\
&\quad + \frac{1}{2}\mu^2\mu' \sin 2(\phi' - \phi),\\
(l, l)(l, r) &= -\mu'(1-\mu^2)^{\frac{1}{2}}(1-\mu'^2)^{\frac{1}{2}} \sin(\phi' - \phi)\\
&\quad - \frac{1}{2}\mu\mu'^2 \sin 2(\phi' - \phi),\\
(l, r)(r, r) &= -\frac{1}{2}\mu' \sin 2(\phi' - \phi),\\
(r, l)(r, r) &= \frac{1}{2}\mu \sin 2(\phi' - \phi),\\
(l, l)(r, r) + (r, l)(l, r) &= (1-\mu^2)^{\frac{1}{2}}(1-\mu'^2)^{\frac{1}{2}} \cos(\phi' - \phi)\\
&\quad + \mu\mu' \cos 2(\phi' - \phi),\\
(l, l)(r, r) - (r, l)(l, r) &= \mu\mu'\\
&\quad + (1-\mu^2)^{\frac{1}{2}}(-\mu'^2)^{\frac{1}{2}} \cos(\phi' - \phi),
\end{aligned}\right\} \qquad (11.4.15)$$

where $\mu = \cos\theta$ and $\mu' = \cos\theta'$. Using equations (11.4.15), the phase matrix can be written as

$$\begin{aligned}
P(\mu, \phi; \mu', \phi') = \mathbf{Q}\Big[&\mathbf{P}^{(0)}(\mu, \mu')\\
&+ \left(1-\mu^2\right)^{\frac{1}{2}}\left(1-\mu'^2\right)^{\frac{1}{2}} \mathbf{P}^{(1)}(\mu, \phi; \mu', \phi') + \mathbf{P}^{(2)}(\mu, \phi; \mu', \phi')\Big],
\end{aligned} \qquad (11.4.16)$$

where

$$\mathbf{P}^{(0)}(\mu, \mu') = \frac{3}{4}\begin{pmatrix} 2\left(1-\mu^2\right)\left(1-\mu'^2\right)+\mu^2\mu'^2 & \mu^2 & 0 & 0 \\ \mu'^2 & 1 & 0 & 0 \\ 0 & 0 & 0 & 0 \\ 0 & 0 & 0 & \mu\mu' \end{pmatrix}, \tag{11.4.17}$$

$$\mathbf{P}^{(1)}(\mu, \phi; \mu'\phi') = \frac{3}{4}\begin{pmatrix} 4\mu\mu'\cos(\phi'-\phi) & 0 & 2\mu\sin(\phi'-\phi) & 0 \\ 0 & 0 & 0 & 0 \\ -2\mu'\sin(\phi'-\phi) & 0 & \cos(\phi'-\phi) & 0 \\ 0 & 0 & 0 & \cos(\phi'-\phi) \end{pmatrix}, \tag{11.4.18}$$

$$\mathbf{P}^{(2)}(\mu, \phi; \mu', \phi') = \frac{3}{4}\begin{pmatrix} \mu^2\mu'^2\cos 2(\phi'-\phi) & -\mu^2\cos 2(\phi'-\phi) & \mu^2\mu'\sin 2(\phi'-\phi) & 0 \\ -\mu'^2\cos 2(\phi'-\phi) & \cos 2(\phi'-\phi) & -\mu'\sin 2(\phi'-\phi) & 0 \\ -\mu\mu'^2\sin 2(\phi'-\phi) & \mu\sin 2(\phi'-\phi) & \mu\mu'\cos 2(\phi'-\phi) & 0 \\ 0 & 0 & 0 & 0 \end{pmatrix} \tag{11.4.19}$$

and

$$\mathbf{Q} = \begin{pmatrix} 1 & 0 & 0 & 0 \\ 0 & 1 & 0 & 0 \\ 0 & 0 & 2 & 0 \\ 0 & 0 & 0 & 2 \end{pmatrix}. \tag{11.4.20}$$

We can see that

$$\tilde{\mathbf{P}}^{(i)}(\mu, \phi; \mu', \phi') = \mathbf{P}^{(i)}(\mu, \phi; \mu', \phi') \quad (i = 0, 1, 2), \tag{11.4.21}$$

where $\tilde{\mathbf{P}}^{(i)}$ stands for the matrix $\mathbf{P}^{(i)}$ after its rows and columns have been interchanged and the angle variables (μ, ϕ) have also been interchanged. The symmetry of the phase matrix for transposition is the mathematical reciprocity for single scattering provided due allowance is made for the polarization of scattered light.

Now we can write the transfer equation in plane parallel geometry as

$$\mu\frac{dI(\tau, \mu, \phi)}{d\tau} = \mathbf{I}(\tau, \mu, \phi) - \frac{1}{4\pi}\int_{-1}^{+1}\int_0^{2\pi} \mathbf{P}(\mu, \phi; \mu', \phi')I(\tau, \mu', \phi')\,d\mu'\,d\phi', \tag{11.4.22}$$

where τ is the normal optical depth, measured in terms of the scattering coefficients.

We shall now write the transfer equation for an electron scattering atmosphere. Rayleigh scattering belongs to the class of conservative scattering. Therefore in the case of axially symmetric and semi-infinite plane parallel atmospheres with a constant net flux the total intensity $(I_l + I_r)$ is quite important. This can be applied to the predominantly electron scattering atmospheres of early type stars whose temperatures exceed 15 000 K. Thompson scattering by free electrons agrees with Rayleigh scattering in the prediction of the angular distribution and the state of polarization.

In a plane parallel atmosphere with no incident radiation, the axial symmetry of the radiation field requires that the plane of polarization is along the meridian plane (or at right angles to it). Then $U = V = 0$ and $I_l(\tau, \mu)$ and $I_r(\tau, \mu)$ are enough to describe the radiation field. The transfer equation in this case becomes

$$\mu \frac{d}{d\tau} \begin{pmatrix} I_l(\tau, \mu) \\ I_r(\tau, \mu) \end{pmatrix} = \begin{pmatrix} I_l(\tau, \mu) \\ I_r(\tau, \mu) \end{pmatrix} - \frac{3}{8} \int_{-1}^{+1} \begin{pmatrix} 2(1-\mu^2)(1-\mu'^2) + \mu^2\mu'^2 & \mu^2 \\ \mu'^2 & 1 \end{pmatrix} \begin{pmatrix} I_l(\tau, \mu) \\ I_r(\tau, \mu) \end{pmatrix} d\mu'. \tag{11.4.23}$$

The boundary conditions are

$$\left.\begin{array}{lll} I_l(0, -\mu) = I_r(0, -\mu) = 0 & \tau = 0 \quad \text{and} \quad 0 < \mu \leq 1, \\ I_l(\tau, \mu) = a \quad \text{and} \quad I_r(\tau, \mu) = b & \text{at} \quad \tau \to \infty, \end{array}\right\} \tag{11.4.24}$$

which correspond to the boundary conditions in a stellar atmosphere.

11.5 Polarization under the assumption of axial symmetry

This theory predicts an 11% polarization in an electron scattering atmosphere. The transfer equation was given in the previous section (see equation (11.4.23) and the boundary conditions (11.4.24)). We rewrite equation (11.4.23) for the components $I_l(\tau, \mu)$ and $I_r(\tau, \mu)$ in the directions parallel and perpendicular respectively to the meridian plane containing the directions μ, thus

$$\mu \frac{dI_l}{d\tau} = I_l - \frac{3}{8}\left[2\int_{-1}^{+1} I_l(\tau, \mu')(1-\mu'^2)\, d\mu' + \mu^2 \int_{-1}^{+1} I_l(\tau, \mu')(3\mu'^2 - 2)\, d\mu' + \mu^2 \int_{-1}^{+1} I_r(\tau, \mu')(\tau, \mu')\, d\mu'\right] \tag{11.5.1}$$

and

$$\mu \frac{dI_r}{d\tau} = I_r - \frac{3}{8}\left[\int_{-1}^{+1} I_l(\tau, \mu')\mu'^2 d\mu' + \int_{-1}^{+1} I_r(\tau, \mu')\, d\mu'\right]. \tag{11.5.2}$$

We replace the integrals by the Gauss sums and rewrite equations (11.5.1) and (11.5.2) as

$$\mu_i \frac{dI_{l,i}}{d\tau} = I_{l,i} - \frac{3}{8}\Big\{2\sum a_j(1-\mu_j^2)I_{l,j} + \mu_i^2\Big[\sum a_j\left(3\mu_j^2 - 2\right)I_{l,j} + \sum a_j I_{r,j}\Big]\Big\}$$
$$(i = \pm 1, \ldots, \pm n) \tag{11.5.3}$$

and

$$\mu_i \frac{dI_{r,i}}{d\tau} = I_{r,i} - \frac{3}{8}\left(\sum a_j I_{r,j} + \sum a_j \mu_j^2 I_{l,j}\right) \quad (i = \pm 1, \ldots, \pm n), \tag{11.5.4}$$

where $I_l(\tau, \mu)$ and $I_r(\tau, \mu)$ are replaced by $I_{l,i}$ and $I_{r,i}$ respectively and the rest of the symbols have their usual meanings. Chandrasekhar (1946) gave a general solution (see chapter 3 for the method) for the discrete equations (11.5.3) and (11.5.4) as follows:

$$I_{l,i} = b\Bigg\{\tau + \mu_i + Q + \left(1 - \mu_i^2\right)\sum_{\beta=1}^{n-1}\left[\frac{L_\beta \exp(-\kappa_\beta \tau)}{1+\mu_i\kappa_\beta} + \frac{L_{-\beta}\exp(\kappa_\beta\tau)}{1-\mu_i\kappa_\beta}\right] + \sum_{\alpha=1}^{n} M_\alpha\,(1 - k_\alpha\mu_i)\exp(-k_\alpha\tau) + \sum_{\alpha=1}^{n-1} M_{-\alpha}\,(1 + k_\alpha\mu_i)\exp(+k_\alpha\tau)\Bigg\},$$
$$(i = \pm 1, \ldots, \pm n) \tag{11.5.5}$$

and

$$I_{r,i} = b\Bigg\{\tau + \mu_i + Q - \sum_{\alpha=1}^{n} M_\alpha \frac{k_\alpha^2 - 1}{1 + \mu_i k_\alpha}\exp(+k_\alpha\tau) - \sum_{\alpha=1}^{n} M_{-\alpha}\frac{k_\alpha^2 - 1}{1 - \mu_i\mu_\alpha}\exp(+k_\alpha\tau)\Bigg\}$$
$$(i = \pm 1, \ldots, \pm n), \tag{11.5.6}$$

where k is a root of

$$\sum_{j=1}^{n} \frac{a_j(1-\mu_j^2)}{1-\mu_j^2 k^2} = \frac{4}{3} \tag{11.5.7}$$

and κ^2 is a root of

$$\sum_{j=1}^{n} \frac{a_j(1-\mu_j^2)}{1-\mu_j^2 k^2} = \frac{2}{3}. \tag{11.5.8}$$

Equation (11.5.5) is of the order n in k^2 and admits $2n$ distinct non-vanishing roots for k which occur in pairs as

$$\pm k_\alpha \quad (\alpha = 1, \ldots, n). \tag{11.5.9}$$

Furthermore, equation (11.5.6) is of the order $2n$ in k^2 and admits only $(2n-1)$ distinct non-vanishing roots for k since $k^2 = 0$ is a root. These $2n-2$ roots also occur in pairs which are denoted by

$$\pm \kappa_\beta \quad (\beta = 1, \ldots, n-1). \tag{11.5.10}$$

In equations (11.5.3) and (11.5.4) $L_{\pm\beta}$ $(\beta = 1, \ldots, n-1)$, $M_{\pm\alpha}$ $(\alpha = 1, \ldots, n)$, b and Q are the $4n$ constants of integration.

We need to impose certain boundary conditions on the solutions (11.5.3) and (11.5.4): none of the I_is should tend to infinity exponentially as $\tau \to \infty$ and there should be no radiation incident at $\tau = 0$. The first condition requires that we can omit terms containing $\exp(+\kappa_\beta \tau)$ and $\exp(k_\alpha \tau)$. We are then left with

$$I_{l,i} = b\left\{\tau + \mu_i + Q + \left(1-\mu_i^2\right)\sum_{\beta=1}^{n-1} \frac{L_\beta \exp(-\kappa_\beta \tau)}{1+\mu_i \kappa_\beta} + \sum_{\alpha=1}^{n} M_\alpha (1 - k_\alpha \mu_i) \exp(-k_\alpha \tau)\right\}$$
$$(i = \pm 1, \ldots, \pm n) \tag{11.5.11}$$

and

$$I_{r,i} = b\left\{\tau + \mu_i + a \sum n_{\alpha=1} \frac{M_\alpha \left(k_\alpha^2 - 1\right)}{1+\mu_i k_\alpha} \exp(-k_\alpha \tau)\right\}$$
$$(1 = \pm 1, \ldots, \pm n). \tag{11.5.12}$$

After some simplification, solutions (11.5.3) and (11.5.4) are in the second and third approximations as follows (see Chandrasekhar (1946) or (1960)):

Second Approximation. The emergent intensities in the two states of polarization are given by,

$$I_l(0,\mu) = \frac{3}{8} F\left\{\mu + 0.696\,38 - (1-\mu^2)\frac{0.192\,65}{1+1.5275\mu} + 0.0218\,30(1 - 2.236\,07\mu) - 0.029\,516(1 - 1.080\,12\mu)\right\}, \tag{11.5.13}$$

$$I_r(0,\mu) = \frac{3}{8}F\left\{\mu + 0.696\,38 - \frac{0.087\,321\,5}{1+2.236\,07\mu} + \frac{0.004\,919\,3}{1+1.080\,12\mu}\right\}, \tag{11.5.14}$$

where F is the constant net flux.

Third approximation. The emergent intensities in the two states of polarization are given by,

$$\begin{aligned} I_l(0,\mu) = \frac{3}{8}F\Big\{\mu &+ 0.705\,927 \\ &- (1-\mu^2)\left[\frac{0.140\,264\,6}{1+2.718\,38\mu} + \frac{0.067\,916\,96}{1+1.118\,216\mu}\right] \\ &+ 0.007\,183\,92\,(1-3.458\,589\mu) \\ &+ 0.018\,612\,55\,(1-1.327\,570\mu) \\ &- 0.032\,866\,4(1-1.046\,766\mu)\Big\}, \end{aligned} \tag{11.5.15}$$

$$\begin{aligned} I_r(0,\mu) = \frac{3}{8}F\Big\{\mu &+ 0.705\,927 - \frac{0.078\,749\,0}{1+3.458\,589\mu} - \frac{0.014\,190\,99}{1+1.325\,70\mu} \\ &+ \frac{0.003\,145\,93}{1+1.046\,765\,9\mu}\Big\}. \end{aligned} \tag{11.5.16}$$

The degree of polarization is

$$\delta = \frac{I_r(0,\mu) - I_l(0,\mu)}{I_r(0,\mu) + I_l(0,\mu)}. \tag{11.5.17}$$

See figure 11.4 and table 11.1 for the laws of darkening in the two states of polarization.

11.6 Polarization in spherically symmetric media

The problem discussed in the previous section has been studied by Sobolev (1963) and Siewart and Fraley (1967). Smith and Siewart (1967) analysed this problem of polarization using the singular eigensolution method developed by Case (1960), which is well known in neutron transport theory (Case and Zweifel 1961) but has not been applied frequently in radiative transfer problems. All the solutions that have been produced use the tabulated Chandrasekhar H-functions in uniform media (for example see Bond and Siewart (1967)).

Grant and Hunt (1968) applied discrete space theory to obtain the polarized components of the intensity at different internal points in an inhomogeneous plane parallel medium (see chapter 6 on discrete space theory). They solved the transfer equation for I_l and I_r with the Rayleigh phase function in plane parallel layers given by

$$\mu \frac{d\mathbf{I}(\tau, \mu)}{d\tau} + \mathbf{I}(\tau, \mu) = [1 - \varpi(\tau)]\,\mathbf{B}(\tau) + \frac{\varpi(\tau)}{2} \int_{-1}^{+1} \mathbf{P}(\tau, \mu, \mu')\mathbf{I}(\tau, \mu')\, d\mu' \tag{11.6.1}$$

and

$$-\mu \frac{d\mathbf{I}(\tau, -\mu)}{d\tau} + \mathbf{I}(\tau, -\mu) = [1 - \varpi(\tau)]\,\mathbf{B}(\tau) + \frac{\varpi(\tau)}{2} \int_{-1}^{+1} \mathbf{P}(\tau, -\mu, \mu')\mathbf{I}(\tau, \mu')\, d\mu', \tag{11.6.2}$$

where

$$\mathbf{I}(\tau, \mu) = \begin{pmatrix} I_l(\tau, \mu) \\ I_r(\tau, \mu) \end{pmatrix}, \quad \mathbf{B}(\tau) = \begin{pmatrix} B_l(\tau) \\ B_r(\tau) \end{pmatrix}, \tag{11.6.3}$$

the phase matrix $\mathbf{P}$ is given by

Table 11.1 The laws of darkening in the two states of polarization given by the third approximation: the degree of polarization of the emergent radiation and a comparison of the total intensities, given by the theory ignoring the polarization of the existing field of radiation but incorporating Rayleigh phase function, are shown. Notice that there is 11% polarization at the limb which is observed in preliminary observations by Hiltner (1947).

μ	$\frac{I_l(0,\mu)}{F}$	$\frac{I_r(0,\mu)}{F}$	$\frac{I_l(0,\mu)}{I_l(0,1)}$	$\frac{I_r(0,\mu)}{I_r(0,1)}$	$\frac{I_r(0,\mu)}{I_l(0,\mu)}$	$\delta(\mu)$	$\frac{I_l(0,\mu)+I_r(0,\mu)}{F}$	$\frac{I(0,\mu)}{F}$
0.0	0.1840	0.2310	0.2914	0.3659	1.2557	0.1134	0.4151	0.4195
0.1	0.2354	0.2767	0.3728	0.4382	1.1753	0.0806	0.5120	0.5175
0.2	0.2832	0.3190	0.4486	0.5053	1.1264	0.0594	0.6023	0.6076
0.3	0.3291	0.3598	0.5213	0.5699	1.0932	0.0445	0.6890	0.6937
0.4	0.3738	0.3997	0.5921	0.6330	1.0691	0.0334	0.7735	0.7773
0.5	0.4178	0.4390	0.6616	0.6953	1.0508	0.0248	0.8567	0.8593
0.6	0.4611	0.4779	0.7303	0.7569	1.0364	0.0179	0.9390	0.9402
0.7	0.5041	0.5165	0.7983	0.8181	1.0247	0.0122	1.0206	1.0203
0.8	0.5467	0.5549	0.8659	0.8789	1.0150	0.0075	1.1017	1.0998
0.9	0.5891	0.5932	0.9331	0.9396	1.0069	0.0034	1.1824	1.1789
1.0	0.6314	0.6314	1.0000	1.0000	1.0000	0.0000	1.2628	1.2576

$$\mathbf{P}(\tau, \mu, \mu') = \frac{3}{4} \begin{pmatrix} 2(1-\mu^2)(1-\mu'^2) + \mu^2\mu'^2 & \mu^2 \\ \mu'^2 & 1 \end{pmatrix} \tag{11.6.4}$$

and the Bs are the Planck functions such that

$$B_l(\tau), \quad B_r(\tau) \geq 0 \quad \text{and} \quad 0 \leq \varpi(\tau) \leq 1, \tag{11.6.5}$$

where $\varpi(\tau)$ is the albedo for single scattering.

The boundary conditions are that $\mathbf{I}(0, \mu)$ and $\mathbf{I}(\tau_{max}, \mu)$ $(0 < \mu \leq 1)$ are specified. The degree of polarization is calculated by the formula

$$P^+ = \frac{I_r^+(\tau_n) - I_l^+(\tau_n)}{I_r^+(\tau_n) + I_l^+(\tau_n)}, \tag{11.6.6}$$

where the $+$ sign corresponds to the intensities $\mathbf{I}(\tau_n, +\mu)$ and n represent the nth layer of the medium which is divided into several plane parallel layers. P^- is defined similarly. Grant and Hunt (1968) solved equations (11.6.1) and (11.6.2) and obtained the polarization given in figure 11.5.

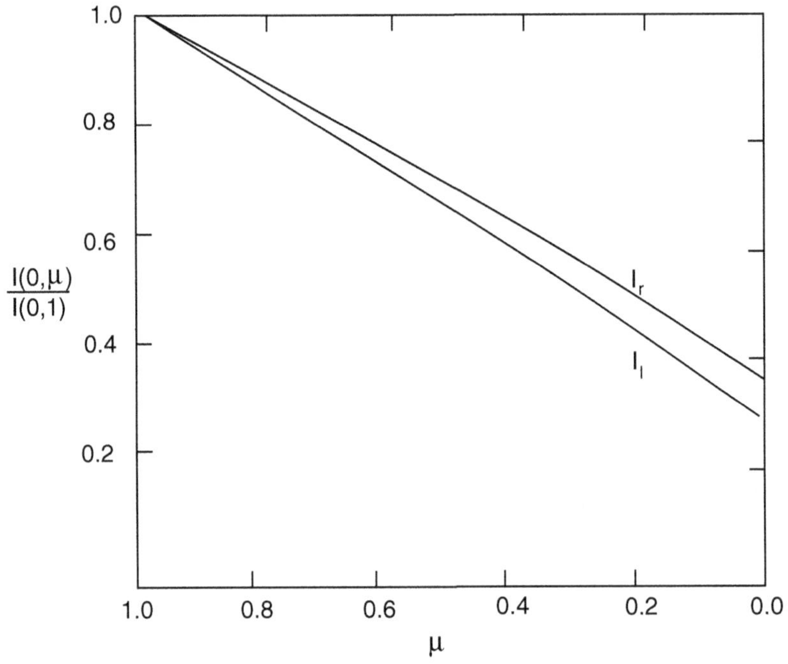

Figure 11.4 The laws of darkening in the two states of polarization for an electron scattering atmosphere. l refers to the component polarized with the electric vector in the meridian plane and r refers to the component with the electric vector at right angles to the meridian plane (from Chandrasekhar (1960), figure 23, page 247, with permission).

Peraiah (1975) computed the polarization in spherically symmetric atmospheres using the Rayleigh phase function. The transfer equation in spherically symmetric geometry is written in the divergence form, which is

$$\frac{\mu}{r^2}\frac{\partial}{\partial r}\left[r^2 I(r,\mu)\right] + \frac{1}{r}\frac{\partial}{\partial\mu}\left[(1-\mu^2)I(r,\mu)\right]$$
$$= \sigma(r)I(r,\mu) - \sigma(r)\left\{[1-\varpi(r)]b(r) + \frac{\varpi}{2}\int_{-1}^{+1} P(r,\mu,\mu')I(r,\mu')\,d\mu'\right\}, \tag{11.6.7}$$

where all symbols have their usual meanings, $\sigma(r)$ is the absorption coefficient and $b(r)$ is the source inside the medium at the radial point r. Let us write that

$$U(r,\mu) = r^2 I(r,\mu). \tag{11.6.8}$$

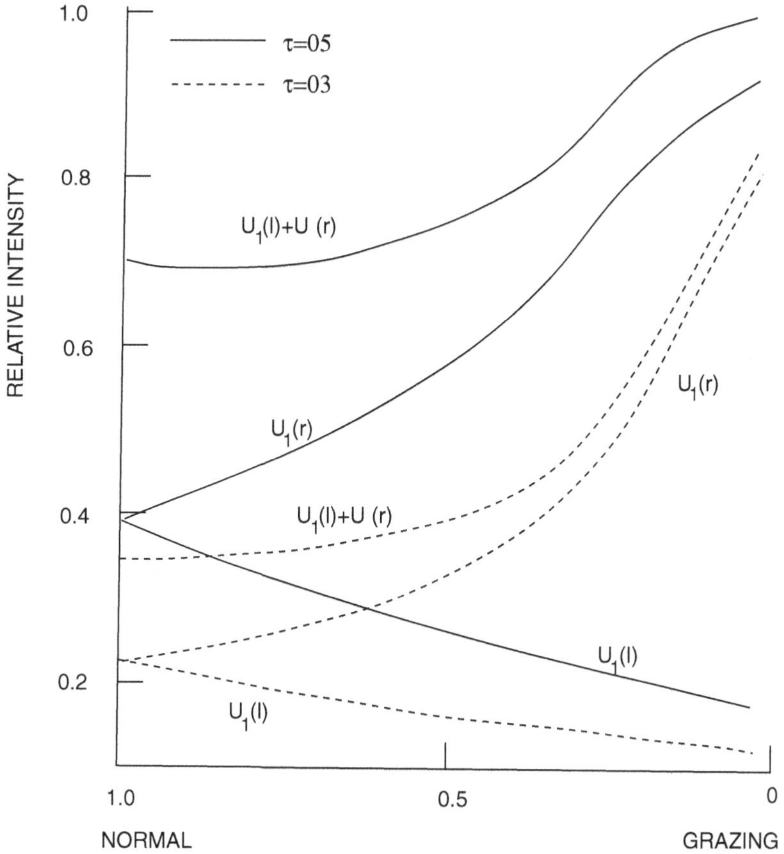

Figure 11.5 The angular distribution of the relative intensity of the total radiation $I_1^-(l) + I_1^-(r)$ and of the intensity components $I_1^-(r)$, $I_1^-(l)$ of the radiation reflected by slabs of optical thickness $\tau = 0.3, 0.5$. The atmospheres are illuminated on their upper surfaces by normally incident beams. $I^-(r) = I_r(-\mu)$ etc. (from Grant and Hunt (1968), with permission).

We set $\varpi(r) = 1$ as we are considering conservative scattering. Then equation (11.6.7) can be written as

$$\mu\frac{\partial U(r,\mu)}{\partial r} + \frac{1}{r}\frac{\partial}{\partial\mu}\left[(1-\mu^2)U(r,\mu)\right] + \sigma(r)U(r,\mu)$$
$$= \frac{\sigma(r)}{2}\int_{-1}^{+1} P(r,\mu,\mu')U(r,\mu')\,d\mu', \quad \mu\in(0,1) \tag{11.6.9}$$

and

$$-\mu\frac{\partial U(r,-\mu)}{\partial r} + \frac{1}{r}\frac{\partial}{\partial\mu}\left[(-\mu^2)U(r,-\mu)\right] + \sigma(r)U(r,-\mu)$$
$$= \frac{\sigma(r)}{2}\int_{-1}^{+1} P(r,-\mu,\mu')U(r,\mu')\,d\mu', \quad \mu\in(0,1) \tag{11.6.10}$$

for the oppositely directed beam.

The radiation field is represented by the two perpendicularly polarized intensity beams $U_L(r,\mu)$ and $U_R(r,\mu)$. Thus

$$U(r,\mu) = \begin{pmatrix} U_L(r,\mu) \\ U_R(r,\mu) \end{pmatrix}, \tag{11.6.11}$$

where U_L and U_R refer respectively to the states of polarization in which the electric vector vibrates along and perpendicular to the principal medium. The phase function is

$$P(r,\mu,\mu') = \frac{3}{4}\begin{pmatrix} 2(1-\mu^2)(1-\mu'^2) & \mu^2 \\ \mu'^2 & 1 \end{pmatrix}$$
$$= \begin{pmatrix} P_{11}(\mu,\mu') & P_{12}(\mu,\mu') \\ P_{21}(\mu,\mu') & P_{22}(\mu,\mu') \end{pmatrix}. \tag{11.6.12}$$

The transfer equation can now be written for each component, U_L and U_R, as

$$\mu\frac{\partial U_L(r,\mu)}{\partial r} + \frac{1}{r}\frac{\partial}{\partial\mu}\left[(1-\mu^2)U_L(r,\mu)\right] + \sigma(r)U_L(r,\mu)$$
$$= \frac{\sigma(r)}{2}\int_{-1}^{+1} \left[P_{11}(\mu,\mu')U_L(r,\mu') + P_{12}(\mu,\mu')U_R(r,\mu')\right]d\mu' \tag{11.6.13}$$

and

$$-\mu\frac{\partial U_L(r,-\mu)}{\partial r} - \frac{1}{r}\frac{\partial}{\partial\mu}\left[(1-\mu^2)U_L(r,-\mu)\right] + \sigma(r)U_L(r,-\mu)$$
$$= \frac{\sigma(r)}{2}\int_{-1}^{+1} \left[P_{11}(-\mu,\mu')U_L(r,\mu') + P_{12}(-\mu,\mu')U_R(r,\mu')\right]d\mu', \tag{11.6.14}$$

with similar equations for $U_R(r,\mu)$.

The discrete representation of equations (11.6.13) and (11.6.14) and the two similar equations for $U_R(r, \mu)$, following the procedure given in chapter 6, are written as

$$M^\star \left(\mathbf{U}^+_{n+1} - \mathbf{U}^+_n\right) + \rho_c \left(\mathbf{\Lambda}^\star_+ \mathbf{U}^+_{n+\frac{1}{2}} + \mathbf{\Lambda}^\star_- \mathbf{U}^-_{n+\frac{1}{2}}\right) + \tau_{n+\frac{1}{2}} \mathbf{U}^+_{n+\frac{1}{2}}$$
$$= \frac{1}{2}\tau_{n+\frac{1}{2}} \left(\mathbf{P}^{++}_{n+\frac{1}{2}} \mathbf{C}^\star \mathbf{U}^+_{n+\frac{1}{2}} + P^{+-}_{n+\frac{1}{2}} \mathbf{C}^\star \mathbf{U}^-_{n+\frac{1}{2}}\right) \tag{11.6.15}$$

and

$$M^\star \left(\mathbf{U}^+_n - \mathbf{U}^+_{n+1}\right) - \rho_c \left(\mathbf{\Lambda}^\star_+ \mathbf{U}^-_{n+\frac{1}{2}} + \mathbf{\Lambda}^\star_- \mathbf{U}^+_{n+\frac{1}{2}}\right) + \tau_{n+\frac{1}{2}} \mathbf{U}^-_{n+\frac{1}{2}}$$
$$= \frac{1}{2}\tau_{n+\frac{1}{2}} \left(\mathbf{P}^{-+}_{n+\frac{1}{2}} \mathbf{C}^\star \mathbf{U}^+_{n+\frac{1}{2}} + P^{--}_{n+\frac{1}{2}} \mathbf{C}^\star \mathbf{U}^-_{n+\frac{1}{2}}\right), \tag{11.6.16}$$

where

$$\mathbf{U}^\pm_n = \begin{pmatrix} \mathbf{U}^\pm(L) \\ \mathbf{U}^\pm(R) \end{pmatrix}, \quad \mathbf{U}^+_n(L) = \begin{pmatrix} U_{n,1}(L) \\ U_{n,2}(L) \\ \vdots \\ U_{n,m} \end{pmatrix}, \quad \mathbf{U}^-_n(L) = \begin{pmatrix} U_{n,-1}(L) \\ U_{n,-2}(L) \\ \vdots \\ U_{n,-m} \end{pmatrix}, \tag{11.6.17}$$

$$U_{n,\pm j}(L) = U_L(r_n, \pm\mu_j), \quad j = 1, 2, \ldots, m. \tag{11.6.18}$$

The vector $\mathbf{U}^\pm(R)$ is defined similarly and

$$\mathbf{M}^\star = \begin{pmatrix} \mathbf{M} & 0 \\ 0 & \mathbf{M} \end{pmatrix}, \quad \mathbf{M} = \left[\mu_j \delta_{jk}\right], \quad C^\star = \begin{pmatrix} \mathbf{C} & 0 \\ 0 & \mathbf{C} \end{pmatrix}, \quad \mathbf{C} = \left[c_j \delta_{jk}\right],$$
$$j = 1, 2, \ldots, m, \tag{11.6.19}$$

where the μs and Cs are the roots and weights of suitable quadrature formulae. Furthermore

$$\left.\begin{aligned} P^{++}_{\alpha\beta,n+\frac{1}{2}} &= \mathbf{P}_{\alpha\beta}(+\mu_j, +\mu_k) = \mathbf{P}_{\alpha\beta,n+\frac{1}{2}} \\ P^{+-}_{\alpha\beta,n+\frac{1}{2}} &= \mathbf{P}_{\alpha\beta}(+\mu_j, -\mu_k) = \mathbf{P}_{\alpha\beta,n+\frac{1}{2}} \quad \alpha, \beta = 1, 2, \quad \mu_j, \mu_k > 0 \end{aligned}\right\} \tag{11.6.20}$$

and

$$\mathbf{P}^{++}_{n+\frac{1}{2}} = \begin{pmatrix} \mathbf{P}^{++}_{11} & \mathbf{P}^{++}_{12} \\ \mathbf{P}^{++}_{21} & \mathbf{P}^{++}_{22} \end{pmatrix}, \tag{11.6.21}$$

with similar expressions for $\mathbf{P}^{+-}_{n+\frac{1}{2}}$, $\mathbf{P}^{-+}_{n+\frac{1}{2}}$, and $\mathbf{P}^{--}_{n+\frac{1}{2}}$. The quantities with the index $n+\frac{1}{2}$ represent the average of the 'cell' bounded by the radii r_n and r_{n+1} (see chapter 6). The optical depth $\tau_{n+\frac{1}{2}}$ is calculated using the formula

$$\tau_{n+\frac{1}{2}} = \int_{r_{n+1}}^{r_n} \sigma(r)\, dr = \sigma_{n+\frac{1}{2}} (r_{n+1} - r_n). \tag{11.6.22}$$

The curvature factor ρ_c is given by

$$\rho_c = \frac{\Delta r}{\bar{r}}, \tag{11.6.23}$$

where Δr is the geometrical thickness of the 'cell' and $\bar{r}$ is the mean radius of the 'cell'. The quantity $\mathbf{\Lambda}^{\star}$ is given by

$$\mathbf{\Lambda}^{\star}_{\pm} = \begin{pmatrix} \mathbf{\Lambda}_{\pm} & 0 \\ 0 & \mathbf{\Lambda}_{\pm} \end{pmatrix}, \tag{11.6.24}$$

where the $\mathbf{\Lambda}$s are the curvature matrices (see chapter 6). The average intensities $U^{\pm}_{n+\frac{1}{2}}$ over the 'cell' are expressed as a weighted mean of the interface intensities. Thus

$$\left(\mathbf{E} - \mathbf{X}^{+}_{n+\frac{1}{2}}\right) \mathbf{U}^{+}_{n} + \mathbf{X}^{+}_{n+\frac{1}{2}} \mathbf{U}^{+}_{n+1} = \mathbf{U}^{+}_{n+\frac{1}{2}}, \tag{11.6.25}$$

$$\left(\mathbf{E} - \mathbf{X}^{-}_{n+\frac{1}{2}}\right) \mathbf{U}^{-}_{n+1} + \mathbf{X}^{-}_{n+\frac{1}{2}} \mathbf{U}^{-}_{n} = \mathbf{U}^{-}_{n+\frac{1}{2}}, \tag{11.6.26}$$

where $\mathbf{E}$ is the unit matrix and $\mathbf{X}^{+}_{n+\frac{1}{2}}$ are $2m \times 2m$ diagonal matrices with the structure

$$\mathbf{X}^{\pm}_{n+\frac{1}{2}} = \begin{pmatrix} \mathbf{X}_{n+\frac{1}{2}}(L) & \\ & \mathbf{X}_{n+\frac{1}{2}}(R) \end{pmatrix}, \tag{11.6.27}$$

where $\mathbf{X}_{n+\frac{1}{2}}$ are diagonal $m \times m$ matrices. Generally we choose $\mathbf{X}_{n+\frac{1}{2}} = \frac{1}{2}\mathbf{E}$ for the diamond scheme.

We can now calculate the transmission and reflection matrices by following the procedure given in chapter 6. For flux conservation we must have

$$\|T(n+1, n) + R(n+1, n)\| = 1, \tag{11.6.28}$$

$$\|T(n, n+1) + R(n, n+1)\| = 1 \tag{11.6.29}$$

and

$$\sum_{j=1}^{m} c_j \left(\mathbf{\Lambda}^{+}_{jk} - \mathbf{\Lambda}^{-}_{jk}\right) = 0 \quad \text{for all } k. \tag{11.6.30}$$

This means that

$$\frac{1}{2} \sum_{j=1}^{m} c_j \big[P_{11}(\mu_j, \mu_k) + P_{21}(\mu_j, \mu_k) + P_{11}(-\mu_j, \mu_k) + P_{21}(-\mu_j, \mu_k)\big] = 1 \quad \text{for all } k \tag{11.6.31}$$

and

$$\frac{1}{2}\sum_{j=1}^{m} C_j\big[P_{21}(\mu_j,\mu_k)+P_{22}(\mu_j,\mu_k)+P_{12}(-\mu_j,\mu_k)$$
$$+P_{22}(-\mu_j,\mu_k)\big]=1 \quad \text{for all } k, \tag{11.6.32}$$

with two similar equations for $P^{--}_{n+\frac{1}{2}}$ and $P^{+-}_{n+\frac{1}{2}}$.

Equations (11.6.31) and (11.6.32) should be satisfied exactly. If they are not satisfied exactly, then renormalization is necessary (Plass *et al.* 1973). We have chosen the roots and weights of the Gauss quadrature formula for the μs and Cs respectively. This problem of polarization is solved by using discrete space theory (chapter 6).

Calculations have been performed for spherical shells whose thicknesses are 1.5 and 5 times their respective stellar radii and whose optical depths are each 10. Fifty discrete points have been chosen along the radial direction (the spherical shell is divided into 50 smaller shells, $N = 50$). The step size $\Delta\tau$ is calculated by using the formula

$$\Delta\tau < \Delta\tau_{crit} = \min_j \left| \frac{\mu_j \pm \frac{1}{2}\rho\Lambda^{+}_{jj}}{\frac{1}{2}(1-\varpi)\,P^{++}_{jj}C_j} \right|, \tag{11.6.33}$$

where $\Delta\tau$ is the optical depth of the 'cell'. This will ensure the positivity of the r and t operators and the stability of the numerical method. If in equation (11.6.33), the quantity $\Delta\tau$ ('cell') is smaller than the optical depth of the shell (which is the one obtained by dividing the spherical atmosphere into N shells), then we need to subdivide the shell until we get a shell whose optical thickness is less than or equal to $\Delta\tau$ ('cell') given in equation (11.6.33). The r and t operators of the composite 'cell' are obtained by using the 'star' algorithm given in chapter 6. The curvature factor of the outermost shell is

$$\rho_{out} = \frac{B-A}{N\cdot B}, \tag{11.6.34}$$

where B and A are the outermost and innermost radii of the spherical atmosphere. The curvature factor of any shell n inside the spherical medium can be calculated in terms of ρ_{out} as follows:

$$\rho_n = \frac{\rho_{out}}{1-(n-1)\rho_{out}}. \tag{11.6.35}$$

The internal field is computed using the boundary conditions given by

1. at $\tau = 0$ $(n = 1)$

$$\left.\begin{aligned} U_1^{+}(L) &= 0 \quad \text{for all } \mu\text{s},\\ U_1^{+}(R) &= 0 \quad \text{for all } \mu\text{s}; \end{aligned}\right\} \tag{11.6.36}$$

2. and an unpolarized radiation incident at $\tau = T$ $(n = N + 1)$

$$\left.\begin{aligned} U_1^+(L) &= 1 \quad \text{for all } \mu\text{s}, \\ U_1^+(R) &= 1 \quad \text{for all } \mu\text{s}. \end{aligned}\right\} \tag{11.6.37}$$

This gives a flux of $\sum_{j=1}^{m} U_{N+1}(L \text{ or } R) = 1$. The degree of polarization is computed, for any shell n, using the formula

$$P_n^{\pm} = \frac{U_n^{\pm}(R) - U_n^{\pm}(L)}{U_n^{\pm}(R) + U_n^{\pm}(L)}. \tag{11.6.38}$$

The results of polarization are shown in figures 11.6, 11.7 and 11.8 below. This theory has been applied to the distorted (due to self radiation and the tidal influence of the secondary component) components of close binary systems (Peraiah 1976), assuming the following model.

An electron (or molecular) scattering atmosphere is assumed. If N_0 is the density at A, then the density $N(r)$ is taken as

$$N(r) = N_0 \left(\frac{A}{r}\right)^2. \tag{11.6.39}$$

Therefore, the total radial optical depth $T(r)$ is given by

$$T(r) = \sigma \int_A^r N(r')\,dr', \tag{11.6.40}$$

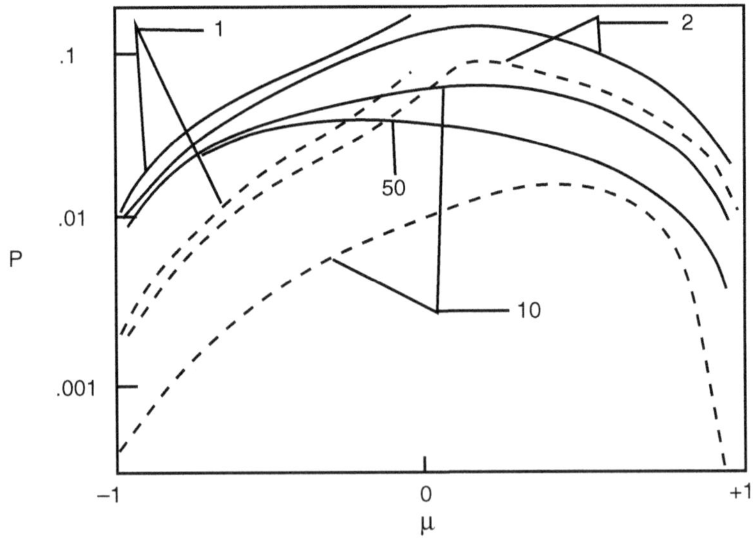

Figure 11.6 The angular distribution of polarization along the radius vector $n = 1, \ldots, 50$ for $\tau = 10$ is shown. Continuous curves represent the spherical case with $B/A = 5$ and the dashed curves represent the plane parallel case with $B/A = 1$. At $n = 50$ for $B/A = 1$, the polarization is too small to be shown here. Numbers refer to n (from Peraiah (1975), with permission).

where σ is the scattering coefficient. The results shown in figures 11.6, 11.7 and 11.8 are obtained by solving the transfer equation in a homogeneous medium. It

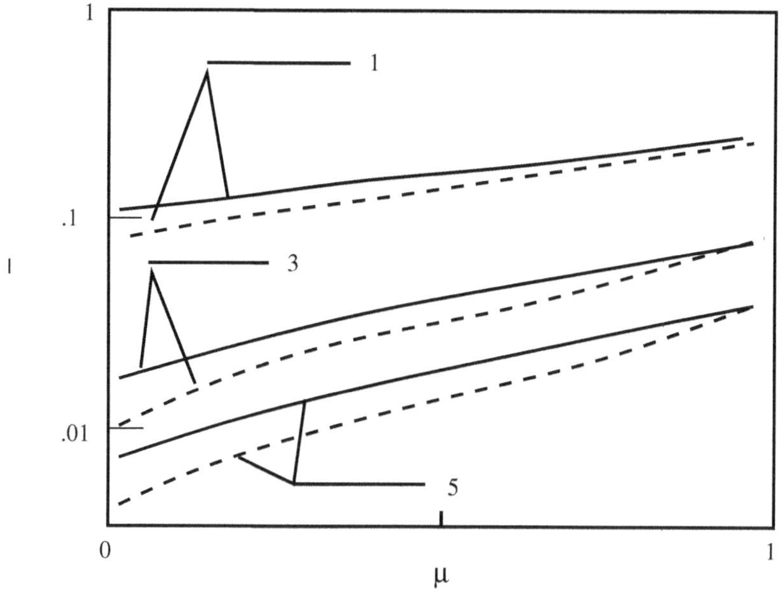

Figure 11.7 Angular distribution of the emergent radiation for the indicated values of B/A: dashed curves represent I_L and the continuous curves represent I_R. Note that at $\mu \approx 1$, $I_L = I_R$ (from Peraiah (1975), with permission).

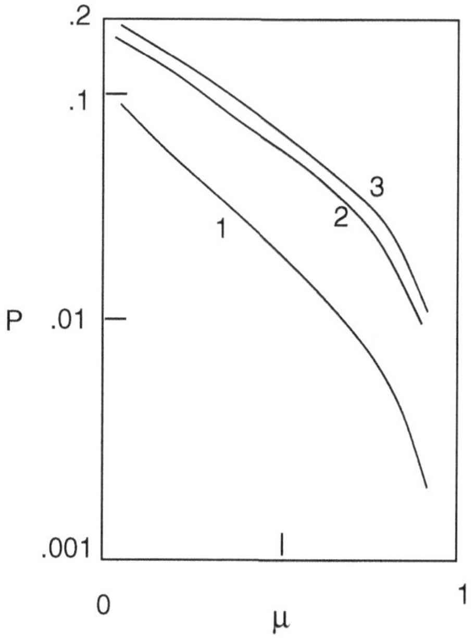

Figure 11.8 Angular distribution of polarization of the emergent radiation for $\tau = 10$ and the indicated parameters of B/A (from Peraiah (1975), with permission).

can be seen from equation (11.6.39) that the density changes as $1/r^2$ and therefore we need to modify the procedure slightly as shells of equal geometrical thickness are not shells of equal optical thickness. Shells of equal optical thickness have an advantage in that the curvature factor decreases towards the centre of the star and the doubling process (or the 'star' addition) needs to be applied only in the first few shells near the top of the atmosphere. The curvature factor ρ ($= \Delta r/\bar{r}$, where Δr is the geometrical thickness of the shell and $\bar{r}$ is the mean radius of the shell – $\bar{r}$ is taken to be the outer radius of the shell in these calculations) and is calculated as follows. If r_n and r_{n+1} are the boundaries of the shell (r_n, r_{n+1}), r_{n+1} is obtained from the recurrence relation

$$r_{n+1} = \frac{p' r_n}{p' + r_n \tau_{n+\frac{1}{2}}}, \tag{11.6.41}$$

where $p' = N_0 A^2 \sigma$ and $\tau_{n+\frac{1}{2}}$ is the optical thickness of the shell bounded by the radii r_n and r_{n+1}. Starting from some value for r_n ($r_1 = B$, the outermost radius of the atmosphere), we can calculate the boundaries r_n of the subsequent shells with the help of equation (11.6.41). Then the curvature factor is calculated from

$$\rho_{n+\frac{1}{2}} = \frac{(r_n - r_{n+1})}{r_n}. \tag{11.6.42}$$

The components of the polarized flux parallel and perpendicular to the polar axis (F_{pol} and F_{eq} respectively) are calculated by integrating over the surface of the apparent disc as described in Harrington and Collins (1968)). The linear polarization P is calculated from the relation

$$P = \frac{\left(F_{eq} - F_{pol}\right)}{\left(F_{eq} + F_{pol}\right)}. \tag{11.6.43}$$

The results for an electron scattering atmosphere are shown in figure 11.9.

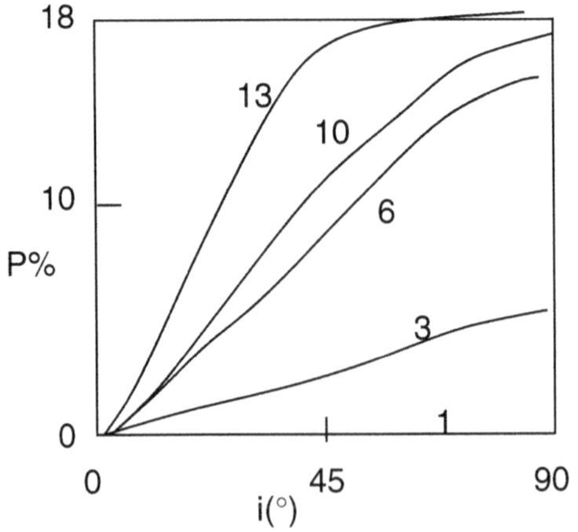

Figure 11.9 Polarization versus i, the angle between the line of sight and the orbital plane; $f = 0.5$ (the ratio of the centrifugal force at the equator to that of gravity). The numbers refer to B_p/A, where B_p is the outermost radius at the pole (from Peraiah (1976), with permission).

Polarization calculations for H_2 molecule scattering atmospheres are shown in figure 11.10. The Rayleigh scattering coefficient for H_2 molecules as a function of λ(Å) is given by

$$\sigma(H_2) = 8.14 \times 10^{-13}\lambda^{-4} + 1.28 \times 10^{-6}\lambda^{-6} + 1.61\lambda^{-8}. \tag{11.6.44}$$

Dumont (1971) obtained a numerical solution of the transfer equation for polarized continuum radiation by using the Feautrier numerical method. Barman and Peraiah (1991) computed models for estimating the linear polarization from extended dusty outer layers of the components of close binary systems. Cassinelli *et al.* (1987) computed the polarization of light scattered from the winds of early type stars.

11.7 Rayleigh scattering and scattering using planetary atmospheres

It is diffuse reflection and transmission of radiation scattered according to the Rayleigh phase function that gives the illumination and polarization of the sky. The same physical processes occur in the planetary atmospheres – the scattering of the sun's radiation by planets such as Jupiter and Venus.

A parallel beam of radiation of the net flux

$$\pi\mathbf{F} = \pi(F_l, F_r, F_U, F_V) \tag{11.7.1}$$

per unit area normal to itself in the four Stokes parameters is incident on a plane parallel atmosphere of optical thickness τ_1 in a given direction $(-\mu_0, \varphi_0)$. We need

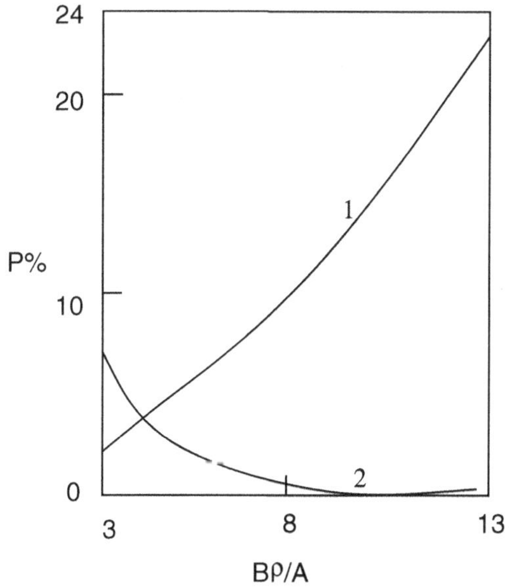

Figure 11.10 The dependence of polarization on the extendedness of the atmosphere for: (1) $\lambda = 3162$ Å and (2) $\lambda = 10\,000$ Å, for $f = 0.7$, $\varpi = 1$ and $i = 90°$ (from Peraiah (1976), with permission).

to find the angular distribution and the state of polarization of the light diffusely reflected by the surface at $\tau = 0$ and that diffusely transmitted at $\tau = \tau_1$. The laws of diffuse reflection and transmission are expressed in terms of a scattering matrix $\mathbf{S}(\tau_1; \mu, \varphi; \mu_0, \varphi_0)$. The reflected and transmitted intensities are given in terms of these functions as

$$\mathbf{I}(0; +\mu, \varphi) = \frac{1}{4\mu}\mathbf{S}(\tau_1, \mu, \varphi; \mu_0, \varphi_0)\mathbf{F} \tag{11.7.2}$$

and

$$\mathbf{I}(0; -\mu, \varphi) = \frac{1}{4\mu}\mathbf{T}(\tau_1, \mu, \varphi; \mu_0, \varphi_0)\mathbf{F}. \tag{11.7.3}$$

In the above equations the factor $1/\mu$ is introduced to maintain the symmetry of $\mathbf{S}$ and $\mathbf{T}$ to transposition as was done for the phase matrix given in equation (11.4.21). We need to distinguish between the reduced incident flux present at different depths and the diffuse radiation field that arises because of multiple scattering. In view of these two aspects of the radiation field, we write the radiative transfer equation in plane parallel geometry as

$$\mu\frac{d\mathbf{I}(\tau, \mu, \varphi)}{d\tau} = \mathbf{I}(\tau, \mu, \varphi) - \frac{1}{4\pi}\int_{-1}^{+1}\int_{0}^{2\pi} \mathbf{P}(\mu, \varphi; \mu'\varphi')\mathbf{I}(\tau, \mu', \varphi')\, d\mu'\, d\varphi'$$
$$- \frac{1}{4}e^{-\tau/\mu_0}\mathbf{P}(\mu, \varphi; -\mu_0, \varphi_0)\mathbf{F}, \tag{11.7.4}$$

with the boundary conditions:

$$\left.\begin{aligned} &\mathbf{I}(0, -\mu, \varphi) = 0 \qquad (0 < \mu \le 1, 0 < \varphi \le 2\pi), \\ &\mathbf{I}(0, +\mu, \varphi) = 0 \qquad (0 < \mu \le 1, 0 < \varphi \le 2\pi). \end{aligned}\right\} \tag{11.7.5}$$

We notice that when the phase matrix (see equations (11.4.16)–(11.4.20)) is reducible with respect to the last row and column, the Stokes parameter V is scattered independently of the others according to phase function $\frac{3}{2}\cos\Theta$. Then the transfer equation for V is given by

$$\begin{aligned} \mu\frac{dV(\tau, \mu, \varphi)}{d\tau} &= V(\tau, \mu, \varphi) \\ &- \frac{3}{8\pi}\int_{-1}^{+1}\int_{0}^{2\pi}\left[\mu\mu' + \left(1-\mu^2\right)^{\frac{1}{2}}\left(1-\mu'^2\right)^{\frac{1}{2}}\cos(\varphi'-\varphi)\right] \\ &\times V(\tau, \mu', \varphi)\, d\mu'\, d\varphi' \\ &- \frac{3}{8}F_V\left[-\mu\mu_0 + \left(1-\mu_0^2\right)^{\frac{1}{2}}\left(1-\mu^2\right)^{\frac{1}{2}}\cos(\varphi_0-\varphi)\right]\exp(-\tau/\mu_0). \end{aligned} \tag{11.7.6}$$

If we write V in the following form:

$$V(\tau,\mu,\varphi)=\frac{3}{8}F_V\Big[-\mu\mu_0V^{(0)}(\tau,\mu)+\left(1-\mu^2\right)^{\frac{1}{2}}\left(1-\mu_0^2\right)^{\frac{1}{2}}V^{(1)}(\tau,\mu)\cos(\varphi_0-\varphi)\Big], \tag{11.7.7}$$

then we obtain the following pair of equations:

$$\mu\frac{dV^{(0)}(\tau,\mu)}{d\tau}=V^{(0)}(\tau,\mu)-\frac{3}{4}\int_{-1}^{+1}\mu'^2V^{(0)}(\tau,\mu')\,d\mu'-\exp(\tau/\mu_0) \tag{11.7.8}$$

and

$$\mu\frac{dV^{(1)}(\tau,\mu)}{d\tau}=V^{(1)}(\tau,\mu)-\frac{3}{8}\int_{-1}^{+1}\mu'^2V^{(1)}(\tau,\mu')\,d\mu'-\exp(\tau/\mu_0). \tag{11.7.9}$$

Now remembering that $\mathbf{I}=[I_l,I_r,U]$, the transfer equation becomes

$$\begin{aligned}\mu\frac{d\mathbf{I}(\tau,\mu,\varphi)}{d\tau}&=I(\tau,\mu,\varphi)\\&=\frac{1}{4\pi}\int_{-1}^{+1}\int_0^{2\pi}\mathbf{P}(\mu,\varphi;\mu',\varphi')\mathbf{I}(\tau,\mu',\varphi')\,d\mu'\,d\varphi'\\&\quad-\frac{1}{4}\mathbf{P}(\mu,\varphi;-\mu_0,\varphi_0)\mathbf{F}\exp(-\tau/\mu_0),\end{aligned} \tag{11.7.10}$$

where

$$\mathbf{F}=[F_l,F_r,U] \tag{11.7.11}$$

and the matrix $\mathbf{P}(\mu,\varphi;\mu',\varphi')$ is the same as that defined in equations (11.4.16)–(11.4.20) but with the last row and column deleted, that is

$$\mathbf{P}(\mu,\varphi;\mu',\varphi')=\mathbf{Q}\Big[\mathbf{P}^{(0)}(\mu,\mu')+(1-\mu'^2)^{\frac{1}{2}}(1-\mu^2)^{\frac{1}{2}}\mathbf{P}^{(1)}(\mu,\varphi;\mu',\varphi')+\mathbf{P}^{(2)}(\mu,\varphi;\mu',\varphi')\Big], \tag{11.7.12}$$

where

$$\mathbf{Q}=\begin{pmatrix}1&0&0\\0&1&0\\0&0&2\end{pmatrix}, \tag{11.7.13}$$

$$\mathbf{P}^{(0)}(\mu,\mu')=\frac{3}{4}\begin{pmatrix}2\left(1-\mu^2\right)\left(1-\mu'^2\right)+\mu^2\mu'^2&\mu^2&0\\\mu'^2&1&0\\0&0&0\end{pmatrix}, \tag{11.7.14}$$

$$\mathbf{P}^{(1)}(\mu,\varphi;\mu',\varphi') = \frac{3}{4}\begin{pmatrix} 4\mu\mu'\cos(\varphi'-\varphi) & 0 & 2\mu\sin(\varphi'-\varphi) \\ 0 & 0 & 0 \\ -2\mu'\sin(\varphi'-\varphi) & 0 & \cos(\varphi'-\varphi) \end{pmatrix}, \tag{11.7.15}$$

$$\mathbf{P}^{(2)}(\mu,\varphi;\mu',\varphi') = \frac{3}{4}\begin{pmatrix} \mu^2\mu'^2\cos 2(\phi'-\varphi) & -\mu^2\cos 2(\varphi'-\varphi) & \mu^2\mu'^2\sin 2(\varphi'-\varphi) \\ -\mu'^2\cos 2(\varphi'-\varphi) & \cos 2(\varphi'-\varphi) & -\mu'\sin 2(\varphi'-\varphi) \\ -\mu\mu'^2\sin 2(\varphi'-\varphi) & \mu\sin 2(\varphi'-\varphi) & \mu\mu'\cos 2(\varphi'-\varphi) \end{pmatrix}. \tag{11.7.16}$$

The law of diffuse reflection is given by

$$\mathbf{I}(0,\mu,\varphi) = \begin{pmatrix} I_l \\ I_r \\ U \end{pmatrix} = \frac{1}{4\mu}\mathbf{QS}(\mu,\phi;\mu_0,\phi_0)\begin{pmatrix} F_l \\ F_r \\ F_U \end{pmatrix}, \tag{11.7.17}$$

(see Chandrasekhar (1960)), where

$$\begin{aligned} &\left(\frac{1}{\mu_0}+\frac{1}{\mu}\right)\mathbf{S}(\mu,\varphi,\mu_0,\varphi_0) \\ &= \frac{3}{4}\begin{pmatrix} \psi(\mu) & 2^{\frac{1}{2}}\phi(\mu) & 0 \\ \chi(\mu) & 2^{\frac{1}{2}}\xi(\mu) & 0 \\ 0 & 0 & 0 \end{pmatrix}\begin{pmatrix} \psi(\mu_0) & \chi(\mu_0) & 0 \\ 2^{\frac{1}{2}}\phi(\mu_0) & 2^{\frac{1}{2}}\xi(\mu_0) & 0 \\ 0 & 0 & 0 \end{pmatrix} \\ &+ \frac{3}{4}\begin{pmatrix} -4\mu\mu_0\cos(\varphi_0-\varphi) & 0 & 2\mu\sin(\varphi_0-\varphi_0) \\ 0 & 0 & 0 \\ 2\mu_0\sin(\varphi_0-\varphi) & 0 & \cos(\varphi_0-\varphi) \end{pmatrix} \\ &\times \left(1-\mu^2\right)^{\frac{1}{2}}\left(1-\mu_0^2\right)^{\frac{1}{2}} H^{(1)}(\mu)H^{(1)}(\mu_0) \\ &+ \frac{3}{4}\begin{pmatrix} \mu^2\mu_0^2\cos 2(\varphi_0-\varphi) & -\mu^2\cos 2(\varphi_0-\varphi) & -\mu^2\mu_0^2\sin 2(\varphi_0-\varphi) \\ -\mu_0^2\cos 2(\varphi_0-\varphi) & \cos 2(\varphi_0-\varphi) & \mu_0\sin 2(\varphi_0-\varphi) \\ \mu\mu_0^2\sin 2(\varphi_0-\varphi) & \mu\sin 2(\varphi_0-\varphi) & -\mu\mu_0\cos 2(\varphi_0-\varphi) \end{pmatrix} \\ &\times H^{(2)}(\mu)H^{(2)}(\mu_0) \end{aligned} \tag{11.7.18}$$

and

$$\left.\begin{aligned} \psi(\mu) &= q\mu H_l(\mu); & \phi(\mu) &= H_l(\mu)(1-c_\mu); \\ \chi(\mu) &= H_r(\mu)(1-c\mu); & \xi(\mu) &= \frac{1}{2}q\mu H_r(\mu). \end{aligned}\right\} \tag{11.7.19}$$

$H_l(\mu)$, $H_r(\mu)$, $H^{(1)}(\mu)$ and $H^{(2)}(\mu)$ are defined in terms of the characteristic functions

$$\frac{3}{4}\left(1-\mu^2\right),\quad \frac{3}{8}\left(1-\mu^2\right),\quad \frac{3}{8}\left(1-\mu^2\right)\left(1+2\mu^2\right) \quad \text{and} \quad \frac{3}{16}\left(1+\mu^2\right)^2 \tag{11.7.20}$$

respectively. The constants q and c are functions of $H_l(\mu)$ and $H_r(\mu)$ (see Chandrasekhar (1960)).

The problem of greatest interest is the diffuse reflection of an incident beam of natural light in which case

$$F_l = F_r = \frac{1}{2}F \quad \text{and} \quad F_V = F_U = 0. \tag{11.7.21}$$

From equation (11.7.18) we obtain

$$\begin{aligned} I_l(0,\mu,\varphi) = \frac{3}{32}\frac{1}{(\mu+\mu_0)}\Big\{&\psi(\mu)\left[\psi(\mu_0)+\chi(\mu_0)\right] \\ &+ 2\phi(\mu)\left[\phi(\mu_0)+\xi(\mu_0)\right] \\ &- 4\mu\mu_0\left(1-\mu^2\right)^{\frac{1}{2}}\left(1-\mu_0^2\right)^{\frac{1}{2}} H^{(1)}(\mu)H^{(1)}(\mu_0)\cos(\varphi_0-\varphi) \\ &- \mu^2\left(1-\mu_0^2\right)H^{(2)}(\mu)H^{(2)}(\mu_0)\cos 2(\varphi_0-\varphi)\Big\}\mu_0 F, \end{aligned} \tag{11.7.22}$$

$$\begin{aligned} I_r(0,\mu,\varphi) = \frac{3}{32(\mu+\mu_0)}\{&\chi(\mu)[\psi(\mu_0)+\chi(\mu_0)] + 2\xi(\mu)[\phi(\mu_0)+\xi(\mu_0)] \\ &+ (1-\mu_0^2)H^{(2)}(\mu)H^{(2)}(\mu_0)\cos 2(\varphi_0-\varphi)\}\mu_0 F, \end{aligned} \tag{11.7.23}$$

and

$$\begin{aligned} U(0,\mu,\varphi) = \frac{3}{16(\mu+\mu_0)}\Big[&2\left(1-\mu^2\right)^{\frac{1}{2}}\left(1-\mu_0^2\right)^{\frac{1}{2}}\mu_0 H^{(1)}(\mu) \\ &\times H^{(1)}(\mu_0)\sin(\varphi_0-\varphi) \\ &+ \mu\left(1-\mu_0^2\right)H^{(2)}(\mu)H^{(2)}(\mu_0)\sin 2(\varphi_0-\varphi)\Big]\mu_0 F. \end{aligned} \tag{11.7.24}$$

The corresponding expressions for the intensities which represent light that has suffered a single scattering process in the atmosphere can be obtained from equations (11.7.22)–(11.7.24) by letting

$$\left.\begin{aligned} &\psi(\mu)\to\mu^2, \quad \phi(\mu)\to 1-\mu^2, \quad \chi(\mu)\to 1, \\ &\xi(\mu)\to 0, \quad H^{(1)}(\mu)\to 1 \quad \text{and} \quad H^{(2)}\to 1. \end{aligned}\right\} \tag{11.7.25}$$

Therefore we obtain $I_l^{(1)}(0,\mu,\varphi)$, $H_l^{(1)}(0,\mu,\varphi)$ and $U^{(1)}(0,\mu,\varphi)$ as

$$I_l^{(1)}(0,\mu,\varphi) = \frac{3}{32(\mu+\mu_0)}\left[\mu^2\left(1+\mu_0^2\right)+2\left(1-\mu^2\right)\left(1-\mu_0^2\right)\right.$$

$$- 4\mu\mu_0 \left(1 - \mu_0^2\right)^{\frac{1}{2}} \left(1 - \mu^2\right)^{\frac{1}{2}} \cos(\varphi_0 - \varphi)$$
$$- \mu^2 \left(1 - \mu_0^2\right) \cos 2(\varphi_0 - \varphi)\big]\mu_0 F, \qquad (11.7.26)$$

$$I_r^{(1)}(0, \mu, \varphi) = \frac{3}{32(\mu + \mu_0)} \left[1 + \mu_0^2 + \left(1 - \mu_0^2\right) \cos 2(\varphi_0 - \varphi)\right] \mu_0 F \qquad (11.7.27)$$

and

$$U^{(1)}(0, \mu, \varphi) = \frac{3}{16(\mu + \mu_0)} \big[2(1 - \mu^2)^{\frac{1}{2}} \left(1 - \mu_0^2\right)^{\frac{1}{2}} \mu_0 \sin(\varphi_0 - \varphi)$$
$$+ \mu \left(1 - \mu_0^2\right) \sin 2\,(\varphi_0 - \varphi)\,\big]\mu_0 F. \qquad (11.7.28)$$

The laws of diffuse reflection given in equations (11.7.22)–(11.7.28) are illustrated in figures 11.11, 11.12 and 11.13 for angles of incidence corresponding to $\mu_0 = 0.8$ (figure 11.11), $\mu_0 = 0.5$ (figure 11.12) and $\mu_0 = 0.2$ (figure 11.13). The variations of I_r, I_l, $I_r + I_l$, $I_r - I_l$ in the principal plane ($\varphi_0 - \varphi = 0$ and π) containing the direction of incidence and in the plane ($\varphi_0 - \varphi = \pm\frac{1}{2}\pi$) at right angles to the directions of incidence are shown. In the principal plane, $U \equiv 0$ which is required by symmetry since, in the plane containing the direction of incidence, the plane of polarization must be along the direction of l and r. However, in the plane $\varphi_0 - \varphi = \pm\frac{1}{2}\pi$, $U \neq 0$ and the variation of U is also shown. The intensity of light which has suffered only a single scattering in the atmosphere is shown in figures 11.11 and 11.12. From the figures, it is seen that I_l is more dependent on angle than I_r. This is particularly true in the meridian plane for moderate angles of incidence because I_l shows strong variation while I_r is nearly independent of angle. One can understand this on the physical grounds because light polarized at right angles to the plane of scattering is isotropically scattered while light polarized parallel to the plane of scattering is scattered in accordance with the phase function $3\cos^2\Theta$. Moreover, in the principal plane, there is a reversal of the sign of polarization, the polarization vanishing at two points – a phenomenon of neutral points which occurs in the polarization of the sky.

In the standard problem, we consider a parallel beam of radiation incident on a plane parallel atmosphere at $\tau = 0$ and no radiation incident at $\tau = T$, from below, i.e. $I(\tau_1, \mu, \varphi) = 0$ $(0 \le \mu \le 1)$. However, in the case of planetary atmospheres, the atmosphere is illuminated by the sun and the atmosphere rests on solid ground or an ocean or a cloud bank which will modify the law of diffuse reflection at $\tau = \tau_1$. Thus 'ground' reflection will modify the boundary condition at $\tau = \tau_1$. One does not have exact knowledge of the reflecting properties of ground but one can use Lambert's law (see figure 11.14) according to which a surface with an 'albedo' λ_0 reflects as follows:

$$I(0, \mu, \varphi; \mu_0, \varphi_0) = \lambda_0 \mu_0 F \tag{11.7.29}$$

(see Chandrasekhar 1960, Grant and Hunt 1968).

Therefore, the planetary problem may be stated as follows. A parallel beam of light of net flux πF (or $\pi \mathbf{F}$ when polarization is taken into account) per unit area normal to itself is incident on a plane parallel atmosphere in a given direction $(-\mu_0, \varphi_0)$. The atmosphere is of optical thickness τ_1. At τ_1 is the other boundary (the 'ground') which reflects according to Lambert's law with an albedo λ_0. We

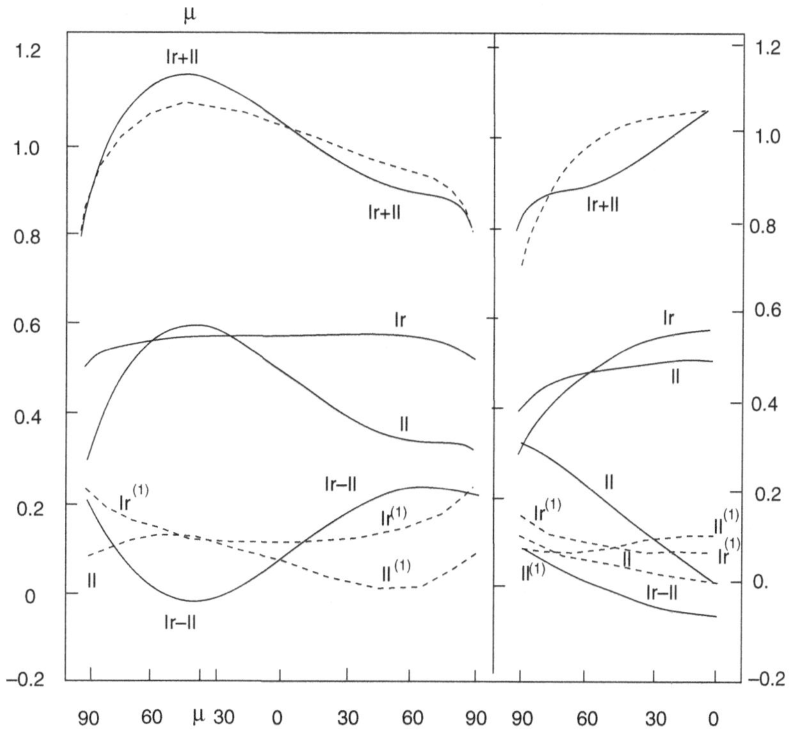

Figure 11.11 The law of diffuse reflection by a semi-infinite atmosphere on Rayleigh scattering. The ordinates represent the intensities in units of $\mu_0 F$ and the abscissae the angle in degrees. An angle of incidence corresponding to $\mu_0 = 0.8$ is considered and the variation of the reflected intensities and the planes $\varphi_0 - \varphi = 0$ (the curves on the LHS of the diagram) and $\varphi_0 - \varphi = 90°$ (the curves on the RHS of the diagram) are illustrated. The intensities I_l and I_r (in directions parallel and perpendicular to the meridian plane containing the directions of the reflected light), the total intensity $I_l + I_r$ and $I_l - I_r$ are all shown and, in the plane $\varphi_0 - \varphi = 90°$, the variation of U is also shown. The intensities $I_l^{(1)}$, $I_r^{(1)}$ and $U^{(1)}$ obtained due to the light which suffered a single scattering are shown as dashed lines. The total intensity $I_l + I_r$ predicted by the present exact theory of Rayleigh scattering is compared to what could be expected (shown by the continuous curves) from a theory which does not take into account the state of polarization of the radiation field but allows for an anisotropy of the scattered radiation according to Rayleigh phase function (from Chandrasekhar (1960), figure 24, page 262, with permission).

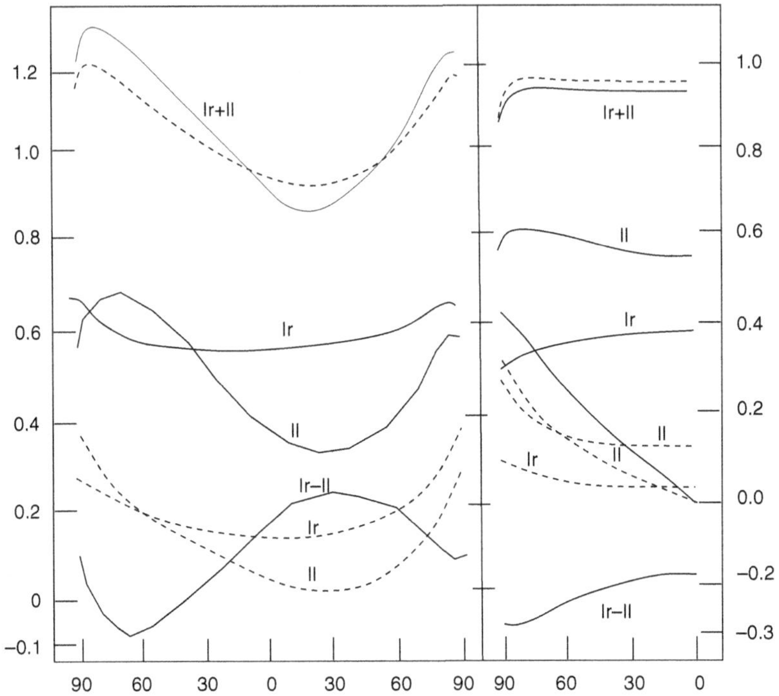

Figure 11.12 Same as in figure 11.11 but for an angle of incidence of $\mu_0 = 0.5$. Notice that the ordinate on the RHS of the diagram is shifted relative to that on the LHS (from Chandrasekhar (1960), figure 25, page 263, with permission).

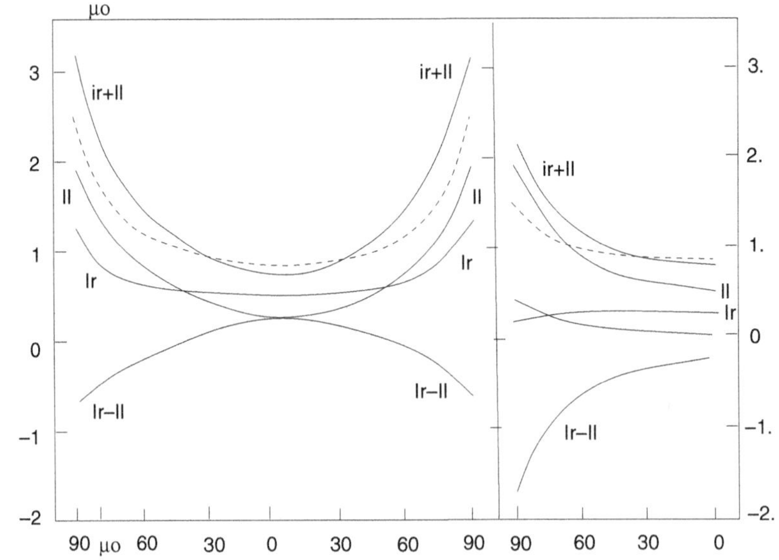

Figure 11.13 Same as in figures 11.11 and 11.12 but for $\mu_0 = 0.2$ but here the results of single scattering are not shown (from Chandrasekhar (1960), figure 26, page 264, with permission).

need to find the state of polarization and the angular distribution of the radiation diffusely reflected from the surface $\tau = 0$ and also to specify the illumination and polarization of the 'sky' as seen by an observer at $\tau = \tau_1$. This is the planetary problem. For details of the solution see Chandrasekhar (1960).

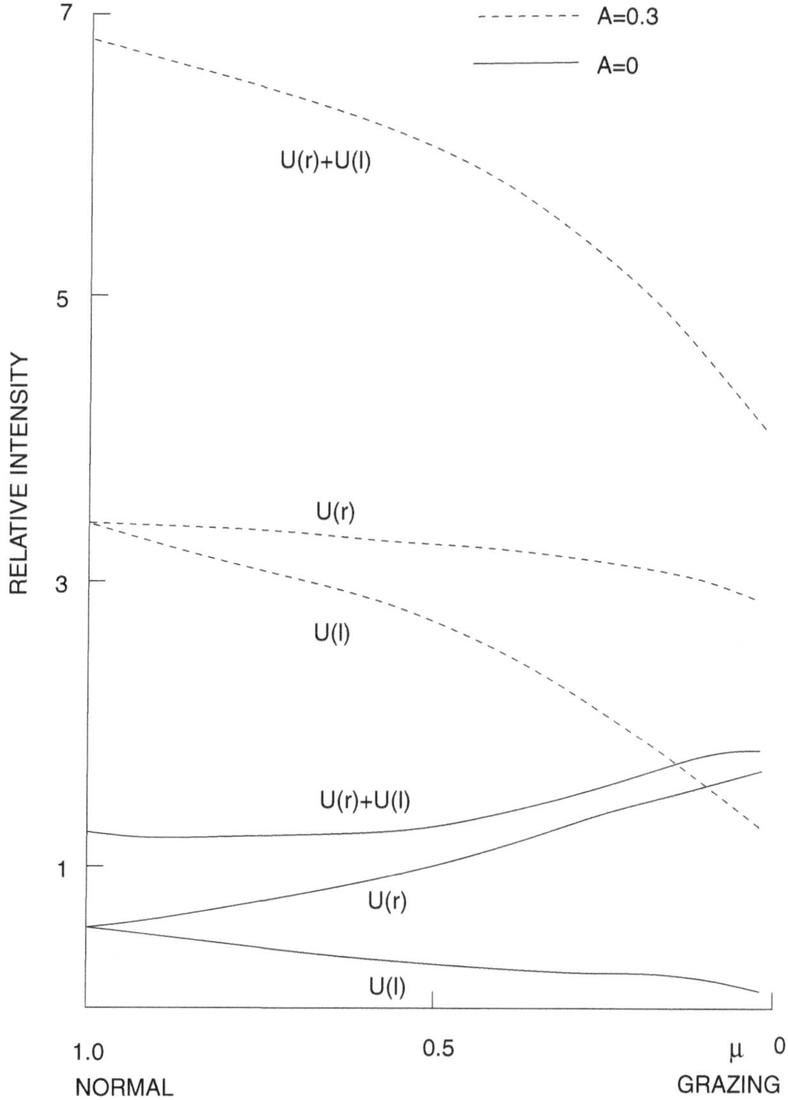

Figure 11.14 The effect of a Lambert surface of albedo A. The angular distribution of the relative intensity of the total radiation $[U_l^-(l) - U_l^-(r)]$ and of the intensity components $U_1^-(r)$ and $U_1^-(l)$ of the radiation reflected by a slab of optical thickness $\tau = 0.5$ is shown. The atmosphere is illuminated on its upper surface by a normally incident beam. These calculations were done using discrete space theory. The Us are the intensities (from Grant and Hunt (1968), with permission).

It is well known that Lord Rayleigh accounted for the principal features of the brightness and polarization of the sky radiation in terms of the molecular scattering that is associated with his name. The deep blue of the sky at the zenith is related to the λ^{-4} dependence of the scattering coefficient on wavelength (see equation (11.4.4)) and to the fact that for a small optical thickness τ_1, the intensity of the transmitted light is proportional to τ_1. Similarly the strong variation in the polarization of the sky radiation is related to the fact that according to Rayleigh's law, the light scattered at right angles to the direction of incidence is completely polarized while the light scattered in the forward or the backward directions has the same polarization as the incident light. However, the law of single scattering cannot account for all the features of the sky radiation. It is known that at right angles to the sun the polarization is not complete ($\delta = 87\%$) and that the polarization in the direction of the sun is not zero, but it is instead weak and negative. Also for general angles of incidence there are two points where the light is unpolarized in the meridian plane. These points of zero polarization are called neutral points. For angles of incidence $< 70°$, the neutral points occur between 10 and $-20°$ above and below the sun. These are the *Babinet* and *Brewster* points respectively. When the sun is low, then near the horizon, opposite the sun and about 20° above the anti-solar point another neutral point occurs. This is called the *Arago* point.

There are occasions when all these neutral points occur simultaneously and persist even after sunset. These may be due to the curvature of the Earth's atmosphere. For the details of obtaining the solution of Earth's curvature problem see Chandrasekhar (1960). Rangarajan *et al.* (1994) have computed the polarization of light with non-conservative Rayleigh scattering using the discrete space theory computational scheme of radiative transfer (Peraiah 1978). They computed the polarization in a finite medium. They derived several interesting results which we shall briefly mention here (Abhyankar 1996):

1. Both I and $Q = |I_r - I_l|$ decrease with ω in all directions as the probability of scattering over absorption decreases.
2. For smaller τs I has double maxima in the plane of the sun's vertical which are close to the horizon and these shift towards the zenith as τ_1 increases.
3. At small values of ω, polarization increases when $I_l < I_r$ and decreases when $I_l > I_r$.
4. The Babinet and Brewster points are bright closer to the sun as ω decreases. Non-conservative Rayleigh scattering brings out the effect of aerosols prominently.
5. The Arago point moves towards the anti-solar point with decreasing ω and vanishes for small values of ω.
6. The polarization is maximum at 90° from the sun, as in the case of conservative scattering, and decreases as the optical thickness increases and

increases with decreasing values of ω.

7. When $\varphi_0 - \varphi = 0$, the absolute value of the polarization has two minima and they move closer as ω decreases. In the opposite direction when $\varphi - \varphi_0 = 180^\circ$, the maximum polarization is shifted towards the horizon as the sun goes towards the zenith, independent of the τ and ω values. When $\varphi - \varphi_0 = 90^\circ$ and 270°, the plane of polarization is the same for all values of ω at the zenith, but at the horizon it depends on ω (see table 11.2).

Hovenier (1987) developed a unified treatment of polarized radiation emerging from a homogeneous plane parallel atmosphere by using exit functions. Daguchi and Watson (1985) studied the circular polarization of interstellar absorption lines.

11.8 Resonance line polarization

In quantum theory, the resonance line scattering arises from the transition from an initial ground state to an excited intermediate state and back to the ground

Table 11.2 Neutral points for the transmitted light. The numbers in columns 3–7 are the zenith angles in degrees (Rangarajan *et al.* 1994).

Optical depth τ	Sun's position $\phi - \phi_0$	ϕ_0	$\omega = 1.0$ (conservative scattering)	$\omega = 0.8$	$\omega = 0.5$	$\omega = 0.2$	Remarks
0.2	0	84°	62.0	65.2	70.0	75.0	Babinet
0.2	180 (anti-solar direction)	84	72.5	76.0	80.0	86.0	Arago
0.2	0	60	43.47	45.4	48.6	52.7	Babinet
			80.5	77.5	73.5	68.0	Brewster
0.2	0	36	26.5	28.0	29.5	32.7	Babinet
			50.0	48.5	45.4	41.9	Brewster
0.5	0	84	56.5	61.4	67.0	74.2	Babinet
0.5	180	84	66.2	71.0	77.4	85.3	Two Arago
0.5	0	60	39.0	42.1	47.0	51.7	Babinet
			86.9	82.2	75.8	68.6	Brewster
0.5	0	36	24.0	25.6	28.0	31.2	Babinet
			54.9	51.7	47.0	42.2	Brewster
2.0	0	60	36.7	44.6	54.0		Babinet
			82.1	70.2	60.6		Brewster
2.0	0	36	21.5	25.6	29.6	32.7	Babinet
			64.6	51.7	44.6	40.6	Brewster

state. In considering these transitions we need to distinguish between the different substates of each level as specified by the magnetic quantum number m (which is the eigenvalue of the z-component of the total angular momentum in units of $\hbar$). Let A_k, B_n and A_p represent the different states of the radiating atom, where A and B refer to the ground (initial or final) and the intermediate (excited) states respectively, and the subscripts refer to the m values of the different substates in question. The probability of a transition $A_k \rightarrow B_n$ between a single pair of m

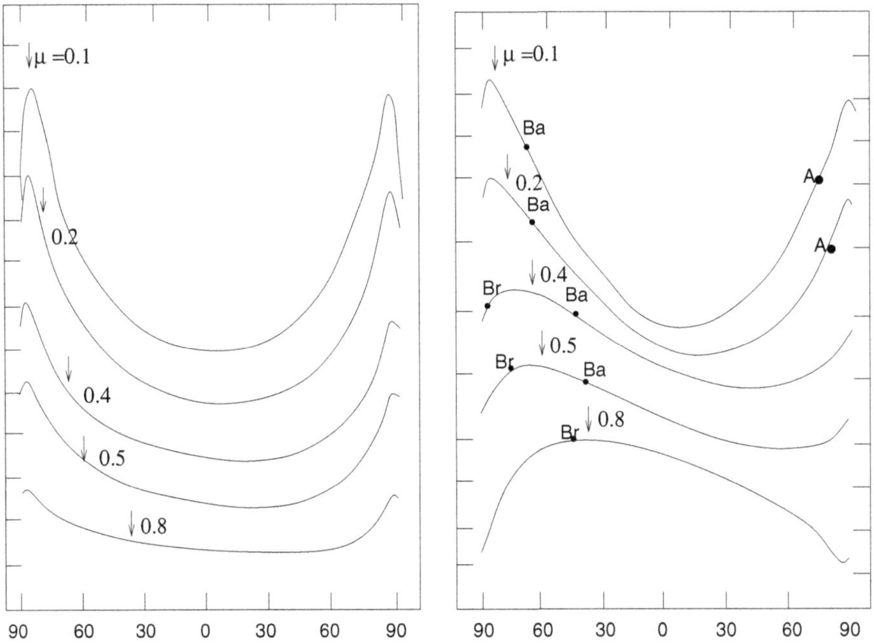

Figure 11.15 The laws of diffuse transmission on Rayleigh scattering by a plane parallel atmosphere. The ordinates represent the intensities in units $\mu_0 F$ and the abscissae the angle in degrees. Angles of incidence corresponding to $\mu_0 = 0, 1, 0.2, 0.4, 0.5$ and 0.8 are considered; also $\tau_1 = 0.2$ (the arrows indicate the directions of incidence). The curves in the left hand diagram illustrate the variation of the net diffusely transmitted light $(I_l + I_r)$ in the meridian plane. For the sake of clarity, curves for different angles of incidence have been displaced with respect to one another: the scales of the ordinates for the successive curves are indicated. The curves in the right hand diagram illustrate the variation of $I_l - I_r$ (also in the meridian plane) for various angles of incidence. The scale of the ordinates has been staggered, as in the left hand diagram. It can be seen that for $\mu_0 = 0.1$ and 0.2, the neutral points occur in positions appropriate for the Babinet (Ba) and the Arago (A) points. For larger angles of incidence, the neutral points occur on either side of the 'sun' in the positions of the Babinet and Brewster (Br) points. The 'setting' of the Arago point therefore coincides with the 'rising' of the Brewster point and conversely (see Chandrasekhar (1960)). Rangarajan *et al.* (1994) explained the shift of the Babinet points with non-conservative scattering using discrete space theory (from Chandrasekhar (1960), page 283, figure 27, with permission).

states is easily calculable from any given stream of incident radiation. Similarly, the angular distribution and the state of polarization of the quantum emitted in a transition $B_n \rightarrow A_p$ between a single pair of m states are also known. However, one must pay attention to the fact that the different sequences of transitions which are possible starting from the same state A_k are not uncorrelated, because when transitions from a given state A_k to different substates B_n are possible, the wave functions of these substates have phases which are related in a definite manner to the phases of the wave functions belonging to A_k. This means that the resulting transitions $B_n \rightarrow A_p$ from the different substates B_n will not be independent to each other. In resonance fluorescence the Stokes parameters I_l, I_r and U are scattered in accordance with the phase matrix of the form (see Chandrasekhar (1960))

$$\frac{3}{2}E_1 \begin{pmatrix} \cos^2\Theta & 0 & 0 \\ 0 & 1 & 0 \\ 1 & 0 & \cos^2\Theta \end{pmatrix} + \frac{1}{2}E_2 \begin{pmatrix} 1 & 1 & 0 \\ 1 & 1 & 0 \\ 0 & 0 & 0 \end{pmatrix}, \tag{11.8.1}$$

where E_1 and E_2 are constants depending on the initial j value and Δj $(= \pm 1$ or $0)$ involved in the transition. In the case of Rayleigh scattering and in the case of scattering by anisotropic particles, the parameter V is scattered independently of the rest and according to a phase function of the form

$$\frac{2}{3}E_3 \cos\Theta, \tag{11.8.2}$$

where E_3 is another constant depending on j and Δj. E_1, E_2, E_3 are given in table 11.3. These constants are due to Hamilton (1947). We can see from table 11.3 that

$$E_1 + E_2 = 1. \tag{11.8.3}$$

This is an essential condition for conservative scattering. Furthermore, when $j = 0$ and $\Delta j = 1$

$$E_1 = 1, \ E_2 = 0, \ E_3 = 1. \tag{11.8.4}$$

This is similar to Rayleigh scattering.

Stenflo (1980) pointed out the importance of the diagnostic potential of scattering polarization measurements in the solar spectrum. The non-magnetic polarization in lines such as NaD_2 is a case in point. Stenflo *et al.* (1980), using the HAO Stokes polarimeter, found a polarization maximum in the core of the line which is of the same order of magnitude as the maxima in the wings. Wiehr's (1981) measurements do not show this maximum, but the resonance lines CaI 4227 Å and CaII K have a similar feature. Stokes's Q maximum is reduced while Stokes's U is enhanced. These characteristics appear to be confined to the Doppler core with the wings being unaffected, which is interpreted as depolarization and the rotation of the plane of linear polarization in the core due to the Hanle effect predicted by Omont *et al.* (1973). Dumont *et al.* (1977) and Rees (1978) have studied the cores of the solar resonance absorption lines. Dumont *et al.* (1973) and Auer *et al.* (1980) assumed coherent scattering and computed polarization profiles which are in a fairly good agreement with observation in the line wings but they found it difficult to obtain the simultaneous existence of polarization of the core and wing maximum. Rees and Saliba (1982) used frequency redistribution with coherent scattering in the rest frame of a two-level atom model with natural broadening of the upper level similar to that of Kneer (1975).

The transfer equation in plane parallel layers for the Stokes vectors $I_l(\tau, \mu)$ and $I_r(\tau, \mu)$ is given by (Rees and Saliba 1982)

$$\mu \frac{d\mathbf{I}(x, \mu)}{d\tau} = [\beta + \phi(x)]\,[\mathbf{I}(\tau, \mu) - \mathbf{S}_T(x, \mu)], \tag{11.8.5}$$

where $\mathbf{I}(\tau, \mu) = [I_l(\tau, \mu), I_r(\tau, \mu)]^T$ and the normalized line absorption profile $\phi(x)$ is taken to be a Voigt profile given by

$$\phi(x) = \frac{H(a, x)}{\pi^{1/2}}, \tag{11.8.6}$$

with constant damping to the Doppler width ratio a. The total source function vector $\mathbf{S}_T$ is given by

$$\mathbf{S}_T(x, \mu) = \frac{\phi(x)\mathbf{S}_L(x, \mu) + \beta\mathbf{B}}{\beta + \phi(x)}, \tag{11.8.7}$$

where $\mathbf{B} = \frac{1}{2}B(1, 1)$ is the unpolarized continuum source vector and B is the Planck function. The line source function $\mathbf{S}_L(x, \mu)$ is given by (Dumont *et al.* (1977)

$$\mathbf{S}_L(x, \mu) = \begin{pmatrix} S_l(x, \mu) \\ S_r(x, \mu) \end{pmatrix} = \frac{(1 - \epsilon)4\pi}{\phi(x)} \int_{-\infty}^{+\infty} dx' \int_{4\pi} d\Omega'\, \mathbf{R}(x', \mathbf{n}'; x, \mathbf{n})\mathbf{I}(x', \mathbf{n}') + \mathbf{B}. \tag{11.8.8}$$

The distribution function $\mathbf{R}(x', \mathbf{n}'; x, \mathbf{n})$ gives the correlation in frequency, angle and polarization between light absorbed at frequency x' in the direction $\mathbf{n}'$ and emitted

Table 11.3 The constants E_1, E_2 and E_3.

Δj	E_1	E_2	E_3
1	$\frac{(2j+5)(j+2)}{10(j+1)(2j+2)}$	$\frac{3j(6j+7)}{10j(j+1)(2j+1)}$	$\frac{j+2}{2(j+1)}$
0	$\frac{(2j-1)(2j+3)}{10j(j+1)}$	$\frac{3(2j^2+2j+1)}{10j(j+1)}$	$\frac{1}{2j(j+1)}$
−1	$\frac{(2j-3)(j-1)}{10j(2j+1)}$	$\frac{3(6j^2+5j-1)}{10j(2j+1)}$	$\frac{j-1}{2j}$

at frequency x in the direction $\mathbf{n}$ and $d\Omega'$ is an element of solid angle about $\mathbf{n}'$. For a two-level atom with coherent scattering in the rest frame (lower level with zero width and a naturally broadened upper width), we have the redistribution function given by (in Hummer's (1962) notation)

$$\mathbf{R}_{II}(x', \mathbf{n}'; x, \mathbf{n}) = \frac{\mathbf{P}(\mathbf{n}', \mathbf{n})}{4\pi} \frac{1}{4\pi^2 |\sin\gamma|} \exp\left[-\left(\frac{x'-x}{2}\right)^2 \operatorname{cosec}^2 \frac{\gamma}{2}\right]$$
$$\times H\left(a \sec\frac{\gamma}{2}, \frac{x'+x}{2} \sec\frac{\gamma}{2}\right), \tag{11.8.9}$$

where γ is the angle between $\mathbf{n}', \mathbf{n}$. To maximize polarization Rees and Saliba assumed a $j = 0 \rightarrow j = 1$ transition so that $\mathbf{P}(\mathbf{n}', \mathbf{n})$ has the form of the well known Rayleigh phase matrix described above.

In terms of the phase matrix equation (11.8.9) can be written as

$$\mathbf{R}_{II}(x', \mathbf{n}'; x, \mathbf{n}) = \frac{1}{4\pi}\mathbf{P}(\mathbf{n}', \mathbf{n})\frac{1}{4\pi}\mathbf{R}_{II}(x', x), \tag{11.8.10}$$

where $\mathbf{R}_{II}(x', x)$ is the angle-averaged redistribution function corresponding to the one in equation (11.8.9). The function in equation (11.8.10) will retain the angular distribution through the phase function $\mathbf{P}(\mathbf{n}', \mathbf{n})$ from which we obtain the polarization and the frequency correlation through the angle-averaged scalar function $\mathbf{R}_{II}(x', x)$. For isotropic scattering (see chapter 1) $\mathbf{R}_{II}(x', x)$ is given by

$$\mathbf{R}_{II}(x', x) = \frac{1}{\pi^{3/2}} \int_{\frac{1}{2}|\bar{x}-\underline{x}|}^{\infty} \exp(-u^2)$$
$$\times \left[\tan^{-1}\left(\frac{\underline{x}-u}{a}\right) - \tan^{-1}\left(\frac{\bar{x}-u}{a}\right)\right] du, \tag{11.8.11}$$

where $\bar{x} = \max(|x'|, |x|)$, $\underline{x} = \min(|x'|, |x|)$. The azimuthal integration is done in equation (11.8.10) and $\mathbf{S}_L(x, \mu)$ is written as

$$\mathbf{S}_L(x, \mu) = \frac{1-\epsilon}{\phi(x)} \int_{-\infty}^{+\infty} dx\, \mathbf{R}_{II}(x', x) \int_{-1}^{+1} \mathbf{P}(\mu', \mu)\mathbf{I}(x', \mu')\, d\mu' + \epsilon\mathbf{B}, \tag{11.8.12}$$

where $P(\mu, \mu')$ is the Rayleigh phase function for the Stokes components I_l and I_r and is given by

$$\mathbf{P}(\mu', \mu) = \frac{3}{8}\begin{pmatrix} 2(1-\mu^2)(1-\mu'^2) + \mu^2\mu'^2 & \mu^2 \\ \mu'^2 & 1 \end{pmatrix}. \tag{11.8.13}$$

Rees and Saliba followed the suggestion of Stenflo (1976) and used the approximation of Jefferies and White (1960) which means scattering is applied in the observer's frame as a CRD in the line core with a gradual transition to coherent scattering (CS) in the wings. They used Kneer's (1975) approximation which is given by

$$R_{II}(x', x) = \langle a \rangle_x \, \delta(x' - x)\phi(x') + \left(1 - a_{x',x}\right) \phi(x')\phi(x), \tag{11.8.14}$$

where

$$\langle a \rangle_x = \int_{-\infty}^{+\infty} a_{x',x}\phi(x')\, dx' \tag{11.8.15}$$

and

$$a_{x',x} = 1 - \exp\left[-\frac{(\hat{x} - 2)^2}{4}\right], \quad \hat{x} = \max\left(\left|x'\right|, |x|\right). \tag{11.8.16}$$

For CRD,

$$a_{x',x} = 0; \quad \text{for CS } \; a_{x',x} = 1. \tag{11.8.17}$$

Equation (11.8.5) is written in terms of $I = I_l + I_r$ and $Q = I_l - I_r$ in the form of two equations as follows:

$$\mu \frac{dI(x, \mu)}{d\tau} = [\beta + \phi(x)]\left[I(x, \mu) - S_T^I(x, \mu)\right] \tag{11.8.18}$$

and

$$\mu \frac{dQ(x, \mu)}{d\tau} = [\beta + \phi(x)]\left[Q(x, \mu) - S_T^Q(x, \mu)\right], \tag{11.8.19}$$

where S_T^I and S_T^Q are given by

$$S_T^I(x, \mu) = \frac{\phi(x)S^I(x, \mu) + \beta B}{\beta + \phi(x)}, \tag{11.8.20}$$

$$S_T^Q(x, \mu) = \frac{\phi(x)(1 - \mu^2)\bar{P}(x)}{\beta + \phi(x)}. \tag{11.8.21}$$

The line source function for I is

$$S^I(x, \mu) = S(x) + \left(\frac{1}{3} - \mu^2\right) \bar{P}(x), \tag{11.8.22}$$

with

$$S(x) = \frac{1 - \epsilon}{2\phi(x)} \int_{-\infty}^{+\infty} R_{II}(x', x)\, dx' \int_{-1}^{+1} I(x, \mu')\, d\mu' + \epsilon B \tag{11.8.23}$$

and

$$\bar{P}(x) = \frac{3}{16} \frac{(1 - \epsilon)}{\phi(x)} \int_{-\infty}^{+\infty} R_{II}(x', x)\, dx' \int_{-1}^{+1} \left[\left(1 - 3\mu^2\right) I(x', \mu')\right.$$

$$+3\left(1-\mu'^2\right) Q(x', \mu')\Big] d\mu'. \tag{11.8.24}$$

The boundary conditions for a finite atmosphere of total optical depth T are that no radiation is incident on either side of the boundary, that is,

$$I(x, \mu) = Q(x, \mu) = 0 \quad (\tau = 0, \mu < 0; \tau = T, \mu > 0). \tag{11.8.25}$$

The boundary conditions for a semi-infinite model at the lower boundary are

$$I(x, \mu) = B, \quad Q(x, \mu) = 0; \quad \mu > 0, \tag{11.8.26}$$

which are applied below the thermalization depth. The upper boundary condition is the same as (11.8.25).

The transfer equations (11.8.18) and (11.8.19) were solved using Auer's (1967) modified Feautrier finite-difference scheme combined with an iterative process and the scheme of Vardavas and Cram (1974). Some of the results of how the polarization varies are given in figures 11.16, 11.17, 11.18, 11.19 and 11.20. In Figures 11.19 and 11.20 we see that the polarization profiles with maxima in both cores and wings. This removes the uncertainties in Auer *et al.* (1980). Nagendra (1988) computed resonance line polarization profiles for a large number of models using the partial frequency redistribution given in equation (11.8.11) in spherically symmetric media. He used an inverse square law for the opacity, that is,

$$K^L(r) = K_0 r^{-2}, \tag{11.8.27}$$

where $K^L(r)$ is the line centre absorption at r, and K_0 is that at r_0 the innermost boundary of the atmosphere. He employed Schuster boundary conditions, planetary nebula type boundary conditions and discrete space theory (chapter 6) and derived the resonance line polarization for PRD shown in figure 11.21. Resonance line polarization in expanding spherical atmospheres can be studied by using the Rayleigh

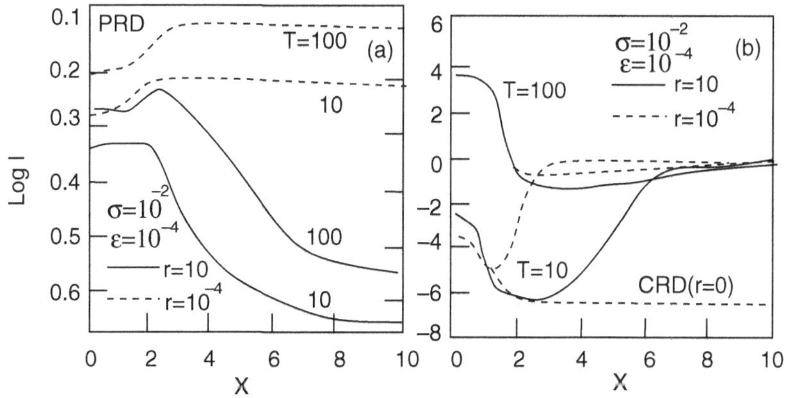

Figure 11.16 PRD profiles for a finite atmosphere of varying thickness T: (a) intensity, (b) polarization. For $T = 10$, the polarization profile for CRD with $\beta = 0$ is also shown (from Rees and Saliba (1982), with permission).

phase function as described in Rees and Saliba (1982). The transfer equation in a spherically symmetric expanding medium in the comoving frame is (see chapter 8)

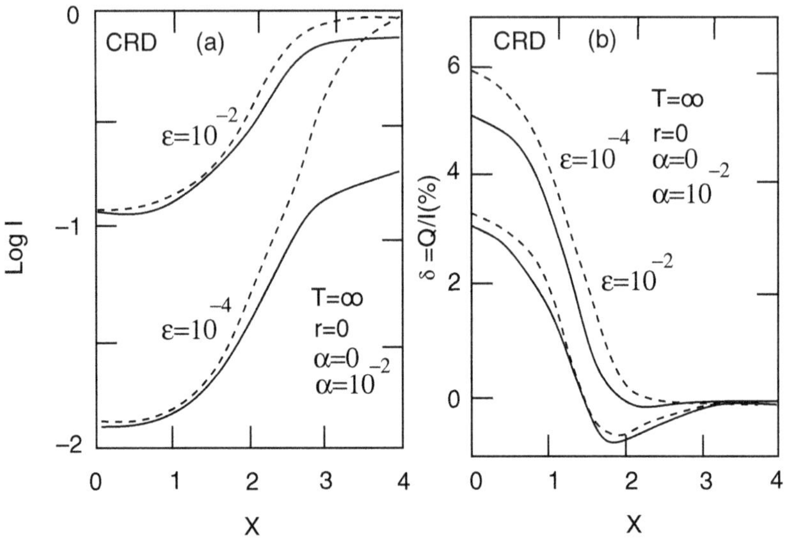

Figure 11.17 A comparison of the CRD profiles for purely Doppler ($a = 0$) and Doppler plus natural ($a = 10^{-2}$) broadening for an isothermal semi-infinite atmosphere with no background continuum: (a) intensity, (b) polarization (from Rees and Saliba (1982), with permission).

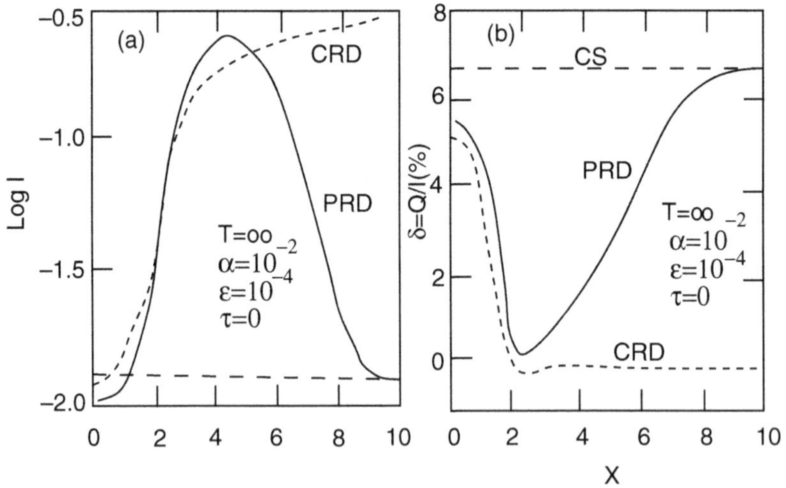

Figure 11.18 A comparison of the CS, CRD and PRD profiles for an isothermal semi-infinite atmosphere with no background continuum: (a) intensity, (b) polarization (from Rees and Saliba (1982), with permission).

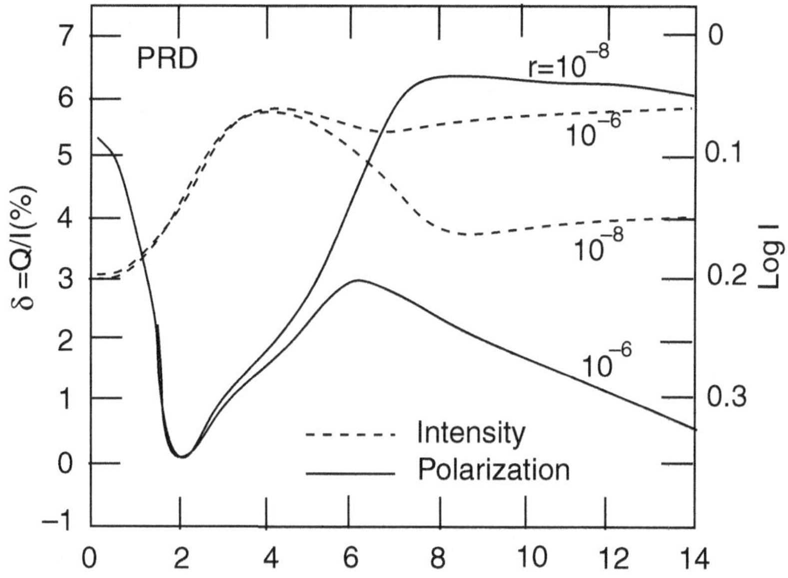

Figure 11.19 The PRD intensity and polarization profiles for a semi-infinite isothermal atmosphere with a background continuum ($\beta = 10^{-6}, 10^{-8}$). Note the core and wing maxima (from Rees and Saliba (1982), with permission).

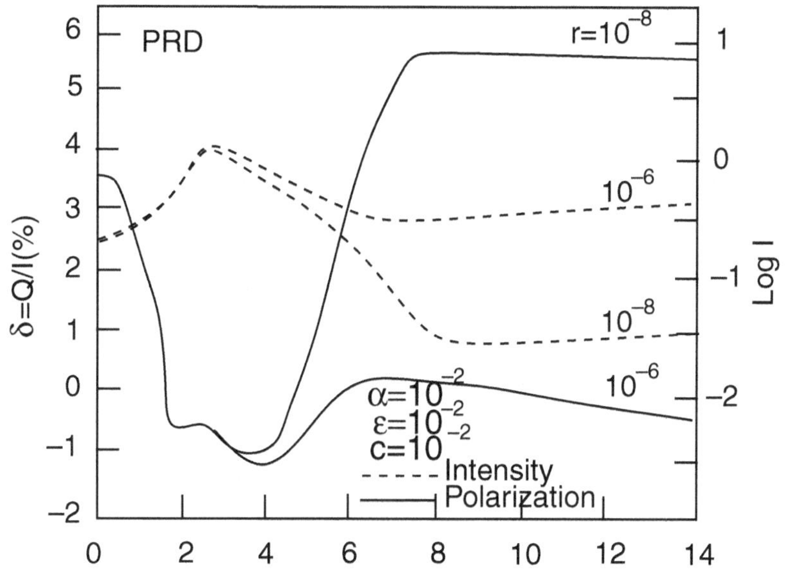

Figure 11.20 The PRD intensity and polarization profiles for the chromospheric model $B = 1 + 100\exp(-\tau)$ ($c = 10^{-2}$) with background continuum ($\beta = 10^{-6}$, 10^{-8}). The polarization profiles have core and wing maxima as in figure 11.19. The wing maximum is barely detectable for $\beta = 10^{-8}$ (from Rees and Saliba (1982), with permission).

$$\mu \frac{\partial}{\partial r} \begin{pmatrix} U_L(x,\mu,r) \\ U_R(x,\mu,r) \end{pmatrix} + \frac{1-\mu^2}{r} \frac{\partial}{\partial \mu} \begin{pmatrix} U_L(x,\mu,r) \\ U_R(x,\mu,r) \end{pmatrix}$$
$$= k_l\,(\beta + \phi(x)) \left\{ \begin{pmatrix} S_L(x,\mu,r) \\ S_R(x,\mu,r) \end{pmatrix} + \begin{pmatrix} U_L(x,\mu,r) \\ U_R(x,\mu,r) \end{pmatrix} \right\}$$
$$+ \left\{ \left(1-\mu^2\right) \frac{V(r)}{r} + \mu^2 \frac{dV(r)}{dr} \right\} \frac{\partial}{\partial x} \begin{pmatrix} U_L(x,\mu,r) \\ U_R(x,\mu,r) \end{pmatrix}, \tag{11.8.28}$$

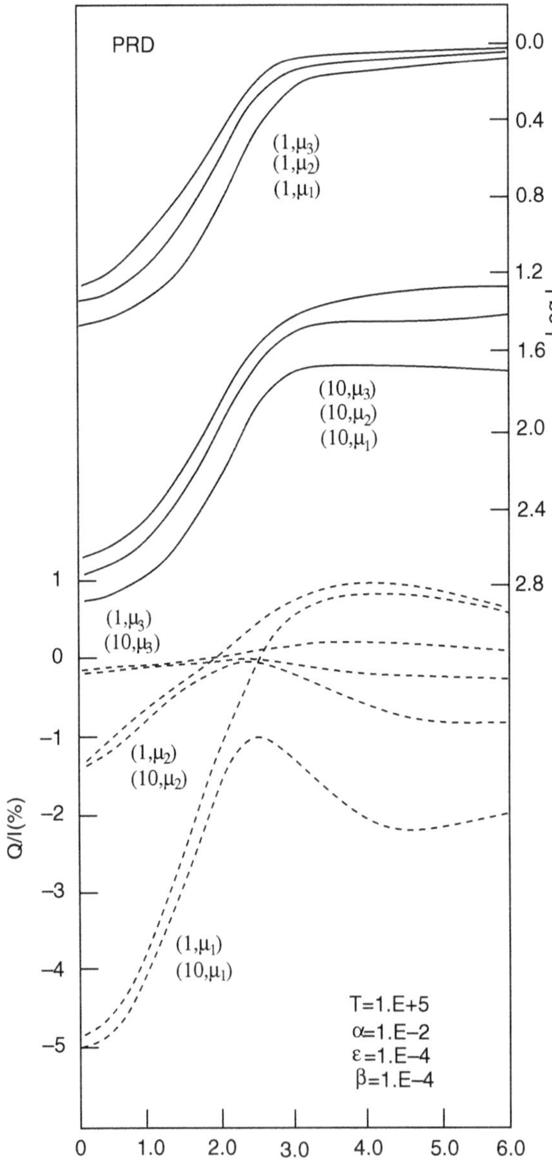

Figure 11.21 The angle dependence of the emergent $I(=I_l+I_r)$ and Q/I $(Q=I_l-I_r)$ profiles corresponding to the PRD scattering mechanism. T is the total optical depth, a is the damping parameter and $R_{II}(x,x')$ is the redistribution function. Three Gauss angle points are used (μ_1, μ_2, μ_3). The first numbers in the brackets refer to B/A, the ratio of the outer to inner radii (from Nagendra (1988), with permission).

where $U_{L,R}(x,\mu,r) = 4\pi r^2 I_{L,R}(x,\mu,r)$ and k_l is the frequency integrated opacity at the line centre, $V(r)$ is the velocity of the medium at radius r in Doppler units or mtu, and all the other symbols have their usual meaning. The normalized absorption profile is represented by a Voigt function

$$\phi(x) = \frac{H(a,x)}{\sqrt{\pi}}. \tag{11.8.29}$$

The line source function $S(x,\mu,r)$ is

$$S(x,\mu,r) = \frac{1-\epsilon}{\phi(x)} \int_{-\infty}^{+\infty} dx' \int_{-1}^{+1} \mathbf{R}(x',\mu';x,\mu) \begin{pmatrix} U_L(x',\mu',r) \\ U_R(x',\mu',r) \end{pmatrix} d\mu' + \frac{1}{2}\epsilon B(r) \begin{pmatrix} 1 \\ 1 \end{pmatrix}. \tag{11.8.30}$$

The redistribution function $R(x',\mu';x,\mu)$ is adopted from Rees and Saliba (1982) as

$$\mathbf{R}(x',\mu';x,\mu) = \mathbf{P}(\mu',\mu)\mathbf{R}(x',x), \tag{11.8.31}$$

where $P(\mu',\mu)$ is the Rayleigh phase matrix for U_L and U_R. The angle-averaged $R(x',x)$ is taken to be

$$R(x',x) = \phi(x')\phi(x). \tag{11.8.32}$$

Now equation (11.8.28) is solved in the frame work of discrete space theory (see chapter 6). The discrete equations for the oppositely directed beams of radiation are

$$\begin{aligned} \mathbf{M}\left[\mathbf{U}_{n+1}^+ - \mathbf{U}_n^+\right] + \rho_c \left[\mathbf{\Lambda}^+\mathbf{U}_{n+\frac{1}{2}}^+ + \mathbf{\Lambda}^-\mathbf{U}_{n+\frac{1}{2}}^-\right] + \tau_{n+\frac{1}{2}}\Phi_{n+\frac{1}{2}}^+\mathbf{U}_{n+\frac{1}{2}}^+ \\ = \tau_{n+\frac{1}{2}}\mathbf{S}_{n+\frac{1}{2}}^+ + \frac{1-\epsilon}{2}\tau_{n+\frac{1}{2}}\left[\mathbf{R}^{++}\mathbf{W}^{++}\mathbf{U}^+ + \mathbf{R}^{+-}\mathbf{W}^{+-}\mathbf{U}^-\right]_{n+\frac{1}{2}} \\ + \mathbf{M}'\mathbf{dU}_{n+\frac{1}{2}}^+, \end{aligned} \tag{11.8.33}$$

$$\begin{aligned} \mathbf{M}\left[\mathbf{U}_n^- - \mathbf{U}_{n+1}^-\right] - \rho_c \left[\mathbf{\Lambda}^+\mathbf{U}_{n+\frac{1}{2}}^- + \mathbf{\Lambda}^-\mathbf{U}_{n+\frac{1}{2}}^+\right] + \tau_{n+\frac{1}{2}}\Phi_{n+\frac{1}{2}}^-\mathbf{U}_{n+\frac{1}{2}}^- \\ = \tau_{n+\frac{1}{2}}\mathbf{S}_{n+\frac{1}{2}}^- + \frac{1-\epsilon}{2}\tau_{n+\frac{1}{2}}\left[\mathbf{R}^{-+}\mathbf{W}^{-+}\mathbf{U}^+ + \mathbf{R}^{--}\mathbf{W}^{--}\mathbf{U}^-\right]_{n+\frac{1}{2}} \\ + \mathbf{M}'\mathbf{dU}_{n+\frac{1}{2}}^-, \end{aligned} \tag{11.8.34}$$

where the different matrices are explained in chapters 6 and 8. Furthermore

$$\mathbf{M} = \lfloor \mu_k \delta_{kk'} \rfloor, \quad \mathbf{U}_n^{\pm} = \lfloor U \rfloor_{k,n}^{\pm}, \quad \left[U^{\pm}\right]_{k,m} = U(x_i, \pm\mu_j; r_n; p),$$

where $k = j + (i-1)J + (p-1)IJ$, $1 \leq k \leq pIJ$, and p is the number of polarization states ($p = 2$ in this case). J and I are the number of angles and frequency points respectively, j and i are the running indices of angle and frequency points respectively (see Varghese (2000)).

$$\left.\begin{aligned} \Phi^{\pm}_{n+\frac{1}{2}} &= (\beta + \phi^{\pm}_{l})_{n+\frac{1}{2}}\, \delta_{kk'}, \\ \phi^{\pm}_{k,n+\frac{1}{2}} &= \phi(x_i, \pm\mu, r_{n+\frac{1}{2}}; p), \\ S^{\pm}_{n+\frac{1}{2}} &= \left[\rho\beta + \epsilon\phi^{\pm}_{k}\right]_{n+\frac{1}{2}} B_{n+\frac{1}{2}} \delta_{kk'}, \\ W^{++}_{n+\frac{1}{2}} &= \left[W^{++}_{k,n+\frac{1}{2}} \delta_{kk'}\right], \\ \phi^{+}_{k,n+\frac{1}{2}} W^{++}_{k,n+\frac{1}{2}} &= a^{++}_{i,n+\frac{1}{2}} c_j, \end{aligned}\right\} \tag{11.8.35}$$

where $a^{++}_{i,n+\frac{1}{2}}$ are the normalized weights given by

$$a^{++}_{i,n+\frac{1}{2}} = \frac{a_i \phi_{kmn+\frac{1}{2}}}{\sum_{k=1}^{2IJ} a_i c_j \sum_{k'=1}^{2IJ} R^{++}_{k,k';n+\frac{1}{2}}}, \tag{11.8.36}$$

with

$$R^{++}_{i,j,i',n+\frac{1}{2}} = R(x_i, \mu_{j'}; x_i, \mu_j; r_n). \tag{11.8.37}$$

The rest of the procedure for obtaining the solution is same as that described in chapter 6 or 8.

Varghese studied the problem of polarization of radiation in a dusty expanding spherical atmosphere. He used the CRD function to find the effects of dust on polarization. The transfer equation in the comoving frame with dust and two states of polarization is written as

$$\begin{aligned} &\mu \frac{\partial}{\partial r}\begin{pmatrix} U_L(x,\mu,r) \\ U_R(x,\mu,r) \end{pmatrix} + \frac{1-\mu^2}{r}\frac{\partial}{\partial \mu}\begin{pmatrix} U_L(x,\mu,r) \\ U_R(x,\mu,r) \end{pmatrix} \\ &= k_l \left[\beta + \phi(x)\left\{\begin{pmatrix} S_L(x,\mu,r) \\ S_R(x,\mu,r) \end{pmatrix} - \begin{pmatrix} U_L(x,\mu,r) \\ U_R(x,\mu,r) \end{pmatrix}\right]\right. \\ &\quad + \left.\left[(1-\mu^2)\frac{V(r)}{r} + \mu^2 \frac{dV(r)}{dr}\frac{\partial}{\partial x}\begin{pmatrix} U_L(x,\mu,r) \\ U_R(x,\mu,r) \end{pmatrix}\right]\right\} \\ &\quad + k_{dust}\begin{pmatrix} S^d_L(x,\mu,r) \\ S^d_R(x,\mu,r) \end{pmatrix} - \begin{pmatrix} U_L(x,\mu,r) \\ U_R(x,\mu,r) \end{pmatrix}, \end{aligned} \tag{11.8.38}$$

where k_{dust} is extinction due to dust (only isotropic scattering due to dust is taken into account), S^d_L and S^d_R are the source functions due to dust and are given by

$$S^d_{L,R}(x,\mu,r) = (1-\varpi)B_{dust} + \frac{1}{2}\varpi \int_{-\infty}^{+1} P(\mu,\mu',r) U_{L,R}(x,\mu',r)\, d\mu', \tag{11.8.39}$$

where B_{dust} is the Planck function for the dust emission, ϖ is the albedo of the dust and P is the isotropic and coherent scattering phase functions. The quantity B_{dust} is normally neglected because the re-emission is far away from the line centre and may not contribute to the line radiation. Therefore we can neglect this term. Equation (11.8.38) has been solved using discrete space theory and some of the results of these calculations are given in figure 11.22.

Mohan Rao and Rangarajan (1993) studied the influence of collisional redistribution on resonance line polarization and found that polarization at the line centre is a monotonic function of the coherence parameter. Rangarajan (1997) studied the resonance line polarization when wave motion is present. He solved the transfer

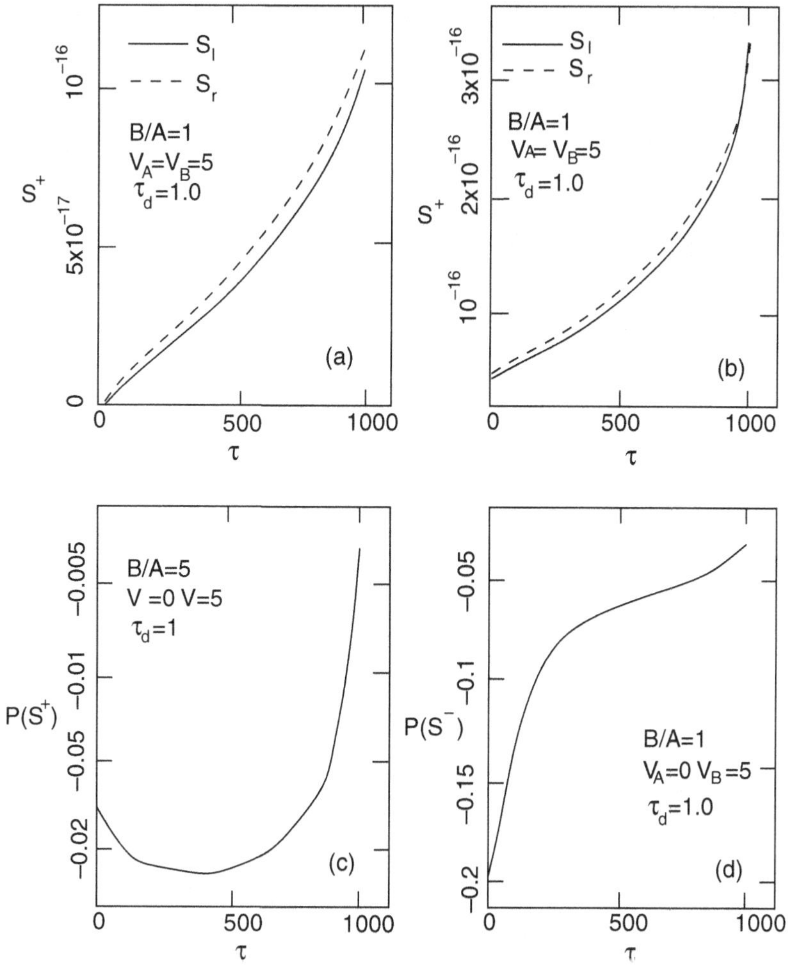

Figure 11.22 Source functions ((a) and (b)) and their degrees of polarization in a spherically symmetric expanding medium with dust ((c) and (d)) (from Varghese (2000), with permission).

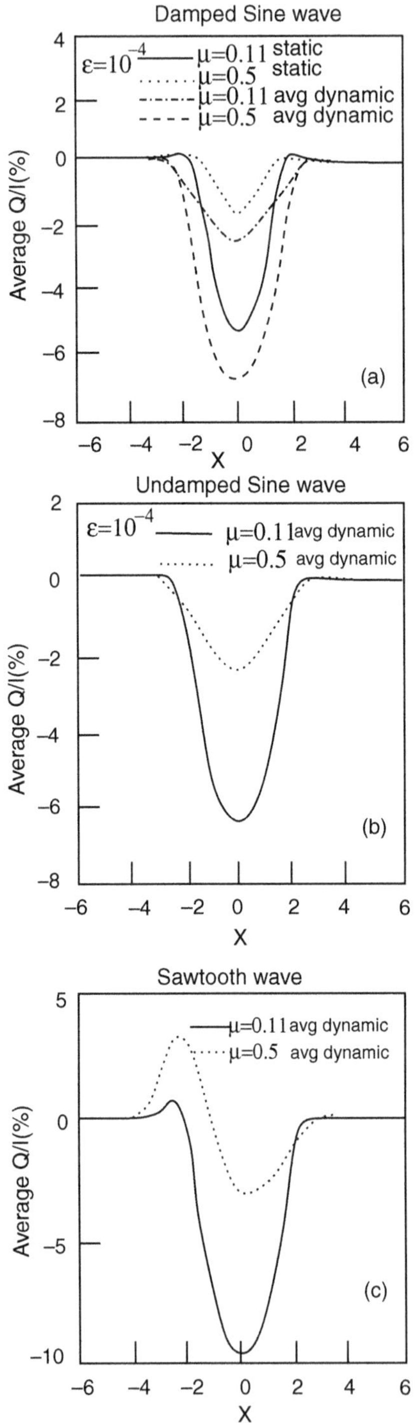

Figure 11.23 Time average of the polarization over one full period: (a) a damped sine wave, (b) an undamped sine wave, (c) a sawtooth wave (from Rangarajan (1997), with permission).

equation for the two states of polarization in plane parallel geometry and in the rest frame. He applied this theory to Ca II K-like lines in a chromosphere, adopting $\beta = 10^{-7}$, $\epsilon = 10^{-4}$ and a depth dependent damping parameter $a = 10^{-3}$. The Planck function used was

$$B(\tau) = 1 + 10\tau_c^{-0.9} + 100\exp(-70.7\tau_c^{1/2}), \qquad (11.8.40)$$

giving a temperature minimum at about $\tau_c = 10^{-2}$, where τ_c is the continuum optical depth. Three types of waves have been used: (1) a damped sine wave, (2) an undamped sine wave and (3) a sawtooth wave. The results of polarization are shown in figures 11.23.

Poutenen *et al.* (1996) computed the Compton spectrum reflected from cold matter. They derived the basic characteristics of the polarized spectra produced by Rayleigh and Compton scattering by using discrete space theory (see figure 11.24). They derived the angular and polarization properties of the fully relativistic Compton scattering cross section together with the photoelectric absorption and the generation of a fluorescent Fe line. The transfer equation for polarized radiation in plane parallel geometry is

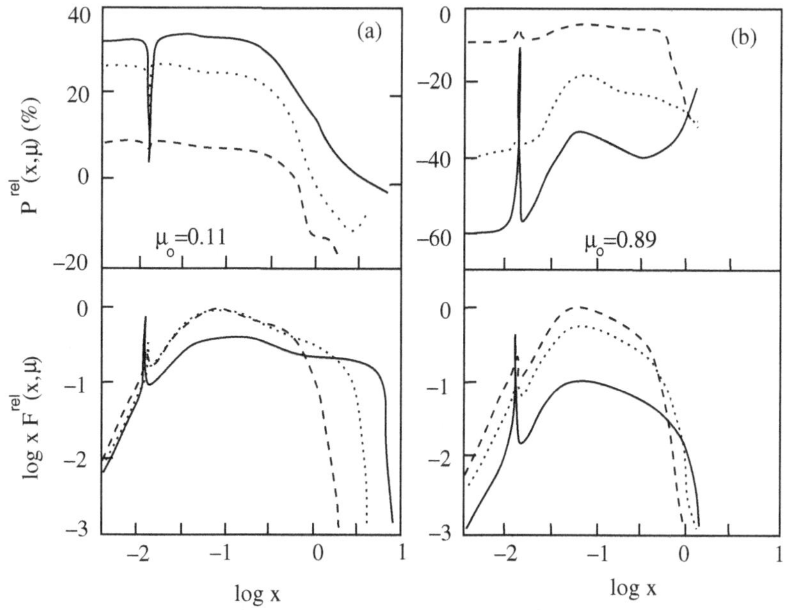

Figure 11.24 The reflected flux $F^{ref}(x, \mu)$ and polarization $P^{ref}(x, \mu)$ at different angles of view μ for an incident cone of unpolarized radiation with opening angle $\cos^{-1}\mu_0$, (a) $\mu_0 = 0.11$ and (b) $\mu_0 = 0.89$. The incident flux has a power law spectrum $F^{in}(x', \mu') = \delta(\mu' - \mu_0)(x')^{-1}$, where x' extends upto $x' = 8$. The solid, dotted and dashed curves corresponds to $\mu = 0, 0.5$ and 0.89 respectively (from Poutenen *et al.* (1996), with permission).

$$\mu \frac{d\bar{\mathbf{I}}(x, \mu, \tau)}{d\tau} = \bar{\mathbf{I}}(x, \mu, \tau) + \varpi(x) \int_0^{\infty} \left(\frac{x}{x'}\right) dx' \int_{-1}^{+1} d\mu' \hat{\mathbf{R}}(x, \mu; x', \mu') \bar{\mathbf{I}}(x', \mu', \tau), \tag{11.8.41}$$

where $\bar{\mathbf{I}} = [I, Q]^T$, $d\tau = n_H \left[\sigma_{ph}(x) + \sigma_{sc}(x)\right] dz$ is the optical depth corresponding to the geometrical depth dz, $\sigma_{ph}(x)$ is the photoelectric absorption cross section, σ_{sc} is the scattering cross section, which is the sum of coherent and incoherent cross sections $\sigma_{coh}(x)$ and $\sigma_{incoh}(x)$, and $x = h\nu/m_e c^2$. The albedo for single scattering $\varpi(x)$ is defined in this case as

$$\varpi(x) = \frac{\sigma_s c(x)}{\sigma_{ph}(x) + \sigma_{sc}(x)}. \tag{11.8.42}$$

$\hat{\mathbf{R}}(x, \mu; x', \mu')$ is the azimuth-averaged polarized 2×2 matrix. The general form of $\hat{\mathbf{R}}$ used by Poutenen *et al.* (1996) is

$$\hat{\mathbf{R}}(x, \mu; x', \mu') = \hat{\mathbf{R}}_{Rayl}(x, \mu; x', \mu') \frac{\sigma_{coh}(x)}{\sigma_{sc}(x)} + \hat{\mathbf{R}}_{Comp}(x, \mu; x', \mu') \frac{\sigma_T \sigma_{incoh}(x)}{\sigma_{KN}(x) \sigma_{sc}(x)} + \hat{\mathbf{R}}_{fluor}(x, \mu; x', \mu') \frac{\sigma_T}{\sigma_{sc}(x)} \tag{11.8.43}$$

where $\hat{\mathbf{R}}_{Rayl}$ is the classical coherent Rayleigh scattering redistribution function, $\hat{\mathbf{R}}_{Comp}$ is the incoherent Compton scattering, $\hat{\mathbf{R}}_{fluor}$ is the fluorescent line production redistribution function, σ_{KN} is the angle integrated Klein–Nishina cross section for Compton scattering and σ_T is the Thompson scattering cross section. The above redistribution is described in chapter 1. Equation (11.8.41) has been solved in the framework of discrete space theory (see chapter 6 and Poutenen *et al.* (1996) for details). Green's matrix is computed numerically by solving the polarized transfer equation in an optically thick plane parallel slab using discrete space theory. The Green's function is given by

$$\mathbf{G} = \mathbf{r}(1, N + 1), \tag{11.8.44}$$

where $\mathbf{r}(1, N+1)$ is the diffuse reflection operator of the radiation diffusely scattered between the layers 1 and $N+1$. This is obtained through internal field computations.

Faurobert (1987, 1988) calculated the linear polarization of the resonance lines in the absence of a magnetic field in finite and semi-infinite media.

Exercises

11.1 Derive the phase matrix in equation (11.4.13) and the relation given in equations (11.4.14) and (11.4.15).

11.2 Derive equations (11.6.15) and (11.6.16).

11.3 In equations (11.6.15) and (11.6.16), if the isotropic scattering phase function is used instead of the Rayleigh phase function, what sort of equation do you obtain? Derive these equations.

11.4 Derive the condition on phase functions to satisfy the conservation of radiant flux in a conservatively scattering medium.

11.5 Obtain equation (11.6.29) from the condition of flux conservation. Apply Gauss, Labotto and trapezoidal points and find which satisfies this relation most accurately.

11.6 Derive the transmission and reflection operators of equations (11.6.13) and (11.6.14) using discrete space theory. Obtain the condition on the optical (critical) depth.

11.7 If

$$\mu \frac{d\begin{pmatrix} I_l(\tau,\mu) \\ I_r(\tau,\mu) \end{pmatrix}}{d\tau} = \begin{pmatrix} I_l(\tau,\mu) \\ I_r(\tau,\mu) \end{pmatrix}$$

$$-\frac{3}{8}\int_{-1}^{+1}\left[\begin{pmatrix} 2\left(1-\mu^2\right)\left(1-\mu'^2\right)+\mu^2\mu'^2 & \mu^2 \\ \mu'^2 & 1 \end{pmatrix}\begin{pmatrix} I_l(\tau,\mu) \\ I_r(\tau,\mu) \end{pmatrix}\right]dt',$$

$$I = (I_l + I_r), \quad Q = (I_l - I_r)$$

and $\hat{\mathbf{I}} = [I\,Q]^T$, write down the transfer equation for $\bar{\mathbf{I}}$ and the corresponding phase matrix.

REFERENCES

Abhyankar, K.D., 1996, *Quarterly J. R. Astron. Soc.*, **37**, 281.

Auer, L.H., 1967, *ApJ*, **150**, L3.

Auer, L.H., Rees, D.E., Stenflo, J.O., 1980, *A&A*, **88**, 302.

Barman, S.K., Peraiah, A., 1991, *Bull. Astron. Soc. India*, **19**, 37.

Bastian, P., 1982, *A&A Suppl. Ser.*, **48**, 153.

Bastian, P., 1985, *A&A Suppl. Ser.*, **59**, 277.

Bond, G.R., Siewart, C.E., 1967, *ApJ*, **150**, 357.

Case, K.M., 1960, *Ann. Phys.*, **9**, 1.

Case, K.M., Zweifel, P.F., 1967, *Linear Transport Theory*, Addison-Wesley, Reading, MA.

Cassinelli, J.P., Nordsiek, K.H., Murison, M.A., 1987, *ApJ*, **317**, 290.

Chandrasekhar, S., 1946, *ApJ*, **103**, 351.

Chandrasekhar, S., 1960, *Radiative Transfer*, Dover, New York.

Daguchi, S., Watson, W.D., 1985, *ApJ*, **289**, 621.

Dumont, S., 1971, *JQSRT*, **11**, 1675.

Dumont, S., Omont, A., Pecker, J.C., 1973, *Sol. Phys.*, **28**, 271.

Dumont, S., Omont, A., Pecker, J.C., Rees, D.E., 1977, *A&A*, **54**, 675.

Faurobert, M., 1987, *A&A*, **178**, 269.

Faurobert, M., 1988, *A&A*, **194**, 268.

Grant, I.P., Hunt, G.E., 1968, *JQSRT*, **8**, 1817.

Hamilton, D.R., 1947, *ApJ*, **106**, 457.

Harrington, J.P., Collins II, G.W., 1968, *ApJ*, **151**, 1051.

Hiltner, W.A., 1947, *ApJ*, **106**, 231.

Hovenier, J.W., 1987, *A&A*, **183**, 363.

Hummer, D.G., 1962, *MNRAS*, **125**, 21.

Jefferies, J.T., White, O.R., 1960, *ApJ*, **132**, 767.

Kneer, F., 1975, *ApJ*, **200**, 367.

Magalhãs, A.M., Coyne, G.V., Codina-Landaberry, S.J., Gneiding, C., 1986, *A&A*, **154**, 1.

Mihalas, D., Kunasz, P.B., Hummer, D.G., 1976, *ApJ*, **206**, 515.

Mohan Rao, D., Rangarajan, K.E., 1993, *A&A*, **274**, 993.

Nadeau, R., Bastian, P., 1986, *ApJ*, **307**, L5-L8.

Nagendra, K.N., 1988, *ApJ*, **335**, 269.

Omont, A., Smith, E.W., Cooper, J., 1973, *ApJ*, **182**, 283.

Peraiah, A., 1975, *A&A*, **40**, 75.

Peraiah, A., 1976, *A&A*, **46**, 237.

Peraiah, A., 1978, *Kodaikanal Obs. Bull. Ser. A*, **2**, 115.

Plass, G.N., Kattawar, G.W., Catchings, F.E., 1973, *Appl. Opt.*, **12**, 314.

Poutenen, J., Nagendra, K.N., Svensson, R., 1996, *MNRAS*, **283**, 892.

Rangarajan, K.E., 1997, *A&A*, **320**, 263.

Rangarajan, K.E., Mohan Rao, D., Abhyankar, K.D., 1994, *Bull. Astron. Soc. India*, **22**, 465.

Rees, D.E., 1978, *Publ. Astron. Soc. Japan*, **30**, 455.

Rees, D.E., Saliba, G.J., 1982, *A&A*, **115**, 1.

Siewart, C.E., Fraley, S.K., 1967, *Ann. Phys.*, **43**, 388.

Smith, O.J., Siewart, C.E., 1967, *J. Math. Phys.*, **8**, 2467.

Sobolev, V.V., 1963, *A Treatise on Radiative Transfer*, translated by S.I. Gaposchkin, Van Nostrand Company Inc., New York.

Stenflo, J.O., 1976, *A&A*, **46**, 61.

Stenflo, J.O., 1980, *Proceeding of Conference: Solar Instrumentation; What next?*, Sacramento Peak Observatory, Sacramento.

Stenflo, J.O., Bauer, T.G., Elmore, D.F., 1980, *A&A*, **84**, 60.

Stokes, Sir George, 1852, *Trans. Camb. Philos. Soc.*, **9**, 399.

Vardavas, I.M., Cram, L.E., 1974, *Sol. Phys.*, **38**, 3677.

Varghese, B.A., 2000, PhD thesis, Bangalore University.

Wiehr, E., 1981, *A&A*, **35**, 54.

Chapter 12

Polarization in magnetic media

12.1 Polarized light in terms of I, Q, U, V

Many books discuss the representation of polarized radiation. We shall start from the basics of the subject from classical electromagnetic theory. We need to understand the density matrix if we want to study polarized transfer of radiation from the quantum mechanical point of view. Polarized light is described by the four Stokes parameters I, Q, U, V, where I denotes intensity, Q and U describe linear polarization and V describes circular polarization.

The properties of light are described in terms of the electric wave vector $\mathbf{E}$ in the plane perpendicular to the direction of propagation (see Rees (1987)). Let us consider a quasi-monochromatic wave which is the superposition of many randomly timed and statistically independent wave trains with a mean frequency ν (and wavelength λ). If the resultant wave has a spectral width $\Delta\nu \ll \nu$, then such a wave propagating in the $+z$-direction will have complex analytic representations of the mutually orthogonal components of $\mathbf{E}$, given by

$$\left.\begin{aligned} E_x(t) &= a_z(t)\exp\bigl[i(\tilde{\phi}_x(t) - 2\pi\nu t + 2\pi z/\lambda)\bigr], \\ E_y(t) &= a_y(t)\exp\bigl[i(\tilde{\phi}_y(t) - 2\pi\nu t + 2\pi z/\lambda)\bigr], \end{aligned}\right\} \tag{12.1.1}$$

where x, y and z form a right handed coordinate system; a_x and a_y are the amplitudes and $\tilde{\phi}_x$ and $\tilde{\phi}_y$ are the phases, which vary slowly with time. They are regarded as approximately constant over any interval short in comparison with the coherence time $(\Delta\nu)^{-1}$ of the wave. The electric vector $\mathbf{E}$ traces an ellipse at any given point z. We need to establish the sign convention. A clockwise rotation as seen by an observer receiving the radiation is called *right handed* circular polarization and a counterclockwise rotation is called *left handed* circular polarization. In terms of phase difference

$$\delta = \tilde{\phi}_x - \tilde{\phi}_y, \tag{12.1.2}$$

where $\delta > 0$ corresponds to right handed circularly polarized light. We shall define the Stokes parameters through the following experiment.

Let light pass through a compensator that subjects E_y to a phase retardation ϵ relative to E_x, followed by a polarizer that transmits linearly polarized light inclined at an angle θ counterclockwise to the x-axis. Then the component of the electric vector of the transmitted light in the θ-direction is

$$E(t; \theta, \epsilon) = E_x \cos\theta + e_y \exp(i\epsilon) \sin\theta. \tag{12.1.3}$$

Here the retardation is represented by the experiment $\exp(i\epsilon)$. The corresponding transmitted intensity (omitting a constant of proportionality) is

$$I_{trans}(\theta, \epsilon) = \left\langle E(t; \theta, \epsilon) E^{\star}(t; \theta, \epsilon) \right\rangle, \tag{12.1.4}$$

where * denotes the complex conjugate and $\langle \cdots \rangle$ denotes a time average over the observation period which is $\gg \nu^{-1}$, the natural period of the wave.

One can make six intensity measurements with following angle settings:

$\epsilon = 0$ (no compensator, $\theta = 0, \pi/4, \pi/2, 3\pi/4$ with the transmission of linear polarization);

$\epsilon = \pi/2$ (compensator is a quarter wave plate);

$\theta = \pi/4$ and $3\pi/4$ with the transmission of right and left circularly polarized light.

We can write these six Is using the symbols $a,\ b,\ c,\ d,\ e,\ f$ as

$$\left.\begin{aligned} I_a &= I_{trans}(0, 0), & I_b &= I_{trans}\left(\frac{\pi}{2}, 0\right), \\ I_c &= I_{trans}\left(\frac{\pi}{4}, 0\right), & I_d &= I_{trans}\left(\frac{3\pi}{4}, 0\right), \\ I_e &= I_{trans}\left(\frac{\pi}{4}, \frac{\pi}{2}\right), & I_f &= I_{trans}\left(\frac{3\pi}{4}, \frac{\pi}{2}\right). \end{aligned}\right\} \tag{12.1.5}$$

Now, we define the Stokes parameters associated with the incident wave $\mathbf{E}$ as follows:

$$\left.\begin{aligned} I &= I_a + I_b, \\ Q &= I_a - I_b, \\ U &= I_c - I_d, \\ V &= I_c + I_d. \end{aligned}\right\} \tag{12.1.6}$$

In terms of the the amplitudes and phase of $\mathbf{E}$ we write

$$\left.\begin{aligned} I &= \langle a_x^2 \rangle + \langle a_y^2 \rangle, \\ Q &= \langle a_x^2 \rangle - \langle a_y^2 \rangle, \\ U &= 2\langle a_x a_y \cos\delta \rangle, \\ V &= 2\langle a_x a_y \sin\delta \rangle. \end{aligned}\right\} \quad (12.1.7)$$

From equations (12.1.6) and (12.1.7), we note that $V > 0$ ($\sin\delta > 0$) implies an excess of right handed (clockwise) polarization and $V < 0$ ($\sin\delta < 0$) implies left handed (counterclockwise) polarization.

12.2 Transfer equation for the Stokes vector

The vector transfer equation for the Stokes vector $\mathbf{I} = (I, Q, U, V)^T$ (T denotes transpose) is written for polarized light (see Rees (1987)) as

$$\frac{d\mathbf{I}}{ds} = \mathbf{K}(\mathbf{I} - \mathbf{S}). \quad (12.2.1)$$

In figure 12.1, the Stokes parameters are defined with reference to the coordinate system X, Y, Z, the Z-axis being in the direction of the observer. $\mathbf{K}$ is the absorption matrix and $\mathbf{S}$ the source vector. The absorption matrix in the case of Zeeman split spectral lines depends on the magnetic field vector $\mathbf{B}$, the inclination γ to the line of sight and χ the azimuthal angle relative to the X-axis (see figure 12.1). The theory formulated below is similar to that of Landi Degl'Innocenti (1976) and Landolfi and Landi Degl'Innocenti (1982).

Assuming LTE in both continuum and lines and neglecting continuum polarization, we have

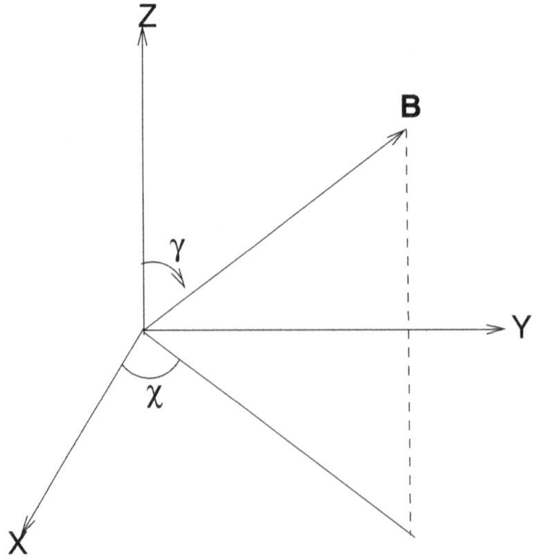

Figure 12.1 Schematic diagram of the Stokes vector **I** and the direction of the emergent vector **B**.

$$\mathbf{S} = B_\nu(T_e)\mathbf{I}_0, \tag{12.2.2}$$

where $\mathbf{I}_0 = [1, 0, 0, 0]^T$, $B_\nu(T_e)$ is the Planck function at the local electron temperature and

$$\mathbf{K} = \mathbf{K}_c\mathbf{1} + \mathbf{K}_0\boldsymbol{\Phi}, \tag{12.2.3}$$

where $\mathbf{K}_c$ is the continuum opacity, $\mathbf{1}$ is the 4×4 unit matrix and $\mathbf{K}_0$ is the line centre opacity for zero damping and zero magnetic field. The line absorption matrix $\boldsymbol{\Phi}$ is given by

$$\boldsymbol{\Phi} = \begin{pmatrix} \phi_I & \phi_Q & \phi_U & \phi_V \\ \phi_Q & \phi_I & \phi'_V & -\phi'_U \\ \phi_U & -\phi'_V & \phi_I & \phi'_Q \\ \phi_V & \phi'_U & -\phi'_Q & \phi_I \end{pmatrix}, \tag{12.2.4}$$

where

$$\left.\begin{aligned}
\phi_I &= \frac{1}{2}\phi_p \sin^2\gamma + \frac{1}{4}(\phi_r + \phi_b)\left(1 + \cos^2\gamma\right), \\
\phi_Q &= \frac{1}{2}\left[\phi_p - \frac{1}{2}(\phi_r + \phi_b)\right]\sin^2\gamma\cos 2\gamma, \\
\phi_U &= \frac{1}{2}\left[\phi_p - \frac{1}{2}(\phi_r + \phi_b)\right]\sin^2\gamma\sin 2\chi, \\
\phi_V &= \frac{1}{2}(\phi_p - \phi_b)\cos\gamma, \\
\phi'_Q &= \frac{1}{2}\left[\phi'_p - \frac{1}{2}(\phi'_r + \phi'_b)\right]\sin^2\gamma\cos 2\chi, \\
\phi'_U &= \frac{1}{2}\left[\phi'_p - \frac{1}{2}(\phi'_r + \phi'_b)\right]\sin^2\gamma\sin 2\chi, \\
\phi'_V &= \frac{1}{2}(\phi'_r - \phi'_b)\cos\gamma
\end{aligned}\right\} \tag{12.2.5}$$

(see also Unno (1956)), where $\phi_{p,b,r}$ are the absorption profiles and $\phi'_{p,b,r}$ are anomalous dispersion profiles. The prime notation is introduced to represent an anomalous dispersion profile which has a shape similar to the negative of the derivative of the corresponding absorption profile. The indices correspond to a normal Zeeman triplet (p = the unshifted π-component, b, r = the blue and red shifted components). If m_{upper} and m_{lower} are the magnetic quantum numbers of the Zeeman sublevels in the upper and lower energy levels of the line forming transition and

$$\delta m = m_{upper} - m_{lower}, \tag{12.2.6}$$

then

$$\Delta m = \begin{cases} +1 \equiv b, \\ 0 \equiv p, \\ -1 \equiv r. \end{cases} \tag{12.2.7}$$

The profiles are

$$\left.\begin{aligned} \phi_p &= H\,(a, v + v_{los}), & \phi'_p &= 2F\,(a, v + v_{los}), \\ \phi_b &= H\,(a, v + v_B + v_{los}), & \phi'_b &= 2F\,(a, v + v_B + v_{los}), \\ \phi_r &= H\,(a, v - v_B + v_{los}), & \phi'_r &= 2F\,(a, v - v_B + v_{los}), \end{aligned}\right\} \tag{12.2.8}$$

where $H(a, v)$ and $F(a, v)$ are the Voigt and Faraday–Voigt functions given by

$$H(a, v) = \frac{a}{\pi} \int_{-\infty}^{+\infty} \frac{\exp(-y^2)}{(v - y)^2 + a^2} dy \tag{12.2.9}$$

and

$$F(a, v) = \frac{1}{2\pi} \int_{-\infty}^{+\infty} \frac{(v - y)\exp(-y^2)}{(v - y)^2 + a^2} dy, \tag{12.2.10}$$

where a is the damping constant, v is the wavelength measured from the line centre in the laboratory frame, v_B is the wavelength of Zeeman splitting and v_{los} is the Doppler shifted wavelength produced by the macroscopic velocity component measured positively in the direction of the observer. These are given as follows:

$$a = \frac{\Gamma \lambda_0^2}{4\pi c \Delta\lambda_D}, \tag{12.2.11}$$

where Γ is the line damping, λ_0 is the laboratory line centre wavelength and c is the velocity of light;

$$v = \frac{(\lambda - \lambda_0)}{\Delta\lambda_D}, \tag{12.2.12}$$

where λ is the wavelength of the line;

$$v_B = \frac{g e \lambda_0^2 B}{4\pi m c^2 \Delta\lambda_D}, \tag{12.2.13}$$

and is the absolute shift of each σ-component of a line with the Landeé g-factor, and with e the electron charge and m the mass;

$$v_{los} = \lambda_0 \frac{\mathbf{v} \cdot \mathbf{n}}{c \Delta\lambda_D}, \tag{12.2.14}$$

and is the Doppler shift caused by the macroscopic velocity field,with $\mathbf{v}$ the velocity vector and $\mathbf{n}$ the unit vector along the direction of the observer.

It should be noted that

$$\lim_{B \to 0} \mathbf{\Phi} = \phi \mathbf{1},$$

where $\phi = H(a, v + v_{los})$. The quantities a, v, v_B and v_{los} are measured in units of Doppler width. For an accurate evaluation of $H(a, v)$ and $F(a, v)$, see Humlicek (1982).

If one wants to work in frequency units rather than wavelength units one should observe the following transformation to equations (12.2.8):

$$v,\, v_B,\, v_{los}(\text{wavelength}) \rightarrow -v,\, -v_B,\, -v_{los}(\text{frequency}). \tag{12.2.15}$$

$H(a, v)$ is a symmetric function, while $F(a, v)$ is an asymmetric function in v, or

$$F(a, -v) = -F(a, v); \quad F(a, v) > 0 \quad \text{if} \quad v > 0. \tag{12.2.16}$$

If we consider a thin layer dz between z and $z + dz$ with no radiation incident at z, the incident and transmitted Stokes vectors $\mathbf{I}(z)$ and $\mathbf{I}(z + dz)$ are approximately related by the following relation:

$$\mathbf{I}(z + dz) = \mathbf{MI}(z), \tag{12.2.17}$$

where $\mathbf{M}$ is the Muller matrix for the layer, given by

$$\mathbf{M} = \begin{pmatrix} 1 & 0 & 0 & 0 \\ 0 & \cos\delta_V & -\sin\delta_V & 0 \\ 0 & \sin\delta_V & \cos\delta_V & 0 \\ 0 & 0 & 0 & 1 \end{pmatrix}, \tag{12.2.18}$$

δ_V being the retarded angle with the layer dz as a circular retarder and

$$\delta_V = K_0\phi_V'\, dz; \tag{12.2.19}$$

and

$$\delta_V > 0 \quad \text{if} \quad v > v_B \quad \text{or} \quad v < -v_B, \tag{12.2.20}$$

which means that the layer acts as a left handed (counterclockwise) circular retarder in these wavelength ranges.

12.3 Solution of the vector transfer equation with the Milne–Eddington approximation

In the Milne–Eddington model the source function varies linearly with the optical depth τ, or

$$B_\nu(T_e) = B_0 + B_1\tau. \tag{12.3.1}$$

Furthermore, $\eta_0 = k_0/k_c$, the opacity ratio, the line damping, the Doppler width, the magnetic vector and the line of sight velocity (and hence $\mathbf{\Phi}$ matrix) are constant. Then from equations (12.2.1) and (12.2.3), we obtain

$$\mu\frac{d\mathbf{I}}{d\tau} = \mathbf{K}'\,(\mathbf{I} - \mathbf{S}), \tag{12.3.2}$$

where

$$\mathbf{K}' = \mathbf{1} + \eta_0 \mathbf{\Phi}. \tag{12.3.3}$$

Equation (12.3.2) has a formal solution given by (at $\tau = 0$)

$$\mathbf{I}(0, \mu) = \int_0^\infty \exp\left(-\frac{\mathbf{K}'\tau}{\mu}\right) \mathbf{K}'\mathbf{S}\frac{d\tau}{\mu}. \tag{12.3.4}$$

Integrating by parts, we obtain

$$\mathbf{I}(0, \mu) = \mathbf{S}(0) + \int_0^\infty \exp\left(-\frac{\mathbf{K}'\tau}{\mu}\right) \frac{d\mathbf{S}}{d\tau} d\tau. \tag{12.3.5}$$

From equations (12.2.2) and (12.2.1), we get

$$\mathbf{S}(0) = B_0 \mathbf{I}_0 \tag{12.3.6}$$

and

$$\frac{d\mathbf{S}}{d\tau} = B_1 \mathbf{I}_0, \tag{12.3.7}$$

then the integral in equation (12.3.5) can be evaluated analytically to give

$$\mathbf{I}(0, \mu) = (B_0 + \mu B_1 \mathbf{K}_1^{-1}) \mathbf{I}_0. \tag{12.3.8}$$

If

$$\eta_{I,Q,U,V} = \eta_0 \phi_{I,Q,U,V} \tag{12.3.9}$$

and

$$\rho_{Q,U,V} = \eta_0 \phi'_{Q,U,V}, \tag{12.3.10}$$

the emergent Stokes parameters are given by

$$\left.\begin{aligned}
I &= B_0 + \frac{\mu B_1}{\Delta} \left\{(1+\eta_I)\left[(1+\eta_I)^2 + \rho_Q^2 + \rho_U^2 + \rho_V^2\right]\right\}, \\
Q &= -\frac{\mu B_1}{\Delta} \left[(1+\eta+I)^2 \eta_Q + (1+\eta_I)(\eta_V \rho_U - \eta_U \rho_V) + \rho_Q W\right], \\
U &= -\frac{\mu B_1}{\Delta} \left[(1+\eta_I)^2 \eta_U + (1+\eta_I)(\eta_Q \rho_V - \eta_V \rho_Q) + \rho_U W\right], \\
V &= -\frac{\mu B_1}{\Delta} \left[(1+\eta_I)^2 \eta_V + \rho_V W\right],
\end{aligned}\right\} \tag{12.3.11}$$

where

$$W = \eta_Q \rho_Q + \eta_U \rho_U + \eta_V \rho_V \tag{12.3.12}$$

and

$$\Delta = (1+\eta_I)^2 \left[(1+\eta_I)^2 - \eta_Q^2 - \eta_U^2 - \eta_V^2 + \rho_Q^2 + \rho_U^2 + \rho_V^2\right] - W^2. \tag{12.3.13}$$

Δ is the determinant of the matrix $\mathbf{K}_1'$.

Unno (1956), Rachkovsky (1962a,b) and Beckers (1969a,b) formulated the LTE transfer equations for Zeeman lines using the classical theory of oscillators of absorption and emission. Rachkovsky (1963) attempted a non-LTE approach for the solution of the Zeeman split lines using a complete redistribution in the lines. A survey of developments is given in Rees (1987) and reference therein. Non-LTE polarized transfer is much more complicated and this can be seen from workshops on solar polarization held at Pittsburgh (Stenflo and Nagendra 1996) and Bangalore (Nagendra and Stenflo 1999).

12.4 Zeeman line transfer: the Feautrier method

Auer *et al.* (1997) applied the Feautrier technique to the Stokes vector problem by omitting the magneto-optical (birefringence) terms (by putting $\phi'_Q = \phi'_U = \phi'_V = 0$). They transformed the first order equation to a second order one. We describe the method following Rees and Murphy (1987).

Let $\mathbf{I}(+v, +\mathbf{n})$ be the Stokes radiation vector at wavelength v travelling in the direction $+\mathbf{n}$ out of the atmosphere and let $\mathbf{I}(-v, -\mathbf{n})$ be its counterpart on the opposite side of the line at wavelength $-v$ travelling in the opposite direction. The quantity $-\mathbf{n}\ \mathbf{I}(-v, -\mathbf{n})$ is to be understood as defined with respect to the right handed reference frame $x\tilde{y}\tilde{z}$ where $\tilde{y} = -y$ and $\tilde{z} = -z$, with the $\tilde{z}$-axis parallel to $-\mathbf{n}$. In this frame $\mathbf{B}$ has an inclination $\pi - \gamma$ to the z-axis and an azimuth $-\chi$ relative to the x-axis. We write the transfer equation for the Stokes vectors as

$$\frac{d\mathbf{I}}{dz} = -\mathbf{KI} + \mathbf{j} = -\mathbf{K}(\mathbf{I} - \mathbf{S}), \tag{12.4.1}$$

where

$$\mathbf{K} = \mathbf{k}_c \mathbf{1} + \mathbf{k}_0 \mathbf{\Phi} \tag{12.4.2}$$

and

$$\mathbf{j} = \mathbf{k}_c B_\nu \mathbf{I}_0 + \mathbf{k}_0 S_L \mathbf{\Phi} \mathbf{I}_0. \tag{12.4.3}$$

The physical assumptions are: (i) the magnetic field is strong enough that there are no quantum interferences between Zeeman sublevels; (ii) collision rates are high and therefore atomic polarization can be neglected (this means that within each atomic level, the Zeeman sublevel populations are equal); (iii) complete frequency redistribution is assumed; and (iv) stimulated emission is treated as negative absorption.

In equation (12.4.3), S_L is the line source function and the other symbols are same as those given in equations (12.2.2) (12.2.3)–(12.2.5) and (12.2.8)–(12.2.10). From equation (12.4.1), we have

$$+\frac{d\mathbf{I}(+v, +\mathbf{n})}{dz} = -\mathbf{K}(+v, +\mathbf{n})\mathbf{I}(+v, +\mathbf{n}) + \mathbf{j}(+v, +\mathbf{n}), \tag{12.4.4}$$

and

$$-\frac{d\mathbf{I}(-v,-\mathbf{n})}{dz}=-\mathbf{K}(-v,-\mathbf{n})\mathbf{I}(-v,-\mathbf{n})+\mathbf{j}(-v,-\mathbf{n}). \tag{12.4.5}$$

From equations (12.2.8)–(12.2.10) we obtain

$$\left.\begin{aligned}
\phi_p(-v,-\mathbf{n})&=+\phi_p(+v,+\mathbf{n}); & \phi'_p(-v,-\mathbf{n})&=-\phi'_p(+v,+\mathbf{n});\\
\phi_b(-v,-\mathbf{n})&=+\phi_r(+v,+\mathbf{n}); & \phi'_b(-v,-\mathbf{n})&=-\phi'_r(+v,+\mathbf{n});\\
\phi_r(-v,-\mathbf{n})&=+\phi_b(+v,+\mathbf{n}); & \phi'_r(-v,-\mathbf{n})&=-\phi'_b(+v,+\mathbf{n});
\end{aligned}\right\} \tag{12.4.6}$$

and

$$H(a,-v)=H(a,v); \quad F(a,-v)=-F(a,+v). \tag{12.4.7}$$

Using these equations in equations (12.2.5) and substituting $\pi-\gamma$ for γ and $-\chi$ in considering the direction $-\mathbf{n}$, we get

$$\left.\begin{aligned}
\phi_I(-v,-\mathbf{n})&=+\phi_I(+v,+\mathbf{n}); & &\\
\phi_Q(-v,-\mathbf{n})&=+\phi_Q(+v,+\mathbf{n}); & \phi'_Q(-v,-\mathbf{n})&=\phi'_Q(+v,+\mathbf{n});\\
\phi_U(-v,-\mathbf{n})&=+\phi_U(+v,+\mathbf{n}); & \phi'_U(-v,-\mathbf{n})&=\phi'_U(+v,+\mathbf{n});\\
\phi_V(-v,-\mathbf{n})&=+\phi_V(+v,+\mathbf{n}); & \phi'_V(-v,-\mathbf{n})&=\phi'_V(+v,+\mathbf{n});
\end{aligned}\right\} \tag{12.4.8}$$

Now, we shall introduce a modified Stokes vector $\tilde{\mathbf{I}}(-v,-\mathbf{n})$ for the inward directed radiation, where

$$\tilde{\mathbf{I}}=(I,Q,-U,V)^T. \tag{12.4.9}$$

Using equations (12.4.2), (12.4.3), (12.4.4) and(12.4.8), equation (12.4.5) can be written as

$$-\frac{d\tilde{\mathbf{I}}(-v,-\mathbf{n})}{dz}=-\mathbf{K}(+v,+\mathbf{n})\tilde{\mathbf{I}}(-v,-\mathbf{n})+\mathbf{j}(+v,+\mathbf{n}). \tag{12.4.10}$$

We shall write,

$$\left.\begin{aligned}
\mathbf{K}&=\mathbf{K}(+v,+\mathbf{n}),\\
\mathbf{j}&=\mathbf{j}(+v,+\mathbf{n}),
\end{aligned}\right\} \tag{12.4.11}$$

and define the generalized Feautrier vectors

$$\left.\begin{aligned}
\mathbf{J}&=\frac{1}{2}\left[\mathbf{I}(+v,+\mathbf{n})+\tilde{\mathbf{I}}(-v,-\mathbf{n})\right],\\
\mathbf{H}&=\frac{1}{2}\left[\mathbf{I}(+v,+\mathbf{n})-\tilde{\mathbf{I}}(-v,-\mathbf{n})\right].
\end{aligned}\right\} \tag{12.4.12}$$

Now equations (12.4.4) and (12.4.10) can be written using equations (12.4.12) as

$$\frac{d\mathbf{J}}{dz}=-\mathbf{K}\mathbf{H}, \tag{12.4.13}$$

$$\frac{d\mathbf{H}}{dz} = -\mathbf{K}\mathbf{H} + \mathbf{j}. \tag{12.4.14}$$

Eliminating **H** from equations (12.4.13) and (12.4.14) we obtain a second order equation for **J** given by

$$\frac{d}{dz}\left(\mathbf{K}^{-1}\frac{d\mathbf{I}}{dz}\right) = \mathbf{K}\mathbf{J} - \mathbf{j}. \tag{12.4.15}$$

The medium can be divided into n layers, that is $z_1, z_2, \ldots, z_n$, where $z_1 = 0$ and z_N is the depth at which the radiation field is thermalized. The boundary condition at z_1 is that there is no incident radiation field which transmits into the atmosphere, or

$$\tilde{\mathbf{I}}(-v, -\mathbf{n}) = 0. \tag{12.4.16}$$

Or from equation (12.4.13) the boundary condition at z_1 as

$$\mathbf{K}^{-1}\frac{d\mathbf{J}}{dz} = -\mathbf{J} \tag{12.4.17}$$

and the boundary condition at z_N is

$$\mathbf{K}^{-1}\frac{d\mathbf{J}}{dz} = -\mathbf{I}(+v, +\mathbf{n}) + \mathbf{J}. \tag{12.4.18}$$

Note that equation (12.4.17) and (12.4.18) are the first order boundary conditions. At the depth z_N, one can give the asymptotic approximation for the Stokes vector (see Auer *et al.* (1977)). Thus, we have

$$\mathbf{I}(+v, +\mathbf{n}) \approx B_\nu \mathbf{I}_0 - \mathbf{K}^{-1}\frac{dB_\nu}{dz}\mathbf{I}_0. \tag{12.4.19}$$

Equations (12.4.15)–(12.4.19) are replaced by their finite-difference equivalents on the z_k grid ($k = 1, \ldots, N$). The subscript denotes that the variable is evaluated at z_k. We thus write

$$\delta_k = |z_{k+1} - z_k|, \tag{12.4.20}$$

$$\Delta_k = \frac{1}{2}\frac{\left(\mathbf{K}_{k+1}^{-1} + \mathbf{K}_k^{-1}\right)}{\delta_k}, \quad k = 1, \ldots, N-1. \tag{12.4.21}$$

Then the difference equations are

$$\left.\begin{aligned} &\mathbf{B}_1\mathbf{J}_1 - \mathbf{C}_1\mathbf{J}_2 = \mathbf{L}_1, \\ &-\mathbf{A}_k\mathbf{J}_{k-1} + \mathbf{B}_k\mathbf{J}_k - \mathbf{C}_k\mathbf{J}_{k+1} = \mathbf{L}_k, \quad k = 2, \ldots, N-1, \\ &-\mathbf{A}_N\mathbf{J}_{N-1} + \mathbf{B}_N\mathbf{J}_N = \mathbf{L}_N, \end{aligned}\right\} \tag{12.4.22}$$

(note: **B** in the above equations is not the magnetic field vector) where

$$\left.\begin{aligned}
&\mathbf{B}_1 = \boldsymbol{\Delta}_1 + \mathbf{1} + \frac{1}{2}\delta_1 \mathbf{K}_1, \quad \mathbf{C}_1 = \boldsymbol{\Delta}_1, \quad \mathbf{L}_1 = \frac{1}{2}\delta_1 \mathbf{j}_1,\\
&\mathbf{A}_k = 2\frac{\boldsymbol{\Delta}_{k-1}}{\delta_k + \delta_{k-1}}, \quad \mathbf{C}_k = \frac{2\boldsymbol{\Delta}_k}{\delta_k + \delta_{k-1}},\\
&\mathbf{B}_k = \mathbf{A}_k + \mathbf{C}_k + \mathbf{K}_k, \quad \mathbf{L}_k = \mathbf{j}_k, \quad k = 2, \ldots, N-1,\\
&\mathbf{A}_N = \boldsymbol{\Delta}_{N-1} - \frac{1}{2}\mathbf{1}, \quad \mathbf{B}_N = \boldsymbol{\Delta}_{N-1} + \frac{1}{2}\mathbf{1},\\
&\mathbf{L}_N = \frac{1}{2}\mathbf{I}_0(B_{\nu,N-1} + B_{\nu,N}) + \boldsymbol{\Delta}_{N-1}\mathbf{I}_0(B_{\nu,N} - B_{\nu,N-1}).
\end{aligned}\right\} \tag{12.4.23}$$

Equations (12.4.22) form a block tri-diagonal system which can be solved using the Gaussian elimination scheme through the recurrence formulae:

$$\mathbf{J}_N = \mathbf{r}_N + \mathbf{J}_k = \mathbf{r}_k + \mathbf{D}_k \mathbf{J}_{k+1}, \quad k = 1, 2, \ldots, N-1, \tag{12.4.24}$$

where

$$\left.\begin{aligned}
&\mathbf{r}_1 = \mathbf{B}_1^{-1}\mathbf{L}_1, \quad \mathbf{D}_1 = \mathbf{B}_1^{-1}\mathbf{C}_1,\\
&\mathbf{r}_k = (\mathbf{B}_k - \mathbf{A}_k\mathbf{D}_{k-1})^{-1}(\mathbf{L}_k + \mathbf{A}_k\mathbf{r}_{k-1}),\\
&\mathbf{D}_k = (\mathbf{B}_k - \mathbf{A}_k\mathbf{D}_{k-1})^{-1}\mathbf{C}_k, \quad k = 2, \ldots, N-1.
\end{aligned}\right\} \tag{12.4.25}$$

The emergent Stokes vector is

$$\mathbf{I}(0) = 2\mathbf{J}_1. \tag{12.4.26}$$

12.5 Lambda operator method for Zeeman line transfer

One can perform a one-way integration (in the Feautrier method, a two-way integration is done) in the line of sight.

From equations (12.2.4) and (12.4.2) we can see that the diagonal elements are all equal to $K_I = K_c + K_0\phi_I$. We can define the line of sight optical depth in terms of K_I as

$$d\tau = -K_I\,dz = -(K_c + K_0\phi_I)\,dz. \tag{12.5.1}$$

We have a modified absorption matrix and a modified total source function defined respectively as

$$\tilde{\mathbf{K}} = \frac{\mathbf{K}}{K_I} - \mathbf{1} \tag{12.5.2}$$

and

$$\tilde{\mathbf{S}} = \frac{\mathbf{j}}{K_I} \tag{12.5.3}$$

(the actual total source function is $\mathbf{S} = \mathbf{K}^{-1}\mathbf{j}$). Now equation (12.4.1) becomes

$$\frac{d\mathbf{I}}{d\tau} - \mathbf{I} = \tilde{\mathbf{K}}\mathbf{I} - \tilde{\mathbf{S}}. \tag{12.5.4}$$

The above equation has a formal solution with lambda transformation with respect to τ and is written as

$$I(\tau) = \int_{\tau}^{\infty} \exp[-(\tau' - \tau)] \left[\tilde{\mathbf{S}}(\tau') - \tilde{\mathbf{K}}(\tau')\mathbf{I}(\tau')\right] d\tau', \tag{12.5.5}$$

which reduces exactly to the formal integral for $I(\tau)$ when the magnetic field vector $\mathbf{B} = 0$ and therefore $\tilde{\mathbf{K}} = 0$.

The above equation can be solved by lambda iteration with an initial estimate of $I(\tau) = 0$ (say). Staude (1969, 1970) used this technique but found that the convergence of the successive iterations is too slow to be practical. Following Kalkofen (1974), we will describe a numerical representation of the lambda integral operator.

Let $z_i (i = 1, \ldots, N)$ be the geometrical grid and τ_i be its corresponding optical depth grid

$$\left.\begin{aligned} \tilde{\delta}_i &= \tau_{i+1} - \tau_i = \frac{1}{2}\left(K_{I,i+1} + K_{I,i}\right)|z_{i+1} - z_i|, \\ E_i &= \exp(-\delta_i), \quad i = 1, \ldots, N-1. \end{aligned}\right\} \tag{12.5.6}$$

On the interval (τ_i, τ_{i+1}), we have from equation (12.5.5)

$$\mathbf{I}_i = E_i \mathbf{I}_{1+1} + \int_{\tau_i}^{\tau_{i+1}} \exp[-\left(\tau' - \tau_i\right)] \left[\tilde{\mathbf{S}}(\tau') - \tilde{\mathbf{K}}(\tau')\mathbf{I}(\tau')\right] d\tau'. \tag{12.5.7}$$

We now have an explicit relation between $\mathbf{I}_i$ and $\mathbf{I}_{i+1}$. On the interval $(i, i+1)$, $(\tilde{\mathbf{S}} - \tilde{\mathbf{K}}\mathbf{I})$ is approximated by a linear function of optical depth and is written as

$$\tilde{\mathbf{S}}(\tau) - \tilde{\mathbf{K}}(\tau)\mathbf{I}(\tau) = \left(\tilde{\mathbf{S}} - \tilde{\mathbf{K}}\mathbf{I}\right)_i + \left[\left(\tilde{\mathbf{S}} - \tilde{\mathbf{K}}\mathbf{I}\right)_{i+1} - \left(\tilde{\mathbf{S}} - \tilde{\mathbf{K}}\mathbf{I}\right)_i\right] \frac{(\tau - \tau_i)}{\tilde{\delta}_i}. \tag{12.5.8}$$

Equation (12.5.7) can be integrated and after some algebraic manipulations, we obtain

$$\mathbf{I}_i = \mathbf{P}_i + \mathbf{Q}_i \mathbf{I}_{i-1}, \tag{12.5.9}$$

where

$$\left.\begin{aligned} \mathbf{P}_i &= \left[\mathbf{1} + (\mathbf{F}_i - G_i)\,\tilde{\mathbf{K}}_i\right]^{-1} \left[(F_i - G_i)\,\tilde{\mathbf{S}}_i + G_i \tilde{\mathbf{S}}_{i-1}\right], \\ \mathbf{Q}_i &= \left[\mathbf{1} + (F_i - G_i)\,\tilde{\mathbf{K}}_i\right]^{-1} \left(E_i \mathbf{1} - G_i \tilde{\mathbf{K}}_{i+1}\right), \\ F_i &= 1 - E_i, \\ G_i &= \frac{[1 - (1 + \delta_i)\,E_i]}{\tilde{\delta}_i}. \end{aligned}\right\} \tag{12.5.10}$$

At the boundary τ_N , the Stokes vector is given by (see also equation (12.4.19))

$$\mathbf{I}_N = B_{\nu,N}\mathbf{I}_0. \tag{12.5.11}$$

Now equation (12.5.9) can be used recursively and the emergent Stokes vector at $\tau = 0$ is given by

$$\mathbf{I}(0) = \mathbf{I}_1. \tag{12.5.12}$$

12.6 Solution of the transfer equation for polarized radiation

An integration method is now presented below which follows Landi Degl'Innocenti (1987). The transfer equation with Stokes vectors along the ray paths is given by

$$\frac{d\mathbf{I}}{ds} = -\mathbf{K}(\mathbf{I} - \mathbf{S}), \tag{12.6.1}$$

where $\mathbf{I}$ is the Stokes vector, $\mathbf{I} = (I, Q, U, V)^T$, $\mathbf{K}$ is a 4×4 matrix describing the absorption and $\mathbf{S} = (S_I, S_Q, S_U, S_V)^T$ is the source function vector. The 4×4 $\mathbf{K}$ absorption matrix is given by

$$\mathbf{K} = \begin{pmatrix} \eta_I & \eta_Q & \eta_U & \eta_V \\ \eta_Q & \eta_I & \rho_V & -\rho_U \\ \eta_U & -\rho_V & \eta_I & \rho_Q \\ \rho_V & \rho_U & -\rho_Q & \eta_I \end{pmatrix}. \tag{12.6.2}$$

Where the ηs are related to the ϕs in equations (12.2.3)–(12.2.5) by

$$\eta_I = k_c + k_0\phi_I, \text{ etc.} \tag{12.6.3}$$

First we consider a homogeneous equation

$$\frac{d}{ds}\mathbf{I}(s) = -\mathbf{K}(s)\mathbf{I}(s). \tag{12.6.4}$$

We define a linear operator $\mathbf{O}(s, s')$ which acts on the Stokes vector at the point s' to give the Stokes vector at s, or

$$\mathbf{I}(s) = \mathbf{O}(s, s')\mathbf{I}(s'). \tag{12.6.5}$$

The limiting conditions on $\mathbf{O}(s, s')$ are

$$\mathbf{O}(s, s') = \mathbf{1}, \tag{12.6.6}$$

$\mathbf{1}$ being the 4×4 identity matrix, and

$$\mathbf{O}(s, s') = \mathbf{O}(s, s')\mathbf{O}(s', s'). \tag{12.6.7}$$

Taking the derivative of equation (12.6.5) with respect to s and comparing this with equation (12.6.4), we obtain the differential equation

$$\frac{d}{ds}\mathbf{O}(s, s') = -\mathbf{K}(s)\mathbf{O}(s, s'). \tag{12.6.8}$$

Differentiating equation (12.6.5) with respect to s' and using equation (12.6.8) we get

$$\frac{d\mathbf{O}(s, s')}{ds'} = \mathbf{O}(s, s')\mathbf{K}(s'). \tag{12.6.9}$$

We can now write the solution of equation (12.6.1) as

$$\mathbf{I}(s) = \int_{s_0}^{s} \mathbf{O}(s, s')\mathbf{K}(s')\, ds' + \mathbf{O}(s, s_0)I(s_0) \tag{12.6.10}$$

and

$$\lim_{s' \to -\infty} \mathbf{O}(s, s')\mathbf{K}(s')\mathbf{S}(s') = 0. \tag{12.6.11}$$

An analytical expression can be obtained for $\mathbf{O}(s, s')$ by first integrating equation (12.6.8), which gives

$$\mathbf{O}(s, s') = 1 - \int_{s'}^{s} \mathbf{K}(s_1)\mathbf{O}(s_1, s')\, ds_1, \tag{12.6.12}$$

and then substituting for the operator $\mathbf{O}(s, s')$ in the RHS its expression as given by the LHS of the same equation. Iterating, we obtain

$$\mathbf{O}(s, s') = 1 + \sum_{n=1}^{\infty}(-1)^n \int_{s'}^{s} ds_1 \int_{s'}^{s_1} ds_2 \cdots \cdots \int_{s'}^{s_{n-1}} ds_n\, p\,[\mathbf{K}(s_1)\mathbf{K}(s_2)\cdots\mathbf{K}(s_n)], \tag{12.6.13}$$

or

$$\mathbf{O}(s, s') = 1 + \sum_{n=1}^{\infty} \frac{(-1)^n}{n!} \int_{s'}^{s} ds_1 \int_{s'}^{s} ds_2 \cdots \cdots \int_{s'}^{s} ds_n P\,[\mathbf{K}(s_1)\mathbf{K}(s_2)\cdots\mathbf{K}(s_n)], \tag{12.6.14}$$

where P is the chronological operator that sets the different matrices according to the following order:

$$P\,\{\mathbf{K}(s_1)\mathbf{K}(s_2)\cdots\mathbf{K}(s_n)\} = \mathbf{K}(s_{j1})\mathbf{K}(s_{j2})\cdots\mathbf{K}(s_{jn}), \quad \text{where } s_{j1} \geq s_{j2} \geq \cdots \geq s_{jn}. \tag{12.6.15}$$

The operator $\mathbf{O}(s, s')$ is the generalization of the attenuation operator

$$\exp\left[-\int_{s'}^{s} K(s'')\, ds''\right] \tag{12.6.16}$$

of the usual transfer equation to the polarized transfer of radiation. The order of the operator in equation (12.6.15) is connected to the physical fact that two different slabs, say a and b, acting differently on the polarization properties of the radiation beam do not 'commute' in the sense that the emerging polarization in general differs depending on whether slab a is located in front of slab b or vice versa.

If $\mathbf{K}$ = constant, equation (12.6.15) is written as

$$\mathbf{O}(s, s') = 1 + \sum_{n=1}^{\infty} (-1)^n \frac{(s - s')^n}{n!} \mathbf{K}^n, \tag{12.6.17}$$

which gives us,

$$\mathbf{O}(s, s') = \exp\left[-(s - s')\mathbf{K}\right]. \tag{12.6.18}$$

The exponential of a matrix is given by its Taylor's expansion. This expression can be computed in terms of the elements of the matrix $\mathbf{K}$, by writing $\mathbf{K}$ as a linear combination of 4×4 matrices. The properties of these matrices allow the reduction of the infinite products of equations (12.6.18) to a closed form. Let

$$\mathbf{K} = \eta_I \mathbf{1} + \vec{a} \cdot \vec{\mathbf{A}} + \vec{b} \cdot \vec{\mathbf{B}}, \tag{12.6.19}$$

where $\vec{a}$ and $\vec{b}$ are the formal vectors given by

$$\left.\begin{aligned} \vec{a} &= \vec{\eta} + i\vec{\rho} = \frac{1}{2}\left(\eta_Q + i\rho_Q, \eta_U + i\rho_U, \eta_V + i\rho_V\right), \\ \vec{b} &= \vec{\eta} - i\vec{\rho} = \frac{1}{2}\left(\eta_Q - i\rho_Q, \eta_U - i\rho_U, \eta_V - i\rho_V\right), \end{aligned}\right\} \tag{12.6.20}$$

and $\vec{\mathbf{A}}$ and $\vec{\mathbf{B}}$ are matrices given by

$$\mathbf{A}_1 = \begin{pmatrix} 0 & 1 & 0 & 0 \\ 1 & 0 & 0 & 0 \\ 0 & 0 & 0 & -i \\ 0 & 0 & i & 0 \end{pmatrix}, \quad \mathbf{A}_2 = \begin{pmatrix} 0 & 0 & 1 & 0 \\ 0 & 0 & 0 & i \\ 1 & 0 & 0 & 0 \\ 0 & -i & 0 & 0 \end{pmatrix}, \quad \mathbf{A}_3 = \begin{pmatrix} 0 & 0 & 0 & 1 \\ 0 & 0 & -i & 0 \\ 0 & i & 0 & 0 \\ 1 & 0 & 0 & 0 \end{pmatrix}, \tag{12.6.21}$$

$$\mathbf{B}_i = \mathbf{A}_i^{\star}. \tag{12.6.22}$$

$\mathbf{A}_i$ and $\mathbf{B}_i$ have the following properties:

$$\left.\begin{aligned} \left[A_i, B_j\right] &= \mathbf{A}_i\mathbf{B}_j - \mathbf{B}_j\mathbf{A}_i = 0, \\ \mathbf{A}_i\mathbf{A}_j &= \delta_{ij}\mathbf{1} + i\sum_k \epsilon_{ik}\mathbf{A}_k, \\ \mathbf{B}_i\mathbf{B}_j &= \delta_{ij}\mathbf{1} - i\sum_k \epsilon_{ijk}\mathbf{B}_k, \end{aligned}\right\} \tag{12.6.23}$$

where ϵ_{ijk} is the complete anti-symmetrical tensor. Also

$$\exp(\mathbf{A} + \mathbf{B}) = \exp(\mathbf{A})\exp(\mathbf{B}), \text{ if } [\mathbf{A}, \mathbf{B}] = 0. \tag{12.6.24}$$

Substituting x for $s - s'$, we get

$$\mathbf{O}(x) = \exp(-\eta_I x \mathbf{1}) \exp\left(-\vec{a} \cdot \vec{\mathbf{A}} x\right) \exp\left(-\vec{b} \cdot \vec{\mathbf{B}} x\right) \quad (12.6.25)$$

Applying Taylor's expansion to the exponentials we get

$$\left.\begin{aligned} \exp(-\eta_I x \mathbf{1}) &= \exp(-\eta_I x)\mathbf{1}, \\ \exp\left[-\vec{a} \cdot \vec{\mathbf{A}} x\right] &= \cosh(ax)\mathbf{1} - \frac{\sinh(ax)}{a} \vec{a} \cdot \mathbf{A}, \\ \exp\left[-\vec{b} \cdot \vec{\mathbf{B}} x\right] &= \cosh(bx)\mathbf{1} - \frac{\sinh(bx)}{b} \vec{b} \cdot \vec{\mathbf{B}}, \end{aligned}\right\} \quad (12.6.26)$$

where $a = (\vec{a} \cdot \vec{a})^{\frac{1}{2}}, b = (\vec{b} \cdot \vec{b})^{\frac{1}{2}}$.

Substituting equation (12.6.26) into equation (12.6.25) invoking the conjugation properties of the matrices **A** and **B**, we obtain the operator **O** (see Landi Degl'Innocenti and Landi Degl'Innocenti (1985))

$$\begin{aligned} O(s, s') &= \exp[-x\mathbf{K}] \\ &= \exp(-\eta_I x)\left\{\frac{1}{2}[\cosh(\Lambda_1 x) + \cosh(\Lambda_2 x)]\right. \\ &\quad - \sin(\Lambda_2 x)\mathbf{M}_2 - \sinh(\Lambda_1 x)\mathbf{M}_3 \\ &\quad \left. + \frac{1}{2}[\cosh(\Lambda_1 x) - \cos(\Lambda_2 x)]\,\mathbf{M}_4\right\}, \end{aligned} \quad (12.6.27)$$

$$M_2 = \frac{1}{\theta}\begin{pmatrix} 0 & \Lambda_2\eta_Q - \sigma\Lambda_1\rho_Q & \Lambda_2\eta_U - \sigma\Lambda_1\rho_U & \Lambda_2\eta_V - \sigma\Lambda_1\rho_V \\ \Lambda_2\eta_Q - \sigma\Lambda_1\rho_Q & 0 & \sigma\Lambda_1\eta_V + \Lambda_2\rho_V & -\sigma\Lambda_1\eta_U - \Lambda_2\rho_U \\ \Lambda_2\eta_U - \sigma\Lambda_1\rho_U & -\sigma\Lambda_1\eta_V - \Lambda_2\rho_V & 0 & \sigma\Lambda_1\eta_Q + \lambda_2\rho_Q \\ \Lambda_2\eta_V - \sigma\Lambda_1\rho_V & \sigma\Lambda_1\eta_U + \Lambda_2\rho_U & -\sigma\Lambda_1\eta_Q - \Lambda_2\rho_Q & 0 \end{pmatrix}, \quad (12.6.28)$$

$$M_3 = \frac{1}{\theta}\begin{pmatrix} 0 & \Lambda_1\eta_Q + \sigma\Lambda_2\rho_Q & \Lambda_1\eta_U + \sigma\Lambda_2\rho_U & \Lambda_1\eta_V + \sigma\Lambda_2\rho_V \\ \Lambda_1\eta_Q - \sigma\Lambda_2\rho_Q & 0 & -\sigma\Lambda_2\eta_V + \Lambda_1\rho_V & \sigma\Lambda_2\eta_U - \Lambda_1\rho_U \\ \Lambda_1\eta_U + \sigma\Lambda_2\rho_U & \sigma\Lambda_2\eta_V - \Lambda_1\rho_V & 0 & -\sigma\Lambda_2\eta_Q + \Lambda_1\rho_Q \\ \Lambda_1\eta_V + \sigma\Lambda_2\rho_V & -\sigma\Lambda_2\eta_U + \Lambda_1\rho_U & \sigma\Lambda_2\eta_Q - \Lambda_1\rho_Q & 0 \end{pmatrix}, \quad (12.6.29)$$

$$M_4 - \frac{2}{\theta}\begin{pmatrix} a_1 & a_2 & a_3 a_4 \\ b_1 & b_2 & b_3 b_4 \\ c_1 & c_2 & c_3 c_4 \\ d_1 & d_2 & d_3 d_4 \end{pmatrix}, \quad (12.6.30)$$

where

$$\left.\begin{aligned}
a_1 &= \frac{\eta^2+\rho^2}{2}, & a_2 &= \eta_V\rho_U - \eta_U\rho_V,\\
a_3 &= \eta_Q\rho_V - \eta_V\rho_Q, & a_4 &= \eta_U\rho_Q - \eta_Q\rho_U,\\
b_1 &= \eta_U\rho_V - \eta_V\rho_U, & b_2 &= \eta_Q^2 + \rho_Q^2 - \frac{(\eta^2+\rho^2)}{2},\\
c_1 &= \eta_V\rho_Q - \eta_Q\rho_V, & c_2 &= \eta_Q\eta_U + \rho_Q\rho_V,\\
c_3 &= \eta_U^2 + \rho_U^2 + \frac{(\eta^2+\rho^2)}{2}, & c_4 &= \eta_U\eta_V + \rho_U\rho_V,\\
d_1 &= \eta_Q\rho_U - \eta_U\rho_Q, & d_2 &= \eta_V\eta_Q + \rho_V\rho_Q,\\
d_3 &= \eta_U\eta_V + \rho_U\rho_V, & d_4 &= \eta_V^2 + \rho_v^2 + \frac{(\eta^2+\rho^2)}{2},
\end{aligned}\right\} \tag{12.6.31}$$

and

$$\Lambda_1 = (\alpha^{\frac{1}{2}} + \beta)^{\frac{1}{2}}, \quad \Lambda_2 = (\alpha^{\frac{1}{2}} - \beta)^{\frac{1}{2}}, \tag{12.6.32}$$

where

$$\begin{aligned}
\alpha &= \frac{(\eta^2-\rho^2)}{4} + (\vec{\eta}\cdot\vec{\rho})^2,\\
\beta &= \frac{(\eta^2-\rho^2)}{2},\\
\sigma &= \frac{(\vec{\eta}\cdot\vec{\rho})}{|\vec{\eta}\cdot\vec{\rho}|} = \mathrm{sign}(\vec{\eta}\cdot\vec{\rho}),\\
\theta &= \Lambda_1^2 + \Lambda_2^2 = 2\left[\frac{(\eta^2-\rho^2)}{4} + (\vec{\eta}\cdot\vec{\rho})^2\right]^{\frac{1}{2}}.
\end{aligned}$$

Equation (12.6.28) can be applied to obtain the solution of the polarized transfer equation, through the following steps: (1) divide the path of integration ($-\infty \le s' \le s$) into n intervals and assume that the matrix is constant and equal in each of these n intervals at the midpoint of the interval. Let $s_0, s_1, s_2, \ldots, s_{n-1}, s_n$ denote the grid points with $s_0 = -\infty$ and $s_n = s$, then in the nth interval we have

$$\mathbf{K}(t) = \mathbf{K}((s_{i-1} - s_i)/2) = \mathbf{K}_i, \quad s_{i-1} \le t \le s_i; \tag{12.6.33}$$

(2) in each interval, calculate the operator $\mathbf{O}$ numerically from equation (12.6.27)

$$\mathbf{O}_i = \exp\left[-(s_i - s_{i-1})\mathbf{K}_i\right], \tag{12.6.34}$$

and calculate the vector $\mathbf{V}_i$ given by

$$\mathbf{V}_i = \int_{s_{i-1}}^{s_i} \exp\left[-(s_i - s')\mathbf{K}_i\right] \mathbf{K}_i\mathbf{S}(s')\,ds', \tag{12.6.35}$$

which is done by using a standard numerical integration formula; and lastly (3) compute $\mathbf{I}(s)$ using the equation

$$\mathbf{I}(s) = \sum_{i=1}^{n} \mathbf{O}_n \mathbf{O}_{n-1} \cdots \mathbf{O}_{i+1} \mathbf{V}_i. \tag{12.6.36}$$

Alternatively, $\mathbf{I}(s)$ can be calculated by using the formula:

$$\mathbf{V}_i = \int_{s_{i-1}}^{s_i} \left\{ \frac{d}{ds} \exp\left[-(s_i - s')\mathbf{K}_i\right] \right\} \mathbf{S}(s')\, ds' = \mathbf{S}_i - \mathbf{O}_i \mathbf{S}_{i-1} + V_i', \tag{12.6.37}$$

where

$$\mathbf{S}_i = \mathbf{S}(s_i)$$

and

$$\mathbf{V}_i' = \int_{s_{i-1}}^{s_i} \exp\left[-(s_i - s')\mathbf{K}_i\right] \left\{ \frac{d}{ds'} \mathbf{S}(s') \right\} ds'. \tag{12.6.38}$$

Inserting equation (12.6.35) into equation (12.6.36) and assuming that the condition of equation (12.6.11) for the source function holds good at large optical depths in the atmosphere, we obtain

$$\mathbf{I}(s) = \mathbf{S}_n + \sum_{i=1}^{n} \mathbf{O}_n \mathbf{O}_{n-1} \cdots \mathbf{O}_{i+1} \mathbf{V}_i'. \tag{12.6.39}$$

The integration method described above has a disadvantage in that the absorption matrix $\mathbf{K}$ is kept constant in each of the n intervals into which the total path of integration is divided. To get a better solution, one should inverse the number of intervals n. The advantage in this method is that one can choose the number of intervals depending upon the variation of the absorption matrix $\mathbf{K}$ along the path of integration which gives better accuracy. This means that the number of grid points can be conveniently adjusted in advance for the desired accuracy.

12.7 Polarization approximate lambda iteration (PALI) methods

The approximate lambda iteration (ALI) methods have been applied to the problems of polarization calculation. These methods have been found to be quite fast and required much less computer memory. We shall describe these methods briefly following a series of papers of Faurobert-Scholl (1991, 1993), Faurobert-Scholl *et al.* (1997), Frisch (1999), Nagendra *et al.* (1998, 1999). For a review of the approximate lambda iterative methods see Hubený (1992). An alternative method has been developed by Ivanov and his coworkers (Ivanov 1995, Ivanov *et al.* 1997).

The ALI methods use the integral formulations of the transfer equation. The source function is used to define the lambda operator. This becomes a vector in the polarized radiation field.

We write the transfer equation in a non-magnetic situation (Faurobert-Scholl *et al.* 1997) in plane parallel geometry as

$$\mu \frac{\partial \mathbf{I}(\tau, x, \mu)}{\partial \tau} = \phi(x)\mathbf{I}(\tau, x, \mu) - \phi(x)\mathbf{S}(\tau, \mu), \tag{12.7.1}$$

where $\phi(x)$ is the absorption profile function at the normalized frequency point x, $\mathbf{I}(I, Q)^T$. $\mathbf{S}(\tau, \mu)$ is the vector source function given (for a two-level atom) by

$$\mathbf{S}(\tau, \mu) = \frac{(1-\epsilon)}{2} \int_{-\infty}^{+\infty} \hat{P}(\mu, \mu')\mathbf{I}(x', \mu')\, d\mu'\, dx' + \epsilon \mathbf{B}(\tau), \tag{12.7.2}$$

$$\mathbf{B}(\tau) = (B, 0)^T, \tag{12.7.3}$$

and ϵ is the probability per scatter that a photon is destroyed by collisional de-excitation. The phase matrix $\hat{P}(\mu, \mu')$ is given by

$$\hat{P}(\mu, \mu') = \hat{P}_{is} + \frac{3}{4} W_2 \hat{P}_0^{(2)}(\mu, \mu'), \tag{12.7.4}$$

where the matrix $\hat{P}_{is}$ is the isotropic phase function (the first element is 1 and the rest are zeros) and $\hat{P}_0^{(2)}$ is given by

$$\hat{P}_0^{(2)}(\mu, \mu') = \frac{1}{2} \begin{pmatrix} \frac{1}{3}(1-3\mu^2)(1-3\mu'^2) & (1-3\mu^2)(1-\mu'^2) \\ (1-\mu^2)(1-3\mu'^2) & 3(1-\mu^2)(1-\mu'^2) \end{pmatrix}. \tag{12.7.5}$$

The quantity W_2 is connected to the quantum numbers J and J' of the lower and upper levels of the normal Zeeman triplet. In the case of a Hanle effect due to a weak microturbulent magnetic field, a phase matrix similar to that of equation (12.7.4) can be used. The Hanle phase matrix can be averaged over the angular distribution of the magnetic field (Stenflo 1982, Landi Degl'Innocenti and Landi Degl'Innocenti 1988, Faurobert-Scholl 1993). The averaged microturbulent Hanle phase matrix is described by equation (12.7.4) with W_2 multiplied by

$$1 - 0.4(s_I^2 + s_{II}^2), \tag{12.7.6}$$

where

$$s_I = \gamma \left(1 + \gamma^2\right)^{-\frac{1}{2}}, \quad s_{II} = 2\gamma \left(1 + 4\gamma^2\right)^{-\frac{1}{2}} \tag{12.7.7}$$

and

$$\gamma = 0.88 g_J \frac{H}{\Gamma_R + D^{(2)} + \Gamma_I}. \tag{12.7.8}$$

g_J is the Landé factor of the upper level and the quantities Γ_R, $D^{(2)}$ and Γ_I are respectively the rates of radiative damping, depolarizing and inelastic collisions in units of 10^7 s^{-1}. H is the magnetic intensity measured in gauss. The phase matrix $\hat{P}(\mu, \mu')$ can be factorized or the variables μ and μ' can be separated (see Sekera (1963), van de Hulst (1980), chapter 16, Ivanov (1995)). Using Ivanov's factorization, we get

$$\hat{P}(\mu, \mu') = \hat{A}(\mu)\hat{A}^T(\mu'), \tag{12.7.9}$$

where

$$\hat{A}(\mu) = \begin{pmatrix} 1 & \left(\dfrac{W_2}{8}\right)^{\frac{1}{2}} (1 - 3\mu^2) \\ 0 & \left(\dfrac{W_2}{8}\right)^{\frac{1}{2}} 3(1 - \mu^2) \end{pmatrix}. \tag{12.7.10}$$

In the case of resonance polarization or polarization by a weak turbulent magnetic field, one can factorize the τ and μ dependence of the source function in the form

$$\mathbf{S}(\tau, \mu) = \hat{A}(\mu)\mathbf{P}(\tau), \tag{12.7.11}$$

where $\mathbf{P}(\tau)$ is a two-level component column vector depending only on τ (Rees 1978). Thus $\mathbf{P}(\tau)$ can be written as

$$P(\tau) = \frac{(1 - \epsilon)}{2} \int_{-\infty}^{+\infty} \phi(x') \int_{-1}^{+1} \hat{A}^T(\mu')\mathbf{I}(\tau, x', \mu')\, d\mu'\, dx' + \epsilon\mathbf{B}(\tau), \tag{12.7.12}$$

where the thermal emission term in $\mathbf{B}(\tau)$ has the same form as in equation (12.7.2) as $\mathbf{B}(\tau) = [B(\tau), 0]^T$, which means that $\mathbf{B}(\tau) = \hat{A}(\mu)\mathbf{B}(\tau)$. Now the two components of $\mathbf{P}(\tau)$ are

$$P_l(\tau) = \frac{(1 - \epsilon)}{2} \int_{-\infty}^{+\infty} \phi(x') \int_{-1}^{+1} I(\tau, x', \mu')\, d\mu'\, dx' + \epsilon B(\tau) \tag{12.7.13}$$

and

$$P_2(\tau) = \left(\frac{W_2}{8}\right)^{\frac{1}{2}} \left(\frac{1 - \epsilon}{2}\right) \int_{-\infty}^{+\infty} \phi(x') \int_{-1}^{+1} \Big[(1 - 3\mu'^2)I(\tau, x', \mu') + 3(1 - \mu'^2)Q(\tau, x', \mu')\, d\mu'\, dx'\Big], \tag{12.7.14}$$

where I and Q are the two Stokes parameters.

One can obtain the integral equation for $\mathbf{P}(\tau)$ by inserting the formal solution of equation (12.7.1) into equation (12.7.12). Thus we obtain $\mathbf{P}(\tau)$ for a plane parallel slab of total optical thickness T with no incident radiation as

$$\mathbf{P}(\tau) = (1 - \epsilon) \int_0^T \hat{K}(\tau - \tau')\, d\tau' + \epsilon\mathbf{B}(\tau), \tag{12.7.15}$$

where the 2×2 kernel matrix $\hat{K}$ is given by

$$\hat{K}(\tau) = \frac{1}{2}\int_{-\infty}^{+\infty} \phi^2(x')\,dx' \int_0^1 \hat{A}^T(\mu')\hat{A}(\mu')\exp\left[-\tau\,|\tau|\,\phi(x')/\mu'\right]\frac{d\mu'}{\mu'}dx'. \tag{12.7.16}$$

Therefore, according to Ivanov (1995), the transfer equation with the matrix formation is

$$\mu\frac{\partial \hat{I}(\tau,\mu,x)}{\partial \tau} = \phi(x)\hat{I}(\tau,\mu,x) - \phi(x)\hat{S}(\tau), \tag{12.7.17}$$

where the source matrix is given by

$$\hat{S}(\tau) = \frac{1-\epsilon}{2}\int_{-\infty}^{+\infty} \phi(x') \int_{-1}^{+1} \hat{A}^T(\mu')\hat{A}(\mu')\hat{I}(\tau,\mu',x')\,dx' + \hat{S}^\star(\tau) \tag{12.7.18}$$

and

$$\mathbf{P}(\tau) = \hat{S}(\tau)\mathbf{e}, \quad \mathbf{e} = (1,1)^T, \tag{12.7.19}$$

and for a plane parallel slab of total optical thickness without incident radiation

$$\hat{S}(\tau) = (1-\epsilon)\int_0^T K(\tau-\tau')\hat{S}(\tau')\,d\tau' + \hat{S}^\star(\tau) \tag{12.7.20}$$

and

$$\epsilon\mathbf{B}(\tau) = \hat{S}^\star(\tau)\mathbf{e}. \tag{12.7.21}$$

The integral equation for the vector $\mathbf{P}$ or the matrix $\hat{S}$ can be written in symbolic form as

$$\mathbf{P}(\tau) = (1-\epsilon)\mathbf{\Lambda}(\mathbf{P}) + \mathbf{Q}(\tau), \tag{12.7.22}$$

where $\mathbf{Q}(\tau)$ is a given primary source and $\mathbf{\Lambda}$ is the integral operator in equations (12.7.15) and (12.7.18) when $T = \infty$. The integral equation for $\mathbf{P}$ can be written explicitly as

$$\begin{aligned} P_I(\tau) = (1-\epsilon)&\int_0^\infty K_{11}(\tau-\tau')P_I(\tau')\,d\tau' + (1-\epsilon) \\ &\times \int_0^\infty K_{12}(\tau-\tau')P_Q(\tau')\,d\tau' + Q_I(\tau) \end{aligned} \tag{12.7.23}$$

and

$$\begin{aligned} P_Q(\tau) = (1-\epsilon)&\int_0^\infty K_{22}(\tau-\tau')P_Q(\tau')\,d\tau' + (1-\epsilon) \\ &\times \int_0^\infty K_{21}(\tau-\tau')P_I(\tau')\,d\tau' + Q_Q(\tau). \end{aligned} \tag{12.7.24}$$

The integral equation (12.7.20) can be written symbolically as

$$\hat{S} = (1-\epsilon)\mathbf{\Lambda}(\hat{S}) + \hat{S}^\star. \tag{12.7.25}$$

With an initial estimate of $\hat{S}$ denoted by $\hat{S}(n)$, iteration $(n+1)$ gives us

$$\hat{S}(n+1) = \hat{S}(n) + \delta\hat{S}. \tag{12.7.26}$$

The correction term is given by the equation

$$\mathbf{A}(\delta\hat{\mathbf{S}}) = -\mathbf{A}(\hat{\mathbf{S}}^{(n)}) + \hat{\mathbf{S}}^\star, \tag{12.7.27}$$

where $\mathbf{A} = [\mathbf{E} - (1-\epsilon)\mathbf{\Lambda}]$, $\mathbf{E}$ being the identity operator. The operator A contains three linear operators: A_{11} and A_{22} defined as

$$A_{\alpha\alpha} f(\tau) = f(\tau) - (1-\epsilon)\int_0^\infty K_{\alpha\alpha}(\tau-\tau') f(\tau')\, d\tau', \quad (\alpha = 1, 2), \tag{12.7.28}$$

and A_{12} defined as

$$A_{12} f(\tau) = -(1-\epsilon)\int_0^\infty K_{12}(\tau-\tau') f(\tau')\, d\tau', \tag{12.7.29}$$

with $K_{12} = K_{21}$, $A_{12} = A_{21}$: τ is discretized ($\tau = \{\tau_I\}$, $i = 1, \ldots, N$). The matrix $\mathbf{A}$ on the LHS of (12.7.27) is replaced by the block diagonal matrix $\mathbf{D} = \{d_{ii}\}$ given by

$$\mathbf{D} = \begin{pmatrix} A_{11}(i,i) & A_{12}(i,i) \\ A_{12}(i,i) & A_{22}(i,i) \end{pmatrix}. \tag{12.7.30}$$

In terms of the operator $\mathbf{\Lambda}$,

$$d_{ii} = \begin{pmatrix} 1-(1-\epsilon)\Lambda_{11}(i,i) & -(1-\epsilon)\Lambda_{12}(i,i) \\ -(1-\epsilon)\Lambda_{12}(i,i) & 1-(1-\epsilon)\Lambda_{22}(i,i) \end{pmatrix}. \tag{12.7.31}$$

The matrix $\mathbf{A}$ on the RHS of equation (12.7.27) can be expressed in terms of $\mathbf{\Lambda}$ giving $\delta\hat{S}(\tau_i)$ at each optical depth grid point:

$$\delta\hat{S}(\tau_i) = d_{ii}^{-1}\left[(1-\epsilon)\Lambda(\hat{S}^{(n)}(\tau_i) - \hat{S}^{(n)}(\tau_I) + \hat{S}^\star(\tau_i))\right]. \tag{12.7.32}$$

In the above equation, each term is a 2×2 square matrix. The matrices $\Lambda(\hat{S}^{(n)})(\tau_i)$ and d_{ii} are obtained by integrating the transfer equation (12.7.17) with matrices $\hat{S}(\tau)$. From equations (12.7.18) and (12.7.25), we can infer that

$$\Lambda(\hat{S}^{(n)})(\tau) = \hat{J}^{(n)}(\tau), \tag{12.7.33}$$

where

$$\hat{J}(\tau) = \int_{-\infty}^{+\infty} \phi(x') \int_{-1}^{+1} \hat{A}^T(\mu')\hat{A}(\mu')\hat{I}(\tau,\mu',x')\frac{d\mu'}{2}dx'. \tag{12.7.34}$$

The matrix $\hat{J}^{(n)}$ is the frequency, and angle-averaged intensity for polarized radiation. We can calculate $\hat{I}^{(n)}(\tau)$ once we know $\hat{S}^{(n)}(\tau)$ from equation (12.7.17) using a

Feautrier formal solution method and then integrate over frequencies and directions as in equation (12.7.34). The matrix in the square bracket of equation (12.7.32) must be calculated accurately for convergence of the solution. The transfer equation must have pre-conditions to avoid the round-off and truncation errors (Rybicki and Hummer 1991) in the calculation of the differences $\hat{J}^{(n)}(\tau_i) - \hat{S}^{(n)}(\tau_i)$. Large optical depths may cause some difficulty. The matrices d_{ii} need to be computed only once by solving equation (12.7.17) using $\hat{S}(\tau)$ given by

$$\hat{S}(\tau) = \delta(\tau - \tau_i) \begin{pmatrix} 1 & 0 \\ 0 & 1 \end{pmatrix}, \tag{12.7.35}$$

where δ is the Dirac delta-function. From equation (12.7.33),

$$\Lambda_{\alpha,\beta}(i, i) = J_{\alpha,\beta}(\tau_i). \tag{12.7.36}$$

The elements of d_{ii} are calculated from equation (12.7.31). We need two different vectorial source terms $\mathbf{P}(\tau) = \delta(\tau - \tau_i)(1, 0)^T$ and $\mathbf{P}(\tau) = \delta(\tau - \tau_i)(0, 1)^T$ to calculate $D_{i,i}$. The former gives $\Lambda_{11}(i, i)$ and $\Lambda_{21}(i, i)$ and the latter gives $\Lambda_{12}(i, i) = \Lambda_{21}(i, i)$ and $\Lambda_{22}(i, i)$. The computational details and some test cases are described in Faurobert-Scholl *et al.* (1997).

We have described the approximate lambda method for polarization (PALI). Ivanov (1995) formalism of the transfer equation has been used for the iterative scheme. Nagendra *et al.* (1998) extended the PALI method described above (Faurobert-Scholl *et al.* 1997) to include the Hanle effect which they called the PALI-H method. In this method they have a Fourier decomposition of the radiation field in the azimuthal angle which can tackle the depth dependent magnetic field. This operator perturbation method can handle arbitrarily large amounts of polarization. For details see Nagendra *et al.* (1998), Nagendra *et al.* (1999) and Frisch (1999).

There have been significant advances in the areas of continuum and line polarization transfer in magnetic white dwarfs (Martin and Wickramasinghe 1979a,b, 1981, 1982; Nagendra and Peraiah 1984, 1985a,b), cyclotron line formation in accreting magnetized neutron stars (Meszaros 1984, Meszaros and Nagel 1985), maser line formation in molecular clouds (Goldrich and Kylafis 1982, Western and Watson 1983, Deguchi and Watson 1985).

Several other methods are available for calculating line transfer in magnetic fields (see for example Domke and Staude (1973)). Faurobert-Scholl (1991) studied lines using the Hanle effect and partial frequency redistribution.

Exercises

12.1 Supply the intermediate steps between equations (12.3.8) and (12.3.11).

12.2 Develop a computer code to obtain $\mathbf{I}(0)$ in equation (12.4.26).

REFERENCES

Auer, L.H., Heasley, J.N., House, L.L., 1977, *ApJ*, **216**, 531.

Avrett, E.H., Hummer, D.G., 1965, *MNRAS*, **130**, 295.

Beckers, J.M., 1969a, *Sol. Phys.*, **9**, 372.

Beckers, J.M., 1969b, *Sol. Phys.*, **10**, 262.

Born, M., Wolf, E., 1959, *Principles of Optics*, Pergamon Press, Oxford.

Deguchi, S., Watson, L.R., 1985, *ApJ*, **289**, 621.

Domke, H., Staude, J., 1973, *Sol. Phys.*, **31**, 279, 291

Faurobert-Scholl, M., 1991, *A&A*, **246**, 469.

Faurobert-Scholl, M., 1993, *A&A*, **268**, 765.

Faurobert-Scholl, M., Frisch, H., Nagendra, K.N., 1997, *A&A*, **322**, 896.

Frisch, H., 1999, in *Solar Polarization*, eds. K.N. Nagendra and J.O. Stenflo, Kluwer, Dordrecht, page 97.

Goldrich, P., Kylafis, N.D., 1982, *ApJ*, **253**, 606.

Hubený, I., 1992, *The Atmospheres of Early Type Stars*: Lectures Notes in Physics 401, eds. U. Heber and C.J. Jeffery, Springer, Berlin, page 377.

Hulst, H.C. van de., 1980, *Multiple Light Scattering*, Academic Press, New York.

Humlicek, J., 1982, *JQSRT*, **27**, 437.

Ivanov, V.V., 1995, *A&A*, **303**, 609.

Ivanov, V.V., Grachev, S.I., Loskutov, V.M., 1997, *A&A*, **318**, 315.

Kalkofen, W., 1974, *ApJ*, **188**, 105.

Landi Degl'Innocenti, E., 1976, *A&A Suppl.*, **25**, 379.

Landi Degl'Innocenti, E., Landi Degl'Innocenti, M., 1972, *Sol. Phys.*, **27**, 219.

Landi Degl'Innocenti, E., 1987, in *Numerical Radiative Transfer*, ed. W. Kalkofen, Cambridge University Press, Cambridge, page 265.

Landi Degl'Innocenti, E., Landi Degl'Innocenti, M., 1985, *Sol. Phys.*, **97**, 239.

Landi Degl'Innocenti, M., Landi Degl'Innocenti, E., 1988, *A&A*, **192**, 374.

Landolfi, M., Landi Degl'Innocenti, E., 1982, *Sol. Phys.*, **78**, 355.

Martin, B., Wickramasinghe, D.T., 1979a, *Proc. Astron. Soc. Australia*, **3**, 351.

Martin, B., Wickramasinghe, D.T., 1979b, *MNRAS*, **189**, 883.

Martin, B., Wickramasinghe, D.T., 1981, *MNRAS*, **196**, 23.

Martin, B., Wickramasinghe, D.T., 1982, *MNRAS*, **200**, 993.

Meszaros, P., 1984, *Space Sci. Rev.*, **38**, 325.

Meszaros, P., Nagel, W., 1985, *ApJ*, **298**, 147.

Mihalas, D., 1978, *Stellar Atmospheres*, 2nd Edition, W.H. Freeman and Company, San Francisco.

Nagendra, K.N., Peraiah, A., 1984, *Astrophys. Space Sci.*, **104**, 61.

Nagendra, K.N., Peraiah, A., 1985a, *MNRAS*, **214**, 203.

Nagendra, K.N., Peraiah, A., 1985b, *Astrophys. Space Sci.*, **117**, 121.

Nagendra, K.N., Stenflo, J.O., 1999, eds., *Solar Polarization*, Kluwer, Dordrecht.

Nagendra, K.N., Frisch, H., Faurobert-Scholl, M., 1998, *A&A*, **332**, 610.

Nagendra, K.N., Paletou, F., Frisch, H., Faurobert-Scholl, M., 1999, in *Solar Polarization*, eds. K.N. Nagendra and J.O. Stenflo, Kluwer, Dordrecht, page 127.

Rachkovsky, D.M., 1962a, *Izv. Krymsk. Astrofiz. Obs.*, **27**, 148.

Rachkovsky, D.M., 1962b, *Izv. Krymsk. Astrofiz. Obs.*, **28**, 259.

Rachkovsky, D.M., 1963, *Izv. Krymsk. Astrofiz. Obs.*, **30**, 267.

Rachkovsky, D.M., 1967, *Izv. Krymsk. Astrofiz. Obs.*, **37**, 56.

Rees, D., 1987, in *Numerical Radiative Transfer*, ed. W. Kalkofen, Cambridge University Press, Cambridge, page 213.

Rees, D., 1978, *Publ. Astron. Soc., Japan*, **30**, 455.

Rees, D., Murphy, G.A., 1987, in *Numerical Radiative Transfer*, ed. W. Kalkofen, Cambridge University Press, Cambridge, page 241.

Rybicki, G.B., Hummer, D.G., 1991, *A&A*, **245**, 171.

Sekera, Z., 1963, Rand Corporation Memo. R-413, DR.

Staude, J., 1969, *Sol. Phys.*, **8**, 264.

Staude, J., 1970, *Sol. Phys.*, **15**, 102.

Stenflo, J.O., 1982, *Sol. Phys.*, **80**, 209.

Stenflo, J.O., Nagendra, K.N., 1996, ed., *Solar Polarization*, Kluwer, Dordrecht.

Unno, W., 1956, *Publ. Astron. Soc. Japan.*, **8**, 108.

Western, L.R., Watson, W.D., 1983, *ApJ*, **268**, 849.

Wittman, A., 1974, *Sol. Phys.*, **35**, 11.

Chapter 13

Multi-dimensional radiative transfer

13.1 Introduction

So far we have seen the problems of radiative transfer mostly in one-dimensional plane parallel or spherically symmetric geometries in the context of astrophysical situations. The book by Sen and Wilson (1990) deals extensively with the basic techniques for solving radiative transfer problems in spherically and cylindrically symmetric media (see also Leong and Sen (1969, 1970, 1971a,b), Uesugi and Tsujita (1969), Kho and Sen (1972)). Taking into account the effects of geometrical convergence and oblique incidence arising out of the sphericity of the medium, Bellman, Kagiwada, Kalaba and Ueno (see the references given in Sen and Wilson (1990)) solved the problems of the diffuse transmission of light from a central point source through an inhomogeneous spherical shell medium. Tsujita used a corresponding method to solve transfer problems in infinite cylindrical media (see Sen and Wilson (1990)). Certain approximate techniques such as ray-by-ray methods may be useful in special circumstances, but we need to explore the solution in multi-dimensional geometries so that any $I(X_1, Y_1, Z_1; t_1)$ can be correlated to any other $I(X_2, Y_2, Z_2; t_2)$ exactly. This is essential especially in scattering media which generate diffuse radiation fields.

However, the developments in multi-dimensional radiative transfer are not as advanced as the one-dimensional (including curved geometries) case. The solution of multi-dimensional radiative transfer is most important and is needed in astrophysical problems. We shall sketch some of the available results in this chapter.

13.2 Reflection effect in binary stars

Radiation transfer in the distorted atmospheres of close binary components is a matter of some difficulty. No symmetric solution of the transfer equation will apply in this case. The atmosphere is distorted by the self rotation and the tidal forces due to the proximity of the companion. This can be treated with strictly three-dimensional radiative transfer. The boundary conditions create another complication since in addition to the self radiation of the component there is incident radiation from the companion which needs a special treatment of the boundary conditions. This requires knowledge of the three-dimensional shape of the surface of the distorted component which can be obtained through an exact knowledge of the dynamics of the system. This is similar to the searchlight problem. Chandrasekhar (1958) studied this problem by applying the principle of invariance to a beam of radiation from a point source incident on a plane parallel layer (see also Richards (1956)). These problems have been reviewed extensively in Vaz (1985), Vaz and Nordlund (1985) and Wilson (1990), Claret and Giménez (1992) amongst others. We shall describe an approximate numerical computation of the reflected radiation from the component's atmosphere.

The radiation field can be divided into two parts: (1) self radiation and (2) radiation incident from the companion. We shall describe the procedure in Peraiah (1982, 1983a,b). Peraiah and Rao (1983, 1998) and Srinivasa Rao and Peraiah (2000).

We assume that the shapes of the components of the binary system are spherical and that the incident radiation is coming from a spherical surface of the secondary. The atmosphere of the primary is divided into several concentric spherical shells. In figure 13.1, we have given the geometrical description of the model. A and B are the centres of the components, separated by the distance AB. The atmosphere of the component with its centre at A receives radiation from that with its centre at B. Radii vector are drawn to intersect the shells at points such as P in the atmosphere of the component with centre A. We calculate the radiation field at points such as P which receives the radiation emitted by the surface SW of the component with centre B. We calculate the ray path lengths $P\tau$, PT_2, $P\tau'$, PT_3 etc., to estimate the optical depths and the radiation field at P due to the incidence of radiation from component B. The quantities AP, AB and $BS(= BW)$ are known in advance. Furthermore, the radii AT_1 $A\tau$, etc. are also assumed to be known quantities. The lengths SE and WF are discretized to calculate the corresponding ray paths along $P\tau$ etc., inside the atmosphere. The segment $P\tau$ corresponding to SB is calculated by the relation

$$P\tau = A\tau \frac{\nu}{\Delta\nu}, \tag{13.2.1}$$

where

$$\nu = \mu'\Delta + \mu\Delta', \qquad \mu = \frac{AP}{A\tau}\Delta,$$

$$\mu' = (1-\mu^2)^{\frac{1}{2}}, \qquad \Delta = \alpha\beta\gamma + \alpha\beta\gamma' + \alpha\beta'\gamma' - \alpha'\beta'\gamma',$$
$$\Delta' = (1-\Delta^2)^{\frac{1}{2}}, \qquad \alpha = \frac{BS}{PB},$$
$$\alpha' = (1-\alpha^2)^{\frac{1}{2}}, \qquad \beta = \frac{AB}{PB}\cos\theta,$$
$$\beta' = (1-\beta^2)^{\frac{1}{2}}, \qquad \gamma = \frac{SE}{PE}, \qquad \gamma' = (1-\gamma^2)^{\frac{1}{2}},$$
$$PB = (AP^2 + AB^2 - 2AP \times AB\sin\theta)^{\frac{1}{2}}, \qquad PE = (SE^2 + PS^2)^{\frac{1}{2}}.$$

Similarly, the segment $P\tau'$ corresponding to BW is given by

$$P\tau' = A\tau'\frac{e}{c}, \tag{13.2.2}$$

where

$$e = cd' + c'd, \qquad d = \frac{AP}{A\tau'}c,$$
$$d' = (1-d^2)^{\frac{1}{2}}, \qquad c = \beta b' - \beta' b,$$
$$c' = (1-c^2)^{\frac{1}{2}}, \qquad b = \alpha a' - \alpha' a,$$
$$b' = (1-b^2)^{\frac{1}{2}}, \qquad a = \frac{WF}{PF},$$

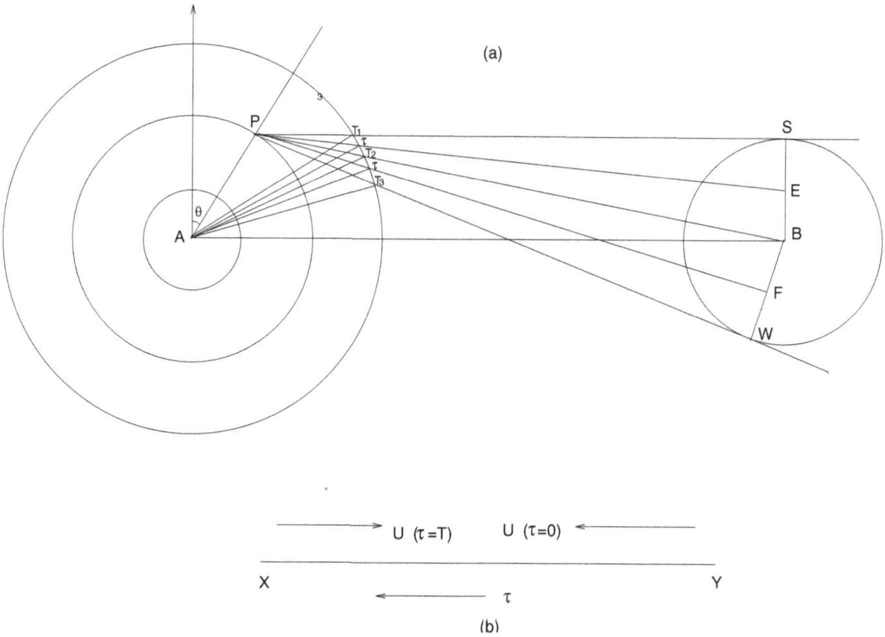

Figure 13.1 (a) Schematic diagram of the incidence of radiation from an extended surface. (b) Schematic diagram of the rod model (from Peraiah (1983b), with permission).

$$a' = (1 - a^2)^{\frac{1}{2}}.$$

We now calculate the specific intensity at points such as P due to the ray paths $P\tau$, $P\tau'$ etc. This is done by using the rod model which is fully described in chapter 6 (see figure 13.1(b)). The source function at P is given by

$$S_d(\tau) = S(\tau) + \varpi(\tau)P(\tau)U_b(\tau), \tag{13.2.3}$$

where ϖ is the albedo for single scattering and the first term on the RHS is due to the diffusely scattered radiation field while the second term is due to the incident radiation. The quantity $S(\tau)$ is given by

$$S(\tau) = (1 - \varpi)B(\tau) + \varpi(\tau)P(\tau)U(\tau), \tag{13.2.4}$$

where

$$\left.\begin{aligned} P(\tau) &= \begin{pmatrix} p & 1-p \\ 1-p & p \end{pmatrix}, \quad B = [B^+, B^-]^T, \\ S &= [S^+, S^-]^T, \quad U(\tau) = [U^+(\tau), U^-(\tau)]^T, \end{aligned}\right\} \tag{13.2.5}$$

with Bs the Planck functions, Us the specific intensities in the opposite directions along the segments $P\tau$ etc., p the probability that a photon is scattered in each of the two directions (for isotropic scattering $p = 0.5$) and

$$S^+ = [1 - \varpi(\tau)]\, B^+ + \varpi(\tau)\left[p(\tau)U^+(\tau) + (1 - p(\tau))U^-(\tau)\right] \tag{13.2.6}$$

and

$$S^- = [1 - \varpi(\tau)]\, B^- + \varpi(\tau)\left[(1 - p(\tau))U^+(\tau) + p(\tau)U^-(\tau)\right]. \tag{13.2.7}$$

The boundary conditions are

$$U^+(\tau = 0) = U_1 \quad \text{and} \quad U^-(\tau = T) = U_2. \tag{13.2.8}$$

The quantity U_b in equation (13.2.3) is given by

$$U_b(\tau) = \begin{pmatrix} U_1 \exp(-\tau) \\ U_2 \exp(-T - \tau) \end{pmatrix}. \tag{13.2.9}$$

The quantities U_1 and U_2 can be specified in advance. The physical meaning of the vector $U_b(\tau)$ is that the intensity at any given point and in a given direction results from the incident radiation at $\tau = 0$ from all the anterior points reduced by the factor $\exp(-\tau)$ and by the factor $\exp[-(T - \tau)]$. In this model, the incident radiation is given at the points T_1, τ etc., and no radiation is given at the point P. Hence, the boundary conditions are given by

$$U_2 = 0 \quad \text{and} \quad U_1 = I \cos\mu', \tag{13.2.10}$$

where I is the ratio of the radiations corresponding to the two components with centres at B and A (and should not be confused with the specific intensity $I(\tau, \pm\mu)$), and μ' is given in equation (13.2.1). The intensities U^+ and U^- are obtained by solving the transfer equation for the rod (see chapter 6):

$$\mathbf{M}\frac{d\mathbf{U}}{d\tau} + \mathbf{U} = \mathbf{S}, \tag{13.2.11}$$

where

$$\mathbf{M} = \begin{pmatrix} 1 & 0 \\ 0 & -1 \end{pmatrix}. \tag{13.2.12}$$

The solution is given by,

$$U^+(\tau) = U_1 \frac{1 + (T - \tau)(1 - p)}{1 + T(1 - p)} \tag{13.2.13}$$

and

$$U^-(\tau) = U_1 \frac{(T - \tau)(1 - p)}{1 + T(1 - p)}, \tag{13.2.14}$$

where

$$\tau(x) = \int_0^x N(x')\sigma \, dx', \tag{13.2.15}$$

with σ the electron scattering coefficient, $N(x')$ the electron density at x' along the ray path and T the total optical depth. Equations (13.2.13) and (13.2.14) reduce to,

$$U^+(\tau = T) = \frac{U_1}{1 + T(1 - p)} \quad \text{and} \quad U^-(\tau = 0) = U_1 \frac{T(1 - p)}{1 + T(1 - p)}. \tag{13.2.16}$$

Using equations (13.2.6)–(13.2.16), we can compute the source functions S_d given in equation (13.2.3). The source function due to self radiation is obtained by solving the transfer equation in spherical symmetry given by

$$\mu \frac{\partial U(r, \mu)}{\partial r} + \frac{1}{r}\frac{\partial}{\partial \mu}\left[(1 - \mu^2)U(r, \mu)\right] + \kappa(r)U(r, \mu) =$$
$$\kappa(r)\left\{[1 - \varpi(r)]\, B^+(r) + \frac{1}{2}\varpi(r)\int_{-1}^{+1} P(r, \mu, \mu')U(r, \mu')\, d\mu'\right\} \tag{13.2.17}$$

and

$$-\mu \frac{\partial U(r, -\mu)}{\partial r} - \frac{1}{r}\frac{\partial}{\partial \mu}\left[(1 - \mu^2)U(r, -\mu)\right] + \kappa(r)U(r, -\mu) =$$
$$\kappa(r)\left\{[1 - \varpi(r)]\, B^-(r) + \frac{1}{2}\varpi(r)\int_{-1}^{+1} P(r, -\mu, \mu')U(r, \mu)\, d\mu'\right\}, \tag{13.2.18}$$

where $U(r, \mu) = 4\pi r^2 I(r, \mu)$, $I(r, \mu)$ is the specific intensity of the ray at the radial point r making an angle $\cos^{-1}\mu$ with the radius vector. $\kappa(r)$ is the absorption coefficient and the quantities B^+ and B^- are the Planck functions (the + and − signs are useful in a moving atmosphere), $P(r, \mu, \mu')$ is the isotropic phase function and $0 \le \mu \le 1$. For a scattering medium, we set $\varpi = 1$ and $B^+ = B^- = 0$, in which case the source function becomes

$$S_s(\tau) = \frac{1}{2}\int_{-1}^{+1} I(r, \mu)\, d\mu. \tag{13.2.19}$$

The details for obtaining the solution of equations (13.2.17) and (13.2.18) are given in chapter 6.

We now obtain the total source functions at P by adding S_d given in equation (13.2.3) and S_s given in equation (13.2.19). Thus

$$S_T = S_d(r, \tau) + S_s(r) \tag{13.2.20}$$

This source function is used to estimate the radiation field in terms of $I(\tau, +\mu)$ and $I(\tau, -\mu)$ which are given by,

$$I(\tau, +\mu) = I(\tau, \mu)\exp[-(\tau_1 - \tau)/\mu] + \int_{\tau}^{\tau_1} S_T(t)\exp[-(t - \tau)/\mu]\, dt/\mu \tag{13.2.21}$$

for the outward intensities and

$$I(\tau, -\mu) = I(0, -\mu)\exp(-\tau/\mu) + \int_0^{\tau} S_T(t)\exp[-(\tau - t)/\mu]\, dt/\mu \tag{13.2.22}$$

for the inward intensities. The intensities $I(\tau, +\mu)$ and $I(\tau, -\mu)$ are shown in figure 13.2.

Thus, by combining the solutions of the rod spherical symmetry and plane parallel models, we can obtain the radiation field of the reflected light from the components of close binary stars. This procedure does not give an exact solution of the problem but by increasing the number of rays such as $P\tau$, etc., the accuracy can be increased and a computationally fast solution can be obtained.

The limb darkening can be obtained by calculating the radiation at infinity, using the line of sight scheme described in chapter 6. In figure 13.3, we give the

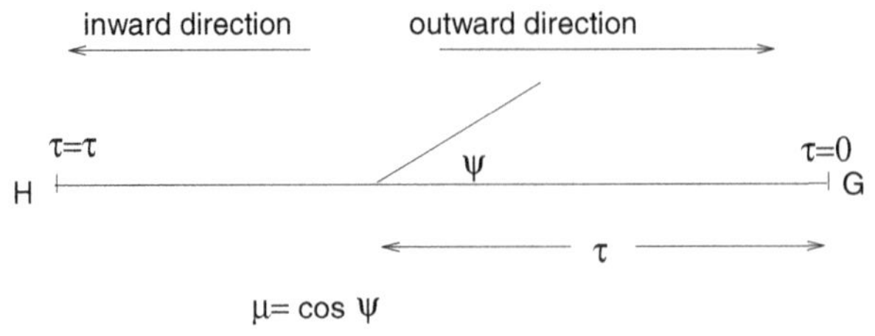

Figure 13.2 Schematic diagram of the rod model.

total source function S_T with respect to the shell numbers, and the corresponding radiation field is shown in figure 13.4.

This method has been applied to the computation of non-LTE lines with the reflection effect (Peraiah and Rao 1983), assuming a purely scattering medium. It

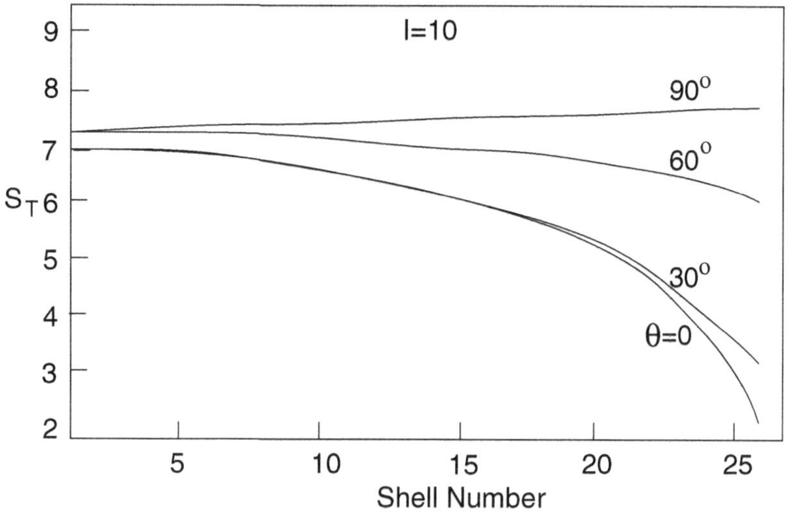

Figure 13.3 Total source function S_T versus shell number for different values of the colatitude (see figure 13.1(a)) $\theta = 0°, 30°, 60°$ and $90°$ for $I = 10$, where I is the ratio of radiations corresponding to the two components with centres at B and A (from Peraiah (1983b), with permission).

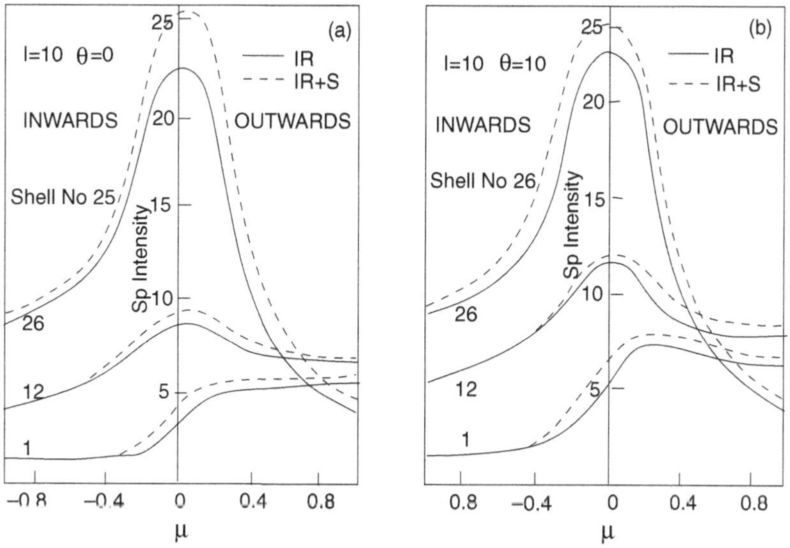

Figure 13.4 Angular distribution of radiation field for $I = 10$ and (a) $\theta = 0°$ and (b) $\theta = 10°$ (from Peraiah (1983b), with permission).

was found that the fluxes in the lines are increased at all frequency points, the cores of the lines receiving more radiation than the wings. The proximity of the secondary component changes the equivalent widths considerably, that is the equivalent widths increase with increasing separation between the components.

This method has been applied to expanding atmospheres of close binary components using the transfer equation in the comoving frame of the gas (Peraiah and Rao 1998) and with the light incident on the atmosphere from the secondary. The transfer equation in the comoving frame is given by

$$\mu \frac{\partial I(x,\mu,r)}{\partial r} + \frac{1-\mu^2}{r}\frac{\partial I(x,\mu,r)}{\partial \mu} = \kappa(x,r)S_L(r) + \kappa_c(r)S_c(r) - [\kappa(x,r)+\kappa_c(r)]\,I(x,\mu,r) + \left[(1-\mu^2)\frac{V(r)}{r} + \mu^2\frac{dV(r)}{r}\right]\frac{\partial I(x,\mu,r)}{\partial x}, \quad 0 \le \mu \le 1 \quad (13.2.23)$$

with a similar equation for $-1 \le \mu \le 0$, where $I(x,\mu,r)$ is the specific intensity making an angle $\cos^{-1}\mu$ with the radius r and x is the normalized frequency $x = (\nu - \nu_0)/\Delta\nu_D$, $\Delta\nu_D$ being the Doppler width and ν and ν_0 being the frequencies at

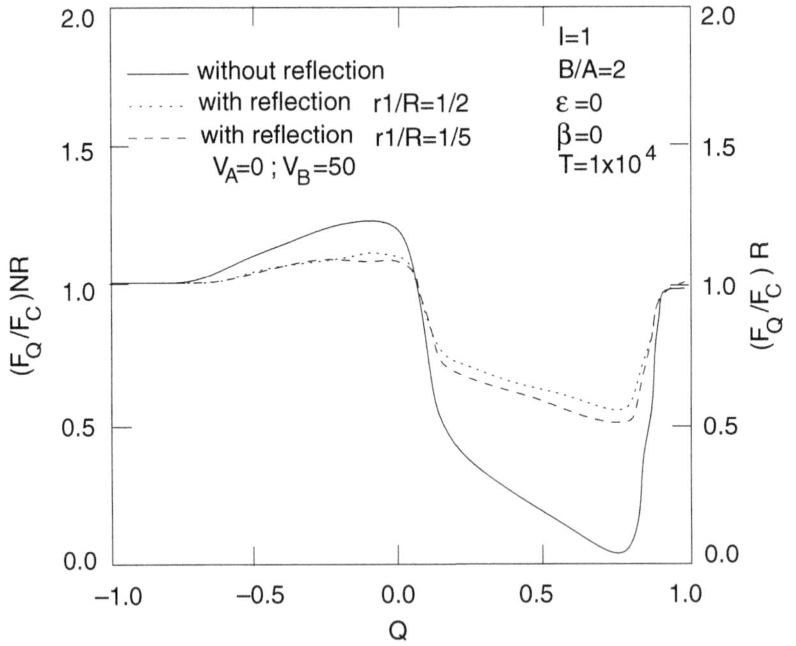

Figure 13.5 Line fluxes with respect to $Q = x/x_{max}$. B/A is the ratio of the outer to inner radii of the atmosphere. r_1/R is the ratio of the radius of the primary to the distance between the centres of stars. V_A and V_B are the velocities at $r = A$ and $r = B$ in mtu. ϵ is a non-LTE parameter and β is the ratio of the absorption in the continuum to that in the line. T is the total optical depth at the line centre (from Peraiah & Rao (1998), with permission).

any point in the line and at the line centre respectively. $V(r)$ is the velocity of the gas at r in mean thermal units (mtu) and $\kappa(x, r)$ and $\kappa_c(r)$ are the absorption coefficients per unit frequency interval in the line and continuum respectively. The quantities S_L and S_c are the line and continuum source functions respectively. Assuming a two-level atom model and complete redistribution, line profiles have been computed in an expanding atmosphere. In figure 13.5, a profile of the spectral lines is given. A linear law was used for the velocity of expansion and this produced a PCygni type profile. In the case of reflection, the core of the line receives more of the photons of the incident light from the companion and has less absorption. However, the wings of the line remain largely unchanged. Srinivasa Rao and Peraiah (2000) extended this method to dusty, irradiated atmospheres of the close binary components.

13.3 Two-dimensional transfer and discrete space theory

We shall now describe a method of solving transfer equation X–Y geometry using discrete space theory following Ellison and Grant (1974). Using the interaction principle, a set of linear equations for the intensities on a discrete grid in a Cartesian X–Y domain are developed. This method gives a solution with non-negative intensities and flux conservation provided the size of the cells of the grid is sufficiently small. The transfer equation in X–Y geometry is

$$\mu \frac{\partial U(x, \theta)}{\partial x} + \eta \frac{\partial U(x, \theta)}{\partial y} + \kappa(x) U(x, \theta) = S(x), \tag{13.3.1}$$

where

$$S(x) = \kappa(x) [1 - \varpi(r)] B(x) + \frac{x(x)\varpi(x)}{2} \int_0^{2\pi} p(x, \theta, \theta') U(x, \theta') \, d\theta'. \tag{13.3.2}$$

Here $U(x, \theta)$ is the intensity of radiation at the point $x \equiv (x, y)$ of the X–Y plane in the direction θ with the x-axis, $\mu = \cos\theta$, $\eta = \sin\theta$, $\varpi(x)$ is albedo for single scattering, $\kappa(x)$ is the absorption coefficient and $B(x)$ is the source at the point x. $P(x, \theta, \theta')$ is the scattering phase function for radiation scattered from direction θ' into the direction θ.

Let us introduce an operator T which relates locally the emergent intensity from a cell to the incident intensity and internal sources, or

$$TU = S, \tag{13.3.3}$$

where the matrix T contains the absorption and scattering information and S contains information about the internal sources.

The determination of the operator T is different from the usual procedure adopted in discrete space theory. In multi-dimensional geometry, we assume some ‘flux shape' within each cell. However, this leads to problems in flux conservation and non-negativity of the intensities. To overcome this difficulty, the operator T is determined by using the relation (see figure 13.6)

$$U_o = U_i \exp(-xs') + \int_0^{s'} \exp\left[-\kappa(s'-s)\right] ds, \tag{13.3.4}$$

where the subscripts i and o refer to input and output quantities.

We now determine the T operator. We solve the transfer equation (13.3.1) in the X–Y plane with $X_0 \leq x \leq X_I$ and $Y_0 \leq y \leq Y_I$ and divide the region into MN cells, not necessarily of the same size. The angular discretization in the first quadrant is $Q_1 \equiv \{\theta_k | k = 1, 2, \ldots, K\}$ so that $0 < \theta_1 < \theta_2 < \theta_3 \cdots < \theta_k < \pi/2$. The angles in the other three quadrants are $\theta_2 \equiv \{\phi_k | \phi_k = \pi - \theta_k\}$, $Q_3 \equiv \{\phi_k\ \phi_k = \pi + \theta_k\}$, $Q_4 \equiv \{\phi_k | \phi_k = 2\pi - \theta_k\}$. We choose $M = [\mu_k \delta_{kk'}]$ and $N = [\eta_k \delta_{kk'}]$, where $k = 1, \ldots, K$ and $\mu_k = \cos\theta_k$ and $\eta_k = \sin\theta_k$.

Now on the cell in the X–Y plane bounded by the lines parallel to the axes at $x = x,\ x + \delta x$ and $y = y,\ y + \delta y$, the averaged intensity on the sides of this cell $U(y + \delta y, \theta)$ is given by

$$\delta x \cdot U(y + \delta y, \theta) = \int_x^{x+\delta x} U^1(x', y + \delta y, \theta)\, dx'. \tag{13.3.5}$$

If $U^i(x, \theta) = \{U(x, \theta,\ \theta \in Q_i\}$, then the averaged output intensities from the cell are the vectors $U^1(x + \delta x)$, $U^1(y + \delta y), \ldots, U^4(x + \delta x)$, where $U^1(y + \delta y) \equiv U^1(y + \delta y, \theta_k) | \theta_k \in Q_1,\ k = 1, \ldots, K\}$. We define $\rho_k = (\eta_k/\mu_k)(\delta x/\delta y)$. There are two types of rays passing through, for which $\rho_k > 1$ and $\rho_k \leq 1$. The symbols $<$ and $>$ are used to represent directions for which $\rho \leq 1$ and $\rho > 1$ respectively. For example $U^1(<, y)$ is the averaged intensity vector on side y for all directions in the first quadrant such that $\eta_k/\mu_k = \tan\theta_k \leq \delta y/\delta x$ (these rays enter at the bottom and emerge at the right), or

$$U^1(<, x + \delta x) \equiv [U^1(x + \delta x, \theta), U^1(x + \delta x, \theta_2), \ldots, U^1(x + \delta x, \delta\bar{k}]^T, \tag{13.3.6}$$

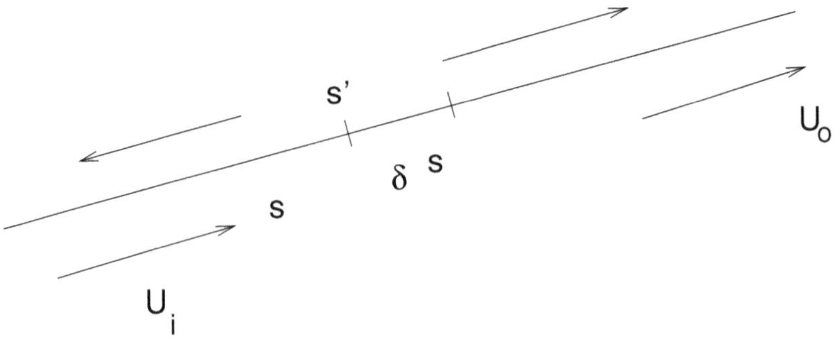

Figure 13.6 Schematic diagram of the input and output intensities.

where T denotes the transpose of the vector. Similarly

$$U^1(>, x+\delta x) = [U^1(x+\delta x, \theta_{\bar{k}+1}), \ldots, U^1(x+\delta x, \theta_k)]^T, \tag{13.3.7}$$

where $1 \le \bar{k} \le K$ and

$$U^1(x+\delta x) = \begin{pmatrix} U^1(<, x+\delta x) \\ U^1(>, x+\delta x) \end{pmatrix}. \tag{13.3.8}$$

We define output vectors

$$U_o^1 = \begin{pmatrix} U^1(x+\delta x) \\ U^1(y+\delta y) \end{pmatrix}, \quad U_o^2 = \begin{pmatrix} U^2(x) \\ U^2(y+\delta y) \end{pmatrix}, \text{ etc.} \tag{13.3.9}$$

with $U_o = [U_o^1, U_o^2, U_o^3, U_o^4]^T$, $U_I = [U_I^1, U_I^2, U_I^3, U_I^4]^T$ (for example $U_I^1(U^1(x), U^1(y)]^T)$.

The interaction principle is written as

$$\begin{pmatrix} U_o^1 \\ U_o^2 \\ U_o^3 \\ U_o^4 \end{pmatrix} = \begin{pmatrix} \Delta_1 + P_{11} & P_{12} & P_{13} & P_{14} \\ P_{21} & \Delta_2 + P_{22} & P_{23} & P_{24} \\ P_{31} & P_{32} & \Delta_3 + P_{23} & P_{34} \\ P_{41} & P_{42} & P_{43} & \Delta_4 + P_{44} \end{pmatrix} \times \begin{pmatrix} U_I^1 \\ U_I^2 \\ U_I^3 \\ U_I^4 \end{pmatrix} + \begin{pmatrix} \Sigma^1 \\ \Sigma^2 \\ \Sigma^3 \\ \Sigma^4 \end{pmatrix}, \tag{13.3.10}$$

where

$$P_{rs} = \frac{\kappa\varpi}{2\pi} \begin{pmatrix} \frac{M^{-1}(<)}{\delta y} & & & \\ & \frac{M^{-1}(>)}{\delta y} & & \\ & & \frac{M^{-1}(<)}{\delta x} & \\ & & & \frac{M^{-1}(<)}{\delta y} \end{pmatrix} \times \begin{pmatrix} XX_{rs}(<<) & XX_{rs}(<>) & XY_{rs}(<<) & XY_{rs}(<>) \\ XX_{rs}(><) & XX_{rs}(>>) & XY_{rs}(><) & XY_{rs}(>>) \\ XX_{rs}(<<) & YX_{rs}(<>) & YY_{rs}(<<) & YY_{rs}(<>) \\ YX_{rs}(><) & YX_{rs}(>>) & YY_{rs}(><) & YY_{rs}(>>) \end{pmatrix},$$

$$\text{for} \quad r, s = 1, \ldots, 4, \tag{13.3.11}$$

$$\Delta_r = \begin{pmatrix} A_{XX}(<) & 0 & A_{XY}(<) & 0 \\ 0 & 0 & 0 & A_{XY}(<) \\ A_{YX}(<) & 0 & 0 & 0 \\ 0 & A_{YX}(<) & 0 & A_{YY}(<) \end{pmatrix},$$
$$r = 1, \ldots, 4 \tag{13.3.12}$$

and

$$\Sigma^r = \left[\Sigma_X^r(<), \Sigma_X^r(>), \Sigma_Y^r(<), \Sigma_Y^r(>)\right]^T . \tag{13.3.13}$$

The various matrices in (13.3.11), (13.3.12) and (13.3.13) are defined below (with similar expressions for N):

$$M(<) = [\mu_k \delta_{kk'}], \quad 1 \le k \le \bar{k}, \quad M(>) = [\mu_k \delta_{kk'}], \ \bar{k} < k \le K. \tag{13.3.14}$$

$$\left.\begin{aligned} A_{XX}(<) &= \left[(1-\rho_k)\left(1-\frac{\kappa\delta x}{\mu_k}\right)\right]\delta_{kk'}, \\ A_{YY}(>) &= \left[\left(1-\frac{1}{\rho_k}\right)\left(1-\frac{\kappa\delta y}{\eta_k}\right)\right]\delta_{kk'}, \\ A_{XY}(<) &= \left[\rho_k\left(1-\frac{\kappa\delta x}{2\mu_k}\right)\right]\delta_{kk'}, \\ A_{YX}(>) &= \left[\frac{1}{\rho_k}\left(1-\frac{\kappa\delta y}{2\eta_k}\right)\right]\delta_{kk'}, \\ A_{XY}(>) &= \left[\left(1-\frac{\kappa\delta y}{2\eta_k}\right)\right]\delta_{kk'}, \\ A_{YX}(<) &= \left[\left(1-\frac{\kappa\delta x}{\mu_k}\right)\right]\delta_{kk'}. \end{aligned}\right\} \tag{13.3.15}$$

The Σs are defined similarly.

The matrices $(XY)_{rs}(>, <)$ etc. are more complicated. They involve the scattering phase function angular quadrature, geometric weighting. For details and further numerical analysis see Ellison and Grant (1974).

13.4 Three-dimensional radiative transfer

Stenholm (1977) extended Rybicki's core-saturation method to multi-dimensional geometry. Stenholm *et al.* (1991) developed a method for solving the problems in two- and three-dimensional geometries using an implicit discretization of the transfer equation in Cartesian coordinates. We shall describe this method briefly below.

The transfer equation in the three-dimensional Cartesian system is given by

$$n_x \frac{\partial I}{\partial x} + n_y \frac{\partial I}{\partial y} + n_z \frac{\partial I}{\partial z} = \kappa(S - I), \tag{13.4.1}$$

where

$$n_x = \sin\theta\cos\phi, \quad n_y = \sin\theta\sin\phi, \quad n_z = \cos\theta \tag{13.4.2}$$

and θ is the angle between the ray direction and the positive z-axis and ϕ is the angle between the positive x-axis and projection of the ray on the x–y plane. S is the source function. It is assumed that the infalling intensities are given as boundary values for all surfaces. Equation (13.4.1) is discretized implicitly in the direction of the rays, that is for positive n_x, n_y and n_z, giving

$$\begin{aligned}&\frac{n_x}{\Delta x}(I_{i+1,j+1,k+1} - I_{i,j+1,k+1}) + \frac{n_y}{\Delta y}(I_{i+1,j+1,k+1} - I_{i+1,j,k+1})\\ &\quad + \frac{n_z}{\Delta z}(I_{i+1,j+1,k+1} - I_{i+1,j+1,k})\\ &\quad = \kappa_{i+1,j+1,k+1}(S_{i+1,j+1,k+1} - I_{i+1,j+1,k+1}).\end{aligned} \tag{13.4.3}$$

$I_{i+1,j+1,k+1}$ is obtained from equation (13.4.3) as

$$\begin{aligned}I_{i+1,j+1,k+1} = &\left(\frac{n_x}{\Delta x}I_{i,j+1,k+1} + \frac{n_y}{\Delta y}I_{i+1,j,k+1} + \frac{n_z}{\Delta z}I_{i+1,j+1,k}\right.\\ &\left.+\kappa_{i+1,j+1,k+1}S_{i+1,j+1,k+1}\right)\left(\frac{n_x}{\Delta x} + \frac{n_y}{\Delta y} + \frac{n_z}{\Delta z} + \kappa_{i+1,j+1,k+1}\right)^{-1}.\end{aligned} \tag{13.4.4}$$

If the source function is prescribed in advance at all grid points, the intensities can be determined recursively starting from the point (1,1,1). If the source function is not known *a priori* it can be obtained by iteration starting with the value

$$S^0 = (1 - \varpi)B, \tag{13.4.5}$$

where ϖ is the albedo for single scattering, and

$$S^{n+1} = (1 - \varpi)B + \varpi J^n, \tag{13.4.6}$$

where J^n is the mean intensity obtained from S^n. This scheme converges slowly. The rate of convergence can be increased by the method of Ng in which a succession of immediate previous iterations are used to give enough information to obtain a better than linear rate of convergence. The accelerated scheme is

$$S = (1 - \alpha_1 - \alpha_2)S_n + \alpha_1 S^{n-1} + \alpha_2 S^{n-2}, \tag{13.4.7}$$

where the α coefficients are to be determined. Equation (13.4.7) can be written as

$$S' = (1 - \alpha_1 - \alpha_2)S^{n-1} + \alpha_1 s^{n-2} + \alpha_2 s^{n-3}. \tag{13.4.8}$$

The αs are determined by minimizing the distance between S and S' by a method similar to that of least squares. The equations chosen to minimize are r^2 =

$\sum \omega_{i,j,k}(S_{i,j,k} - S'_{i,j,k})^2$, the $S_{i,j,k}$s being the components of the estimates of the solution vector and the ωs the weights. The normal equations

$$\mathbf{A}\boldsymbol{\alpha} = \mathbf{b}, \tag{13.4.9}$$

form a system of two symmetric linear equations that determine the two unknown acceleration coefficients. The elements are given by

$$A_{l,n} = \sum_d \omega_{i,j,k} \left(\Delta S^n_{i,j,k} - \Delta S^{n-1}_{i,j,k}\right) \left(\Delta S^n_{i,j,k} - \Delta S^{n-m}_{i,j,k}\right) \tag{13.4.10}$$

and

$$b_l = \sum_d \omega_{i,j,k} \Delta S^n_{i,j,k} (\Delta S^n_{i,j,k} - \Delta S^{n-1}_{i,j,k}), \tag{13.4.11}$$

with

$$\Delta S^k = S^k - S^{k-1}. \tag{13.4.12}$$

This acceleration scheme is used in the iteration procedure to obtain four successive estimates of the solution error. Equation (13.4.9) is then solved to obtain the two αs. The weights are taken to be

$$\omega_d = 1/S^n_d. \tag{13.4.13}$$

Cannon's iteration scheme (see chapter 10) can be used. It is given by

$$S^{n+1} = \varpi \Lambda^\star[S^{n+1}] + \varpi(\Lambda - \Lambda^\star)[S^n] + (1 - \varpi)B \tag{13.4.14}$$

or

$$S^{n+1} = (1 - \varpi\Lambda^\star)^{-1} \left\{(\varpi(\Lambda - \Lambda^\star)[S^n] + (1 - \varpi)B\right\}. \tag{13.4.15}$$

The quantity $\Lambda[S^n]$ represents the $(n + 1)$th iteration of the basic method. The approximate diagonal is derived analogously to the operator (see Olson *et al.* (1986)), which is done by using the following relations:

$$S_{i+1,j+1,k+1} = 1, \tag{13.4.16}$$

$$I^\star_{i,j+1,k+1} = I^\star_{i+1,j+1,k+1} \exp[-\tau_{i+1,j+1,k+1} - \tau_{i,j+1,k+1}], \tag{13.4.17}$$

$$I^\star_{i+1,j,k+1} = I^\star_{i+1,j+1,k+1} \exp[-\tau_{i+1,j+1,k+1} - \tau_{i+1,j,k+1}], \tag{13.4.18}$$

$$I^\star_{i+1,j+1,k} = I^\star_{i+1,j+1,k+1} \exp[-\tau_{i+1,j+1,k+1} - \tau_{i+1,j+1,k}], \tag{13.4.19}$$

Substituting equations (13.4.16)–(13.4.19) into equation (13.4.4) we obtain the local algebraic equation for approximate $I^\star$, which gives the diagonal approximate lambda operator:

$$\Lambda_{i+1,j+1,k+1} = \frac{1}{4\pi}\int_{\Omega} \kappa_{i+1,j+1,k+1}\left\{\kappa_{i+1,j+1,k+1} + \frac{n_x}{n_y}\left[1-\exp(-\Delta\tau_x)\right]\right.$$
$$\left. + \frac{n_y}{\Delta y}\left[1-\exp(-\Delta\tau_y)\right] + \frac{n_z}{\Delta z}\left[1-\exp(-\Delta\tau_z)\right]\right\}^{-1} d\Omega. \qquad (13.4.20)$$

The $\Delta\tau$s are the same as those in the exponents of equations (13.4.17)–(13.4.19). Equations similar to equations (13.4.17)–(13.4.19) are used at the boundaries in the difference equations for the boundary conditions. $\Lambda^\star[S^n]$ can be calculated and then equation (13.4.15) is used to obtain new estimates of the source function.

Wehrse *et al.* (2000a,b) developed the diffusion of radiation in moving media in a three-dimensional geometry using the solutions of Baschek *et al.* (1997a,b).

13.5 Time dependent radiative transfer

Time dependent radiative transfer problems involve the complex interaction of matter and radiation. When a photon interacts with quantized atomic states one needs to know the time spent by the photon in the absorbed state or the time spent by it in two successive scatterings (see Mohan Rao *et al.* (1990)). There are several areas in which the time dependence of the transfer of radiation plays an important role. Most of the research in these problems has used the analytical approach. Latko and Pomraning (1972) used the synthesis method to solve time dependent two-dimensional non-linear radiative transfer. The synthesis method gives a logical consistent technique for constructing two-dimensional solutions from a small number of one-dimensional calculations. Leong and Sen (1972) developed a probabilistic approach to time dependent transfer in spherically symmetric inhomogeneous isotropically scattering media. They derived an integro-differential equation through four probability functions. They obtained emergent intensities in terms of scattering and transmission in both directions. Munier (1987, 1988) developed the integral form of the time dependent transfer equation in plane parallel, spherical, moving boundaries and three-dimensional geometry.

We shall describe a numerical solution of the time dependent inertial frame transfer equation in moving media to the order of (v/c). For a full discussion of this method see Mihalas and Klein (1982). We start with the equations

$$p_k(\mu) = \frac{1}{2}[I_k(+\mu) + I_k(-\mu)] \qquad (13.5.1)$$

and

$$q_k(\mu) = \frac{1}{2}[I_k(+\mu) - I_k(-\mu)]. \qquad (13.5.2)$$

Introducing these into the transfer equation, we obtain

$$\frac{1}{c}\frac{\partial I_k}{\partial t} + \mu\frac{\partial I_k}{\partial z} = \kappa_k(B_k - I_k) + \mu\beta\kappa_k(I_k + 3\bar{B}_k), \qquad (13.5.3)$$

where $\beta = v/c$ and

$$I_k = \int_{\nu_k}^{\nu_{k+1}} I_\nu \, d\nu, \qquad B_\nu = \int_{\nu_k}^{\nu_{k+1}} B_\nu \, d\nu,$$

$$\begin{aligned} 3\kappa_k \bar{B}_k &= \bar{\eta}_k \\ &= 3\kappa_k B_k + \frac{1}{2}[\nu_{k+1}(\kappa_k B_k + \kappa_{k+1} B_{k+1}) - \nu_k(\kappa_{k-1} B_{k-1} + \kappa_k B_k)], \end{aligned}$$

B being the Planck function. Then in terms of equations (13.5.1) and (13.5.2) for $\pm\mu$, we obtain from equation (13.5.3)

$$\frac{1}{c}\frac{\partial p_k}{\partial t} + \mu \frac{\partial q_k}{\partial z} = \kappa_k (B_k - p_k) + \mu \beta \kappa_k q_k \tag{13.5.4}$$

and

$$\frac{1}{c}\frac{\partial q_k}{\partial t} - \mu \frac{\partial p_k}{\partial z} = -\kappa_k q_k + \mu \beta \kappa_k (p_k + 3\bar{B}_k). \tag{13.5.5}$$

We shall describe the first order schemes. The z-axis is discretized into the grid $\{z_d, \ d = 1, 2, \ldots, D+1\}$ with $z_d > z_{d+1}$, which means that z_{d-1} is at the 'top' of the medium and $z_{d=D+1}$ is at the bottom of the medium. The angle grid points μ_m are chosen from a quadrature formula, while the frequencies are discretized over the mesh $\{\nu_k\}$. All quantities are specified at the centre of the cell $z_{d+\frac{1}{2}}$, $d = 1, \ldots, D$, while velocities and the flux-like vector $q(\nu)$ are specified at the cell boundaries, z_d. We use only depth levels and time levels t^n and suppress reference to angles μ_m and frequencies ν_k in assigning superscript and subscripts.

(A) An explicit scheme

We centre $p^n_{d+\frac{1}{2}}$ at time t^n and $q_d^{n+\frac{1}{2}}$ at times $t^{n+\frac{1}{2}} = \frac{1}{2}(t^n + t^{n+1})$. Equations (13.5.4) and (13.5.5) can then be written in a standard leapfrog manner as

$$\begin{aligned} &\left(p^{n+1}_{d+\frac{1}{2}} - p^n_{d+\frac{1}{2}}\right)(c\Delta t^{n+\frac{1}{2}})^{-1} + \mu(q_d^{n+\frac{1}{2}} q_{d+1}^{n+\frac{1}{2}})(\Delta z_d)^{-1} = \\ &\qquad \kappa^{n+\frac{1}{2}}_{d+\frac{1}{2}}\left[B^{n+\frac{1}{2}}_{d+\frac{1}{2}} - \frac{1}{2}\left(p^n_{d+\frac{1}{2}} + p^{n+1}_{d+\frac{1}{2}}\right)\right] \\ &\quad + \frac{1}{2}\mu\kappa^{n+\frac{1}{2}}_{d+\frac{1}{2}}\left(\beta_d^{n+\frac{1}{2}} q_d^{n+\frac{1}{2}} + \beta_{d+1}^{n+\frac{1}{2}} q_{d+1}^{n+\frac{1}{2}}\right), \quad d = 1, \ldots, D, \end{aligned} \tag{13.5.6}$$

$$\begin{aligned} &\left(q_d^{n+\frac{1}{2}} - q_d^{n-\frac{1}{2}}\right)(c\Delta t^n)^{-1} - \mu(p^n_{d-\frac{1}{2}} - p^n_{d+\frac{1}{2}})(\Delta z_{d-\frac{1}{2}})^{-1} = \\ &\qquad -\frac{1}{2}\kappa_d^n\left(q_d^{n-\frac{1}{2}} + q_d^{n+\frac{1}{2}}\right) + \mu\beta_d^n\left\{(3\kappa B)_d^n + (\kappa p)_d^n\right\}, \quad d = 2, \ldots, D, \end{aligned} \tag{13.5.7}$$

where $\delta t^n = \frac{1}{2}(\Delta t^{n-\frac{1}{2}} + \Delta t^{n+\frac{1}{2}})$, $\kappa^{n+\frac{1}{2}}_{d+\frac{1}{2}} = \frac{1}{2}(\kappa^n_{d+\frac{1}{2}} + \kappa^{n+1}_{d+\frac{1}{2}})$, $B^{n+\frac{1}{2}}_{d+\frac{1}{2}} = \frac{1}{2}(B^n_{d+\frac{1}{2}} + B^{n+1}_{d+\frac{1}{2}})$ and $\beta^n_d = \frac{1}{2}(\beta^{n-\frac{1}{2}}_d + \beta^{n+\frac{1}{2}}_d)$. The quantity Δz_d is the thickness of the dth slab and $\Delta z_{d-\frac{1}{2}} = \frac{1}{2}(\Delta z_{d-1} + \Delta z_d)$. The quantities $(\kappa B)^n_d$ and $(\kappa p)^n_d$ are interpreted similarly.

To obtain the boundary condition, we use equation (13.5.7) over the half intervals from the cell edge to the centre for the first and last slabs and the relations

$$p^n_1 = I^n_1 + q^n_1 = I^n_- + \frac{1}{2}(q^{n-\frac{1}{2}}_1 + q^{n+\frac{1}{2}}_1) \tag{13.5.8}$$

and

$$p^n_{D+1} = I^n_+ - q^n_{D+1} = I^n_+ - \frac{1}{2}(q^{n-\frac{1}{2}}_{D+1} + q^{n+\frac{1}{2}}_{D+1}). \tag{13.5.9}$$

Then,

$$\begin{aligned}(q^{n+\frac{1}{2}}_1 - q^{n-\frac{1}{2}}_1)(c\Delta t^n)^{-1} &+ \mu\left[I^n_- + (q^{n-\frac{1}{2}}_1 + q^{n+\frac{1}{2}}_1) - p^n_{\frac{3}{2}}\right](\Delta z_{\frac{1}{2}})^{-1} \\ &= -\kappa^n_{\frac{3}{2}}(q^{n-\frac{1}{2}}_1 + q^{n+\frac{1}{2}}_1)/2 + \mu\beta^n_1\kappa^n_{\frac{3}{2}}\left[3B^n_{\frac{3}{2}} + I^n_- + (q^{n-\frac{1}{2}}_1 + q^{n+\frac{1}{2}}_1)/2\right],\end{aligned} \tag{13.5.10}$$

$$\begin{aligned}(q^{n+\frac{1}{2}}_{D+1} - q^{n-\frac{1}{2}}_{D+1})(c\Delta t^n)^{-1} &+ \mu\left[p^n_{D+\frac{1}{2}} - I^n_+ + (q^{n-\frac{1}{2}}_{D+1} + q^{n+\frac{1}{2}}_{D+1})/2\right](\Delta z_D/2)^{-1} \\ &= -\kappa^n_{D+\frac{1}{2}}(q^{n-\frac{1}{2}}_{D+1} + q^{n+\frac{1}{2}}_{D+1})/2 \\ &+ \mu\beta^n_{D+1}\kappa^n_{D+\frac{1}{2}}\left[3B^n_{D+\frac{1}{2}} + I^n_+ - (q^{n-\frac{1}{2}}_{D+1} + q^{n+\frac{1}{2}}_{D+1})^2\right].\end{aligned} \tag{13.5.11}$$

We obtain D equations from equation (13.5.6) for D values of $p^{n+1}_{d+\frac{1}{2}}$ and equations (13.5.7), (13.5.10) and (13.5.11) give $D+1$ equations for $D+1$ values of $q^{n+\frac{1}{2}}_d$. One can update the p and q values by vectorizing the solution either over the depth grid or over all angles and frequencies.

Equations (13.5.6) and (13.5.7) are subject to von Neumann's local stability condition which gives the Courant condition:

$$c\Delta t < \Delta z/\mu,$$

which gives the required time steps. This is quite useful in the case of radiation flow but not as useful in the case of fluid flow in which one requires time steps of the order of $\Delta t \approx \Delta z/v$, where $v(\ll c)$ is the fluid velocity. In the diffusion limit a more restricted time step limitation of the form $\Delta t \leq k(\Delta z)^2$ may be necessary. This indicates that we should turn to implicit schemes.

(B) Two-level implicit schemes

We adopt the same spatial centring as in (A) but put all variables at a common time level. Then equations (13.5.4) and (13.5.5) can be replaced by following equations:

$$
\begin{aligned}
&(p^{n+1}_{d+\frac{1}{2}} - p^{n}_{d+\frac{1}{2}})(c\Delta t^{n+\frac{1}{2}})^{-1} + \epsilon\mu(q^{n+1}_{d} - q^{n+1}_{d+1})(\Delta z_d)^{-1} \\
&\quad + (1-\epsilon)\mu(q^{n}_{d} - q^{n}_{d+1})(\Delta z_d)^{-1} = \epsilon\kappa^{n+1}_{d+\frac{1}{2}}(B^{n+1}_{d+\frac{1}{2}} - p^{n+1}_{d+\frac{1}{2}}) \\
&\quad + (1-\epsilon)\kappa^{n}_{d+\frac{1}{2}}(B^{n}_{d+\frac{1}{2}} - p^{n}_{d+\frac{1}{2}}) \\
&\quad + \frac{1}{2}\epsilon\mu\kappa^{n+1}_{d+\frac{1}{2}}(\beta^{n+1}_{d} q^{n+1}_{d} + \beta^{n+1}_{d+1} q^{n+1}_{d+1}) \\
&\quad + \frac{1}{2}(1-\epsilon)\mu\kappa^{n}_{d+\frac{1}{2}}(\beta^{n}_{d} q^{n}_{d} + \beta^{n}_{d+1} q^{n}_{d+1}), \quad d = 1, \ldots, D
\end{aligned}
\tag{13.5.12}
$$

and

$$
\begin{aligned}
&(q^{n+1}_{d} - q^{n}_{d})(c\Delta t^{n+\frac{1}{2}})^{-1} + \epsilon\mu(p^{n+1}_{d-\frac{1}{2}} - p^{n+1}_{d+\frac{1}{2}})(\Delta z_{d-\frac{1}{2}})^{-1} \\
&\quad + (1-\epsilon)\mu(p^{n}_{d-\frac{1}{2}} - p^{n}_{d+\frac{1}{2}})(\Delta z_{d-\frac{1}{2}})^{-1} = -\epsilon\kappa^{n+1}_{d} q^{n+1}_{d} - (1-\epsilon)\kappa^{n}_{d} q^{n}_{d} \\
&\quad + \epsilon\mu\beta^{n+1}_{d}[3(\kappa B)^{n+1}_{d} + (\kappa p)^{n+1}_{d}] + (1-\epsilon)\mu\beta^{n}_{d}[3(\kappa B)^{n}_{d} + (\kappa p)^{n}_{d}], \\
&\qquad d = 2, \ldots, D.
\end{aligned}
\tag{13.5.13}
$$

Quantities such as κ_d, $(\kappa B)_d$ and $(\kappa p)_d$ are defined as in the explicit scheme. Boundary conditions over half intervals from equation (13.5.13) are applied over half intervals of the two end cells. We thus obtain

$$
\begin{aligned}
&(q^{n+1}_{1} - q^{n}_{1})(c\Delta t^{n+\frac{1}{2}})^{-1} + 2\epsilon\mu(I^{n+1}_{-} + q^{n+1}_{1} - p^{n+1}_{\frac{3}{2}})(\Delta z_1)^{-1} \\
&\quad + 2(1-\epsilon)\mu(I^{n}_{-} + q^{n}_{1} - p^{n}_{\frac{3}{2}})(\Delta z_1)^{-1} = -\mu\kappa^{n+1}_{\frac{3}{2}} q^{n+1}_{1} - (1-\epsilon)\kappa^{n}_{\frac{3}{2}} q^{n}_{1} \\
&\quad + \epsilon\mu\beta^{n+1}_{1}\kappa^{n+1}_{\frac{3}{2}}(3B^{n+1}_{\frac{3}{2}} + I^{n+1}_{-} + q^{n+1}_{1}) \\
&\quad + (1-\epsilon)\mu\beta^{n}_{1}\kappa^{n}_{\frac{3}{2}}(3B^{n}_{\frac{3}{2}} + I^{n}_{-} + q^{n}_{1})
\end{aligned}
\tag{13.5.14}
$$

and

$$
\begin{aligned}
&(q^{n+1}_{D+1} - q^{n}_{D+1})(c\Delta t^{n+\frac{1}{2}})^{-1} + 2\epsilon\mu(p^{n+1}_{D+\frac{1}{2}} - I^{n+1}_{+} + q^{n+1}_{D+1})(\Delta z_D)^{-1} \\
&\quad + 2(1-\epsilon)\mu(p^{n}_{D+\frac{1}{2}} - I^{n}_{+} + q^{n+1}_{D+1})(\Delta z_D)^{-1} = -\epsilon\kappa^{n+1}_{D+\frac{1}{2}} q^{n+1}_{D+1} \\
&\quad - (1-\epsilon)\kappa^{n}_{D+\frac{1}{2}} q^{n}_{D+1} + \epsilon\mu\beta^{n+1}_{D+1}\kappa^{n+1}_{D+\frac{1}{2}}(3B^{n+1}_{D+\frac{1}{2}} + I^{n+1}_{+} \\
&\quad - q^{n+1}_{D+1}) + (1-\epsilon)\mu\beta^{n}_{D+1}\kappa^{n}_{D+\frac{1}{2}}(3B^{n}_{D-1} + I^{n}_{+} - q^{n}_{D+1}).
\end{aligned}
\tag{13.5.15}
$$

When $\epsilon = 1$ we have a fully implicit or backward Euler scheme and when $\epsilon = \frac{1}{2}$ we have the Crank–Nicholson scheme.

If we introduce the solution vector **X** given by

$$\mathbf{X} = \{q_1, p_{3/2}, q_2, \ldots, q_D, p_{D+1/2}, q_{D+1}\}, \tag{13.5.16}$$

then for each angle μ_m and frequency ν_k equations (13.5.12)–(13.5.15) can be written as

$$\mathbf{T}_{mk}\mathbf{X}_{mk} = \mathbf{R}_{mk}, \tag{13.5.17}$$

where $\mathbf{T}_{mk}$ is a tri-diagonal matrix of dimension $(2D + 1)$ and $\mathbf{R}_{mk}$ is a vector of length $(2D + 1)$. As the solution is recursive, it cannot be vectorized over the depth grid. Both the setup and the solution can be vectorized over all angles and frequencies which are treated in parallel.

Equation (13.5.12) can be written in the form

$$D_d q_d^{n+1} + E_d q_{d+1}^{n+1} = F_d p_{d+\frac{1}{2}}^{n+1} + G_d, \quad d = 1, \ldots, D, \tag{13.5.18}$$

and equation (13.5.13) can be written as

$$A_d p_{d-\frac{1}{2}}^{n+1} + B_d p_{d+\frac{1}{2}}^{n+1} = C_d q_d^{n+1} + L_d, \quad d = 2, \ldots, D \tag{13.5.19}$$

The upper and lower boundary conditions are obtained by putting $A_1 = 0$ and $B_{d+1} = 0$ respectively in equation (13.5.19). The matrices A_d, B_d, C_d, D_d, E_d and F_d are matrices of order $(M \times M)$ if scattering is included. The quantities G_D and L_D are vectors and contain all known information from the previous time step t^n. Equations (13.5.18) and (13.5.19) can be solved using the following recursion relations:

$$q_{d+1}^{n+1} = V_{d+1} p_{d+\frac{3}{2}}^{n+1} + W_{d+1}, \tag{13.5.20}$$

$$p_{d+\frac{1}{2}}^{n+1} = U_{d+\frac{1}{2}} q_{d+1}^{n+1} + T_{d+\frac{1}{2}}, \tag{13.5.21}$$

where

$$U_{d+\frac{1}{2}} = (F_d - D_d V_d)^{-1} E_d, \tag{13.5.22}$$

$$V_d = (C_d - A_d U_{d-\frac{1}{2}})^{-1} B_d, \tag{13.5.23}$$

$$W_d = (C_d - A_d U_{d-\frac{1}{2}})^{-1} (A_d T_{d-\frac{1}{2}} - L_d), \tag{13.5.24}$$

$$T_{d+\frac{1}{2}} = (F_d - D_d V_d)^{-1} (D_d W_d - G_d). \tag{13.5.25}$$

The solution begins at $d = 1$, with $A_1 = 0$, then the quantities V, W, U, T are computed recursively for all depths $B_{d+1} = 0$. At the bottom $q_{D+1} = W_d + 1$. We then back substitute using equations (13.5.20) and (13.5.21) and obtain $P_{d+\frac{1}{2}}^{n+1}$ $(d = 1, 2, \ldots, D)$ and q_d^{n+1} $(d = 1, \ldots, D+1)$. For applications of this method see Mihalas and Klein (1982) and for automatic flux limiting see Mihalas and Weaver (1982).

13.6 Radiative transfer, entropy and local potentials

Certain interactions between matter and radiation are irreversible. These were studied by Wildt (1956). The connection between the theory of the thermodynamics of irreversible process and radiative transfer through the possibility of the applicability of a 'striking theorem' was suggested by Wildt (1972). According to Prigogine's theorem the rate of entropy production has its minimum value in the steady state. The application of such an extremum principle to radiative transfer would be important. It could lead to the application of variational techniques in obtaining the solution of radiative transfer problems. For a comprehensive discussion see Essex (1984).

Oxenius (1966) treated a two-level atom without continuum using an irreversible process of emission and Sen (1967) treated this problem by adding the background continuum. Glansdorff and Prigogine (1964, 1965) introduced the concept of 'local potential' in the studies of non-equilibrium chemical and hydrodynamical processes. Sen (1972) studied the role of a 'local potential' in connection with the non-LTE radiative transfer. In the above paper, he studied the possibility of introducing a variational principle based on the entropy of the system to obtain necessary conditions for the existence of a stationary state in the time dependent radiation and matter interaction.

We shall describe a simple model of the time dependent transfer equation with two discrete levels due to Sen (1972).

The time dependent transfer equation in a plane parallel medium is

$$\frac{1}{c}\frac{dI_\nu}{dt} + \frac{dI_\nu}{ds} = \frac{1}{c}\frac{\partial I_\nu}{\partial t} + \mu\frac{\partial I_\nu}{\partial z} = \kappa_\nu(S_\nu^\star - I_\nu) \tag{13.6.1}$$

and, assuming that the number densities in the two discrete levels are time dependent, we can write

$$\frac{dn_2}{dt} = v_z\frac{\partial n_2}{\partial z} + \frac{\partial n_z}{\partial t} = n_1\left(B_{12}\int J_\nu\phi_\nu\,d\nu + \Omega_{12}\right)$$
$$- n_2\left(A_{21} + B_{21}\int J_\nu\phi_\nu\,d\nu + \Omega_{21}\right). \tag{13.6.2}$$

$I_\nu(z, t, \mu)$ is the specific intensity of the unpolarized radiation with frequency ν over a distance ds, n_1 and n_2 are the number densities of the ground and upper states of the two-level atom, B_{12}, B_{21} and A_{21} are the Einstein coefficients of absorption, induced emission and spontaneous emission respectively of the two states, ϕ_ν is the absorption profile (assumed to be the same as the emission profile) and κ_ν and $S_\nu^\star$ are the absorption coefficient and source function respectively. J_ν is the mean intensity. Ω_{12} and Ω_{21} are the rate coefficients for collisional excitation and de-excitation and are given by the relation

$$\Omega_{12}/\Omega_{21} = (g_2/g_1)\exp(-h\nu/kT), \tag{13.6.3}$$

where h is Planck's constant, g_1, g_2 are the statistical weights, k is the Boltzmann constant and T is the kinetic temperature which is assumed to be independent of z and t. The quantities κ_ν and $S^\star_\nu$ are given for a two-level atom by

$$\kappa_\nu = \frac{h\nu}{4\pi}\phi_\nu(n_1 B_{12} - n_2 B_{21}) \tag{13.6.4}$$

and

$$S^\star_\nu = n_2 A_{21}/(n_1 B_{12} - n_2 B_{21}). \tag{13.6.5}$$

From the principle of conservation, we have

$$n = n_1 + n_2 = \text{constant}. \tag{13.6.6}$$

This leads to

$$\frac{dn_2}{dt} = -\frac{dn_1}{dt}. \tag{13.6.7}$$

We define (Kröll, 1967) two temperatures T_ν and T_j by the relations

$$\frac{1}{T_\nu} = \frac{k}{h\nu}\ln\left(1 + \frac{2h\nu^3}{c^2 I_\nu}\right), \tag{13.6.8}$$

$$\frac{1}{T_j} = \left(k/\varepsilon_j\right)\ln(g_j/n_j), \quad j = 1, 2, \tag{13.6.9}$$

where ε is the energy of the jth level and g_j is the statistical weight of the jth level. We write T_{12} as

$$\frac{1}{T_{12}} - (k/h\nu)\ln(g_2 n_1/g_1 n_2), \tag{13.6.10}$$

$$n_j = g_j \exp(-\varepsilon_j/kt). \tag{13.6.11}$$

From equations (13.6.10) and (13.6.11), we have

$$\frac{h\nu}{T_{12}} = \frac{\varepsilon_2}{T_2} - \frac{\varepsilon_1}{T_1}. \tag{13.6.12}$$

From equations (13.6.6), (13.6.11) and (13.6.12), we have

$$n_2 = n[(g_1/g_2)\exp(h\nu/kT_{12}) + 1]^{-1}, \tag{13.6.13}$$

$$n_1 = n[(g_1/g_2)\exp(h\nu/kT_{12})][(g_1/g_2)\exp(h\nu/kT_{12}) + 1]^{-1}. \tag{13.6.14}$$

It is known that $A_{21}/B_{21} = 2h\nu^3/c^2$ and $g_1 B_{12} = g_2 B_{21}$. Using these relations, we get κ_ν and $S^\star$ as,

$$\kappa_\nu = \frac{h\nu}{4\pi}\phi_\nu n[B_{12}(g_1/g_2)\exp(h\nu/kT_{12}) - B_{12}]$$
$$\times [(g_2/g_1)\exp(h\nu/kT_{12}) + 1]^{-1} \tag{13.6.15}$$

and

$$S_\nu^\star = \frac{2h\nu^3}{c^2} \exp[(h\nu/kT_1) - 1]^{-1}. \tag{13.6.16}$$

We shall now consider a variation principle based on entropy considerations. We assume a Maxwell–Boltzmann distribution for the matter and a Bose–Einstein distribution for the radiation field, and define $s_\nu(\mu)$ as the local entropy density for the radiation in a given direction and s_j for the gas, giving

$$s_\nu(\mu) = \frac{2h\nu}{c^2}[(1+a)\ln(1+a) - a\ln a], \tag{13.6.17}$$

where

$$a = c^2 I_\nu / 2h\nu^3 \tag{13.6.18}$$

and

$$s_j = kn_j[\ln(g_j/n_j) + 1], \quad j = 1, 2. \tag{13.6.19}$$

From equations (13.6.17), (13.6.19), (13.6.1) and (13.6.2), we get

$$\frac{\partial s_\nu}{\partial t} = \frac{1}{cT_\nu}\frac{\partial I_\nu}{\partial t}, \quad \frac{\partial s_j}{\partial t} = \frac{\varepsilon_j}{T_j}\frac{n_j}{\partial t} \tag{13.6.20}$$

and

$$\frac{\partial s_\nu}{\partial z} = \frac{1}{cT_\nu}\frac{\partial I_\nu}{\partial z}, \quad \frac{\partial s_j}{\partial z} = \frac{\varepsilon_j}{T_j}\frac{\partial n_j}{\partial z}. \tag{13.6.21}$$

Equations (13.6.1) and (13.6.2) give us

$$\begin{aligned} \frac{\partial s_\nu}{\partial t} &= \frac{\kappa_\nu}{T_\nu}\frac{2h\nu^3}{c^2}\left\{\left[\exp(h\nu/kT_{12}) - 1\right]^{-1} - \left[\exp(h\nu/kT_\nu) - 1\right]^{-1}\right\} \\ &\quad + \frac{\mu}{T_\nu}\frac{2h^2\nu^4}{kc^2}\exp(h\nu kT_\nu)\left[\exp(h\nu/kT_\nu) - 1\right]^{-2}\frac{\partial}{\partial z}\left(\frac{1}{T_\nu}\right), \end{aligned} \tag{13.6.22}$$

$$\begin{aligned} \frac{ds_1}{dt} = \frac{\varepsilon_1}{T_1}\frac{dn_1}{dt} &= \frac{\varepsilon_1}{T_1}\left[n_1\left(B_{12}\int J_\nu\phi_\nu + \Omega_{12}\right)\right. \\ &\quad \left. - n_2\left(A_{21} - B_{21}\int J_\nu\phi_\nu\, d\nu + \Omega_{21}\right)\right], \end{aligned} \tag{13.6.23}$$

$$\begin{aligned} \frac{ds_2}{dt} = \frac{\varepsilon_2}{T_2}\frac{dn_2}{dt} &= \frac{\varepsilon_2}{T_1}\left[n_2\left(A_{21} + B_{21}\int J_\nu\phi_\nu + \Omega_{21}\right)\right. \\ &\quad \left. - n_1\left(B_{12}\int J_\nu\phi_\nu\, d\nu + \Omega_{12}\right)\right]. \end{aligned} \tag{13.6.24}$$

Therefore,

$$\frac{d(s_1 + s_2)}{dt} = \frac{h\nu}{kT_{12}}\left[n_1\left(B_{12}\int J_\nu\phi_\nu\, d\nu + \Omega_{12}\right)\right.$$

$$-n_2\left(A_{21}+B_{21}\int J_\nu\phi_\nu\,d\nu+\Omega_{21}\right)\Bigg] \tag{13.6.25}$$

and

$$\frac{\partial(s_1+s_2)}{\partial t}=\frac{d(s_1+s_2)}{dt}-\left[\frac{\varepsilon_1 v_1}{T_1}\frac{\partial n_1}{\partial z}-\frac{\varepsilon_2 v_2}{T_2}\frac{\partial n_2}{\partial z}\right], \tag{13.6.26}$$

where n_1 and n_2 are given by equations (13.6.13) and (13.6.14). Let

$$s'_\nu = s_\nu(\mu)+\sum_1^2 s_j, \tag{13.6.27}$$

then from equations (13.6.22)–(13.6.26), we get

$$\frac{\partial s'}{\partial t}=\frac{\partial s}{\partial t}+\frac{\partial(s_1+s_2)}{\partial t}. \tag{13.6.28}$$

If $v_1(\partial n_1/\partial z)$ and $v_2(\partial n_2/\partial z)$ are small, then

$$\frac{\partial s'}{\partial t}=\frac{\partial s}{\partial t}+\frac{h\nu}{kT_{12}}\left[n_1\left(B_{12}\int J_\nu\phi_\nu\,d\nu+\Omega_{12}\right)\right.$$
$$\left.-n_2\left(A_{21}+B_{21}\int J_\nu\phi_\nu\,d\nu+\Omega_{21}\right)\right]=F\quad\text{say}. \tag{13.6.29}$$

We have

$$s'(z,\mu)=\int_{t_0(z)}^{T_1(z)}F\,dt. \tag{13.6.30}$$

The global entropy S can be obtained by interpolating over the whole volume and averaging over the solid angle. Thus,

$$S=\frac{1}{2}\int_t\int_{-1}^{+1}\int_A\int_z F(u(s,t),\omega(z,t),U_z,z,t)\,dz\,dA\,d\mu\,dt, \tag{13.6.31}$$

where $U=(1/T_\nu)$, $\omega=(1/T_{12})$ and $U_z=(\partial U/\partial z)$. The above analysis has been done with the azimuthal independent radiation field.

During the evolution of the system $\delta S\geq 0$, which leads to a stationary state. Using the calculus of variations, the necessary condition for the existence of a stationary value can be expressed in the form of Euler equations, that is

$$\frac{\partial F}{\partial U}-\frac{\partial}{\partial z}\frac{\partial F}{\partial U_z}=0 \tag{13.6.32}$$

and

$$\frac{\partial F}{\partial \omega}=0. \tag{13.6.33}$$

The quantity F in equation (13.6.29) can be written in the form

$$F = \sum_{j=1}^{4} P_j(U)Q(\omega) + P_s(U)\frac{\partial U}{\partial z}, \tag{13.6.34}$$

and the Euler equations then become

$$\sum_{j+1}^{4} Q_j(\omega)\frac{\partial P_j(U)}{\partial U} = 0, \tag{13.6.35}$$

$$\sum_{j+1}^{4} P_j(U)\frac{\partial Q_j(\omega)}{\partial \omega} = 0, \tag{13.6.36}$$

where

$$P_1(U) = U, \quad P_2(U) = \frac{2h\nu^3}{c^2}U[\exp(h\nu U/k) - 1]^{-1},$$

$$P_3(U) = \left[B_{12}\int J_\nu\phi_\nu\, d\nu + \Omega_{12}\right],$$

$$P_4(U) = \left[A_{21} + B_{21}\int J_\nu\phi_\nu\, d\nu + \Omega_{21}\right]$$

and

$$P_5(U) = \frac{\mu U 2h^2\nu^4}{kc^2}\exp(h\nu U/k)[\exp(h\nu U/k) - 1]^{-2},$$

$$Q_1(\omega) = \kappa_\nu\frac{2h\nu^3}{c^2}[\exp(h\nu\omega/k) - 1]^{-1},$$

$$Q_2(\omega) = \kappa_\nu, \quad Q_3(\omega) = (h\nu/k)\omega n_1, \quad Q_4(\omega) = (h\nu/k)\omega n_2.$$

The quantities U and ω can be obtained from equations (13.6.35) and (13.6.36). From equations (13.6.31)–(13.6.36) one can obtain $S_\nu^\star$ and κ_ν. The steady state condition can be obtained by setting $(1/c)\partial I_\nu/\partial t = 0$, $\partial n_2/\partial t = \partial n_1/\partial t = 0$, $\partial s_\nu(\mu)/\partial t = \partial s_1/\partial t = \partial s_2/\partial t = 0$.

We now consider the local potentials. The change in the signs of δs and $\partial s_\nu/\partial t$ are to be ascertained following Glansdorff and Prigogine (1964, 1965). From equations (13.6.22) and (13.6.8), we can write

$$\frac{1}{c}\frac{\partial}{\partial t}\left(\frac{1}{T_\nu}\right) = -\frac{\kappa_\nu k}{h\nu}[1 - \exp(-h\nu/kT_\nu)]\{[\exp(h\nu/kT_\nu) - 1]$$
$$\times\left[\exp(h\nu/kT_{12}) - 1\right]^{-1} - 1\} - \mu\frac{\partial}{\partial z}\left(\frac{1}{T_\nu}\right). \tag{13.6.37}$$

From equations (13.6.26) and (13.6.13) and assuming that $v(\partial n_1/\partial z)$ and $v_2(\partial n_2/\partial z)$ are negligibly small, we get

$$\frac{\partial}{\partial t}\left(\frac{1}{T_{12}}\right) = -\frac{k}{h\nu}\left\{[(g_1/g_2)\exp(h\nu/kT_{12}) + 1]\left(B_{12}\int J_\nu\phi_\nu\, d\nu + \Omega_{12}\right)\right.$$

$$-\left[(g_2/g_1)\exp(-h\nu/kT_{12})+1\right]\left(A_{21}+B_{21}\int J_\nu\phi_\nu\,d\nu+\Omega_{21}\right)\Big\}. \tag{13.6.38}$$

Multiplying equation (13.6.37) by $(\partial/\partial t)(1/T_\nu)$ and equation (13.6.38) by $(\partial/\partial t)(1/T_{12})$ and adding we get the local potential φ:

$$\varphi=-\left\{\frac{1}{c}\left[\frac{\partial}{\partial t}\left(\frac{1}{T_\nu}\right)\right]^2+\left[\frac{\partial}{\partial t}\left(\frac{1}{T_{12}}\right)\right]^2\right\}. \tag{13.6.39}$$

φ is made negative and semi-definite. Integrating over the entire volume of the atmosphere (plane parallel), we get

$$\psi=\int\varphi dv=\int_A\int_z\varphi\,dz\,dA. \tag{13.6.40}$$

We can use the variational principle by assuming a quantity E_ν such that

$$\psi=\frac{\partial}{\partial t}E_\nu, \tag{13.6.41}$$

or

$$E_\nu=\int\psi\,dt=\int_t\int_A\int_z\varphi\,dz\,dA\,dt. \tag{13.6.42}$$

Assuming that φ is negative and semi-definite and the ranges of integration are all positive, E_ν can only diminish during the process of evolution and will take a minimum value at the stationary state. Now

$$E_\nu=\int_t\int_A\int_z\varphi(U(z,t),\omega(z,t),U_z,U_t,\omega_t,z,t)\,dz\,dA\,dt, \tag{13.6.43}$$

with $U=1/\tau_\nu$, $\omega=1/T_{12}$, $U+z=\partial U/\partial z$, $U_t=\partial U/\partial t$, $\omega_t=\partial\omega/\partial t$.

The necessary conditions for the existence of a minimum can be written as

$$\frac{\partial\varphi}{\partial U}-\frac{\partial}{\partial z}\frac{\partial\varphi}{\partial U_z}-\frac{\partial}{\partial t}\frac{\partial\varphi}{\partial U_t}=0 \tag{13.6.44}$$

and

$$\frac{\partial\varphi}{\partial\omega}-\frac{\partial}{\partial t}\frac{\partial\varphi}{\partial\omega_t}=0. \tag{13.6.45}$$

φ in equation (13.6.39) can be written as

$$\varphi=\left[\mu\frac{\partial U}{\partial z}+P_1(U)Q_1(\omega)+P_2(U)Q_2(\omega)\right]\frac{\partial U}{\partial t}$$
$$+[P_3(U)Q_3(\omega)+P_4(U)Q_4(\omega)]\frac{\partial\omega}{\partial t}, \tag{13.6.46}$$

where the P and Q coefficients are similar to those appearing in equations (13.6.35) and (13.6.36).

From equations (13.6.44), (13.6.45) and (13.6.46), we get

$$2\mu\frac{\partial^2 U}{\partial z\partial t} + [P_1(U)Q_1'(\omega) + P_2(U)Q_2'(\omega) - Q_3(\omega)P_3'(U) - Q_4(\omega)P_4'(U)]\frac{\partial\omega}{\partial t} = 0 \tag{13.6.47}$$

and

$$\{[P_1(U)Q_1'(\omega) + P_2(U)Q_2'(\omega)] - [Q_3(\omega)P_3'(U) + Q_4(\omega)P_4'(U)]\}\frac{\partial U}{\partial t} = 0, \tag{13.6.48}$$

where

$$\frac{\partial P_j}{\partial t} = \frac{\partial P_j}{\partial U}\frac{\partial U}{\partial t} = P_j'(U)\frac{\partial U}{\partial t}$$

and

$$\frac{\partial Q_j}{\partial t} = Q_j'(\omega)\frac{\partial\omega}{\partial t}. \tag{13.6.49}$$

As $\partial U/\partial t \neq 0$, equation (13.6.48) gives the connection between U and ω or between I_ν and $S_\nu^\star$. Eliminating one of them, one can obtain the differential equation for solving them.

13.7 Radiative transfer in masers

High resolution interferometry gives the structure of maser sources. The study of maser radiation gives information regarding small scale structure. For early reviews of masers see Litwak *et al.* (1966) and Kegel (1975) and for later developments see Alcock and Ross (1985) and Elitzur (1990).

A variational technique was used by Sen (1982) on the radiation stability in a cylindrical homogeneous maser with steady state pumping. He employed time dependent transfer for the two-level atom of Deguchi (1974).

Exercises

13.1 Assuming that the secondary component is a point source (instead of an extended source as in section 13.2), describe the distribution of the reflected radiation field of the atmosphere of the primary component.

13.2 Develop a computer scheme for obtaining the limb darkening when the secondary component is: (a) a point source and (b) an extended source.

13.3 Using the partial redistribution function R_{I-A} (isotropic) compute the line profiles in the expanding irradiated atmospheres of the close binary components.

13.4 Substitute equation (13.5.21) into equation (13.5.20) and solve the resulting tri-diagonal system for $P^{n+1}_{d+\frac{1}{2}}$, and using equation (13.5.21) obtain the equation

$$q_d^{n+1} = C_d^{-1}\left(A_d p^{n+1}_{d-\frac{1}{2}} + B_d p^{n-1}_{d+\frac{1}{2}} - L_d\right).$$

13.5 From the solution obtained in exercise 13.4, derive the moments.

13.6 Derive equations (13.6.20) and (13.6.21).

13.7 Expand equation (13.6.39) and write down the full expression for φ.

13.8 Derive the P and Q coefficients in equation (13.6.46).

13.9 Derive equations (13.6.47) and (13.6.48).

REFERENCES

Alcock, C., Ross, R.R., 1985, *ApJ*, **290**, 433.

Baschek, B., Efimov, G.V., von Waldenfels, W., Wehrse, R., 1997a, *A&A*, **317**, 630.

Baschek, B., Grüber, C., von Waldenfels, W., Wehrse, W., 1997b, *A&A*, **320**, 920.

Chandrasekhar, S., 1958, *Proc. NAS*, **44**, 933.

Claret, A., Giménez, A., 1992, *A&A*, **256**, 572.

Deguchi, S., 1974, *Publ. Astron. Soc. Japan*, **26**, 437.

Elitzur, D., 1990, *ApJ*, **363**, 628 and 638.

Ellison, D., Grant, I.P., 1974, *Comp. Phys. Commun*, **8**, 257.

Essex, C., 1984, *ApJ*, **285**, 279.

Glansdorff, P., Prigogine, I., 1964, *Physica*, **30**, 351.

Glansdorff, P., Prigogine, I., 1965, *Physica*, **31**, 1242.

Kegel, W.H., 1975, in *Problems in Stellar Atmospheres*, eds. B. Baschek, W.H. Kegel, G. Traving, Springer, Berlin, page 257.

Kho, T.H., Sen, K.K., 1972, *Astrophys. Spa. Sci.*, **16**, 151.

Kröll, W., 1967, *JQSRT*, **7**, 715.

Latko, R.J., Pomraning, G.C., 1972, *JQSRT*, **12**, 1.

Leong, T.K., Sen, K.K., 1969, *Publ. Astron. Soc. Japan*, **21**, 167.

Leong, T.K., Sen, K.K., 1970, *Publ. Astron. Soc. Japan*, **22**, 57.

Leong, T.K., Sen, K.K., 1971a, *Publ. Astron. Soc. Japan*, **23**, 99.

Leong, T.K., Sen, K.K., 1971b, *Publ. Astron. Soc. Japan*, **23**, 247.

Leong, T.K., Sen, K.K., 1972, *MNRAS*, **160**, 21.

Litwak, M.M., McWhirter, A.L., Meeks, M.L., Zeiger, H.J., 1966, *Phys. Rev. Lett.*, **17**, 821.

Mihalas, D., Klein, R.I., 1982, *J. Comp. Phys.*, **46**, 97.

Mihalas, D., Weaver, R., 1982, *JQSRT*, **28**, 213.

Mohan Rao, D., Rangarajan, K.E., Peraiah, A., 1990, *ApJ*, **358**, 622.

Munier, A., 1987, *JQSRT*, **38**, 447, 457, 475.

Munier, A., 1988, *JQSRT*, **39**, 43.

Olson, G.L., Auer, L.H., Buchler, J.R., 1986, *JQSRT*, **35**, 431.

Oxenius, J., 1966, *JQSRT*, **6**, 65.

Peraiah, A., 1982, *J. Astrophys. Astr.*, **3**, 485.

Peraiah, A., 1983a, *J. Astrophys. Astr.*, **4**, 11.

Peraiah, A., 1983b, *J. Astrophys. Astr.*, **4**, 151.

Peraiah, A., Rao, M.S., 1983, *J. Astrophys. Astr.*, **4**, 183.

Peraiah, A., Rao, M.S., 1998, *A&A Suppl. Ser.*, **132**, 45.

Richards, P.I., 1956, *J. Opt. Soc. Amer.*, **46**, 927.

Srinivasa Rao, M., Peraiah, A., 2000, *A&A Suppl. Ser.*, **145**, 525.

Sen, K.K., 1967, *JQSRT*, **7**, 517.

Sen, K.K., 1972, *JQSRT*, **12**, 1487.

Sen, K.K., 1982, *Astrophys. Spa. Sci.*, **86**, 477.

Sen, K.K., Wilson, S.J., 1990, *Radiative Transfer in Curved Media*, World Scientific, Singapore.

Stenholm, L.G., 1977, *A&A*, **54**, 577.

Stenholm, L.G., Störzer, H., Wehrse, R., 1991, *JQSRT*, **45**, 47.

Uesugi, A., Tsujita, J., 1969, *Publ. Astron. Soc. Japan*, **21**, 370.

Vaz, L.P.R., 1985, *Astrophys. Spa. Sci.*, **113**, 349.

Vaz, L.P.R., Nordlund, Å., 1985, *A&A*, **147**, 281.

Wehrse, R., Baschek, B., von Waldenfele, W., 2000a Paper I, preprint.

Wehrse, R., Baschek, B., von Waldenfele, W., 2000b Paper II, preprint.

Wildt, R., 1956, *ApJ*, **123**, 107.

Wildt, R., 1972, *ApJ*, **174**, 69.

Wilson, R.E., 1990, *ApJ*, **356**, 613.

Wilson, S.J., Sen, K.K., 1975, *A&A*, **44**, 377.

Symbol index

a = ratio of the damping width to the Doppler width ($\Gamma/4\pi\Delta\nu_D$, $\Delta\nu_D$ being the Doppler width), 16

a_j = weights of Gauss–Legendre quadrature formula, 65

$\mathbf{b}$ = diagonal matrix of quadrature weights, 181

c = velocity of light, 11

$\mathbf{d} = (x_{i-1} - x_i)^{-1}$, 231

dE_ν = amount of radiant energy, 1

$d\omega$ = element of solid angle, 3

$\mathrm{erfc}(x)$ = error function, 18

$f(\mathbf{r}, \mathbf{p}, t)$ = distribution function, 35

f^r = radial function in the adaptive mesh, 255

f^s = structure function, 255

$f_\nu(r, t)$ = variable Eddington factor, 25

g_1, g_2 = statistical weights, 15

h = Planck constant, 11

j_ν = emission coefficient, 10

Index

Printed in the United States
By Bookmasters